The Centrosome

Heide Schatten
Editor

The Centrosome

Cell and Molecular Mechanisms of Functions
and Dysfunctions in Disease

 Humana Press

Editor
Heide Schatten
Department of Veterinary Pathobiology
University of Missouri-Columbia
Columbia MI
USA

ISBN 978-1-62703-034-2   ISBN 978-1-62703-035-9 (eBook)
DOI 10.1007/978-1-62703-035-9
Springer New York Heidelberg Dordrecht London

Library of Congress Control Number: 2012942022

© Humana Press, a part of Springer Science+Business Media, LLC 2012
This work is subject to copyright. All rights are reserved by the Publisher, whether the whole or part of the material is concerned, specifically the rights of translation, reprinting, reuse of illustrations, recitation, broadcasting, reproduction on microfilms or in any other physical way, and transmission or information storage and retrieval, electronic adaptation, computer software, or by similar or dissimilar methodology now known or hereafter developed. Exempted from this legal reservation are brief excerpts in connection with reviews or scholarly analysis or material supplied specifically for the purpose of being entered and executed on a computer system, for exclusive use by the purchaser of the work. Duplication of this publication or parts thereof is permitted only under the provisions of the Copyright Law of the Publisher's location, in its current version, and permission for use must always be obtained from Springer. Permissions for use may be obtained through RightsLink at the Copyright Clearance Center. Violations are liable to prosecution under the respective Copyright Law.
The use of general descriptive names, registered names, trademarks, service marks, etc. in this publication does not imply, even in the absence of a specific statement, that such names are exempt from the relevant protective laws and regulations and therefore free for general use.
While the advice and information in this book are believed to be true and accurate at the date of publication, neither the authors nor the editors nor the publisher can accept any legal responsibility for any errors or omissions that may be made. The publisher makes no warranty, express or implied, with respect to the material contained herein.

Printed on acid-free paper

Humana Press is a brand of Springer

Springer is part of Springer Science+Business Media (www.springer.com)

Preface

The discovery of centrosomes well over 100 years ago has been called as important as the discovery of the nucleus but it was only recently that research on centrosomes has moved this central organelle to the forefront of modern research and has exploded as a result of renewed appreciation, new enthusiasm, and new methods that are now available to study centrosomes on cellular, molecular, and genetic levels. Centrosome functions are critically important for cell cycle regulation while centrosome dysfunctions have been implicated in numerous diseases such as cancer and in disorders such as infertility and other reproductive disorders.

While most frequently described as microtubule organizing centers (MTOCs), the recent recognition that centrosomes are critical for cell signaling coordination and cellular protein degradation, and regulated proteolysis has revolutionized studies into disorders and in the pathogenesis and progression of diseases. To cover the wealth of new data on the significant role of centrosomes in cellular functions and implications in disease, a number of topic-focused review articles have been written and published in various specialized journals, as the demand for understanding the direct and indirect functions of this important organelle has increased in various areas of basic and biomedical research.

This book features a variety of different aspects on classic and modern centrosome research to cover topics of current interest. Each chapter is written by internationally recognized experts in their respective fields who have contributed their unique expertise in specific research areas and include comprehensive and concise reviews of key topics in the field as well as cell and molecular details that are important for the specific subtopics. Cutting edge new information is balanced with background information that is readily understandable for the newcomer and the experienced centrosome researcher alike. In addition, several articles will raise awareness of centrosomes in areas that have not yet considered centrosomes associated with disease including aspects of misguided signal transduction and several others that may find centrosomes as new targets for therapeutic intervention.

The book includes chapters on

- Centriole duplication and inheritance.
- Sperm centrioles and abnormalities underlying sperm pathology and infertility.
- Centrosomal functions and dysfunctions in cat spermatozoa.
- Nuclear-centrosome interactions during fertilization and cell division.
- Human centrosomal dynamics during gametogenesis, fertilization, and embryogenesis.
- Asymmetric centrosome behavior in stem cell divisions.
- Functional associations between the Golgi apparatus and the centrosome.
- The centrosome and its role in regulated proteolysis.
- Regulation of the centrosome cycle by protein degradation.
- Molecular links between centrosome duplication and other cell cycle associated events.
- Regulation of centrosomes by cyclin-dependent kinases.
- Disruption of centrosome duplication control and induction of mitotic instability by the high-risk human papillomavirus oncoproteins E6 and E7.
- Centrosomes, DNA damage, and aneuploidy.
- Centrosome regulation and breast cancer.
- The role of centrosomes in multiple myeloma.
- Centrosomal amplification and related abnormalities.
- Mechanisms and consequences of centrosome clustering in cancer cells.
- The neuronal centrosome as a generator of microtubules for axons and dendrites.
- Centrosomes and cell division in Apicomplexa.
- The centrosome life story in *Xenopus laevis*.
- The role of centrosomes in T cells, and concludes with.
- Thoughts on progress in the centrosome field.

The topics addressed are selected to be of interest to scientists, students, teachers, and to all who are interested in expanding their knowledge related to centrosomes. The volume is intended for a large audience as a reference book on the subject.

It has been a great pleasure and timely to edit this book on centrosomes and I would like to sincerely thank all contributors for their outstanding chapters and for sharing their unique expertise with the centrosome community. I hope that this book will stimulate further advances in centrosome research and contribute new insights and appreciation for the role of centrosomes in the basic and biomedical sciences.

Heide Schatten

Contents

Part I Centrosomes in Reproduction

1. **Centriole Duplication and Inheritance in *Drosophila melanogaster*** 3
   Tomer Avidor-Reiss, Jayachandran Gopalakrishnan,
   Stephanie Blachon and Andrey Polyanovsky

2. **Sperm Centrioles and Their Dual Role in Flagellogenesis and Cell Cycle of the Zygote** 33
   Hector E. Chemes

3. **Centrosomal Functions and Dysfunctions in Cat Spermatozoa** ... 49
   Pierre Comizzoli and David E. Wildt

4. **Nuclear–Centrosome Relationships During Fertilization, Cell Division, Embryo Development, and in Somatic Cell Nuclear Transfer Embryos** 59
   Heide Schatten and Qing-Yuan Sun

5. **Human Centrosomal Dynamics During Gametogenesis, Fertilization, and Embryogenesis and Its Impact on Fertility: Ultrastructural Analysis** 73
   A. Henry Sathananthan

6. **Asymmetric Centrosome Behavior in Stem Cell Divisions** 99
   Therese M. Roth, Yukiko M. Yamashita and Jun Cheng

Part II Cell and Molecular Biology of Centrosomes

7 Functional Associations Between the Golgi Apparatus and the Centrosome in Mammalian Cells 113
Breanne Karanikolas and Christine Sütterlin

8 Many Pathways to Destruction: The Role of the Centrosome in, and Its Control by Regulated Proteolysis. 133
Harold A. Fisk

9 Regulation of the Centrosome Cycle by Protein Degradation 157
Suzanna L. Prosser and Andrew M. Fry

10 Molecular Links Between Centrosome Duplication and Other Cell Cycle-Associated Events. 173
Kenji Fukasawa

11 Regulation of Centrosomes by Cyclin-Dependent Kinases 187
Rose Boutros

Part III Centrosome Abnormalities in Cancer

12 Disruption of Centrosome Duplication Control and Induction of Mitotic Instability by the High-Risk Human Papillomavirus Oncoproteins E6 and E7 201
Nina Korzeniewski and Stefan Duensing

13 Centrosomes, DNA Damage and Aneuploidy 223
Chiara Saladino, Emer Bourke and Ciaran G. Morrison

14 Centrosome Regulation and Breast Cancer 243
Zeina Kais and Jeffrey D. Parvin

15 The Role of Centrosomes in Multiple Myeloma 255
Benedict Yan and Wee-Joo Chng

16 Centrosomal Amplification and Related Abnormalities Induced by Nucleoside Analogs. 277
Ofelia A. Olivero

17 Mechanisms and Consequences of Centrosome Clustering in Cancer Cells. 285
Alwin Krämer, Simon Anderhub and Bettina Maier

Part IV Centrosomes in Other Systems

**18 Re-evaluation of the Neuronal Centrosome as a Generator
of Microtubules for Axons and Dendrites** 309
Peter W. Baas and Aditi Falnikar

19 Centrosomes and Cell Division in Apicomplexa 327
Leandro Lemgruber, Marek Cyrklaff and Freddy Frischknecht

20 The Centrosome Life Story in *Xenopus laevis* 347
Jacek Z. Kubiak and Claude Prigent

21 Role of the MTOC in T Cell Effector Functions 365
Martin Poenie, Laura Christian, Sarah Tan and Yuri Sykulev

22 Thoughts on Progress in the Centrosome Field 385
Jeffrey L. Salisbury

About the Author 393

Index 395

Contributors

Simon Anderhub Clinical Cooperation Unit Molecular Hematology/Oncology, German Cancer Research Center and Department of Internal Medicine V, University of Heidelberg, Heidelberg, Germany

Tomer Avidor-Reiss Department of Cell Biology, Harvard Medical School, Boston, MA, USA

Peter W. Baas Department of Neurobiology and Anatomy, Drexel University College of Medicine, Philadelphia, PA, USA

Stephanie Blachon Department of Cell Biology, Harvard Medical School, Boston, MA, USA

Emer Bourke Institute of Technology Sligo, Sligo, Ireland

Rose Boutros Children's Medical Research Institute, The University of Sydney, Westmead, NSW, Australia

Hector E. Chemes División Endocrinología, Centro de Investigaciones Endocrinológicas, CEDIE-CONICET, Buenos Aires, Argentina

Jun Cheng Department of Bioengineering, University of Illinois, Chicago, IL, USA

Wee-Joo Chng Haematology and Oncology, National University Cancer Institute of Singapore, National University Health System, Singapore; Department of Medicine, Yong Loo Lin School of Medicine, National University of Singapore

Laura Christian Department of Molecular Cell and Developmental Biology, University of Texas at Austin, Austin, TX, USA

Pierre Comizzoli Center for Species Survival, Smithsonian Conservation Biology Institute, Washington, DC, USA

Marek Cyrklaff Department of Infectious Diseases, Parasitology, University of Heidelberg Medical School, Heidelberg, Germany

Stefan Duensing Section of Molecular Urooncology, Department of Urology, University of Heidelberg, School of Medicine, Heidelberg, Germany

Aditi Falnikar Department of Neurobiology and Anatomy, Drexel University College of Medicine, Philadelphia, PA, USA

Harold A. Fisk Department of Molecular Genetics, The Ohio State University, Columbus, OH, USA

Friedrich Frischknecht Department of Infectious Diseases, Parasitology, University of Heidelberg Medical School, Heidelberg, Germany

Andrew M. Fry Department of Biochemistry, University of Leicester, Leicester, UK

Kenji Fukasawa Molecular Oncology Program, H. Lee Moffitt Cancer Center & Research Institute, Tampa, FL, USA

Jayachandran Gopalakrishnan Department of Cell Biology, Harvard Medical School, Boston, MA, USA

Zeina Kais Department of Biomedical Informatics, Ohio State University Comprehensive Cancer Center, Columbus, OH, USA

Breanne Karanikolas Department of Developmental and Cell Biology, University of California, Irvine, CA, USA

Nina Korzeniewski Section of Molecular Urooncology, Department of Urology, University of Heidelberg, School of Medicine, Heidelberg, Germany

Alwin Krämer Clinical Cooperation Unit Molecular Hematology/Oncology, German Cancer Research Center and Department of Internal Medicine V, University of Heidelberg, Heidelberg, Germany

Jacek Kubiak CNRS, UMR, Institut de Génétique et Développement de Rennes, Rennes, France; Faculté de Médecine, Université Rennes, UEB, IFR, Rennes, France

Leandro Lemgruber Department of Infectious Diseases, Parasitology, University of Heidelberg Medical School, Heidelberg, Germany

Bettina Maier Clinical Cooperation Unit Molecular Hematology/Oncology, German Cancer Research Center and Department of Internal Medicine V, University of Heidelberg, Heidelberg, Germany

Ciaran G. Morrison Centre for Chromosome Biology, School of Natural Sciences, National University of Ireland Galway, Galway, Ireland

Ofelia A. Olivero Laboratory of Cancer Biology and Genetics, Carcinogen-DNA Interactions Section, National Cancer Institute, NIH, Bethesda, MD, USA

Jeffrey D. Parvin Department of Biomedical Informatics, Ohio State University Comprehensive Cancer Center, Columbus, OH, USA

Martin Poenie Department of Molecular Cell and Developmental Biology, University of Texas at Austin, Austin, TX, USA

Andrey Polyanovsky Russian Academy of Sciences, Sechenov Institute, St. Petersburg, Russia

Claude Prigent CNRS, UMR, Institut de Génétique et Développement de Rennes, Rennes, France; Faculté de Médecine, Université Rennes, UEB, IFR, Rennes, France

Suzanna L. Prosser Department of Biochemistry, University of Leicester, Leicester, UK

Therese M. Roth Life Sciences Institute, Center for Stem Cell Biology and Department of Cell and Developmental Biology, University of Michigan, Ann Arbor, MI, USA

Chiara Saladino Centre for Chromosome Biology, School of Natural Sciences, National University of Ireland Galway, Galway, Ireland

Jeffrey L. Salisbury Tumor Biology Program, Biochemistry and Molecular Biology, Mayo Clinic, Rochester, MN, USA

A. Henry Sathananthan Monash Immunology and Stem Cell Laboratories, Melbourne, Australia

Heide Schatten Department of Veterinary Pathobiology, University of Missouri-Columbia, Columbia, MO, USA

Qing-Yuan Sun State Key Laboratory of Reproductive Biology, Institute of Zoology, Chinese Academy of Sciences, Beijing, China

Christine Sütterlin Department of Developmental and Cell Biology, University of California, Irvine, CA, USA

Yuri Sykulev Department of Microbiology and Immunology, Kimmel Cancer Center and Jefferson Vaccine Center, Thomas Jefferson University, Philadelphia, PA, USA

Sarah Tan Department of Molecular Cell and Developmental Biology, University of Texas at Austin, Austin, TX, USA

David E. Wildt Center for Species Survival, Smithsonian Conservation Biology Institute, Washington, DC, USA

Yukiko M. Yamashita Life Sciences Institute, Center for Stem Cell Biology and Department of Cell and Developmental Biology, University of Michigan, Ann Arbor, MI, USA

Ben Yan Department of Pathology, National University Health System, Singapore

Part I
Centrosomes in Reproduction

Chapter 1
Centriole Duplication and Inheritance in *Drosophila melanogaster*

Tomer Avidor-Reiss, Jayachandran Gopalakrishnan,
Stephanie Blachon and Andrey Polyanovsky

Abstract Centrosomes are conserved microtubule-based organelles that are essential for animal development. In this chapter, we highlight key centrosomal proteins and describe the centrosome in the context of several developmental processes in *Drosophila melanogaster*. These processes include fertilization, during which the centrosome mediates the fusion of male and female pronuclei; development of the embryonic syncytium, where centrosomes act as microtubule-organizing centers and participate in nuclear division; and the formation of sensory and motile cilia in the adult, where the centrosome's centrioles template axoneme assembly. The study of these processes in *Drosophila* provides a unique experimental system where classical approaches in genetics and biochemistry can be used to dissect centrosome biology.

1.1 What are the Challenges in Studying the Centrosome and Why Use *Drosophila*?

Like chromosomes and yeast spindle pole bodies (SPB), centrosome numbers in the cell is strictly controlled. Control of centrosome numbers is achieved by a process of duplication in which the preexisting structure is used as a means to

T. Avidor-Reiss (✉) · J. Gopalakrishnan · S. Blachon
Department of Cell Biology, Harvard Medical School,
Seeley G. Mudd Building, Room 509A, 250 Longwood Avenue,
Boston 02115, MA, USA
e-mail: tomer_avidor-reiss@hms.harvard.edu

A. Polyanovsky
Sechenov Institute, Russian Academy of Sciences, St. Petersburg, Russia

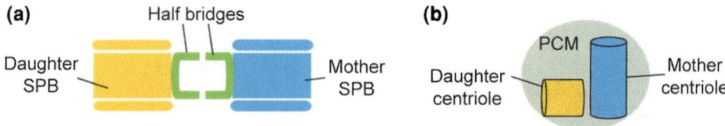

Fig. 1.1 Models for duplication **a** Yeast spindle pole bodies (SPB Duplication); the half bridge of the preexisting (mother) SPB serves as a template and nucleation site for the new (daughter) SPB that is formed in parallel. **b** Centriole Duplication; the new (daughter) centriole is formed perpendicularly to the preexisting (mother) centriole and at a significant distance from the surface of the preexisting centriole

ensure that only a single new structure is formed. Centrosomes consist of two centrioles surrounded by PCM. Centrosome duplication starts after the two preexisting centrioles separate slightly and a new centriole forms near each of the preexisting centrioles. Centrosome duplication is concluded when each centriole pair completely separates, along with some of the PCM of the original centrosome. In each centrosome, the older centriole is also known as the mother centriole and the new centriole is known as the daughter centriole.

The process of centrosome duplication is conceptually similar to how DNA and yeast SPB duplicate (Fig. 1.1). DNA duplication starts by splitting into two strands, which then serve as a template to create a new strand. Yeast SPB duplicate by splitting into two halves. Each half contains a structure known as the half bridge, which then serves to template the formation of another half bridge (Jaspersen and Winey 2004; Jones and Winey 2006). Since, like DNA and SPB duplication, centrosome duplication maintains one preexisting element and creates one new element, it is thought that centrosomes duplicate in a semi–conservative manner.

Although many of the proteins involved in centrosome duplication have been recently identified, the critical question of how the centrioles duplicate remains elusive. Very little is known about the overall mechanism of centriole duplication (Azimzadeh and Marshall 2010; Nigg and Raff 2009). It has become increasingly accepted that centriole duplication does not involve a templating mechanism like in DNA and SPB duplication.

Several lines of observation suggest that de novo formation of the new centrioles takes place at the vicinity of the preexisting centriole.

i) The new centriole is formed perpendicularly to the preexisting centriole.
ii) The new centriole forms at a distance of about 30 nm from the surface of the preexisting centriole (Anderson and Brenner 1971; Phillips 1967),
iii) The new centriole can have a very different structure from the preexisting centriole (Phillips 1967).
iv) Under certain conditions, centrioles can form in the absence of a preexisting centriole (Fulton and Dingle 1971; Rodrigues-Martins et al. 2007b)
v) Several new centrioles can be induced to form simultaneously around the preexisting centriole by overexpressing centriolar components (Kleylein-Sohn et al. 2007).

vi) No proof is available of an intrinsic asymmetry around the preexisting centriole before the onset of centriole duplication

These observations suggest that a yet unidentified mechanism inherent to the centrosome assures that only one new centriole is formed near a preexisting centriole (Sluder and Khodjakov 2010).

Analysis of centriole duplication is challenged by a combination of factors. Centrosomes are essential for development in animals. Centrosomes and centrioles are in low-abundance, found only in one or two copies per cell, thus challenging biochemical approaches. Furthermore, new centriole intermediates are few, small, short-lived, and form too close to the preexisting centriole to be observed as a distinct entity, making it extremely challenging to study centriole intermediates by traditional light microscopy (the internal structure of the centriole is beyond the resolution of standard light microscopy).

Despite these challenges, many proteins involved in centriole duplication have been identified over the past 15 years. The recent identification of many proteins involved in centriole duplication opens new ways to overcome these barriers. One commonly used way is to overexpress the centriolar protein, usually in immortalized cells (in vitro), and observe the consequences using microscopy (Dzhindzhev et al. 2010; Gopalakrishnan et al. 2010; Tang et al. 2009). While sometimes informative, interpreting overexpression data is problematic due to the fact that proteins are studied at non-physiological levels. Also, immortalized cells often have abnormal centrioles, suggesting they already carry mutations that prevent normal centriole duplication. Therefore, to balance these limitations, it is important to develop approaches to study centriole duplication in vivo, when proteins are expressed at physiological levels.

For a number of reasons, *Drosophila* is ideal for developing such a balanced approach for studying centriole duplication and centrosome biogenesis:

First: mutants defective in centrosome biogenesis are available (see Table 1.1). In flies, even null mutations in essential centrosomal proteins are not embryonic lethal and the fly can often develop to maturity. This is due to maternal contribution which allows the embryo to form centrosomes when they are critical for development, namely during early embryonic development. Later during pupal development when the adult fly is forming, maternal contribution becomes depleted but centrosomes are no longer essential for development. This allows extensive characterization of defective centrosome biogenesis in the testes and sensory neurons of pupae (Basto et al. 2006; Blachon et al. 2009; Blachon et al. 2008; Mottier-Pavie and Megraw 2009).

Second: techniques are available to introduce newly-engineered proteins with modified capabilities into a null mutant background, allowing their specific function to be studied with expression at near physiological levels and in the absence of the wild-type protein (Blachon et al. 2008; Gopalakrishnan et al. 2011). This is especially useful in the study of centrosomes, as many centrosomal proteins form multiprotein complexes and may have more than one function. The ability to engineer a mutant that is deficient in one or limited interactions is very insightful

Table 1.1 *Drosophila* genes involved in centrosome biogenesis

Name	CG ID	Ortholog	Localization in *Drosophila*	Phenotype	Additional information	References
Asl	CG2919	Cep152 (MCPH4)	Centriole PCM interphase	Uncoordinated Meiosis defect Nonmotile sperm	Complete block of centriole and centrosome formation in mutant	(Blachon et al. 2008; Bonaccorsi et al. 1998; Varmark et al. 2007)
Ana1	CG6631	KIAA1731	Centrosome	Uncoordinated Meiosis defect Nonmotile sperm	Essential for centrosome formation	(Blachon et al. 2009; Goshima et al. 2007)
Ana2	CG8262	STIL	Centrosome	Unknown	Induces cartwheel-like structures together with Sas-6	(Goshima et al. 2007; Stevens et al. 2010a)
Ana3	CG13162	Rttn	Centrosome	Uncoordinated	Required for centriole structural integrity	(Goshima et al. 2007; Stevens et al. 2009)
Sak	CG7186	Plk4	Centrosome	Uncoordinated Meiosis defect Nonmotile sperm	Essential for centrosome formation	(Bettencourt-Dias et al. 2005)
Sas-4	CG10061	CPAP (MCPH6) TCP10	Centriole and PCM	Uncoordinated Meiosis defect Nonmotile sperm	Essential for centrosome formation	(Basto et al. 2006; Blachon et al. 2009; Gopalakrishnan et al. in press)
Sas-6	CG15524	Sas-6	Cartwheel	Uncoordinated Meiosis defect Nonmotile sperm	Short mc-giant centriole in mutants Centrioles that lack symmetry in mutant; Central tubule protein	(Gopalakrishnan et al. 2010; Rodrigues-Martins et al. 2007a; Stevens et al. 2010b)
D-Plp	CG6735	Pericentrin AKAP450	Centrosome	Uncoordinated Nonmotile sperm	Essential for PCM formation	(Martinez-Campos et al. 2004)
Cnn	CG4832	CDK5RAP2 (MCPH3) Myomegalin	Centrosome, PCM	Meiosis defect Nonmotile sperm	Essential for PCM formation	(Heuer et al. 1995; Li et al. 1998; Megraw et al. 1999; Vaizel-Ohayon and Schejter 1999)

(continued)

Table 1.1 (continued)

Name	CG ID	Ortholog	Localization in Drosophila	Phenotype	Additional information	References
Spd-2	CG17286	Spd-2	Centrosome	Uncoordinated Meiosis defect Nonmotile sperm	PCM formation	(Dix and Raff 2007; Giansanti et al. 2008)
Poc1	CG10191	Poc1 Pix1 Pix2	Centrosome	Mail sterile	A short giant centriole in mutant	(Blachon et al. 2009)
Unc	CG1501	OFD1?	Basal body	Uncoordinated Nonmotile sperm	Mutants result in a short giant centriole	(Baker et al. 2004)
Bld10	CG17081	Cep135	Centriole wall	Nonmotile sperm	Mutants result in a Short giant centriole	(Blachon et al. 2009; Mottier-Pavie and Megraw 2009)
CP190	CG6384	Not found	Centrosome	Pupal lethal	Nuclear function	(Butcher et al. 2004)
CP60	CG6384	Not found	Centrosome	Unknown	Forms a complex with CP190	(Kellogg et al. 1995)

in identifying the separate functions mediated by a centrosomal protein (Gopalakrishnan et al. 2011).

Third: it is possible to biochemically isolate centrosomes and centrosomal complexes from *Drosophila* embryos to study them ex vivo (Gopalakrishnan et al. 2010; Gopalakrishnan et al. 2011; Kellogg and Alberts 1992; Moritz et al. 1995). This allows one to study protein interactions under near physiological conditions. This also opens a window to use purified centrosomal proteins, structures, and complexes in cell-free experiments that can investigate the individual steps in centrosome duplication. Ultimately, this can theoretically allow centrosome duplication and function to be reconstituted using purified components.

Finally: *Drosophila* centrosomes are formed using conserved proteins and the overall structure of *Drosophila* centrosomes is very similar to that of other organisms, suggesting that the basic mechanisms of centrosome duplication used in *Drosophila* are similar to those used in other organisms.

1.2 Centrosomes in *Drosophila* Development

Centrosomes in *Drosophila* were studied in some detail in a context of several developmental processes. In this chapter, we will focus on four processes. The first two processes take place during early embryonic development: (1) Fertilization, and (2) Syncytial blastoderm formation. The next two processes occur in differentiated cell types and can be studied during pupal development: (3) Sensory neuron differentiation, and (4) Spermatogenesis. We will summarize key features of the centrosome in each of these developmental processes, highlighting unique properties that have provided insight into the biology of the centrosome.

1.2.1 Fertilization

Fertilization is the process by which the sperm (male gamete) and oocyte (female gamete) are fused to form a zygote, the first cell of a new organism. In general, a key step in fertilization is the migration of sperm and oocyte pronuclei toward each other and their subsequent fusion (Fig. 1.2).

It is generally recognized that in most animals, including *Drosophila*, the oocyte does not contain centrioles (Krioutchkova and Onishchenko 1999; Manandhar et al. 2005; Sun and Schatten 2007). Instead, oocytes have acentriolar centrosomes or microtubule organization centers that participate in female meiosis and in the formation of the female pronucleus (Megraw and Kaufman 2000). While the oocyte does not appear to have centrioles, it does contain a large amount of centriolar and PCM proteins within its cytoplasm, enough to form 2^{13} centrosomes (Rodrigues-Martins et al. 2007b). These proteins, contributed by the mother via the oocyte (maternal contribution), are sufficient to support centrosome

1 Centriole Duplication and Inheritance in *Drosophila melanogaster* 9

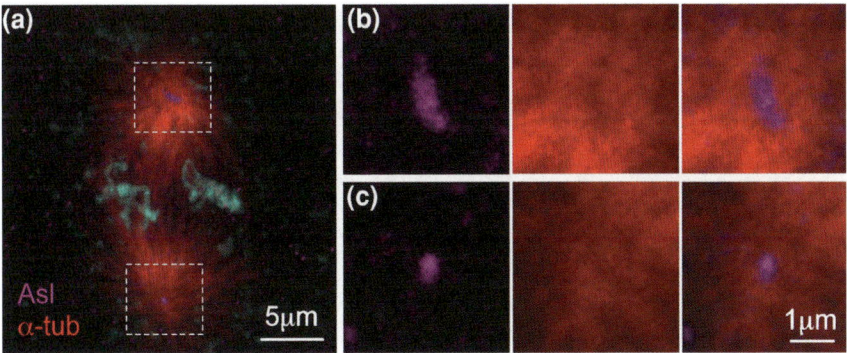

Fig. 1.2 Fertilization during mitosis of the zygote (**a**), the giant centriole (**b**) and a second smaller centriolar structure (**c**) are observed at the two poles. Note that the two pronuclei (*blue*) are not mixed and they divide separately in parallel. A low magnification image with small inset squares outlines the approximate positions of the centriolar structures, which are shown under a higher magnification in (**b** and **c**). Embryos were stained with rat anti-α-tubulin (*red*), and anti–N-ter-Asl (*purple*); 4'-6-diamidino-2-phenylindole (DAPI) stains DNA

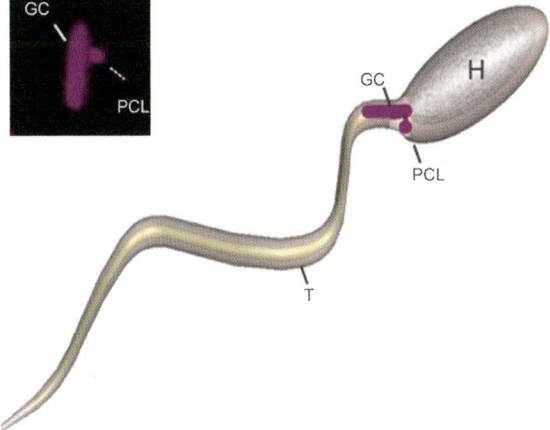

Fig. 1.3 Spermatid with a giant centriole (GC) and a PCL. *Left*, giant centriole and PCL labeled by Ana-1-GFP (*purple*). *Right*, cartoon depicting the relative location of the sperm centriolar structures relative to the sperm head (H) and tail (T). (The panel on the *Left* is modified from Fig. 1.2b in (Blachon et al. 2009)

duplication in early embryonic development, a time when the embryonic genome is not yet fully involved in producing the proteins necessary for development.

On the other hand, the *Drosophila* sperm contains two centriolar structures. The first, termed the "giant centriole" (due to its exceptional length) resembles the distal centriole found in vertebrate sperm and functions to nucleate the sperm flagellum (Friedlander and Wahrman 1966; Fuller 1993; Krioutchkova and Onishchenko 1999; Manandhar et al. 2005; Sun and Schatten 2007). A second centriolar structure associates with the giant centriole and is termed the proximal centriole-like (PCL) structure due to the fact that, like the vertebrate sperm proximal centriole, it does not form a flagellum (Blachon et al. 2009) (Fig. 1.3).

Upon fusion of the sperm and oocyte, the sperm giant centriole recruits PCM proteins from the surrounding cytoplasm and forms a centrosome. The zygote centrosome acts as a microtubule organization center and assembles an aster—a star-like structure consisting of microtubules. These asters are thought to play a role in bringing together the female and male pronuclei and in orchestrating the first cell division (Callaini and Riparbelli 1996). In support for the critical role of centrosomes in zygote biology, it has been reported that interfering with centrosome biogenesis after fertilization inhibits zygote development (Dix and Raff 2007; Stevens et al. 2007; Varmark et al. 2007).

The role of the PCL after fertilization is currently not known; however, one attractive hypothesis is that the PCL later becomes the second centrosome of the zygote. The observation that two centrosomes can be observed after fertilization in mutant oocytes under conditions that block centrosome duplication supports this hypothesis (Stevens et al. 2007).

1.2.2 Syncytial Development

Drosophila early embryonic development takes place in a syncytial blastoderm, a large cell containing many nuclei. In this developmental stage, the embryo undergoes 13 rounds of nuclear duplication and division without forming individual cells surrounded by a plasma membrane; each of these rounds is called a "nuclear cycle". For simplicity, we will refer to each dividing unit that includes a nucleus and its associated centrosomes as a "cell". At the end of syncytial blastoderm development, the nuclei are partitioned into separate cells where they are surrounded by a plasma membrane (cellular blastoderm stage). This partitioning, termed cellularization, is mediated by actin, which forms a cleavage furrow (the structure that mediates the separation of daughter cells).

The syncytial blastoderm is an excellent system where one can do live imaging of centriole duplication in real time (Fig. 1.4). In the embryonic syncytium, the nuclear cycle is very rapid (~ 10 min) and many centrioles duplicate synchronously. The nuclear cycle is comprised of two phases: synthesis and mitosis. Early in synthesis phase, each cell has two centrosomes, each containing one centriole surrounded by PCM (Callaini and Marchini 1989; Callaini and Riparbelli 1990, 1996; Riparbelli et al. 1997). During synthesis phase, the centriole within each of the two centrosomes duplicates, generating a new centriole at the vicinity of the older centriole. At the onset of mitosis, the centrosomes move to opposite poles such that each future daughter "cell" will inherit one of the of the mother "cell" centrosomes. During late mitosis, each of the centrosomes that are associated with one of the nuclei splits into two centrosomes. As a result, each of the daughter "cells" inherently contains two centrosomes, each with one of the centrioles from the original centrosome (Fig. 1.5).

The centriole in the *Drosophila* syncytial blastoderm has a very intriguing structure. Unlike classic centrioles, which are made of nine triplets of

1 Centriole Duplication and Inheritance in *Drosophila melanogaster* 11

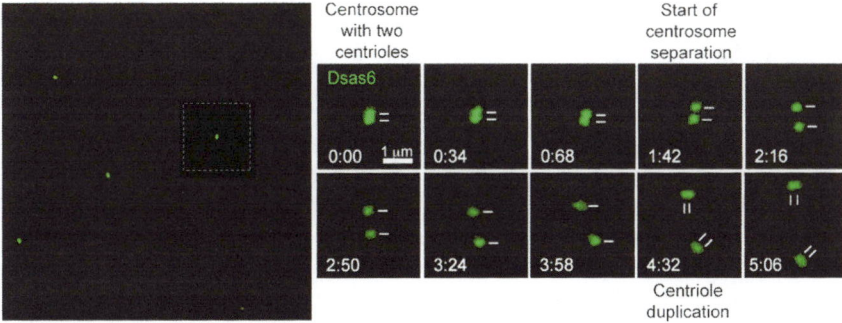

Fig. 1.4 Live imaging of centriole duplication in the early development of an embryo expressing the centriolar marker Sas-6-GFP. The start of centriole separation marks the splitting of the centrosome. The *left* panel shows low magnification of the centrioles, organized in the *right* panels by time. The white square in the panel on the *right* corresponds to the centrioles displayed in the *left* panels. A line distinguishes each centriole. S. B produced this picture

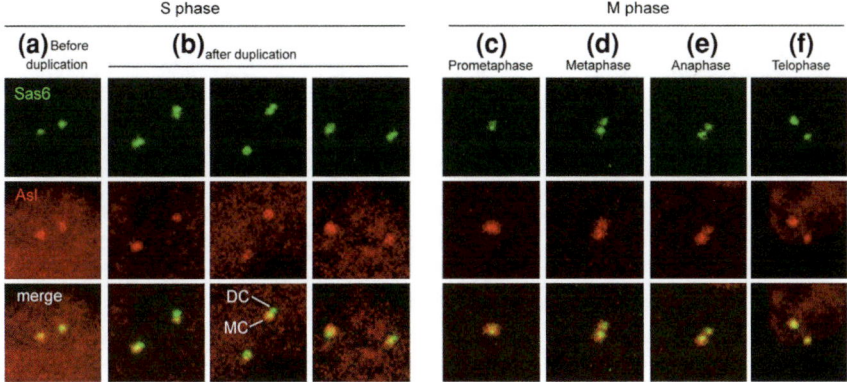

Fig. 1.5 Centriole duplication and centrosome separation in the syncytial blastoderm. The embryo expressed the centriolar marker Sas-6-GFP (*green*) and was stained with an anti-Asl antibody; Asl is a component of the PCM that is found near the centriole (*red*). **a** In early S phase, each of the two centrosomes contain the PCM protein Asl and have a centriole labeled by Sas-6-GFP. **b** During S phase each of the centrioles duplicates to form a daughter centriole (Dc). The daughter centriole is marked with Sas6-GFP but not Asl. **c–f** Only one of the centriole pairs is shown. The mother and daughter centrioles start to separate and the daughter centriole accumulates Asl, finally leading to the formation of two centrosomes. Note that anti-Asl antibody also lightly labeled the nucleus. S. B produced this picture

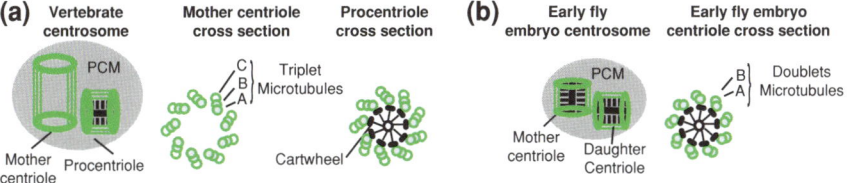

Fig. 1.6 The fly early embryonic centriole resembles a procentriole. (The fig is modified from Fig. 1 in (Avidor-Reiss 2010)

microtubules, centrioles in the syncytial blastoderm are made of nine doublets of microtubules. In addition, syncytial blastoderm centrioles are shorter than classic centrioles: ~200 nm long instead of 400–500 nm long observed in vertebrates. Finally, the centrioles of the syncytial blastoderm have a structure known as the "cartwheel" within their core, which in vertebrate cells is characteristic of a young, developing centriole (procentriole) and is absent from mature centrioles. This raises the hypothesis that syncytial blastoderm centrioles are centriolar structures arrested in the procentriole stage. It is possible that since the syncytial blastoderm nuclei divide rapidly and there is no need for cilia formation (see below), the centrioles do not have the time, nor the need, to develop into their mature states (Fig. 1.6).

1.2.3 The Drosophila Zygote and Syncytial Blastoderm Develop Using Proteins Generated in the Mother

Protein deposition in the oocyte that supports early embryonic development is called maternal contribution. An important implication of the presence of maternal contribution in *Drosophila* development is that, despite the fact that an embryo may be genetically homozygous for a mutation in an essential centrosomal gene, it will still contain the wild-type protein, allowing it to produce normal centrosomes as long as the maternal contribution persists. As a result, studying those mutants cannot reveal the role of centrosomes in early embryogenesis. Indeed, studies using homozygous mutants for an essential centrosomal gene, have demonstrated that the fly embryo can develop normally (Basto et al. 2006; Blachon et al. 2008; Rodrigues-Martins et al. 2007a).

Investigating the role of centrosomes in early embryogenesis requires the study of an embryo that is produced from an oocyte generated in an environment that is also mutated. Since flies with mutations in essential centrosomal proteins are unable to walk, mate, or lay eggs (see below), it is not possible to use embryos produced by homozygous females. However, this obstacle can be overcome using other approaches.

One way is to study centrosomal proteins that are essential for aspects of centrosome function but are not necessary to produce a fertile female. For example, mutations in centrosomin (Cnn) result in flies that are viable but female sterile (Megraw et al. 1999; Vaizel-Ohayon and Schejter 1999). Cnn is a PCM protein that plays an important role in PCM formation and is required for the centrosome's activity as a microtubule organization center (Li and Kaufman 1996). Studies of *cnn* mutants in early embryogenesis reveal an impairment of several aspects of embryo development that depend on the function of the cytoskeleton (Megraw et al. 1999; Vaizel-Ohayon and Schejter 1999). In particular, it appears that Cnn is essential for the organization of actin into cleavage furrows. New information suggests that the centrosome functions as a site where Cnn interacts with Centrocortin, a protein that is required for cleavage furrow formation and is localized both to centrosomes and to cleavage furrows (Kao and Megraw 2009). It is therefore possible that the centrosome functions as a signaling hub within the cell. At this signaling hub, proteins can interact in order to integrate information and later move to other domains in the cell where they execute their function (Alieva and Uzbekov 2008; Wang et al. 2009).

Another way by which to study centrosomes in early embryogenesis is to study hypomorphic mutations of centrosomal proteins that are essential for centrosome formation and produce a female that can mate. In this case, the studied protein is partially functional and missing an activity that is essential specifically for centrosome function during embryogenesis. An example of one such mutation is asl^1, in which the C-terminus of the essential centrosomal protein Asterless (Asl) is truncated (Blachon et al. 2008). While flies homozygous for the severe loss-of-function allele asl^{mecD} are unable to walk and mate, asl^1 generates viable females that can lay eggs. Analysis of embryonic centrosomes generated from an asl^1 female finds that they initially form asters, but these asters are not stable and later fall apart ((Varmark et al. 2007) and S. B. unpublished data). In addition, pronuclei fusion does not take place and embryo development is arrested at the zygote stage. Similar results were obtained when a hypomorphic mutation of the *Drosophila spd*-2 gene was studied, while severe loss-of-function *spd*-2 mutants are unable to walk and mate (Dix and Raff 2007; Giansanti et al. 2008).

One can also study centrosomal proteins that are essential for centrosome formation and produce a female that can mate using genetic tricks. One way to do so is to make germline clones that lack any particular essential centrosomal protein from maternal contribution (Stevens et al. 2007). This is done using the dominant female sterile (DFS) technique (Chou and Perrimon 1996), a variant of FLP-FRT recombination. In this case, recombination is induced in larvae via a heat shock-inducible flippase (FLP); as adults, the fly produces homozygous mutant oocytes that lack the essential centrosomal protein. Study of oocytes that are mutant for *sas*-4 or *sas*-6 after they are fertilized with a wild-type male finds that they possess two centriolar structures (presumably the giant centriole and PCL) (Stevens et al. 2007). These centriolar structures can nucleate centrosomes and can undergo few nuclear cycles before embryogenesis is arrested presumably because the centrosomes cannot duplicate. This suggests that the PCL can form an independent centrosome and demonstrates that centrosomes are essential for syncytial blastoderm development (after their role in zygote pronuclei fusion).

An alternative method to study centrosomes in the syncytial blastoderm is to inject it with an antibody for a particular centrosomal protein and observe the consequence (Conduit et al. 2010). A potential problem with this approach is that when a centrosomal protein forms a complex, binding of an antibody may not only inactivate its intended protein target, but may also inactivate other proteins found in the complex.

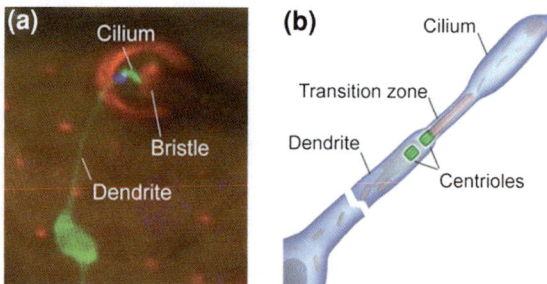

Fig. 1.7 Drosophila mechanosensory neuron morphology. **a** The neuron cell body is filled with tubulin-GFP. The dendrite is lightly labeled by tubulin-GFP. At the dendrite tip, the cilium is strongly labeled by tubulin-GFP. Red labels cuticle structures, including the bristle. Blue labels the Transition zone vicinity. **b** Diagram of sensory cell dendrite and cilium. (The a panel is modified from Fig. 7c in (Avidor-Reiss et al. 2004)

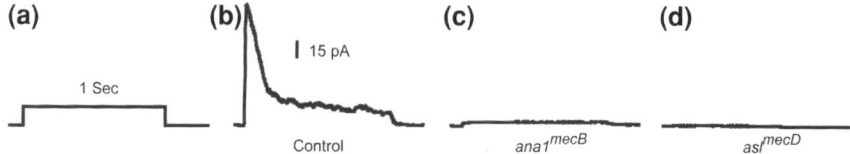

Fig. 1.8 *Drosophila* mutants with centriolar or cilia defects are mechanosensory defective. Displacing a bristle 30 um for one second (**a**) generates a mechanoreceptor current in control flies that adapts over the course of the stimulus (**b**). In contrast, mutations that affect centriole formation (**c** and **d**) and thus cannot form mechanosensory cilia have no mechanoreceptor current. (The a and b panels are modified from Fig. 1.4c in (Avidor-Reiss et al. 2004)

1.3 Centrosomes in Differentiated Cells

1.3.1 Sensory neuron differentiation

In *Drosophila*, the first cells that develop cilia are the type I sensory neurons (Fig. 1.7). These neurons mediate the reception of mechano- and chemo-sensory information and are found in both larvae and the adult fly. One subtype of these sensory neurons is found on the cuticle of the adult fly and mediate touch sensation (Fig. 1.7) (Keil 1997). These neurons are bipolar sensory neurons that extend a dendrite with sensory cilia at their tip in one direction. The sensory cilia are attached to a cutaneous structure termed the bristle. When the bristle moves due to mechanical stimuli, the mechanosensory transduction machinery found in the cilia is activated and a mechanoreceptor current is generated (Fig. 1.8). This current produces an action potential that is delivered to the brain, transmitting information regarding touch sensation or proprioception (Avidor-Reiss et al. 2004; Kernan et al. 1994; Walker et al. 2000).

The sensory neuron is a product of asymmetrical cell division and it inherits a centrosome with two centrioles (Gomes et al. 2009; Keil 1997; Seidl 1991). After the sensory neuron generates a long dendrite, the two centrioles are reorganized and are found in tandem, one of which is attached to a vesicle. This reorganized structure appears to migrate along the dendrite until it reaches the distal end, where the associated vesicle fuses with the plasma membrane to form the sensory cilium (Seidl 1991). The sensory cilium is composed of a transition zone, also called the connecting cilium, and the sensory cilia proper, also called the outer segment (Fig. 1.7b).

1.3.2 Spermatogenesis

Spermatogenesis is the process that takes place in the testes to form mature male gametes and begins when a sperm stem cell divides asymmetrically to form another stem cell and a progenitor spermatogonium. The spermatogonium divides 4 times to form 16 spermatocytes. These spermatocytes grow to ~ 30 times their original size and ultimately undergo two cycles of meiosis to generate 64 spermatids. The spermatids, which are first round, undergo a dramatic differentiation program, called spermiogenesis. The completion of this differentiation program results in the formation of a sperm cell that is ~ 2 mm long, a length comparable to that of the fly itself.

Centrosomes in the *Drosophila* testes have several interesting properties:

First, unlike the syncytial blastoderm, the centrosome and centrioles of the adult testes are similar to their vertebrate counterparts (Tates 1971; Tokuyasu 1975). These centrioles have nine triplet microtubules (Fig. 1.9k). This normal centriolar structure correlates with the ability of the spermatogenic centrioles to form cilia and suggests that in ciliated cells, the centriole needs to develop into its mature state. In this regard, sensory neurons that have cilia also contain centrioles with triplet microtubules (Keil 1997).

Second, in spermatogenesis, the centrosomes form two types of cilia. During spermatocyte growth, each of the spermatocyte's four centrioles forms a primary cilium-like structure of unknown function (Fig. 1.9b) (Tates 1971). Later in spermiogenesis, each of these primary cilia is modified to form a motile cilium—the sperm flagellum.

Third, male meiosis is absolutely dependent on centrosomes. Flies that do not have functional centrosomes fail to accurately separate genetic material and mitochondria (Fig. 1.10), one of the reasons why centrosomal mutants are male sterile. It is currently unclear why centrosomal defects cause abnormalities in male meiosis but do not disrupt mitosis. However, it is possible that male-specific meiotic defects are due to the fact that the centrosomes are associated with a ciliary-like structure that requires centrosomal components for their formation.

Fourth, during spermatocyte growth and spermiogenesis, the centriole elongates to ~ 2.5 um, much larger than centrioles found in any other *Drosophila* tissue or

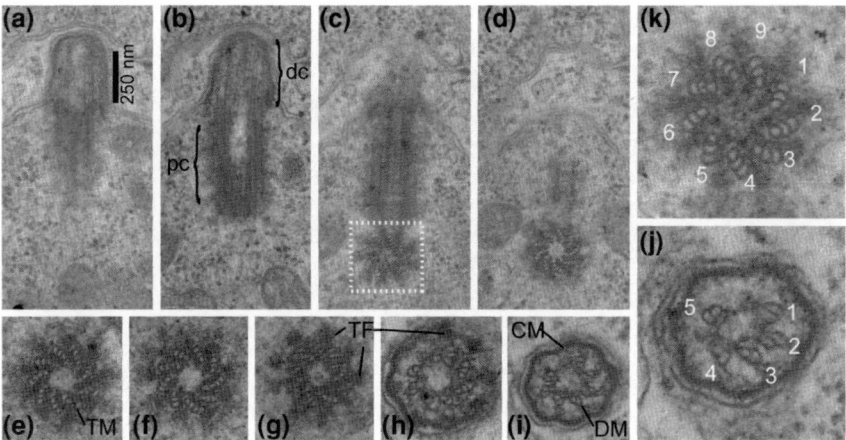

Fig. 1.9 The giant centriole (proximal centriole) and primary cilium (distal centriole) of fly spermatocytes: A-J serial section electron microscopy analysis of a pair of giant centrioles organized in an orthogonal relationship. **k** The cross-section of the daughter centriole from **c** is magnified to demonstrate triplet microtubules. **j** The last cross-section is highlighted to depict the presence of irregular numbers of doublet microtubules in the primary cilium. TF, transitional fibers connecting the centriole to the plasma membrane; TM, triplet microtubules; DM, doublet microtubules; CM, cilium membrane; pc and dc according to Tates: pc, proximal centriole or basal body and dc, distal centriole

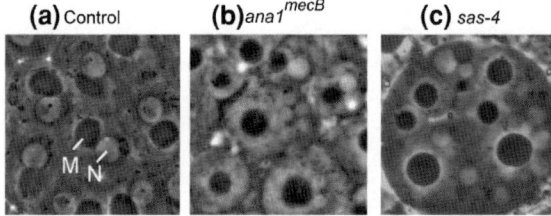

Fig. 1.10 Round spermatids formed imminently after meiosis, contain a white nucleus (N) and dark mitochondria (M) of similar size (**a**). Centriolar mutants in the spermatid state form nuclei and mitochondria of variable size (**b** and **c**)

those of other organisms (Blachon et al. 2009; Mottier-Pavie and Megraw 2009; Tates 1971). At particular stage of spermiogenesis these giant centrioles are surrounded by a long and thick PCM (also referred to as the "centriolar adjunct"). These centrioles provide a very convenient model to study centriole elongation and several mutants that have shortened giant centrosomes have been described. Proteins essential for giant centriole elongation include: Bld10, Poc1, and Sas-4 (Blachon et al. 2009; Mottier-Pavie and Megraw 2009).

Fifth, during spermiogenesis, the spermatid cell forms a centriole precursor-like structure called the PCL (Fig. 1.11a). The PCL has been proposed to be a centriole intermediate that arrests at the stage before the centriolar microtubules are assembled. Therefore, by studying how the PCL forms, one can study the early

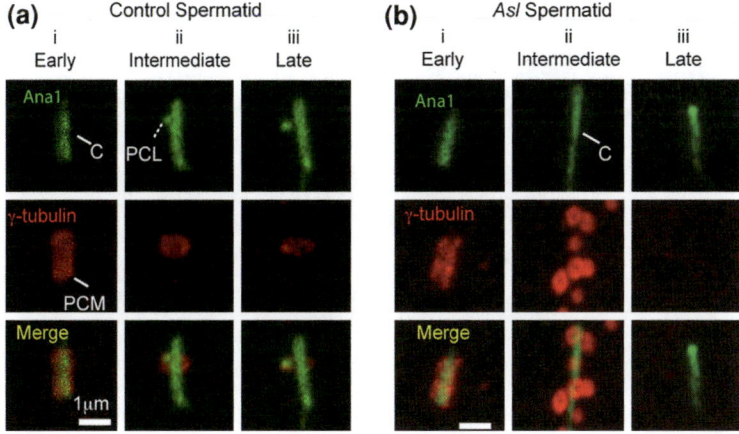

Fig. 1.11 Asl is essential for PCL formation. **a** To determine the relationship between the PCL, the giant centriole (**c**), and the PCM, the localization of Ana1-GFP relative to the centriolar adjunct protein γ-tubulin (Wilson et al. 1997) was analyzed. In early spermatids, γ-tubulin labels the vicinity of the giant centriole along most of its length (i). During the time the Ana1 labelled PCL appears, γ-tubulin assembles a half ring structure around the giant centriole, which touches the PCL (ii). At a later stage, the PCL migrates distally and γ-tubulin labels the PCL. **b** To determine if Asl plays a role in PCL formation we followed maternally contributed giant centrioles in spermatids of the asl^{mecD} mutant. No PCL is observed near the giant centriole, demonstrating an essential role for Asl in PCL formation. However, while in early asl^{mecD} spermatids γ-tubulin localization appears to be normal, many abnormal γ-tubulin rings are found at the vicinity of the maternally contributed giant centrioles in intermediate asl^{mecD} spermatids. (The **a** panel is modified from Fig. 1.2c in (Blachon et al. 2009)

events in centriole biogenesis (Blachon et al. 2009). PCL formation depends on the function of the centrosomal proteins Plk4, Sas-6, and Asl and it contains the following centrosomal proteins: Sas-6, Ana2, Ana1, Sas-4, Bld10, Cnn, and Asl (Blachon et al. 2009; Mottier-Pavie and Megraw 2009; Stevens et al. 2010b) and (Fig. 1.11b).

1.3.3 The Drosophila Adult Sensory Neuron and Testes Develop Using Proteins Generated During Metamorphosis

Unlike early embryogenesis that utilizes maternally contributed proteins, the adult fly develops during metamorphosis by utilizing proteins synthesized from the genome of the fly itself. Therefore, by studying sensory neurons and testes in the pupa or adult, one can study the full impact of a mutation in a centrosomal protein. Interestingly, flies that have mutations in essential centrosomal proteins and have no centrosomes can develop to adulthood but die soon after leaving the pupa

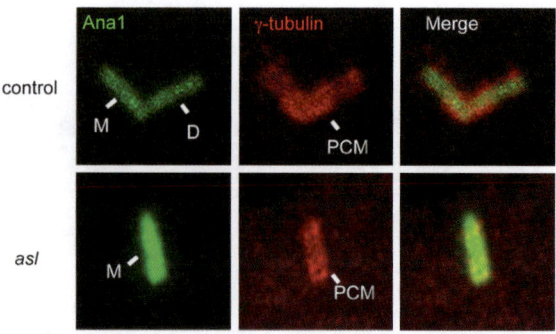

Fig. 1.12 Asl is essential for centriole duplication in vivo. Wild-type cells (control) contain both a mother centriole (M) and its daughter centriole (D). Centrioles are labeled by Ana1-GFP (green); PCM assembled around these centrioles are labeled for γ-tubulin (*red*). In *asl* loss-of-function mutant cells, the maternally contributed centrioles elongate but its duplication is blocked. Modified from (Blachon et al. 2008). (The fig is modified from Fig. 1.5d in (Blachon et al. 2008)

because they cannot stand on their legs, walk, or fly [defects collectively referred to as uncoordination (Kernan et al. 1994)]. This uncoordination results from the fact that flies with centrosomal defects have no mechanosensory cilia and cannot sense the environment or their body parts.

The germline stem cells found in the testes originate from the first group of cells that are generated early in embryonic development (pole cells) (Okada 1998). When germline stem cells divide to form a new stem cell and a spermatogonium, the older centriole (the maternally contributed centriole) stays in the stem cells while the newer centriole is inherited by the spermatogonium (Blachon et al. 2008; Yamashita et al. 2007). A fly that is homozygous mutant for an essential centriole component and cannot form new centrioles will have functional centrioles in the germline stem cells that are made using maternal contribution, but will lack these centrioles in the later progenitors.

Since germline stem cells in the developing *Drosophila* testes have two maternally contributed centrioles, some of the first spermatogonium to form can each inherit one maternally contributed centriole. These centrioles then duplicate and elongate during spermatogenesis and end up in the first spermatids to form. By that time maternal contribution of wild-type proteins becomes depleted. Following maternally contributed centrioles of the spermatogonium and later in spermatogenesis allows one to dissect the role of a particular protein under circumstances where a centriole is present (Fig. 1.12). Using this approach, it was found that maternally contributed centrioles require Sas-4 to elongate but not Asl (Blachon et al. 2009; Blachon et al. 2008). On the other hand, formation of the PCL can form in Sas-4 mutants but not in Asl mutants (Blachon et al. 2009) and Fig. (1.11b).

1.3.4 Phenotypic Characterization of Centrosomal Mutants

The spermatid flagellum is formed in a unique way from that of sensory cilia (Avidor-Reiss et al. 2004; Han et al. 2003). Sensory cilia are formed within a distinct compartment separated from the rest of the cell by a transition zone that is thought to function as the ciliary gate (Betleja and Cole 2010; Craige et al. 2010; Omran 2010; Williams et al. 2011). As a result, sensory cilium formation requires a complex machinery known as the intra-flagellar transport (IFT) (Rosenbaum 2002; Scholey and Anderson 2006). On the other hand, the spermatid flagellum is formed in the cytoplasm and does not depend on IFT, a process called cytoplasmic or non-compartmentalized ciliogenesis (Avidor-Reiss et al. 2004; Han et al. 2003). In flies, this distinction allows phenotypic analysis to rapidly determine if a protein is an essential centriolar protein, if it is important for compartmentalized ciliogenesis, or if it is important for the conversion of a centriole into a basal body.

Mutants in essential centriolar proteins will have the following phenotype that reflects the role of centrioles in their development. Meiosis will be abnormal in males due to the importance of centrosomes in meiosis, sperm will be nonmotile as a result of defects in non-compartmentalized ciliogenesis, and adult flies will be uncoordinated due to defects in compartmentalized ciliogenesis. Such mutants include *Sas-6*, *Sas-4*, *Asl*, *Ana-1*, *Spd-2*, and *Plk4* (Basto et al. 2006; Bettencourt-Dias et al. 2005; Blachon et al. 2009; Blachon et al. 2008; Giansanti et al. 2008; Martinez-Campos et al. 2004; Rodrigues-Martins et al. 2007a).

Mutations in proteins that are essential for compartmentalized ciliogenesis will not affect meiosis or sperm motility, but will still result in adult uncoordination due to defects in compartmentalized ciliogenesis. Examples include mutation in IFT proteins such as Oseg 1 and 2 as well as IFT88 (Avidor-Reiss et al. 2004; Han et al. 2003).

A third group of centriolar proteins seems to affect compartmentalized and non-compartmentalized ciliogenesis, but do not have an impact on meiosis. This group seems to function in the conversion of a centriole into a basal body. Such mutants include *unc* (Baker et al. 2004).

1.4 The Role of the Centrosome

Immediately after the discovery of the centrosome by Flemming (1875) and van Beneden (1876), two major hypotheses regarding its function were postulated. Boveri hypothesized that the centrosome is a cellular organelle found close to the nucleus and with paramount importance in cell division. On the other hand, the Henneguy-Lenhossek theory (1898) claimed that the centrosome and basal body are the same organelle located in two distinct sites, with the centrosome located at the cell center near the nucleus, and the basal body existing at the base of the cilia at the plasma membrane. This theory was the first to emphasize the importance of the centrosome in cilia formation.

These two hypotheses were based only on observations of the centrosome and have remained untested for many years. Recent studies in *Drosophila* have allowed these theories to be tested directly. A genetic approach that has been employed is elimination of the centrosome using mutants and the study of its effect on cilia formation and cell division. Ideally, one would employ a null mutation in a centrosome-specific component that is absolutely essential for centrosome formation. Studying the phenotype of these mutants can test if the centrosome is essential for cell division and/or cilia formation.

Study of fly mutants that block centrosome formation (*sas-4 and asl^{mecD}*) has suggested that flies lacking centrosomes die due to mechanosensory defects caused by a failure to form cilia (Basto et al. 2006; Blachon et al. 2008). These flies also exhibit a failure in male meiosis and their sperm form without an axoneme. However, it is important to note that during larval and pupal development, the centrosome does not play an essential role in cell division; division, though delayed, still takes place. This argues that, at least in larvae, pupae, and adult files, centrosomes and centrioles behave in a way consistent with the Henneguy-Lenhossek theory.

In flies, the oocyte carries a large amount of maternally derived proteins that support the early development of the embryo. Therefore, analyzing the role of the centrosome in the early embryo requires the embryo's mother or oocyte to be mutant (see above). Interestingly, such studies suggest that centrosomes are vital in the zygote and early embryo. However, whether this is because centrosomes are essential to mediate nuclear division is not clear. Regardless, in this developmental stage, fly cells do not have cilia and it is thus possible that the early embryonic syncytium requires the centrosome for "cell" division in a way that is consistent with the Boveri hypothesize. However, an important caveat in this idea is that centrosomes may have an important function in the early embryo that is neither related to cilia formation nor cell division. For example, it appears that centrosome function is essential for the migration and fusion of the gamete pronuclei and in acting as a signaling hub (see above).

1.5 Two Pathways of Centriole Formation

A centriole forms by one of two pathways (Anderson and Brenner 1971; Delattre and Gonczy 2004; Loncarek and Khodjakov 2009; Loncarek et al. 2007). In the "acentriolar pathway", a daughter centriole forms de novo without a preexisting centriole. This occurs when multiple centrioles are required in a cell or under the unusual situation when preexisting centrioles are absent. The acentriolar pathway produces an imprecise number of centrioles. In flies, centrioles can form de novo when certain centrosomal proteins are overexpressed in oocytes or fertilized embryos (e.g., PLK4, Asl, and Ana-2) (Peel et al. 2007; Rodrigues-Martins et al. 2007b; Stevens et al. 2010a). To what extent these centrioles are similar to centrioles that normally form is not clear, but in some cases, they appear to have normal centriolar structures, have the capacity to recruit PCM proteins, and can form astral microtubules. Whether, these induced centrosomes can form cilia or mediate meiosis is unknown. However, overexpression of centrosomal proteins can also form centriolar structures that are clearly abnormal (Gopalakrishnan et al. 2010; Rodrigues-Martins et al. 2007a; Stevens et al. 2010a). The capacity to induce centriole formation by overexpression seems to be tissue-specific and is more prevalent in the oocyte/embryo than in the testes (Peel et al. 2007; Stevens et al. 2009).

Whether centrioles can form de novo in *Drosophila melanogaster* without overexpression of centriolar components is not clear. It was shown that in other *Drosophila* species, oocytes assemble a number of cytoplasmic asters after activation with centrioles and centrosomal proteins (Ferree et al. 2006; Riparbelli and Callaini 2003). In addition, in *sas-4* null mutants that cannot form centrioles, foci containing centriolar markers appear transiently, suggesting that nascent procentrioles form de novo but fail to develop further in the absence of *Sas-4* (Gopalakrishnan et al. 2011).

In the "centriolar pathway", a preexisting centriole acts as a template to ensure that only a single daughter centriole is formed per cell cycle. However, it does not appear to impart structural information to the daughter (Phillips 1967; Rodrigues-Martins et al. 2007b). Before cell division, centriole duplication under the "centriolar pathway" results in four centrioles; each mother/daughter centriole pair forms a centrosome that migrates to one of the cell's two poles where it helps orient the mitotic spindle, a structure essential for cell division. Having precisely one centriole pair in interphase and two centriole pairs during mitosis is believed to be critical for proper cell division and having too many centrioles can initiate tumorigenesis (Basto et al. 2008; Cunha-Ferreira et al. 2009; Fukasawa 2007; Ganem et al. 2009).

How the mother centriole ensures that only a single daughter centriole forms is not clear. However, it is readily observed that there is already only one daughter centriole at the vicinity of the mother centriole by the time a procentriole is present. Therefore, an approach to address this question is to study proteins that are involved early in centriole formation when the formation of the procentriole is initiated. In recent years, extensive proteomic, genetic, and bioinformatic studies have identified many of the key proteins critical for centriole formation (Andersen et al. 2003; Avidor-Reiss et al. 2004; Fritz-Laylin and Cande 2010; Gonczy et al. 2000; Keller et al. 2005; Kilburn et al. 2007; Li et al. 2004; Mahoney et al. 2006). Some of these proteins were analyzed in flies and were shown to function early in procentriole initiation (Table 1.1). Some of the important discoveries regarding these proteins that were made using flies are summarized below.

1.6 Identification of Centrosomal Proteins and Mutants in *Drosophila melanogaster*

Identification of centrosomal components that play a specific role in centrosome have been hampered by initial difficulties in obtaining sufficient quantities of material for biochemical isolation and a lack of clear phenotypes expected from mutation of essential centrosomal proteins. Over time, several approaches have resulted in the identification of *Drosophila* proteins important for centrosome biogenesis and function (Table 1.1). This includes the use of anti-centrosomal antibodies, the use of reverse genetic approaches, and the use of forward genetic approaches such as the cloning of genes from sterile, mechanosensory, or PCL mutants.

1.6.1 Anti-Centrosomal Antibodies

Analysis of the centrosomal gene CP190 (Whitfield et al. 1988) was facilitated by the use of an antibody raised against isolated centrosomes and that were later found to bind CP190. CP60 was identified as a protein that interacts with CP190 (Kellogg and Alberts 1992). CP190 and CP60 localized to the centrosome in a cell cycle-dependent manner and their amount at the centrosome was shown to be maximal during mitosis and was barely detected during interphase. However, CP190 and CP60 are not centrosome-specific and are also found in the nucleus during interphase. A mutant for the CP190 gene was identified using standard genetic approaches and it was found that CP190 is not essential for centrosome biogenesis or function, but its function in the nucleus is essential for fly viability (Butcher et al. 2004). RNAi study of CP60 suggests it is also not essential for centrosome biogenesis or function (Butcher et al. 2004). Orthologs of CP190 and CP60 have not been identified in vertebrates.

1.6.2 Reverse Genetics

Another way to obtain mutants in centrosomal proteins is to study the *Drosophila* ortholog of known centriolar proteins in other organisms. This approach benefits from the availability of large collections of mutant flies in identified genes. For example, *Sas-4* (Kirkham et al. 2003), *Sas-6* (Dammermann et al. 2004; Leidel et al. 2005), *Spd-2* (Kemp et al. 2004; Pelletier et al. 2004), *Bld10* (Matsuura et al. 2004) and *PLK4* (Habedanck et al. 2005) are all conserved centriolar proteins first identified in other model organisms.

1.6.3 Bioinformatic subtractive screen for ciliary and centrosomal proteins

Several centrosomal proteins were identified in a bioinformatic subtractive screen for ciliary and centrosomal proteins (Avidor-Reiss et al. 2004; Li et al. 2004). Such a screen is based on the idea that only organisms that have cilia should have ciliary genes and is used to identify genes that are conserved in organisms with cilia but are missing in organisms that lack cilia. Since centrioles are only observed in ciliated organisms, this screen can also be used to identify centriolar proteins (Carvalho-Santos et al.; Hodges et al.). Examples of *Drosophila* centrosomal genes identified using such approaches are *poc1* and *poc18* (Keller et al. 2005).

1.6.4 Male Sterile Mutants

The centrosomal mutant asterless (asl) was identified in a study of male sterile mutants (Bonaccorsi et al. 1998). Unfortunately, for nearly 10 years, it was incorrectly believed that Asl was solely important for centrosomal function and aster formation (hence its name: Asterless) while its essential role in centriole formation remained unknown (Bonaccorsi et al. 1998; Giansanti et al. 2001; Oliferenko and Balasubramanian, 2002; Varmark et al. 2007; Wakefield et al. 2001). Nonetheless, it was later demonstrated that Asl is instead essential for centriole duplication and that the absence of aster formation in *asl* mutants is mainly due to a lack of centrioles (Blachon et al. 2008). This finding came from analysis of asl^{mecD}, a new mutant that was discovered in a mechanosensory mutant screen (see below). Asl is a conserved centrosomal protein known as *Cep152* in vertebrates (Blachon et al. 2008; Varmark et al. 2007).

Male sterility is a common phenotype in centrosomal mutants and is a valuable method to identify centriolar mutants. Male sterility in centriolar mutants can arise by several distinct mechanisms:

First, centrosomes are essential for male meiosis and flies with abnormal meiosis will fail to form functional sperm. This type of defect is most commonly diagnosed by examining spermatids immediately after meiosis by light microscopy and observing abnormal numbers and shapes of nuclei and mitochondria (Fig. 1.11) (Bonaccorsi et al. 1998; Li et al. 1998).

Second, the centriole acts in templating the sperm flagellum. As a result, abnormalities or absences of the centriole translate to abnormalities or absences of the sperm flagellum. This type of defect is diagnosed by observing that the fly sperm is not motile, or by electron microscopy analysis where fly sperm cross-section shows a missing or abnormal axoneme (Baker et al. 2004; Blachon et al. 2008; Mottier-Pavie and Megraw, 2009; Rodrigues-Martins et al. 2007a).

Third, centrosomes are essential for mechanosensory cilia formation (Blachon et al. 2009; Blachon et al. 2008; Martinez-Campos et al. 2004). In the absence of normal centrioles, flies have abnormal proprioception and cannot court the female and mate with her.

Fourth, it is possible that an additional mechanism of male sterility is the failure to form a normal PCL. If the PCL becomes one of the zygotic centrosomes, it is expected that a nonfunctional PCL will cause male sterility even if the sperm is delivered to the oocyte and fertilization takes place. In this case, genes whose mutations cause PCL failure are expected to fall into a class of interesting mutants referred to as paternal effect genes (Fitch and Wakimoto 1998).

In some mutants of centrosomal proteins, such as *bld10*, sterility is caused by an abnormal sperm flagellum (Mottier-Pavie and Megraw, 2009). In other mutants, such as *cnn*, sterility is a result of meiosis defects followed by abnormal differentiation of the motile sperm (Li et al. 1998). However, in the most severe centrosomal mutants (for example, *asl, sas-4, sas-6, plk4, ana1*) sterility is caused by a combination of all of these mechanisms; the flies are uncoordinated, fail in meiosis, and do not form flagella (Basto et al. 2006; Bettencourt-Dias et al. 2005; Blachon et al. 2009; Blachon et al. 2008; Martinez-Campos et al. 2004; Rodrigues-Martins et al. 2007a). While, defects in sperm motility, meiosis, or proprioception are not restricted to centrosomal mutations, the combination of these three phenotypes is a very strong indication of a mutation in a centrosomal protein (see below).

1.6.5 Mechanosensory Mutants

Several centrosome and basal body-specific proteins were identified by a screen for mechanosensory mutants. This screen led to the identification of the basal body protein *unc* (Baker et al. 2004), the *mecB* allele of *ana1* (Blachon et al. 2009), and the *mecD* allele of *asl* (Blachon et al. 2008).

A screen for mechanosensory mutants looks for adult lethal mutations that have no or an abnormal mechanoreceptor current. Since the mechanosensory apparatus is located in cilia, loss of the centrosome results in a loss of the mechanoreceptor current. Therefore, as in mutants of genes involved in various aspects of mechanotransduction (Chung et al. 2001; Walker et al. 2000), centriolar mutants are adult lethal and have no mechanoreceptor current (Fig. 1.8).

1.6.6 PCL Mutants

The PCL is a centriolar structure resembling an early intermediate in centriole biogenesis in its composition and in the genetic pathway that underlies its formation (Blachon et al. 2009). Since the PCL is similar to an early centriolar intermediate, it was postulated that it should be possible to identify mutants of genes required in early centriole formation by screening for mutants that do not form a normal PCL.

Use of this approach led to the discovery of a mutant in the *Drosophila* ortholog of Poc1. Since it is thought that the PCL is related to male fertility, male sterile mutants were screened for PCL defects. PCL presence was scored using a centriolar protein that labels the PCL strongly (Ana1). Poc1 is a conserved centrosomal protein found in protists and mammals and its localization suggests that it plays a role in early centriole formation (Blachon et al. 2009; Keller et al. 2009; Keller et al. 2005; Pearson et al. 2009).

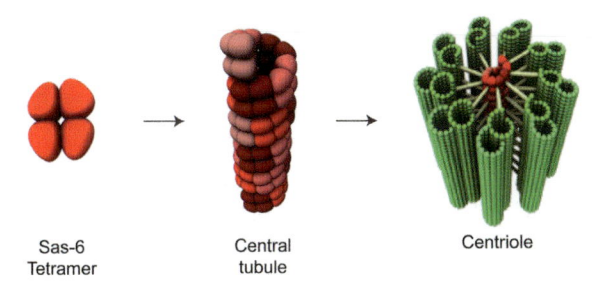

1 Centriole Duplication and Inheritance in *Drosophila melanogaster*

Fig. 1.13 Sas-6 homomers as the building block of centriole central tubule. *Sas-6* forms homomers in the cytosol. These tetramers are hypothesized to polymerize and form the central tubule of the centriole (Gopalakrishnan et al. 2010). (Illustrated by Iwasa, Janet Haru)

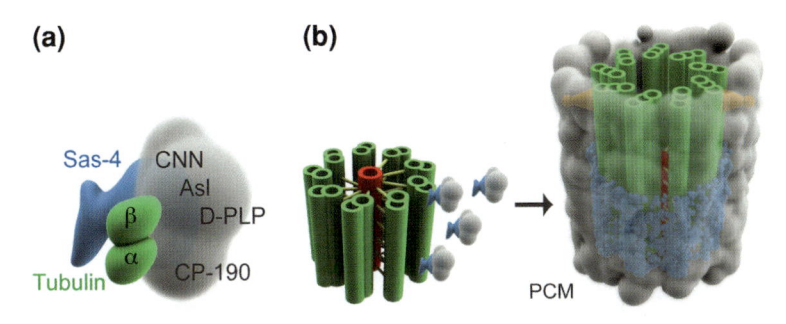

Fig. 1.14 S-CAP complex. **a** *Sas-4* forms a complex known as S-CAP together with Cnn, Asl, and D-PLP. This complex also contains CP190 and Tubulin. **b** S-CAP complexes are tethered to the centrosome via Sas-4 and contribute to the formation of the PCM (Gopalakrishnan et al. in press) (The fig is modified from Fig. 1.7 (Gopalakrishnan et al. 2011)

1.7 Identification of Centrosomal Complexes

Drosophila embryos are a rich source for centrosomal protein complexes (Gopalakrishnan et al. in press; Kellogg and Alberts 1992; Moritz et al. 1995). Originally, this system was used to isolate CP190 and CP60 complexes, which are nuclear and centrosomal proteins (Kellogg and Alberts 1992; Kellogg et al. 1995). Later, this system was used to purify γ-TuRC and γ-TuSC complexes, which are found in the cytosol and reside in the PCM (Moritz et al. 1998; Oegema et al. 1999). More recently, *drosophila* embryos were used to isolate complexes of specific centrosomal proteins such Sas-6 (Gopalakrishnan et al. 2010) and Sas-4 (Gopalakrishnan et al. in press). Sas-6 forms homomers that are hypothesized to be the building block of the centriole central tubule (Gopalakrishnan et al. 2010; Kitagawa et al. 2011; van Breugel et al. 2011) (Fig. 1.13).

Sas-4 forms a complex (named S-CAP) with the centrosomal proteins CNN, Asl, and D-PLP (Gopalakrishnan et al. 2011) (Fig. 1.14). Interestingly, mutations in the orthologs of these proteins results in microcephaly, a developmental disorder where brain size is severely reduced (Al-Dosari et al. 2010; Bond et al. 2005;

Kalay et al. 2011; Thornton and Woods 2009). The finding that microcephaly linked proteins form a common complex may explain why mutations in any of these lead to the same disorder. The S-CAP complex also contains CP190 and tubulin (Gopalakrishnan et al. 2011).

Drosophila embryos are also a rich source for centrosomes. These centrosomes can nucleate microtubules in a cell-free system, allowing the study of the mechanisms of astral microtubules formation (Moritz et al. 1995). The function of centrosomal complexes can be studied further by stripping the centrosome from its PCM using high salt and then adding embryo extract and purified γ-TuRC complexes (Moritz et al. 1998). Stripping the centrosome is also useful to identify the mechanism of recruiting other centrosomal complexes and was used to show that Sas-4 is the S-CAP component that tethers the complex to the centrosome (Gopalakrishnan et al. 2011).

Acknowledgments This material is based upon work supported by the National Science Foundation under Grant No. MCB-1121176. Andrey Polyanovsky was supported by RFBR grant 10-04-01027

References

Al-Dosari MS, Shaheen R, Colak D, Alkuraya FS (2010) Novel CENPJ mutation causes Seckel syndrome. J Med Genet 47:411–414

Alieva IB, Uzbekov RE (2008) The centrosome is a polyfunctional multiprotein cell complex. Biochemistry (Mosc) 73:626–643

Andersen JS, Wilkinson CJ, Mayor T, Mortensen P, Nigg EA, Mann M (2003) Proteomic characterization of the human centrosome by protein correlation profiling. Nature 426:570–574

Anderson RG, Brenner RM (1971) The formation of basal bodies (centrioles) in the Rhesus monkey oviduct. J Cell Biol 50:10–34

Avidor-Reiss T (2010) The cellular and developmental program connecting the centrosome and cilium duplication cycle. Semin Cell Dev Biol 21:139–141

Avidor-Reiss T, Maer AM, Koundakjian E, Polyanovsky A, Keil T, Subramaniam S, Zuker CS (2004) Decoding cilia function: defining specialized genes required for compartmentalized cilia biogenesis. Cell 117:527–539

Azimzadeh J, Marshall WF (2010) Building the Centriole. Curr Biol 20:R816–R825

Baker JD, Adhikarakunnathu S, Kernan MJ (2004) Mechanosensory-defective, male-sterile unc mutants identify a novel basal body protein required for ciliogenesis in Drosophila. Development 131:3411–3422

Basto R, Brunk K, Vinadogrova T, Peel N, Franz A, Khodjakov A, Raff JW (2008) Centrosome amplification can initiate tumorigenesis in flies. Cell 133:1032–1042

Basto R, Lau J, Vinogradova T, Gardiol A, Woods CG, Khodjakov A, Raff JW (2006) Flies without Centrioles. Cell 125:1375–1386

Betleja E, Cole DG (2010) Ciliary trafficking: CEP290 guards a gated community. Curr Biol 20:R928–R931

Bettencourt-Dias M, Rodrigues-Martins A, Carpenter L, Riparbelli M, Lehmann L, Gatt MK, Carmo N, Balloux F, Callaini G, Glover DM (2005) SAK/PLK4 is required for centriole duplication and flagella development. Curr Biol 15:2199–2207

Blachon S, Cai X, Roberts KA, Yang K, Polyanovsky A, Church A, Avidor-Reiss T (2009) A proximal centriole-like structure is present in Drosophila spermatids and can serve as a model to study centriole duplication. Genetics 182:133–144

Blachon S, Gopalakrishnan J, Omori Y, Polyanovsky A, Church A, Nicastro D, Malicki J, Avidor-Reiss T (2008) Drosophila asterless and vertebrate Cep152 Are orthologs essential for centriole duplication. Genetics 180:2081–2094

Bonaccorsi S, Giansanti MG, Gatti M (1998) Spindle self-organization and cytokinesis during male meiosis in asterless mutants of Drosophila melanogaster. J Cell Biol 142:751–761

Bond J, Roberts E, Springell K, Lizarraga SB, Scott S, Higgins J, Hampshire DJ, Morrison EE, Leal GF, Silva EO et al (2005) A centrosomal mechanism involving CDK5RAP2 and CENPJ controls brain size. Nat Genet 37:353–355

Butcher RD, Chodagam S, Basto R, Wakefield JG, Henderson DS, Raff JW, Whitfield WG (2004) The Drosophila centrosome-associated protein CP190 is essential for viability but not for cell division. J Cell Sci 117:1191–1199

Callaini G, Marchini D (1989) Abnormal centrosomes in cold-treated Drosophila embryos. Exp Cell Res 184:367–374

Callaini G, Riparbelli MG (1990) Centriole and centrosome cycle in the early Drosophila embryo. J Cell Sci 97(Pt 3):539–543

Callaini G, Riparbelli MG (1996) Fertilization in Drosophila melanogaster: centrosome inheritance and organization of the first mitotic spindle. Dev Biol 176:199–208

Carvalho-Santos Z, Machado P, Branco P, Tavares-Cadete F, Rodrigues-Martins A, Pereira-Leal JB, Bettencourt-Dias M (2010) Stepwise evolution of the centriole-assembly pathway. J Cell Sci 123:1414–1426

Chou TB, Perrimon N (1996) The autosomal FLP-DFS technique for generating germline mosaics in Drosophila melanogaster. Genetics 144:1673–1679

Chung YD, Zhu J, Han Y, Kernan MJ (2001) nompA encodes a PNS-specific, ZP domain protein required to connect mechanosensory dendrites to sensory structures. Neuron 29:415–428

Conduit PT, Brunk K, Dobbelaere J, Dix CI, Lucas EP, Raff JW (2010) Centrioles regulate centrosome size by controlling the rate of Cnn incorporation into the PCM. Curr Biol 20:2178–2186

Craige B, Tsao CC, Diener DR, Hou Y, Lechtreck KF, Rosenbaum JL, Witman GB (2010) CEP290 tethers flagellar transition zone microtubules to the membrane and regulates flagellar protein content. J Cell Biol 190:927–940

Cunha-Ferreira I, Bento I, Bettencourt-Dias M (2009) From zero to many: control of centriole number in development and disease. Traffic 10:482–498

Dammermann A, Muller-Reichert T, Pelletier L, Habermann B, Desai A, Oegema K (2004) Centriole assembly requires both centriolar and pericentriolar material proteins. Dev Cell 7:815–829

Delattre M, Gonczy P (2004) The arithmetic of centrosome biogenesis. J Cell Sci 117:1619–1630

Dix CI, Raff JW (2007) Drosophila Spd-2 Recruits PCM to the Sperm Centriole, but Is Dispensable for Centriole Duplication. Curr Biol.

Dzhindzhev NS, Yu QD, Weiskopf K, Tzolovsky G, Cunha-Ferreira I, Riparbelli M, Rodrigues-Martins A, Bettencourt-Dias M, Callaini G, Glover DM (2010) Asterless is a scaffold for the onset of centriole assembly. Nature 467:714–718

Ferree PM, McDonald K, Fasulo B, Sullivan W (2006) The origin of centrosomes in parthenogenetic hymenopteran insects. Curr Biol 16:801–807

Fitch KR, Wakimoto BT (1998) The paternal effect gene ms(3)sneaky is required for sperm activation and the initiation of embryogenesis in Drosophila melanogaster. Dev Biol 197:270–282

Friedlander M, Wahrman J (1966) Giant centrioles in neuropteran meiosis. J Cell Sci 1:129–144

Fritz-Laylin LK, Cande WZ (2010) Ancestral centriole and flagella proteins identified by analysis of Naegleria differentiation. J Cell Sci 123:4024–4031

Fukasawa K (2007) Oncogenes and tumour suppressors take on centrosomes. Nat Rev Cancer 7:911–924

Fuller MT (1993) Spermatogenesis. In: Bate M and Martinez-Arias A (eds) The development of Drosophila melanogaster, Cold Spring Harbor, New York, Cold Spring Harbor Laboratory Press), pp. 71–174

Fulton C, Dingle AD (1971) Basal bodies, but not centrioles, in Naegleria. J Cell Biol 51:826–836

Ganem NJ, Godinho SA, Pellman D (2009) A mechanism linking extra centrosomes to chromosomal instability. Nature 460:278–282

Giansanti MG, Bucciarelli E, Bonaccorsi S, Gatti M (2008) Drosophila SPD-2 Is an essential centriole component required for PCM recruitment and astral-microtubule nucleation. Curr Biol

Giansanti MG, Gatti M, Bonaccorsi S (2001) The role of centrosomes and astral microtubules during asymmetric division of Drosophila neuroblasts. Development 128:1137–1145

Gomes JE, Corado M, Schweisguth F (2009) Van Gogh and Frizzled act redundantly in the Drosophila sensory organ precursor cell to orient its asymmetric division. PLoS ONE 4:e4485

Gonczy P, Echeverri C, Oegema K, Coulson A, Jones SJ, Copley RR, Duperon J, Oegema J, Brehm M, Cassin E et al (2000) Functional genomic analysis of cell division in C. elegans using RNAi of genes on chromosome III. Nature 408:331–336

Gopalakrishnan J, Guichard P, Smith AH, Schwarz H, Agard DA, Marco S, Avidor-Reiss T (2010) Self-assembling SAS-6 multimer is a core centriole building block. J Biol Chem 285:8759–8770

Gopalakrishnan J, Mennella V, Blachon S, Zhai B, Smith AH, Megraw TL, Nicastro D, Gygi SP, Agard DA, Avidor-Reiss T (2011) Sas-4 Scaffolds cytoplasmic complexes and tethers them in a centrosome. Nat Commun 2, 359, doi:ncomms1367 [pii] 10.1038/ncomms1367.

Goshima G, Wollman R, Goodwin SS, Zhang N, Scholey JM, Vale RD, Stuurman N (2007) Genes required for mitotic spindle assembly in Drosophila S2 cells. Science 316:417–421

Habedanck R, Stierhof YD, Wilkinson CJ, Nigg EA (2005) The Polo kinase Plk4 functions in centriole duplication. Nat Cell Biol 7:1140–1146

Han YG, Kwok BH, Kernan MJ (2003) Intraflagellar transport is required in Drosophila to differentiate sensory cilia but not sperm. Curr Biol 13:1679–1686

Heuer JG, Li K, Kaufman TC (1995) The Drosophila homeotic target gene centrosomin (cnn) encodes a novel centrosomal protein with leucine zippers and maps to a genomic region required for midgut morphogenesis. Development 121:3861–3876

Hodges ME, Scheumann N, Wickstead B, Langdale JA, Gull K (2010) Reconstructing the evolutionary history of the centriole from protein components. J Cell Sci 123:1407–1413

Jaspersen SL, Winey M (2004) The budding yeast spindle pole body: structure, duplication, and function. Annu Rev Cell Dev Biol 20:1–28

Jones MH, Winey M (2006) Centrosome duplication: is asymmetry the clue? Curr Biol 16: R808–R810

Kalay E, Yigit G, Aslan Y, Brown KE, Pohl E, Bicknell LS, Kayserili H, Li Y, Tuysuz B, Nurnberg G et al (2011) CEP152 is a genome maintenance protein disrupted in Seckel syndrome. Nat Genet 43:23–26

Kao LR, Megraw TL (2009) Centrocortin cooperates with centrosomin to organize Drosophila embryonic cleavage furrows. Curr Biol 19:937–942

Keil TA (1997) Functional morphology of insect mechanoreceptors. Microsc Res Tech 39:506–531

Keller LC, Geimer S, Romijn E, Yates J 3rd, Zamora I, Marshall WF (2009) Molecular architecture of the centriole proteome: the conserved WD40 domain protein POC1 is required for centriole duplication and length control. Mol Biol Cell 20:1150–1166

Keller LC, Romijn EP, Zamora I, Yates JR 3rd, Marshall WF (2005) Proteomic analysis of isolated chlamydomonas centrioles reveals orthologs of ciliary-disease genes. Curr Biol 15:1090–1098

Kellogg DR, Alberts BM (1992) Purification of a multiprotein complex containing centrosomal proteins from the Drosophila embryo by chromatography with low-affinity polyclonal antibodies. Mol Biol Cell 3:1–11

Kellogg DR, Oegema K, Raff J, Schneider K, Alberts BM (1995) CP60: a microtubule-associated protein that is localized to the centrosome in a cell cycle-specific manner. Mol Biol Cell 6:1673–1684

Kemp CA, Kopish KR, Zipperlen P, Ahringer J, O'Connell KF (2004) Centrosome maturation and duplication in C. elegans require the coiled-coil protein SPD-2. Dev Cell 6:511–523

Kernan M, Cowan D, Zuker C (1994) Genetic dissection of mechanosensory transduction: mechanoreception-defective mutations of Drosophila. Neuron 12:1195–1206

Kilburn CL, Pearson CG, Romijn EP, Meehl JB, Giddings TH Jr, Culver BP, Yates JR 3rd, Winey M (2007) New Tetrahymena basal body protein components identify basal body domain structure. J Cell Biol 178:905–912

Kirkham M, Muller-Reichert T, Oegema K, Grill S, Hyman AA (2003) SAS-4 is a C. elegans centriolar protein that controls centrosome size. Cell 112:575–587

Kitagawa D, Vakonakis I, Olieric N, Hilbert M, Keller D, Olieric V, Bortfeld M, Erat MC, Fluckiger I (2011) Structural basis of the 9-fold summetry of centrioles. Cell 144(3):364–375

Kleylein-Sohn J, Westendorf J, Le Clech M, Habedanck R, Stierhof YD, Nigg EA (2007) Plk4-induced centriole biogenesis in human cells. Dev Cell 13:190–202

Krioutchkova MM, Onishchenko GE (1999) Structural and functional characteristics of the centrosome in gametogenesis and early embryogenesis of animals. Int Rev Cytol 185:107–156

Leidel S, Delattre M, Cerutti L, Baumer K, Gonczy P (2005) SAS-6 defines a protein family required for centrosome duplication in C. elegans and in human cells. Nat Cell Biol 7:115–125

Li JB, Gerdes JM, Haycraft CJ, Fan Y, Teslovich TM, May-Simera H, Li H, Blacque OE, Li L, Leitch CC et al (2004) Comparative genomics identifies a flagellar and basal body proteome that includes the BBS5 human disease gene. Cell 117:541–552

Li K, Kaufman TC (1996) The homeotic target gene centrosomin encodes an essential centrosomal component. Cell 85:585–596

Li K, Xu EY, Cecil JK, Turner FR, Megraw TL, Kaufman TC (1998) Drosophila centrosomin protein is required for male meiosis and assembly of the flagellar axoneme. J Cell Biol 141:455–467

Loncarek J, Khodjakov A (2009) Ab ovo or de novo? Mechanisms of centriole duplication. Mol Cells 27:135–142

Loncarek J, Sluder G, Khodjakov A (2007) Centriole biogenesis: a tale of two pathways. Nat Cell Biol 9:736–738

Mahoney NM, Goshima G, Douglass AD, Vale RD (2006) Making microtubules and mitotic spindles in cells without functional centrosomes. Curr Biol 16:564–569

Manandhar G, Schatten H, Sutovsky P (2005) Centrosome reduction during gametogenesis and its significance. Biol Reprod 72:2–13

Martinez-Campos M, Basto R, Baker J, Kernan M, Raff JW (2004) The Drosophila pericentrin-like protein is essential for cilia/flagella function, but appears to be dispensable for mitosis. J Cell Biol 165:673–683

Matsuura K, Lefebvre PA, Kamiya R, Hirono M (2004) Bld10p, a novel protein essential for basal body assembly in Chlamydomonas: localization to the cartwheel, the first ninefold symmetrical structure appearing during assembly. J Cell Biol 165:663–671

Megraw TL, Kaufman TC (2000) The centrosome in Drosophila oocyte development. Curr Top Dev Biol 49:385–407

Megraw TL, Li K, Kao LR, Kaufman TC (1999) The centrosomin protein is required for centrosome assembly and function during cleavage in Drosophila. Development 126:2829–2839

Moritz M, Braunfeld MB, Sedat JW, Alberts B, Agard DA (1995) Microtubule nucleation by gamma-tubulin-containing rings in the centrosome. Nature 378:638–640

Moritz M, Zheng Y, Alberts BM, Oegema K (1998) Recruitment of the gamma-tubulin ring complex to Drosophila salt-stripped centrosome scaffolds. J Cell Biol 142:775–786

Mottier-Pavie V, Megraw TL (2009) Drosophila Bld10 is a centriolar protein that regulates centriole, basal body, and motile cilium assembly. Mol Biol Cell

Nigg EA, Raff JW (2009) Centrioles, centrosomes, and cilia in health and disease. Cell 139:663–678

Oegema K, Wiese C, Martin OC, Milligan RA, Iwamatsu A, Mitchison TJ, Zheng Y (1999) Characterization of two related Drosophila gamma-tubulin complexes that differ in their ability to nucleate microtubules. J Cell Biol 144:721–733

Okada M (1998) Germline cell formation in Drosophila embryogenesis. Genes Genet Syst 73:1–8

Oliferenko S, Balasubramanian MK (2002) Astral microtubules monitor metaphase spindle alignment in fission yeast. Nat Cell Biol 4:816–820

Omran H (2010) NPHP proteins: gatekeepers of the ciliary compartment. J Cell Biol 190:715–717

Pearson CG, Osborn DP, Giddings TH Jr, Beales PL, Winey M (2009) Basal body stability and ciliogenesis requires the conserved component Poc1. J Cell Biol 187:905–920

Peel N, Stevens NR, Basto R, Raff JW (2007) Overexpressing centriole-replication proteins in vivo induces centriole overduplication and de novo formation. Curr Biol 17:834–843

Pelletier L, Ozlu N, Hannak E, Cowan C, Habermann B, Ruer M, Muller-Reichert T, Hyman AA (2004) The Caenorhabditis elegans centrosomal protein SPD-2 is required for both pericentriolar material recruitment and centriole duplication. Curr Biol 14:863–873

Phillips DM (1967) Giant centriole formation in Sciara. J Cell Biol 33:73–92

Riparbelli MG, Callaini G (2003) Drosophila parthenogenesis: a model for de novo centrosome assembly. Dev Biol 260:298–313

Riparbelli MG, Whitfield WG, Dallai R, Callaini G (1997) Assembly of the zygotic centrosome in the fertilized Drosophila egg. Mech Dev 65:135–144

Rodrigues-Martins A, Bettencourt-Dias M, Riparbelli M, Ferreira C, Ferreira I, Callaini G, Glover DM (2007a) DSAS-6 organizes a tube-like centriole precursor, and its absence suggests modularity in centriole assembly. Curr Biol 17:1465–1472

Rodrigues-Martins A, Riparbelli M, Callaini G, Glover DM, Bettencourt-Dias M (2007b) Revisiting the role of the mother centriole in centriole biogenesis. Science 316:1046–1050

Rosenbaum J (2002) Intraflagellar transport. Curr Biol 12:R125

Scholey JM, Anderson KV (2006) Intraflagellar transport and cilium-based signaling. Cell 125:439–442

Seidl S (1991) Structure and differentiation of the sensilla of the ventral sensory field on the maxillary palps ofPeriplaneta americana (Insecta, Blattodea), paying special attention to the ciliogenesis of the sensory cells. Zoomorphology 111:35–47

Sluder G, Khodjakov A (2010) Centriole duplication: analogue control in a digital age. Cell Biol Int 34:1239–1245

Stevens NR, Dobbelaere J, Brunk K, Franz A, Raff JW (2010a) Drosophila Ana2 is a conserved centriole duplication factor. J Cell Biol 188:313–323

Stevens NR, Dobbelaere J, Wainman A, Gergely F, Raff JW (2009) Ana3 is a conserved protein required for the structural integrity of centrioles and basal bodies. J Cell Biol 187:355–363

Stevens NR, Raposo AA, Basto R, St Johnston D, Raff JW (2007) From stem cell to embryo without centrioles. Curr Biol 17, 1498–1503

Stevens NR, Roque H, Raff JW (2010b) DSas-6 and Ana2 Coassemble into Tubules to Promote Centriole Duplication and Engagement. Dev Cell 19:913–919

Sun QY, Schatten H (2007) Centrosome inheritance after fertilization and nuclear transfer in mammals. Adv Exp Med Biol 591:58–71

Tang CJ, Fu RH, Wu KS, Hsu WB, Tang TK (2009) CPAP is a cell-cycle regulated protein that controls centriole length. Nat Cell Biol 11:825–831

Tates AD (1971) Cytodifferentiation during Spermatogenesis in Drosophila melanogaster: An Electron Microscope Study. Rijksuniversiteit de Leiden, Leiden

Thornton GK, Woods CG (2009) Primary microcephaly: do all roads lead to Rome? Trends Genet 25:501–510

Tokuyasu KT (1975) Dynamics of spermiogenesis in Drosophila melanogaster. V. Head-tail alignment. J Ultrastruct Res 50:117–129

Vaizel-Ohayon D, Schejter ED (1999) Mutations in centrosomin reveal requirements for centrosomal function during early Drosophila embryogenesis. Curr Biol 9:889–898

van Breugel M, Hirono M, Andreeva A, Yanagisawa HA, Yamaguchi S, Nakazawa Y, Morgner N, Petrovich M, Ebong IO, Robinson CV et al (2011) Structures of SAS-6 suggest its organization in centrioles. Science 331:1196–1199

Varmark H, Llamazares S, Rebollo E, Lange B, Reina J, Schwarz H, Gonzalez C (2007) Asterless is a centriolar protein required for centrosome function and embryo development in Drosophila. Curr Biol 17:1735–1745

Wakefield JG, Bonaccorsi S, Gatti M (2001) The drosophila protein asp is involved in microtubule organization during spindle formation and cytokinesis. J Cell Biol 153:637–648

Walker RG, Willingham AT, Zuker CS (2000) A Drosophila mechanosensory transduction channel. Science 287:2229–2234

Wang Y, Ji P, Liu J, Broaddus RR, Xue F, Zhang W (2009) Centrosome-associated regulators of the G(2)/M checkpoint as targets for cancer therapy. Mol Cancer 8:8

Whitfield WG, Millar SE, Saumweber H, Frasch M, Glover DM (1988) Cloning of a gene encoding an antigen associated with the centrosome in Drosophila. J Cell Sci 89(Pt 4):467–480

Williams CL, Li C, Kida K, Inglis PN, Mohan S, Semenec L, Bialas NJ, Stupay RM, Chen N, Blacque OE et al (2011) MKS and NPHP modules cooperate to establish basal body/transition zone membrane associations and ciliary gate function during ciliogenesis. J Cell Biol 192:1023–1041

Wilson PG, Zheng Y, Oakley CE, Oakley BR, Borisy GG, Fuller MT (1997) Differential expression of two gamma-tubulin isoforms during gametogenesis and development in Drosophila. Dev Biol 184:207–221

Yamashita YM, Mahowald AP, Perlin JR, Fuller MT (2007) Asymmetric inheritance of mother versus daughter centrosome in stem cell division. Science 315:518–521

Chapter 2
Sperm Centrioles and Their Dual Role in Flagellogenesis and Cell Cycle of the Zygote

Structure, Function, and Pathology

Hector E. Chemes

Abstract This chapter examines the current knowledge on the role of the spermatid centrosome. The dual role of the centrosome as a spermatid basal body that generates the sperm flagellum and as the nucleation site for sperm aster formation in the zygote is mirrored in different sperm pathologies in infertile men that are reviewed in detail. Information is discussed on different sperm centriolar and centrosomal anomalies that are involved in failed fertilizations or abnormal development of the embryo. Particular attention is paid to specific centrosomal anomalies of genetic origin that cause dysfunction of the sperm centrosome with abnormal assembly of the sperm aster and failed pronuclear apposition and cleavage of the zygote. The studies highlight the key role played by sperm centrosomes in flagellogenesis and early zygote development and encourage further investigation on the physiopathology of sperm centrosome-related fertility failures to fully expose the basic mechanisms involved.

2.1 Centrosomes, Centrioles, and Basal Bodies

Centrosomes are ubiquitous organelles found in eukaryotic cells. They are composed of a pair of barrel-shaped centrioles, hollow cylindrical structures with their walls composed of nine triplet microtubules in a "pinwheel" arrangement. The two centrioles, perpendicular to each other (diplosomes), are surrounded by a dense fibro-granular "cloud" of pericentriolar material (PCM) that constitutes the

---

H. E. Chemes (✉)
División Endocrinología, Centro de Investigaciones Endocrinológicas,
CEDIE-CONICET, Gallo 1330, C1425SEFD Buenos Aires, Argentina
e-mail: hchemes@cedie.org.ar; hechemes@yahoo.com.ar

Fig. 2.1 Schematic representation of the centrosome. A pair of barrel-shaped centrioles formed by nine triplet microtubules (diplosome) is surrounded by the PCM where γ-tubulin ring complexes (γTuRCs) serve as nucleation sites for microtubules. Distal and subdistal appendages can be observed at the *upper* end of the mother centriole. Reproduced with permission from J. Lüders

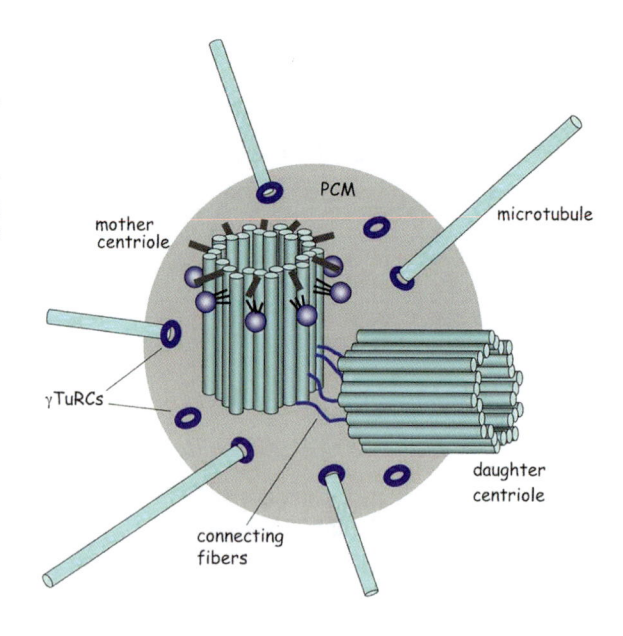

microtubule organizing center (MOCT) of the cell (Fig. 2.1). The PCM is organized as a framework that supports microtubular motor proteins like kinesins and dynein, coiled coil proteins, centrin, pericentrin, speriolin, Cdc20 (spindle checkpoint protein), and NuMA (Nuclear Mitotic Apparatus protein) among about 100 other proteins (reviewed by Schatten and Sun 2009). Centrosomes are involved in numerous cell functions, among them translocation of signal transduction molecules, movement of cell organelles along microtubules and organization of the cytoskeleton, mitotic spindle, and zygote sperm aster. Microtubules do not originate from centrioles themselves but from the γ-tubulin ring complex, a collection of annular structures contained in discrete densities of the PCM (Figs. 2.1, 2.2). The γ-tubulin ring complex serves as nucleation site for tubulin, the main component of microtubules, polarized structures with a minus end anchored to the PCM and a distal plus end where microtubules elongate by polymerization of α- and β-tubulin heterodimers.

Centrioles and basal bodies are structurally similar and functionally interconvertible. In dividing cells, centrosomes organize the mitotic spindle for chromosome alignment, duplication, and partition between daughter cells (Fig. 2.2). During generation of cilia and flagella, centrosomes migrate to the cell periphery where distal centrioles dock to the cell membrane to become basal bodies from which ciliary and flagellar axonemes originate. When these cells enter mitosis, basal bodies move back to the cytoplasm and reconstitute centrosomes. This alternating dual role is essential to understand the functioning of spermatids and spermatozoa.

Mammalian spermatozoa are the end product of spermiogenesis, a complex differentiation process in which organelles of round spermatids undergo a series of

Fig. 2.2 Mitosis of a guinea pig spermatogonium. At one pole of the mitotic spindle a centriole (*C*) is surrounded by dense PCM (inset detail). Note that the proximal ends of spindle microtubules (*m*) converge toward nucleation sites on the area surrounding the centriole (*). The distal ends of spindle microtubules (*m*) are anchored to metaphase chromosomes (*Chr*). Bar represents 3 μm (This figure belongs to the author and is originally reproduced in the present text)

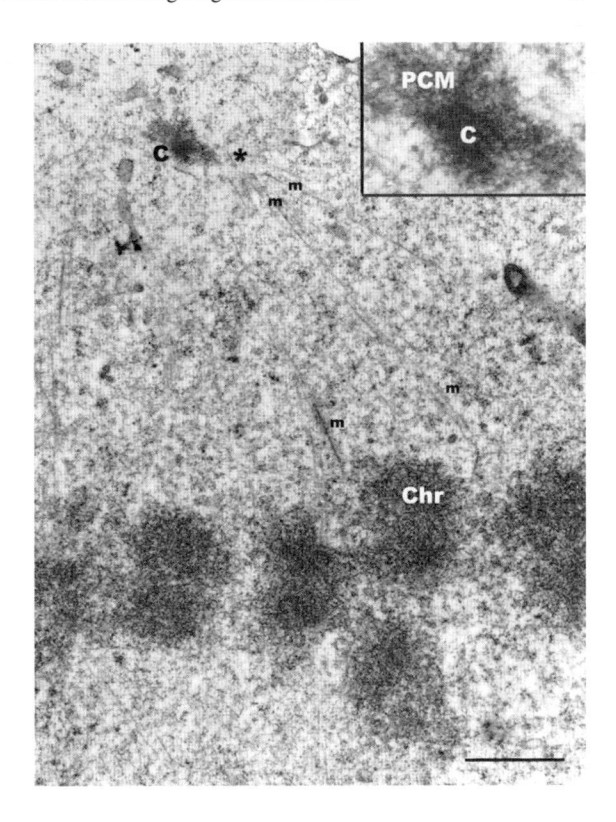

modifications that result in the elaborated structure of mature spermatozoa. The Golgi complex develops into the sperm acrosome and mitochondria organize around the sperm axoneme giving rise to the midpiece (mitochondrial sheath). Spermatids derive from meiosis II of secondary spermatocytes, the last cells to divide in spermatogenesis. Since spermatids will not enter a new mitotic cycle, their centrosomes undergo a functional shift to basal bodies that serve as templates for the assembly of axoneme doublet microtubules by direct tubulin nucleation on subunits a and b of distal centriolar triplets. As flagellar axonemes grow, basal bodies migrate to the cell periphery where distal centrioles dock perpendicular to the plasma membrane as the axoneme sprouts toward extracellular spaces (Fig. 2.4a). In successive steps the basal body-flagellar complex invaginates and attaches to the nuclear envelope at the concave implantation fossa (Fawcett 1981; Holstein and Roosen Runge 1981). ODF2 (a protein component of sperm outer dense fibers) is involved in the initial docking of centrioles to membranes (reviewed by Hoyer-Fender 2010). As spermatid nuclei elongate, acrosomes occupy their cranial pole while basal bodies take up the caudal pole. This topographical arrangement is critical for normal sperm development since it defines the bipolarity of spermatid nuclei and the alignment of heads, midpieces, and tails along the sperm longitudinal axis. As we will discuss later, alterations in this

Fig. 2.3 Human sperm connecting piece. The proximal centriole (*), sectioned at right angle, is enclosed laterally by segmented columns (*SC*) and cranially by the capitulum (*arrow, C*) which is lodged in the implantation fossa at the caudal pole of the sperm head. A dense basal plate (*BP*) lines the outer leaflet of the nuclear envelope at the implantation fossa. Distal ends of SC are continuous with outer dense fibers (*ODF*) of the sperm axoneme. Axonemal microtubules (*Mt*) end cranially in a rarefied area formerly occupied by the distal centriole (**). Mitochondria (*Mi*). Bar represents 0.1 μm (Figure 3 was originally published by Chemes et al. (1999) and reproduced, modified from the original, with permission from the publisher)

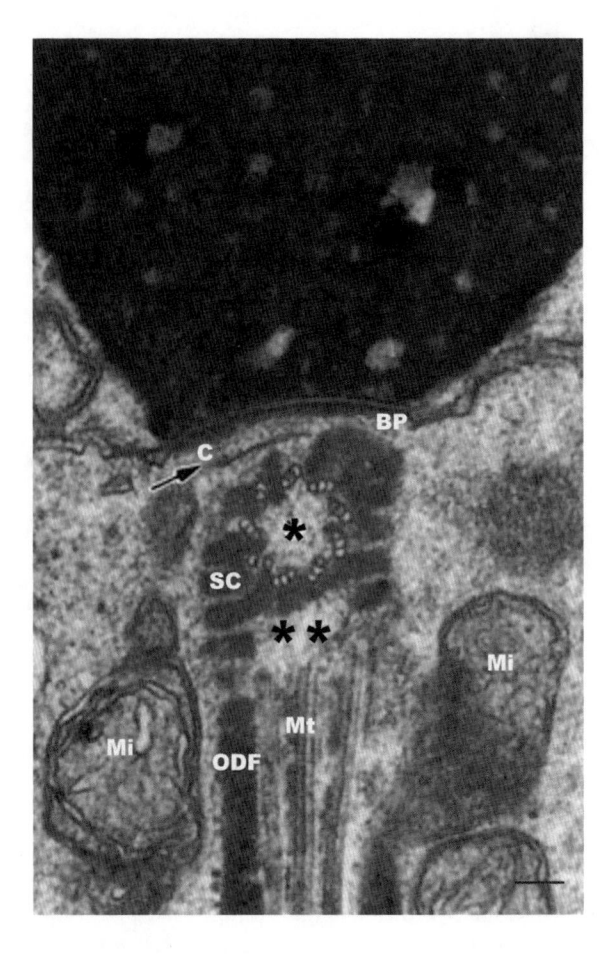

migration–attachment of basal bodies–flagella will result in misalignments of the tail and serious structural and functional sperm anomalies.

The growth of the sperm axoneme is accompanied by complex modifications in the dense PCM. In its place, new proteins organize in nine longitudinal segmented columns (SC) and the capitulum (C) of the connecting piece (Fig. 2.3) (Fawcett 1981). SC and C constitute a dense shield that lodges and encloses both centrioles. The SC are nine cylindrical structures with periodic densities that fuse cranially to form the capitulum, a curved plate-like disk that links connecting pieces to sperm heads by its association to basal plates, dense structures that line the outer nuclear membrane at the implantation fossa. At their caudal end each SC is continuous with one of the nine outer dense fibers (ODF) that associate to peripheral microtubular doublets of the growing axoneme. In many mammals, including humans, the distal centriole vanishes after giving rise to the sperm axoneme, leaving few remnants in mature spermatozoa.

A phosphorylated protein complex has been reported that localizes to ODFs, SC, and C. This complex may regulate sperm centrosomal function through ODF dephosphorylation and connecting piece disassembly since it has been reported that dephosphorylation of sperm midpiece antigens initiate aster formation in rabbit zygotes (Pinto-Correia et al. 1994; Long et al. 1997; Schalles et al. 1998). Long et al. (1997), Rawe et al. (2008) and Hoyer-Fender (2010) have noted that the sperm basal body-centriole must first disengage from the connecting piece to be able to organize the zygote centrosome by recruiting oocyte-derived PCM. Proteasomes localized to the sperm neck are probably necessary for normal centriolar release (Wójcik et al. 2000; Wójcik and DeMartino 2003; Rawe et al. 2008). The organization of centrioles (basal bodies) docking to cell membranes and giving rise to microtubular axonemes is an evolutionary conserved mechanism common to ciliated and flagellated cells. Vashishtha et al. (1996) have studied in *Chlamydomonas* the role of KHP1, a kinesin-homologous protein that localizes to basal bodies and centrioles and possibly acts as a transporter of protein components to their distal site of assembly in axonemes or aster microtubules. Prior to mitosis, flagella are resorbed and basal bodies duplicate to become centrosomes that occupy *Chlamydomonas* spindle poles from where aster microtubules radiate. These observations point to the dual function of basal bodies/centrioles in flagellar assembly and mitotic spindle formation. Similar phenomena occur after fertilization in humans: flagella detach from sperm heads, and basal bodies (proximal centrioles) recruit PCM to become the zygote MTOC, from which the sperm aster and mitotic spindle will assemble. Sutovsky et al. (1996) have reported that after sperm incorporation into oocytes connecting pieces break down and microtubules first associate with proximal centrioles to form sperm asters that direct pronuclear migration and fusion. During this process, capitulum and SC move away and disintegrate in the cytoplasm. After syngamy, sperm centrioles form the zygote centrosome that subsequently duplicates and migrates to both poles of the cell to assemble the mitotic spindle as the embryo enters its first cell cycle.

The need for a functional centriolar complement was demonstrated by Palermo et al. (1997) and Colombero et al. (1999) who showed that injection of separated sperm components (head only, separated head and tail, isolated tail) is followed by oocyte activation and bipronuclear formation, but ultimately results in abnormal centrosomal function and embryonic mosaicism. They concluded that the integrity of the sperm head–neck region is essential for human early embryogenesis. Experimental evidence presented by Comizzoli et al. (2006) points to the importance of complete centriolar maturation, since aster formation was reduced after injection of testicular immature spermatozoa when compared to that obtained with fully mature ejaculated sperm. Recent investigations have shown that the pericentrosomal area is enriched in proteasomes and may function as a proteolytic center of the cell. Under conditions of cell stress or proteasome inhibition increased numbers of proteasomes and ubiquitinated proteins concentrate around the centrosome forming "aggresomes" (Wójcik 1997a, b; Fabunmi et al. 2000; Wójcik and DeMartino 2003; Rawe et al. 2008). These evidences support an active proteasome involvement in centrosomal function during early zygote development. Alterations in these mechanisms are essential events in the physiopathology of some sperm-related fertilization failures.

2.2 Sperm Pathologies in Infertile Men with Special Reference to Those Related to Centrioles and Centrosomes

Teratozoospermia has been reported as an important cause of male infertility. Two main forms of sperm anomalies can be identified in teratozoospermia (Chemes 2000). The first and more frequent variety consists of heterogeneous combinations of sperm anomalies randomly distributed in different patients. These alterations are referred to as *non-specific* or *non-systematic sperm anomalies*. They are usually secondary to andrological conditions of diverse etiologies that affect the testis or the seminal pathway. No genetic component is present. The second variety is characterized by a specific phenotype that affects most spermatozoa in all patients suffering from the same condition. These alterations may be called *systematic anomalies* because the sperm phenotype involves specific organelles and repeats in most spermatozoa. Systematic alterations show family clustering and have proven or suspected genetic origin. To this variety belong *acephalic spermatozoa* (Perotti et al. 1981; Chemes et al. 1987b, 1999), *round head acrosomeless spermatozoa* (Holstein et al. 1973; Nistal et al. 1978), the *miniacrosome sperm defect* (Baccetti et al. 1991), Dysplasia of the Fibrous Sheath (DFS or *stump tail defect*, Chemes et al. 1987a, 1998), and the dynein-deficient axonemes of Primary Ciliary Diskinesia (PCD, Azfelius et al. 1975). Each of these phenotypes is the consequence of distinctive pathologic mechanisms involving different sperm organelles.

Headless sperm flagella, loose heads, and abnormal head–tail alignment are the distinguishing features of a human syndrome of genetic origin characterized by abnormalities in sperm centrioles and the head–neck junction. Later, we will review what is currently known on this interesting sperm pathology.

Various kinds of defects in centrosomes and cilia have been reported in patients suffering from "ciliopathies", a group of disorders of ciliated cells caused by mutations in different genes (Tammachote et al. 2009). These comprise syndromes affecting CNS, eyes, kidneys, biliary ducts, respiratory tract, etc. Among them, lack of dynein arms or other axonemal components is the structural basis of immotility in respiratory cilia and sperm flagella in patients with PCD (Afzelius et al. 1975; reviewed by Chemes and Rawe 2003).

As pointed out by Schatten and Sun (2009), even though genetic components most likely play a role in centrosome pathologies, these can also have acquired origins, including exposure to a variety of environmental factors or toxic compounds that can disrupt centrosomal function.

Examining the ultrastructure of zygotes and aster development after fertilizations with abnormal spermatozoa, Sathananthan (1994) and Van Blerkom (1996) reported sperm centrosome dysfunctions as a cause of infertility or abortive embryonic development. More recently, Sathananthan et al. (2001) identified structural alterations in sperm centrioles of infertile men, including disorganization or loss of centriolar triplets, loss or abnormal positioning of proximal centrioles, and intrusion of mitochondria within centrioles. Hewitson et al. (1997) and Rawe et al. (2008) have also suggested that centrosomal anomalies are responsible for defective sperm aster formation or microtubule elongation in human post-ICSI fertilization failures.

Injection of isolated sperm tails (containing the proximal centriole) into oocytes results in the formation of sperm asters (Van Blerkom and Davis 1995). In later studies it was demonstrated that the use of heterologous ICSI systems (human–bovine, human–rabbit) provide objective information on the capacity of spermatozoa to elicit normal aster development and constitute a novel tool to examine sperm centrosomal function of infertile men (Terada et al. 2002, 2004; Yoshimoto-Kakoi et al. 2008). Using this technique sperm centrosomal failures were reported in teratozoospermia and globozoospermia (Nakamura et al. 2002; Terada et al. 2010), the rate of sperm aster formation from infertile men was found to be lower than that from fertile individuals (Rawe et al. 2002), and Hinduja et al. (2010) communicated that centrosome proteins centrin, α and γ-tubulin, were reduced in oligoasthenozoospermic patients. As a consequence of these observations efforts to develop in vitro methods to restore defective sperm centrosomal function in humans are underway (Nakamura et al. 2005; Terada et al. 2010).

In summary, centriolar and centrosomal abnormalities are involved in failed fertilizations or abnormal development of the embryo. However, all these reports derive from experimental observations or laboratory studies after post-ICSI fertilization failures and do not identify diagnostic categories of clinical value in human infertility. We and others have described a human syndrome of genetic origin in infertile men with systematic teratozoospermia.

In 1987 we published a paper entitled *Lack of a Head in Human Spermatozoa from Sterile Patients: a Syndrome Associated with Impaired Fertilization* (Chemes et al. 1987b). Three adult males were reported who suffered from primary sterility and presented a characteristic sperm defect that repeated in all semen samples examined. Most spermatozoa (75–100 %) presented with minute "heads", no larger than 1 µm in diameter and negative for the Feulgen reaction, which indicated lack of sperm heads. Two main abnormal configurations could be observed. Some forms had cephalic ends with minute spherical thickenings containing sperm centrioles and connecting pieces followed by normally structured midpieces and flagella (Fig. 2.4d). The other type was similar but without midpieces. In a second publication, we documented the findings in 10 patients, the largest series published to date (Chemes et al. 1999). A third abnormal variety could be observed in which heads were present but abnormally attached to midpieces, with no linear alignment with the sperm axis (Figs. 2.4c, e). The angles between heads and tails were up to 180°. Normally formed spermatozoa amounted to no more than 1 %. Immature spermatids in semen showed their flagellum-middle piece complexes abnormally related or completely divorced from nuclei.

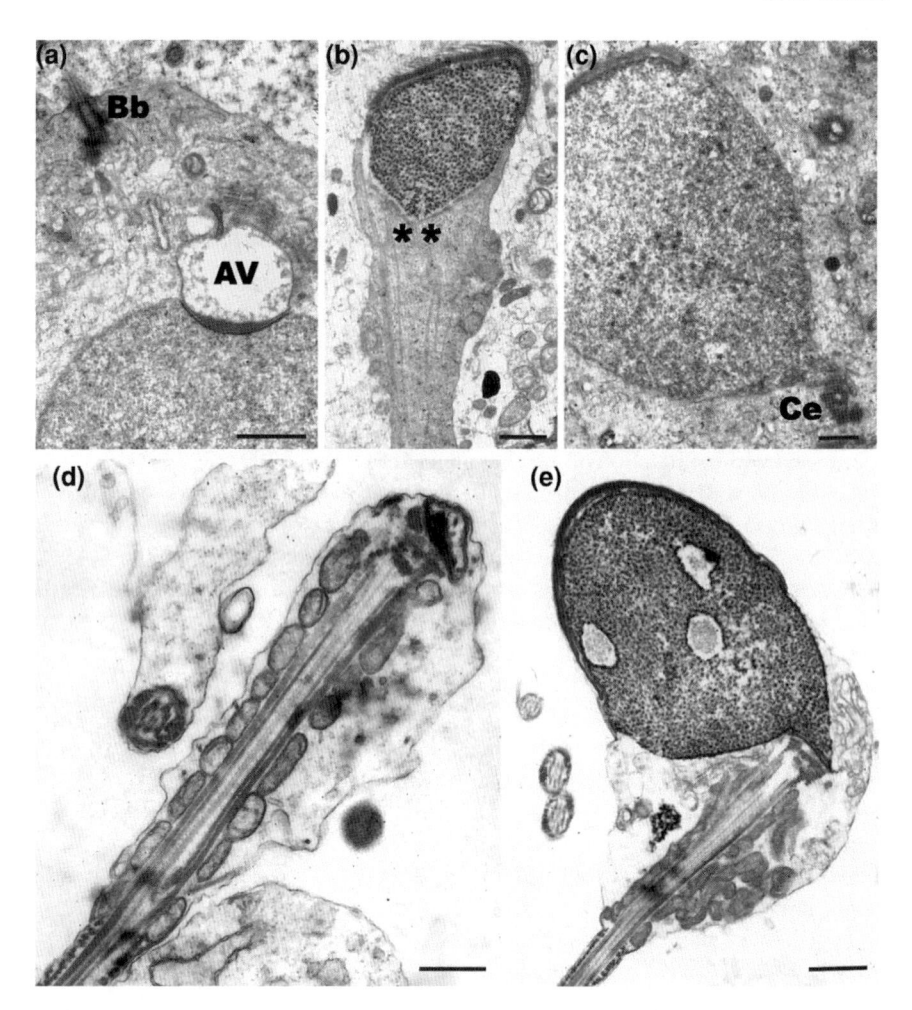

Fig. 2.4 Ultrastructure of neck attachment misalignments and acephalic spermatozoa. **a** At an early stage of spermiogenesis this spermatid still shows a normally positioned basal body (*Bb*) anchored to the cell membrane. Acrosomic vesicle (*AV*). **b** This elongating spermatid lacks a nuclear implantation fossa (**), and the flagellar anlage is not attached to the nucleus. **c** Another spermatid showing centrioles (*Ce*) with abnormal implantation into the nucleus. **d** An acephalic spermatozoon. There is no head. Centrioles, mitochondrial sheath, and flagellum are normal. **e** Spermatid with abnormal angle between head and midpiece. Flagellar attachment is similar to that depicted in panel **c**. Bars represent 1 μm (Figure 4 was originally published by Chemes et al. (1999) and Chemes and Rawe (2010) and reproduced, modified from the original, with permission from the publishers)

Various earlier publications had reported single patients with similar sperm phenotypes and identified them as "microcephalic", "pin-head", or "decapitated" spermatozoa (Zaneveld and Polakoski 1977; Nistal et al. 1978; LeLannou 1979; Perotti and Gioria 1981; Perotti et al. 1981; Baccetti et al. 1984). We introduced the term "acephalic spermatozoa" (Chemes et al. 1999) and, in agreement with Perotti and Gioria (1981), proposed that abnormal head-midpiece alignments originated in the testis because centrioles failed to attach normally to spermatid nuclei. This failure could also result from a nuclear defect that interferes with the formation of the implantation fossa, normal lodging site for the sperm proximal centriole. Nuclei and flagella develop independently and become separated within the seminiferous tubules or in the seminal pathway as a consequence of increased instability of the head-midpiece junction. This interpretation is supported by observations that the separation or abnormal relations between heads and tails increase due to mechanical stress in centrifugation or sperm in vitro manipulation (Chemes et al. 1999; Kamal et al. 1999). The admixture of acephalic spermatozoa and abnormal head-middle piece connections expresses different degrees of abnormalities of the head–neck junction with acephalic forms representing the most extreme situation. In most cases the sperm neck was the preferred region where cleavage between heads and midpieces took place. In occasional reports separation resulted from dissociation between proximal and distal centrioles (Holstein et al. 1986) or due to other sperm defects at more distal locations. The study of a testicular biopsy in one of our patients confirmed that alterations started very early during testicular spermiogenesis with abnormal relations between spermatid nuclei and tails (Fig. 2.4) that resulted in abnormal lateral implantations or completely independent development. When present, heads implanted at abnormal angles on the middle piece. The caudal nuclear pole of elongating spermatids appeared as a protruding area without an implantation fossa to lodge the proximal centriole.

We had previously shown (Chemes et al. 1978) that in early human spermiogenesis the spermatid nucleus differentiates a cranial pole where the Golgi complex attaches to form the acrosome. Shortly after, the centriole-flagellum complex approaches the opposite pole of the nucleus and attaches to it. Acephalic spermatozoa derive from the failure of this caudal migration, while some acrosomeless spermatozoa result from the lack of proper attachment of the Golgi complex to the cranial pole of the spermatid nucleus (Zamboni 1992). The unusual case described by Aughey and Orr (1978), with round acrosomeless heads and acephalic spermatozoa in the same patient indicate that these two abnormal mechanisms have combined, suggesting that there are different pathologies derived from an abnormal differentiation of the bipolar nature of spermatid nuclei. In very recent studies (Alvarez Sedo et al. 2012) we have found that failures of proper Golgi attachment to nuclei are indeed accompanied by frequent failures in head–tail connections.

In one of the reported patients, that had around 1 % normal spermatozoa in his ejaculate, it was possible to follow the evolution of seminal profiles over an extended period, before, during, and after pharmacologic suppression of spermatogenesis. Testosterone propionate treatment was instituted to achieve oligoazoospermia in an attempt to promote expansion of the clone of normal spermatozoa during spermatogenic recovery that follows testosterone administration. However, sperm morphology did not change along the course of spermatogenic regression-recovery, the percentage of normal spermatozoa remained very low, and about 99 % of all newly formed spermatozoa were again acephalic.

The uniform pathologic phenotype, its origin as a consequence of a systematic alteration during spermiogenesis, the fact that seminal characteristics remain constant along clinical evolution and even after pharmacologic induction of germ cell depletion-repopulation and the familial incidence reported in men and bulls (Bloom and Birch 1970; Baccetti et al. 1989, Chemes et al. 1999) indicate that this distinctive phenotype is a centrosome-related primary sperm defect that results from an abnormal spermatogenic programming of genetic origin. Very recently, Liska et al. (2009) and Kierszenbaum et al. (2011) reported mutations in Centrobin and IFT88, two sperm proteins that localize to spermatid centrioles and manchettes. Both phenotypes show separation of centrioles from their normal nuclear attachment site, disruption of head–tail coupling, and spermatid decapitation. No communication of similar mutations in humans is available to date.

All reported patients suffered from long standing primary sterility. In some cases acephalic forms predominate, which makes impossible any attempt at assisted reproduction (LeLannou 1979; Perotti et al. 1981; Holstein et al. 1986; Chemes et al. 1987b, 1999; Baccetti et al. 1989; Toyama et al. 2000). However, in other patients there were good numbers of nucleated forms with alterations in the head-midpiece alignment. This opened the way to consider their use in oocyte microinjections. In the first reported attempt, nucleated spermatozoa were microinjected into four good quality metaphase II oocytes (Chemes et al. 1999). All of them fertilized and formed pronuclei, but zygotes remained at the pronuclear stage and degenerated before syngamy and cleavage (Fig. 2.5). Comparable results were communicated by Saias Magnan et al. (1999) and Rawe et al. (2002), but this last report also documented high βhCG plasma levels followed by preclinical abortions when microinjected spermatozoa were rigorously selected avoiding anomalies of the head–neck junction. When these abnormal spermatozoa were used in a heterologous bovine–human ICSI system, sperm asters either failed to form or had an arrested development (Fig. 2.5, Rawe et al. 2002). The first births in this condition were reported by Porcu et al. (2003) and, more recently, two successful ICSI attempts in one of our patients were followed by pregnancies and births of healthy children (Coco et al., manuscript in preparation). These dissimilar results indicate variations in the degree of abnormalities of the head–neck junction, some of them compatible with normal centrosomal function.

Various observations have demonstrated the ultrastructural integrity of proximal centrioles in spermatozoa with defects of the head–tail attachment (Baccetti et al. 1989; Chemes et al. 1999). In the search for the nature of this centrosome abnormality we realized that there was dissociation between the function of both centrioles. While distal centrioles successfully completed development of flagellar axonemes, proximal centrioles were unable to attach normally to spermatid nuclei and failed to reconstitute zygotic centrosomes. This type of functional dissociation between both centrioles has its counterpart in PCD (Primary Ciliary Diskinesia or Immotile Cilia Syndrome) where the function of proximal centrioles is preserved (immotile PCD spermatozoa fertilize oocytes when microinjected) while distal centrioles generate abnormal sperm axonemes. This double function and dissociated pathology is an interesting dualistic model underscoring a high degree of autonomy between proximal and distal centrioles.

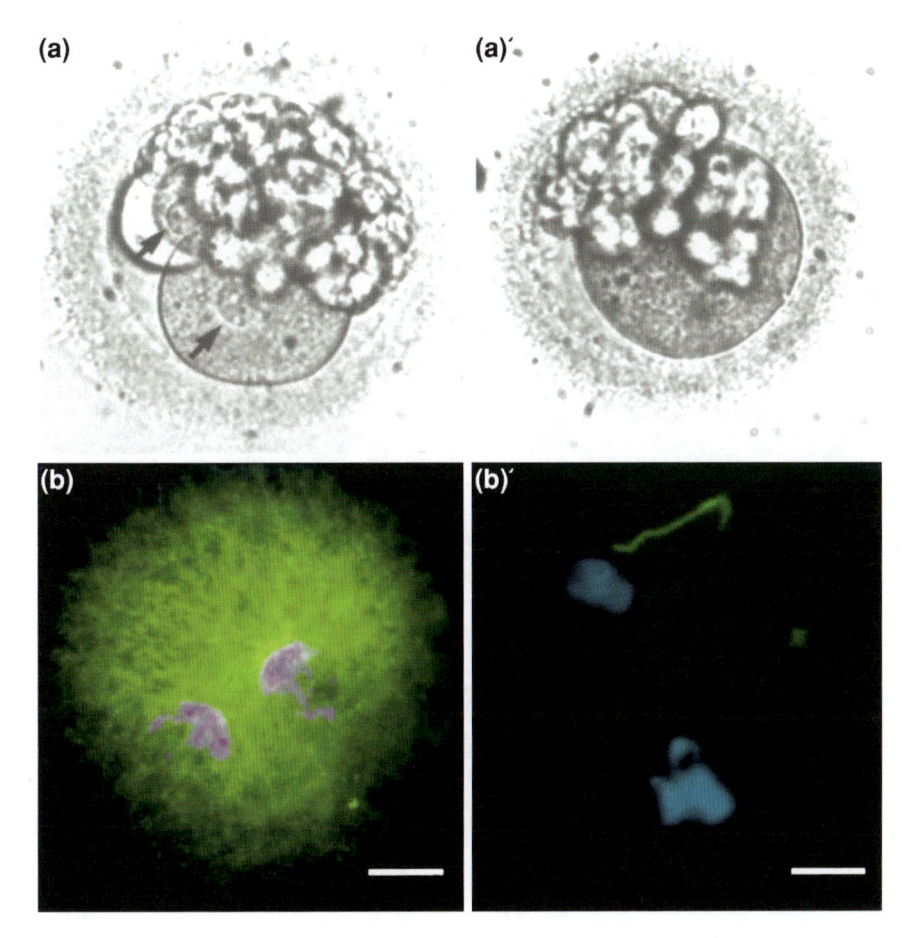

Fig. 2.5 a, a′ Spontaneous post-ICSI fragmented zygotes after failure of syngamy that followed microinjection of spermatozoa with neck anomalies. In A two pronuclei are clearly seen (*arrows*). **b** heterologous ICSI. Normal sperm aster formation (*tubulin green fluorescence*) after injection of a normal human spermatozoon into a bovine oocyte. **b′** When a spermatozoon with neck anomalies is microinjected the sperm aster fails to form. The sperm flagellum (*tubulin green fluorescence*) is still associated with the male pronucleus. **a, a′**, original magnification X400; **b, b′**, bars represent 25 μm (These figures were originally produced by Chemes et al. (1999) and Rawe et al. (2002) and are reproduced, modified from the original, with permission from the publishers)

What impaired mechanisms can account for centrosomal dysfunction? The proximal centriole must disengage from the connecting piece to be able to reconstitute the zygote centrosome. Long et al. (1997) characterized a phosphorylated protein complex from sperm ODF and connecting pieces that may be involved in the regulation of sperm centrosomal activity after connecting piece disassembly, and Pinto Correia et al. (1994) reported that dephosphorylation of sperm connecting piece antigens is required for initiation of aster formation in rabbit oocytes. Centriole release after fertilization

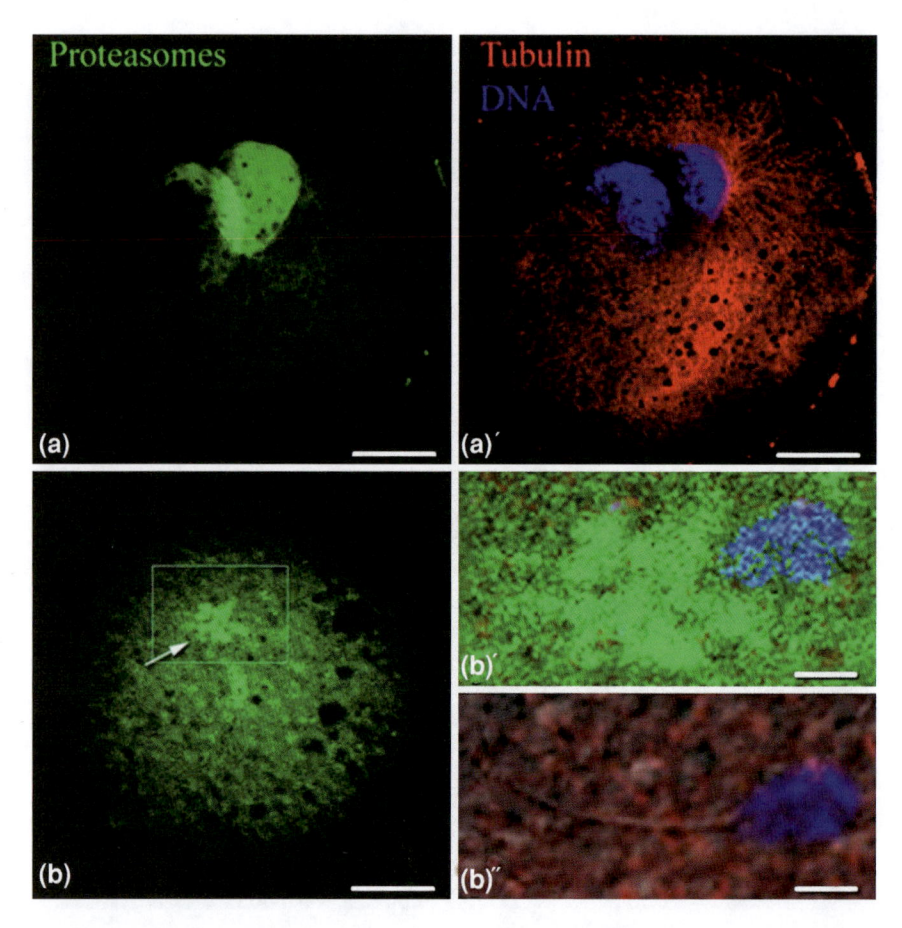

Fig. 2.6 Proteasomes and fertilization. **a, a′** Zygote obtained after in vitro fertilization of bovine oocytes in control conditions. **a** Proteasomes (*green fluorescence*) are concentrated on both pronuclei. **a′** Complete development of sperm aster (*tubulin red fluorescence*) and fully apposed pronuclei (*blue fluorescence*). **b** Bovine oocyte after IVF in the presence of E446 anti proteasome antibodies delivered by the Chariot reagent (Rawe et al. 2008). An intense labeling of proteasomes (*green fluorescence*) is seen on the cytoplasm of the zygote and concentrated on the sperm connecting piece (*arrow, boxed area*). B′: Detail at higher magnification. Strong proteasome concentration (*green fluorescence*) covers the connecting piece close to sperm nucleus (*blue fluorescence*). **b″** Same area as **b′**. Failure of microtubule polymerization and sperm aster formation (*tubulin red fluorescence*). The sperm connecting piece and flagellum are clearly seen and the nucleus (*blue fluorescence*) has not decondensed. Bars represent 20 μm (**a, a′, b**) and 5 μm (**b′, b″**) (Figure 2.6was originally produced by Rawe et al. (2008) and modified by Chemes and Rawe (2010), and is reproduced with permission from the publishers)

may involve various mechanisms including ubiquitin-mediated proteolysis of selected targets by 26S proteasomes recently localized near the centrosome in the neck region of human spermatozoa (Wójcik et al. 2000; Rawe et al. 2008). A reduction below 20 and 40 % of control values was found in the activities of proteasome enzymes

Chymotrypsin and Peptidylglutamyl peptidase indicating that proteasomes of spermatozoa with neck abnormalities were endowed with deficient proteolytic machineries (Morales et al. 2004; Rawe et al. 2008). The important role of these organelles is also supported by Platts et al. (2007) who reported that the major cellular system negatively disrupted in teratozoospermia was the ubiquitin–proteasome pathway. We hypothesized that the failure of centriolar release after sperm penetration was due to insufficiency of proteasome-dependent proteolytic disassembly of the sperm connecting piece. In bovine IVF experiments with pharmacologic and immunologic neutralization of proteasomes, aster development and pronuclear apposition were markedly inhibited (Fig. 2.6), (Rawe et al. 2008). Proteasomes and polyubiquitinated proteins were recruited around the sperm connecting piece. These conglomerates ("aggresomes", Johnston et al. 1998, see Fig 2.6b, b') may represent failed attempts to overcome proteasome insufficiency when their capacity to degrade ubiquitinated proteins is exceeded. These findings point to the male complement of proteasomes as probably involved in the release of a functional centriole after proteolytic degradation of the sperm connecting piece. Similar features have been reported by Rawe et al. (2008) in zygotes from couples with spontaneous post-ICSI fertilization failure.

The assembly of such a complex structure as the sperm neck, with centrioles encased by a shield of dense proteins organized in the connecting piece and its sequential disassembly into the zygote are processes for which pathways still have to be successfully worked out. The studies summarized in this chapter highlight the central role played by the sperm neck ubiquitin–proteasome system in early zygote development and encourage further investigation on the physiopathology of sperm-related fertility failures to fully expose the basic mechanisms involved.

Acknowledgments This chapter was the result of extensive review of the literature on sperm centrosomes and centrosome-related sperm pathologies, including previous publications of our group, in particular Chemes et al. 1978, 1987b, 1999; Rawe et al. 2002, 2008.

References

Afzelius BA, Eliasson R, Johnsen O, Lindholmer C (1975) Lack of dynein arms in immotile human spermatozoa. J Cell Biol 66:225–232

Aughey E, Orr PS (1978) An unusual abnormality of human spermatozoa. J Reprod Fert 53:341–342

Alvarez Sedó C, Rawe VY, Chemes HE (2012) Acrosomal biogenesis in human globozoospermia: immunocytochemical, ultrastructural and proteomic studies. Hum Reprod [Epub ahead of print]

Baccetti B, Selmi MG, Soldani P (1984) Morphogenesis of "decapitated spermatozoa" in a man. J Reprod Fertil 70:395–397

Baccetti B, Burrini AG, Collodel G, Magnano AR, Piomboni P, Renieri T, Sensini C (1989) Morphogenesis of the decapitated and decaudated sperm defect in two brothers. Gamete Res 23:181–188

Baccetti B, Burrini AG, Collodel G, Piomboni P, Renieri T (1991) A "miniacrosome" sperm defect causing infertility in two brothers. J Androl 12:104–111

Bloom E, Birch Andersen A (1970) Ultrastructure of the "decapitated sperm defect" in Guernsey bulls. J Reprod Fertil 23:67–72

Chemes H (2000) Phenotypes of sperm pathology: genetic and acquired forms in infertile men. J Androl 21:799–808

Chemes H, Fawcett DW, Dym M (1978) Unusual features of the nuclear envelope in human spermatogenic cells. Anat Rec 192:493–511

Chemes H, Brugo Olmedo S, Zanchetti F, Carrere C, Lavieri JC (1987a) Dysplasia of the fibrous sheath. An ultrastructural defect of human spermatozoa associated with sperm immotility and primary sterility. Fertil Steril 48:664–669

Chemes HE, Carizza C, Scarinci F, Brugo Olmedo S, Neuspiler N, Schwarsztein L (1987b) Lack of a head in human spermatozoa from sterile patients: a syndrome associated with impaired fertilization. Fertil Steril 47:310–316

Chemes H, Brugo Olmedo S, Carrere C, Oses R, Carizza C, Leisner M, Blaquier J (1998) Ultrastructural Pathology of the sperm flagellum. Association between flagellar pathology and fertility prognosis in severely asthenozoospermic men. Hum Reprod 13:2521–2526

Chemes HE, Puigdomenech ET, Carizza C, Brugo Olmedo S, Zanchetti F, Hermes R (1999) Acephalic spermatozoa and abnormal development of the head-neck attachment. A human syndrome of genetic origin. Hum Reprod 14:1811–1818

Chemes HE, Rawe VY (2003) Sperm pathology: a step beyond descriptive morphology origin, characterization and fertility potential of abnormal sperm phenotypes in infertile men. Hum Reprod Update 9:405–428

Chemes HE, Rawe VY (2010) The making of abnormal spermatozoa: cellular and molecular mechanisms underlying pathological spermiogenesis. Cell Tissue Res 34:349–357

Colombero LT, Moomjy M, Sills ES, Rosenwaks Z, Palermo GD (1999) The role of structural integrity of the fertilising spermatozoon in early human embryogenesis. Zygote 7:157–163

Comizzoli P, Wildt DE, Pukazhenthi BS (2006) Poor centrosomal function of cat testicular spermatozoa impairs embryo development in vitro after intracytoplasmic sperm injection. Biol Reprod 75:252–260

Fabunmi RP, Wigley WC, Thomas PJ, DeMartino GN (2000) Activity and regulation of the centrosome-associated proteasome. J Biol Chem 275:409–413

Fawcett DW (1981) The Cell, 2nd ed. WB Saunders Co. Philadelphia, London, Toronto. ISBN 0-7216-3584-9

Hewitson L, Simerly C, Schatten G (1997) Inheritance defects of the sperm centrosome in humans and its possible role in male infertility. Int J Androl 20(Suppl 3):35–43

Hinduja I, Baliga NB, Zaveri K (2010) Correlation of human sperm centrosomal proteins with fertility. J Hum Reprod Sci 3:95–101

Holstein AF, Roosen Runge EC (1981) Atlas of human spermatogenesis. Grosse Verlag, Berlin

Holstein AF, Schirren C, Schirren CG (1973) Human spermatids and spermatozoa lacking acrosomes. J Reprod Fertil 35:489–491

Holstein AF, Schill WB, Breucker H (1986) Dissociated centriole development as a cause of spermatid malformation in man. J Reprod Fertil 78:719–725

Hoyer-Fender S (2010) Centriole maturation and transformation to basal bodies. Semin Cell Dev Biol 21:142–147

Johnston JA, Ward CL, Kopito RR (1998) Aggresomes: a cellular response to misfolded proteins. J Cell Biol 143:1883–1898

Kamal A, Mansour R, Fahmy I, Serour G, Rhodes C, Aboulghar M (1999) Easily decapitated spermatozoa defect: a possible cause of unexplained infertility. Hum Reprod 14:2791–2795

Kierszenbaum AL, Rivkin E, Tres LL, Yoder BK, Haycraft CJ, Bornens M, Rios RM (2011) GMAP210 and IFT88 are present in the spermatid golgi apparatus and participate in the development of the acrosome-acroplaxome complex, head-tail coupling apparatus and tail. Dev Dyn 240:723–736

LeLannou D (1979) Teratozoospermie consistant en läbsence de tete spermatique par defaut de conexión. J Gynecol Obstet Biol Reprod (Paris) 8:43–45

Liska F, Gosele C, Rivkin E, Tres L, Cardoso MC, Domaing P, Krejcí E, Snajdr P, Lee-Kirsch MA, de Rooij DG, Kren V, Krenová D, Kierszenbaum AL, Hubner N (2009) Rat hd mutation reveals an essential role of centrobin in spermatid head shaping and assembly of the head-tail coupling apparatus. Biol Reprod 81:1196–1205

Long CR, Duncan RP, Robl JM (1997) Isolation and characterization of MPM-2 reactive sperm proteins: homology to components of the outer dense fibers and segmented columns. Biol Reprod 57:246–254

Morales P, Diaz ES, Pizarro E, Rawe VY, Chemes HE (2004) Decreased proteasome enzymatic activity in sperm from patients with genetic abnormalities of the head-tail junction and acephalic spermatozoa. J Androl 25:41

Nakamura S, Terada Y, Horiuchi T, Emuta C, Murakami T, Yaegashi N, Okamura K (2002) Analysis of the human sperm centrosomal function and the oocyte activation ability in a case of globozoospermia, by ICSI into bovine oocytes. Hum Reprod 17:2930–2934

Nakamura S, Terada Y, Rawe VY, Uehara S, Morito Y, Yoshimoto T, Tachibana M, Murakami T, Yaegashi N, Okamura K (2005) A trial to restore defective human sperm centrosomal function. Hum Reprod 20:1933–1937

Nistal M, Herruzo A, Sanchez Corral F (1978) Teratozoospermia absoluta de presentación familiar: espermatozoides microcéfalos irregulares sin acrosoma. Andrologia 10:234–240

Palermo GD, Colombero LT, Rosenwaks Z (1997) The human sperm centrosome is responsible for normal syngamy and early embryonic development. Rev Reprod 2:19–27

Perotti ME, Gioria M (1981) Fine structure and morphogenesis of "headless" human spermatozoa associated with infertility. Cell Biol Int Rep 5:113

Perotti ME, Giarola A, Gioria M (1981) Ultrastructural study of the decapitated sperm defect in an infertile man. J Reprod Fertil 63:543–549

Pinto-Correia C, Poccia DL, Chang T, Robl JM (1994) Dephosphorylation of sperm mid piece antigens initiates aster formation in rabbi oocytes. Proc Natl Acad Sci U S A 91:7894–7898

Platts AE, Dix DJ, Chemes HE, Thompson KE, Goodrich R, Rockett JC, Rawe VY, Quintana S, Diamond MP, Strader LF, Krawetz SA (2007) Success and failure in human spermatogenesis as revealed by teratozoospermic RNAs. Hum Mol Genet 16:763–773

Porcu G, Mercier G, Boyer P, Achard V, Banet J, Vasserot M, Melone C, Saias-Magnan J, D'Ercole C, Chau C, Guichaoua MR (2003) Pregnancies after ICSI using sperm with abnormal head-tail junction from two brothers: case report. Hum Reprod 18:562–567

Rawe VY, Terada Y, Nakamura S, Chillik C, Brugo Olmedo S, Chemes HE (2002) A pathology of the sperm centriole responsible for defective sperm aster formation, syngamy and cleavage. Hum Reprod 17:2344–2349

Rawe VY, Díaz ES, Abdelmassih R, Wójcik C, Morales P, Sutovsky P, Chemes HE (2008) The role of sperm proteasomes during sperm aster formation and early zygote development: implications for fertilization failure in humans. Hum Reprod 23:573–580

Saias-Magnan J, Metzler-Guillemain C, Mercier G, Carles-Marcorelles F, Grillo JM, Guichaoua MR (1999) Failure of pregnancy after intracytoplasmic sperm injection with decapitated spermatozoa: case report. Hum Reprod 14:1989–1992

Sathananthan AH (1994) Functional competence of abnormal spermatozoa. Baillieres Clin Obstet Gynaecol 8:141–156

Sathananthan AH, Ratnasooriya WD, de Silva PK, Menezes J (2001) Characterization of human gamete centrosomes for assisted reproduction. Ital J Anat Embryol 106(Suppl 2):61–73

Schalles U, Shao X, van der Hoorn FA, Oko R (1998) Developmental expression of the 84-kDa ODF sperm protein: localization to both the cortex and medulla of outer dense fibers and to the connecting piece. Dev Biol 199:250–260

Schatten H, Sun QY (2009) The functional significance of centrosomes in mammalian meiosis, fertilization, development, nuclear transfer ad stem cell differentiation. Environ Mol Mutagen 50:620–636

Sutovsky P, Navara CS, Schatten P (1996) Fate of the sperm mitochondria, and the incorporation, conversion, and disassembly of the sperm tail structures during bovine fertilization. Biol Reprod 55:1195–1206

Tammachote R, Hommerding CJ, Sinders RM, Miller CA, Czarnecki PG, Leightner AC, Salisbury JL, Ward CJ, Torres VE, Gattone VH 2nd, Harris PC (2009) Ciliary and centrosomal defects associated with mutation and depletion of the Meckel syndrome genes MKS1 and MKS3. Hum Mol Genet 18:3311–3323

Terada Y, Nakamura S, Hewitson L, Simerly CR, Horiuchi T, Murakami T, Okamura K, Schatten G (2002) Human sperm aster formation after intracytoplasmic sperm injection with rabbit and bovine eggs. Fertil Steril 77:1283–1284

Terada Y, Nakamura S, Simerly C, Hewitson L, Murakami T, Yaegashi N, Okamura K, Schatten G (2004) Centrosomal function assessment in human sperm using heterologous ICSI with rabbit eggs; a new male factor infertility assay. Mol Reprod Dev 67:360–365

Terada Y, Schatten G, Hasegawa H, Yaegashi N (2010) Essential roles of the sperm centrosome in human fertilization: developing the therapy for fertilization failure due to sperm centrosomal dysfunction. Tohoku J Exp Med 220:247–258

Toyama Y, Iwamoto T, Yajima M, Baba K, Yuasa S (2000) Decapitated and decaudated spermatozoa in man, and pathogenesis based on the ultrastructure. Int J Androl 23:109–115

Van Blerkom J (1996) Sperm centrosome dysfunction: a possible new class of male factor infertility in the human. Mol Hum Reprod 2:349–354

Van Blerkom J, Davis P (1995) Evolution of the sperm aster microinjection of isolated human sperm centrosomes into meiotically mature human oocytes. Hum Reprod 10:2179–2182

Vashishtha M, Walther Z, Hall JL (1996) The kinesin-homologous protein encoded by the Chlamydomonas FLA10 gene is associated with basal bodies and centrioles. J Cell Sci 109:541–549

Yoshimoto-Kakoi T, Terada Y, Tachibana M, Murakami T, Yaegashi N, Okamura K (2008) Assessing centrosomal function of infertile males using heterologous ICSI. Syst Biol Reprod Med 54:135–142

Wójcik C (1997a) On the spatial organization of ubiquitin-dependent proteolysis in HeLa cells. Folia Histochem Cytobiol 35:117–118

Wójcik C (1997b) An inhibitor of the chymotrypsin-like activity of the proteasome (PSI) induces similar morphological changes in various cell lines. Folia Histochem Cytobiol 35:211–214

Wójcik C, DeMartino GN (2003) Intracellular localization of proteasomes. Int J Biochem Cell Biol 35:579–589

Wójcik C, Benchaib M, Lornage J, Czyba JC, Guerin JF (2000) Proteasomes in human spermatozoa. Int J Androl 23:169–177

Zaneveld LJD, Polakoski KL (1977) Collection and physical examination of the ejaculate. In: Hafez ESE (ed) Techniques of human andrology. Elsevier North Holland Biomedical Press, Amsterdam, pp 147–172

Zamboni L (1992) Sperm structure and its relevance to infertility. An electron microscopic study. Arch Pathol Lab Med 116:325–344

Chapter 3
Centrosomal Functions and Dysfunctions in Cat Spermatozoa

Pierre Comizzoli and David E. Wildt

Abstract The cat sperm centrosome is a paternally inherited organelle essential to successful fertilization and early embryo development. While there are structural and functional commonalities with other mammals, we have discovered original traits in the cat sperm centrosome that have offered new insights into sperm maturation as well as possible treatments for certain types of infertility. In contrast to ejaculated counterparts, cat testicular spermatozoa contain an immature centrosome preventing the formation of a large sperm aster post fertilization, which increases the incidence of early arrest in embryonic development. Emerging techniques that involve cellular desiccation or preservation in a liquid environment can adversely impact centrosomal properties of normal, mature spermatozoa. Most importantly, our investigations of the sperm centrosome in the cat model (1) re-emphasize the significance of this organelle in achieving successful reproduction and (2) are laying groundwork for overcoming centrosomal immaturity or dysfunctions using gamete micromanipulations or alternative, new in vitro sperm maturation systems.

3.1 Introduction

Our laboratory has strongly advocated for the need to expand reproductive knowledge beyond traditional animal models, especially laboratory and livestock species (Comizzoli et al. 2010; Wildt et al. 2010). Such a comparative approach is

P. Comizzoli (✉) · D. E. Wildt
Center for Species Survival, Smithsonian Conservation Biology Institute,
National Zoological Park, 3001 Connecticut Ave NW,
Washington, DC 20008, USA
e-mail: comizzolip@si.edu

essential to appreciate the diversity of reproductive mechanisms and to speed the discovery of therapies that allow the more efficient treatment of infertility, including in humans. We, and others, have found a striking number of morphological and physiological phenomena in the domestic cat that are relevant to addressing fertility issues in both men and women. Among the commonalities between the cat and human are the temporal profiles of folliculogenesis and oocyte maturation (Pelican et al. 2006; Comizzoli et al. 2011) as well as the expression of teratospermia, a condition whereby >60 % of ejaculated spermatozoa are pleiomorphic (Pukazhenthi et al. 2006).

More recently, we have both characterized and examined the significance of cat sperm centrosome—the paternally inherited organelle essential to successful fertilization (via sperm aster formation) and early embryo development (Schatten and Sun 2011). The first centrosomal study in the cat occurred almost 30 years ago with transmission electron microscopy confirming the presence of a pair of sperm centrioles in the neck region with the more distal centriole clearly undergoing various degrees of degeneration compared to the intact, proximal counterpart (Sato and Oura 1984; Schmehl and Graham 1989). This ultrastructure was similar to other observations made in non-rodent mammals, including the sheep, bull, rhesus monkey, and human (for review see Manandhar et al. 2005). Interestingly, there have been few studies about centrosomal function in felids or in carnivores in general. In our recent studies (Comizzoli et al. 2006), we first confirmed that the cat sperm centrosome played a critical role in fertilization and early embryo development. Specifically, we determined that proper sperm aster formation is mainly directed by the paternal centrosome and is crucial for pronuclear migration and apposition as well as first mitotic spindle formation, as has been shown in other species (Schatten and Sun 2011). We also discovered that a larger size cat sperm aster correlates well with early embryo developmental success to the blastocyst stage (Comizzoli et al. 2006), as has been observed in bovine (Navara et al. 1996) and human (Terada et al. 2002, 2010) systems.

Besides increasing our fundamental base of knowledge, the characterizations of normal centrosomal structure and function in the cat can have applied relevance to addressing certain fertility issues. These conditions are significant in that sperm centrosomal dysfunctions have been identified to contribute to infertility in men (Van Blerkom and Davis 1995). There have been a few investigations of centrosomal functions involving heterologous human sperm injections into bovine or rabbit oocytes (Terada et al. 2002). However, the etiology of centrosomal dysfunctions remains poorly understood, and means for mitigation are still lacking (Schatten and Sun 2011; Terada et al. 2010).

The objective of this review is to highlight the value and detailed findings of sperm centrosomal studies, specifically in the domestic cat model. Such studies are providing new insights into the role of this organelle, a better understanding of its dysfunction, and potential means for overcoming such biological anomalies that may have relevance to improving human reproductive health.

3.2 Centrosomal Immaturity in Cat Testicular Spermatozoa

Spermatozoa extracted from the seminiferous tubules of testicular tissue as well as epididymal and ejaculated cat spermatozoa have a similar-shaped head region that contains highly compacted chromatin with a low incidence (<5 %) of DNA damage (Comizzoli et al. 2006). Labeling of the long midpieces with Mitotracker Green reveals that mitochondria occupy most of the area between the head and the flagellum (Fig. 3.1a). The sperm centrosome is easily detectable by the presence of centrin, a common centrosomal protein, in both centrioles located between the head and the midpiece. Centrin appears equally abundant in testicular (Fig. 3.1b) as well as in ejaculated (Fig. 3.1c) spermatozoa.

The first role of the sperm centrosome after fertilization is to organize the formation of microtubules into a sperm aster that enables both maternal and paternal pronuclei to migrate and undergo syngamy (Schatten and Sun 2011). The formation of a large sperm aster occurs about 5 h after sperm penetration into oocytes in vitro (Comizzoli et al. 2006; Jin et al. 2011; Xu et al. 2011). As previously observed in the bovine system (Navara et al. 1996), cat sperm aster morphology is highly reflective of developmental potential. Specifically, pronuclear migration is accelerated in the presence of a larger size sperm aster that, in turn, promotes the first cleavage division (no later than 26 h post-penetration) that eventually encourages embryonic development to the blastocyst stage (Table 3.1; Comizzoli et al. 2006). More specifically, we measured a 50 % increase in advanced embryo formation in the presence of a large diameter sperm aster (Comizzoli et al. 2006). The source of spermatozoa influences the capacity to develop asters of varying size. For example, a primary reason that testicular spermatozoa fail to fertilize or experience delayed first cleavage and compromised embryo development is due to inability to produce an aster or one of normal, large size (Table 3.1; Comizzoli et al. 2006). Additionally, we have routinely observed that about 25 % of all motile cat spermatozoa from ejaculates or the epididymis consistently produced small asters. Given the clear immaturity of centrosomes in cat testicular spermatozoa, it is expected that earlier sperm stages also contain immature centrosomes, as has been observed in the rabbit (Tachibana et al. 2009).

Based on studies in the porcine system, it is known that nucleation activity of the sperm centrosome influences microtubule length by attracting γ-tubulin (Sun et al. 2001). But this mechanism is complex and not well understood. Dysfunctional centrosomes in human and non-human primate testicular or ejaculated spermatozoa lead to blocked pronuclear stage formation post insemination (Hewitson et al. 1996; Palazzo et al. 2000; Nakamura et al. 2001). By contrast, centrosomal dysfunctions in cat testicular spermatozoa do not result in complete impediments as illustrated by some minimal embryo development (Comizzoli et al. 2006). For the cat, this condition appears to be more of a centrosomal immaturity (with poor nucleation capacity) rather than a true dysfunction. According to previous studies, centrosomal maturation has been defined as the change in microtubule nucleation potential occurring as cells generally pass

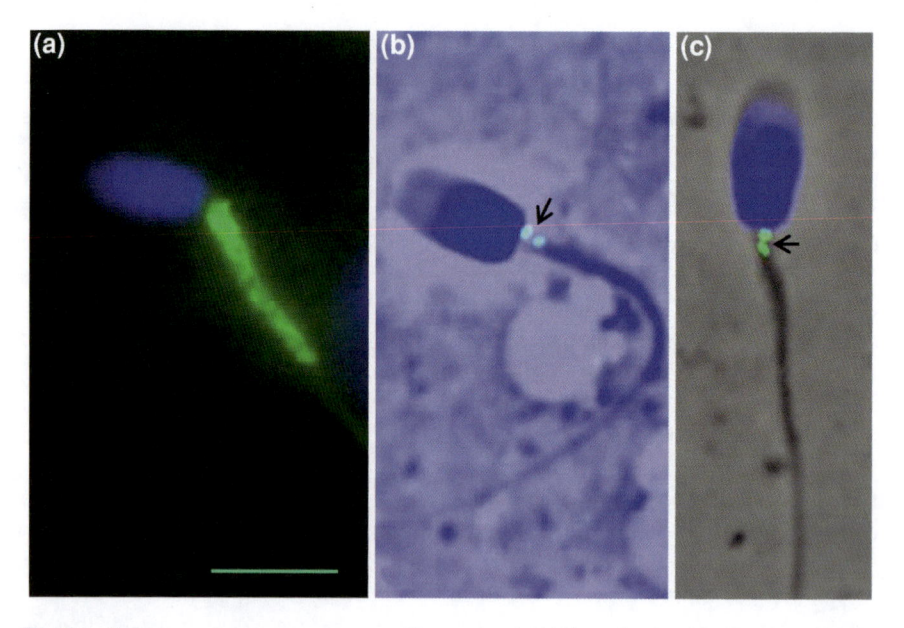

Fig. 3.1 Structure of cat spermatozoa (Sperm head DNA stained with Hoechst 33342). Mitotracker Green staining of the midpiece (**a**), centrin immunostaining (*arrow*) in testicular (**b**), and ejaculated (**c**) spermatozoa. $Bar = 5$ µm

through specific phases of the cell cycle (Palazzo et al. 2000). Although this theory remains to be tested in the cat, we suspect that sperm centrosomes in this species do not contain essential proteins, which have been argued to provoke complete maturation of this organelle in other species (Manandhar and Schatten 2000; Palazzo et al. 2000; Goto et al. 2010). Rather, our functional comparisons of spermatozoa recovered from the testis, epididymis, and ejaculate point to a full centrosomal maturation being acquired during epididymal transit. This maturational phenomenon likely is associated with the accumulation of new cytosolic proteins and/or protein phosphorylations in this region of the male reproductive tract (Axnér 2006).

3.3 Causes of Centrosomal Dysfunctions in Cat Spermatozoa

3.3.1 Impact of Teratospermia

The high proportion (>60 %) of abnormal spermatozoa in the ejaculates of certain felid species, populations, or individuals can be a significant cause of infertility through a host of functional failures, including from cells with apparently normal structure (Pukazhenthi et al. 2006). Nonetheless, Penfold et al. (2003) have demonstrated that morphologically normal cat spermatozoa recovered from a

Table 3.1 Presence of centrosome (centriole pairs detected by centrin immunostaining) and assessment of centrosomal functions (sperm aster formation after sperm injection and success of embryo development) for different types of fresh spermatozoa or after various preservation approaches

	Centrin labeling	Proportion of large sperm aster	Embryo development	Source
Epididymal spermatozoa (from cauda or ductus deferens)	+	>75 %	~35 % blastocysts	Pers. obs.
Ejaculated spermatozoa	+	>75 %	~35 % blastocysts	Comizzoli et al. (2006)
Testicular spermatozoa	+	0 %	Arrest at the 16-cell stage	Comizzoli et al. (2006)
Abnormal spermatozoa from teratospermic ejaculate	+	0 %	Arrest at the 16-cell stage	Pers. obs.
Normal spermatozoa from teratospermic ejaculate	+	>75 %	~35 % blastocysts	Pers. obs.
Epididymal or ejaculated spermatozoa post-freezing	+	>75 %	~35 % blastocysts	Comizzoli et al. (2006)
Epididymal or ejaculated spermatozoa post-drying	+	0 %	Arrest at the 16-cell stage	Pers. obs.; Ringleb et al. (2011)
Epididymal or ejaculated spermatozoa preserved in liquid	+	>75 %	Arrest at the 16-cell stage	Pers. obs.; Murakami et al. (2005)

teratospermic ejaculate can undergo normal fertilization and early embryo developmental events when microinjected into a conspecific oocyte. We confirmed these results by detecting the presence of normal sperm aster formations after injecting structurally normal spermatozoa from teratospermic ejaculates into cat oocytes (Table 3.1). Simultaneously, we observed that malformed spermatozoa from the same teratospermic semen had dysfunctional centrosomes based on the formation of small sperm asters and arrested early embryonic development (Table 3.1). This malfunction occurred even though a normally appearing centriole pair was detected in these pleiomorphic cells (Table 3.1). These observations in the cat are analogous to what occurs in some cases of human teratospermia where poor centrosomal function is due, for instance, to abnormal alignment of the head-flagellum junction (Manandhar et al. 2005; Nakamura et al. 2005). However, human centrosomal aberrations in teratospermic human ejaculate have been associated with lower expression of pericentriolar proteins, including centrin (Manandhar et al. 2005; Nakamura et al. 2005). This finding has not been apparent in the cat, although a search for alternative centrosomal proteins that might be involved in cat teratospermia is a worthy target for future study.

3.3.2 Impact of Classical and Emerging Preservation Techniques

Spermatozoa are commonly exposed to a variety of perturbations for the purposes of evaluation, processing, freezing, and practical use. Classical freeze-preservation approaches that rely on exposing these cells to cryoprotectant, osmotic, and cooling/freezing/thawing stressors are well known to adversely affect the quality of cat spermatozoa (Pukazhenthi et al. 2006). However, the use of a slow freezing (circa −1°C/min) method has been shown to retain both normal centrosomal structure and function (Comizzoli et al. 2006; Table 3.1). There has been growing interest in the scientific community in preserving male gametes via freeze-drying, desiccation, or in a liquid environment at supra-zero temperatures. While promising, these methods easily compromise sperm motility, thereby complicating the ability to select spermatozoa that have functional centrosomes. Interestingly, freeze-drying apparently is not detrimental to centrosomal functions of non-human primate (Sánchez-Partida et al. 2008) and bovine (Hara et al. 2011) spermatozoa. However, primate spermatozoa that are simply desiccated in trehalose appear to lose fertilization potential (Klooster et al. 2011). These important and contradictory findings deserve more thorough validation, especially as Ringleb et al. (2011) recently determined that the injection of freeze-dried cat spermatozoa into oocytes leads to limited early embryo development. We have also confirmed the compromised sperm aster formation using desiccated spermatozoa (dried at ambient temperature in trehalose) that were injected into conspecific oocytes (Table 3.1). We suspect that both phenomena are due to altered centrosomal functions. One study has also demonstrated loss in centrosomal capacity after storing cat spermatozoa in alcohol (Murakami et al. 2005). However, recently, we have effectively preserved cooled (4 °C) cat spermatozoa for up to 2 weeks in a 2 M trehalose solution while successfully retaining DNA integrity, centrosomal structure (presence of centrin), and function (sperm aster formation; Table 3.1).

3.4 Mitigating Centrosomal Immaturity and Dysfunctions

3.4.1 Sperm Micromanipulations and Selections

It is possible to use sonication and micromanipulation to replace an immature centrosome from a non-functioning testicular spermatozoon with a mature counterpart from an ejaculated cell. These reconstructed cat spermatozoa have biological viability, at least in capacity to develop into blastocysts in vitro (Comizzoli et al. 2006). Interestingly, such re-built spermatozoa are comprised of a head and centrosome/midpiece that are sufficiently proximate to ensure adequate interaction, pronuclear alignment, and linear migration no different from that observed after microinjecting an intact (un-reconstructed) spermatozoon (Van Blerkom and

Davis 1995). Confirmation that the sperm head remains adjacent to the centrosome/midpiece is also demonstrated by absence of arrest during the first cell cycle (Palazzo et al. 2000). These observations are important in illustrating that it is possible to replace the centrosome of one spermatozoon with that of another while ensuring the effective reorientation and elongation of the microtubule array toward the female pronucleus, which is known to be a good indicator of sperm aster quality (Navara et al. 1996). This observation naturally leads to considering the potential of this approach as a therapeutic remediation strategy in certain cases of male infertility. This prospect has already been proposed in humans with positive results (Emery et al. 2004). In the latter study, sperm heads detached from the respective flagella were co-injected into oocytes which resulted in a normal pregnancy. In our laboratory, we also envision substantial promise in centrosomal replacement across spermatozoa from different species. For example, we have studied sperm form and function in more than 25 felid species, many of which are endangered and teratospermic (Pukazhenthi et al. 2006). It would be intriguing to determine if centrosomes transferred from a normospermic species to spermatozoa from males that produce high proportions of malformed cells can be used to boost fertility or genetic management of rare individuals or populations.

Lastly, improving the sperm selection before microinjection might be another solution to avoid injecting cells with centrosomal issues. It is now possible to predict centrosomal function on the basis of midpiece morphometry as measured using a high magnification device. This process known as intracytoplasmic, morphologically selected sperm injection (IMSI) has been reported in humans (Ugajin et al. 2010) and is under development in cats.

3.4.2 Centrosomal Maturation In Vitro

Given that centrosomal maturity and normal function are acquired as testicular spermatozoa pass through the epididymis, then a priority is to learn significantly more about maturational processes within this region. Our knowledge on this subject is rudimentary for all species. For the cat, epididymal epithelial cells secrete factors (including hypotaurine or alkaline phosphatase) that impart physiological changes and could permit the sperm centrosome to complete maturation (Axnér 2006). However, a lack of information on specific mechanisms and proteins has made it impossible to artificially mature testicular spermatozoa in vitro for subsequent assisted reproduction use in any species. We are currently investigating centrosomal function and the presence of centriole maturation markers, such as cenexin (Lange BM and Gull K 1995) and speriolin (Goto et al. 2010), in spermatozoa isolated from different regions of the cat epididymis. We hope that resulting knowledge can be used to develop in vitro maturation protocols that could promote centrosomal maturation in testicular gametes and/or overcome functional deficits in spermatozoa regardless of source.

3.5 Conclusions

Clearly, the centrosome is a sperm organelle critical to both fertilization and embryogenesis and is a significant factor in male-related infertility. To date, the sperm centrosome in most species studied (including the domestic cat) share similar structural properties. Primarily, these include a pair of centrioles with centrin labeling and the distal centriole degenerating. Our studies reveal that epididymal transit is essential to normal functionality of this organelle in cats. In the absence of appropriate maturational conditions, an adequately sized sperm aster fails to develop after fertilization which leads to embryo developmental arrest. Results also reveal that the centrosome can be highly sensitive to environmental alterations. For example, some but not all methods of sperm preservation can adversely affect centrosome functionality. The cat and felids in general appear to be particular useful models for centrosomal studies due to the tendency of individuals (or species) to produce high numbers of malformed spermatozoa, as in men. Centrosomal dysfunctions are prevalent in these pleiomorphisms, thereby offering interesting models for testing remediation approaches. Based on preliminary findings, we believe it will be possible to provoke centrosomal maturation of immature (testicular) spermatozoa or overcome deficits in abnormally shaped cells using micromanipulation-cellular reconstruction and/or new in vitro culture systems. These approaches could have widespread application to animal models, endangered species, and human reproductive health.

Competing Interests
The authors declare that they have no competing interests.

Authors' Contributions
PC and DW equally contributed to the conception and writing of the review article.
Both authors read and approved the final manuscript.

References

Axnér E (2006) Sperm maturation in the domestic cat. Theriogenology 66:14–24
Comizzoli P, Wildt DE, Pukazhenthi BS (2006) Poor centrosomal function of cat testicular spermatozoa impairs embryo development in vitro after intracytoplasmic sperm injection. Biol Reprod 75:252–260
Comizzoli P, Songsasen N, Wildt DE (2010) Protecting and extending fertility for females of wild and endangered mammals. Cancer Treat Res 156:87–100
Comizzoli P, Pukazhenthi BS, Wildt DE (2011) The competence of germinal vesicle oocytes is unrelated to nuclear chromatin configuration and strictly depends on cytoplasmic quantity and quality in the cat model. Hum Reprod 26:2165–2177
Emery BR, Thorp C, Malo JW, Carrell DT (2004) Pregnancy from intracytoplasmic sperm injection of a sperm head and detached tail. Fertil Steril 81:686–688
Goto M, O'Brien DA, Eddy EM (2010) Speriolin is a novel human and mouse sperm centrosome protein. Hum Reprod 25:1884–1894

Hara H, Abdalla H, Morita H, Kuwayama M, Hirabayashi M, Hochi S (2011) Procedure for bovine ICSI, not sperm freeze-drying, impairs the function of the microtubule-organizing center. J Reprod Dev 57:428–432

Hewitson LC, Simerly CR, Tengowski MW, Sutovsky P, Navara CS, Haavisto AJ, Schatten G (1996) Microtubule and chromatin configurations during rhesus intracytoplasmic sperm injection: successes and failures. Biol Reprod 55:271–280

Jin YX, Cui XS, Yu XF, Lee SH, Wang QL, Gao WW, Xu YN, Sun SC, Kong IK, Kim NH (2011) Cat fertilization by mouse sperm injection. Zygote 27:1–8

Klooster KL, Burruel VR, Meyers SA (2011) Loss of fertilization potential of desiccated rhesus macaque spermatozoa following prolonged storage. Cryobiology 62:161–166

Lange BM, Gull K (1995) A molecular marker for centriole maturation in the mammalian cell cycle. J Cell Biol 130:919–927

Manandhar G, Schatten G (2000) Centrosome reduction during Rhesus spermiogenesis: gamma-tubulin, centrin, and centriole degeneration. Mol Reprod Dev 56:502–511

Manandhar G, Schatten H, Sutovsky P (2005) Centrosome reduction during gametogenesis and its significance. Biol Reprod 72:2–13

Murakami M, Karja NW, Wongsrikeao P, Agung B, Taniguchi M, Naoi H, Otoi T (2005) Development of cat embryos produced by intracytoplasmic injection of spermatozoa stored in alcohol. Reprod Domest Anim 40:511–515

Nakamura S, Terada Y, Horiuchi T, Emuta C, Murakami T, Yaegashi N, Okamura K (2001) Human sperm aster formation and pronuclear decondensation in bovine eggs following intracytoplasmic sperm injection using a Piezo-driven pipette: a novel assay for human sperm centrosomal function. Biol Reprod 65:1359–1363

Nakamura S, Terada Y, Rawe VY, Uehara S, Morito Y, Yoshimoto T, Tachibana M, Murakami T, Yaegashi N, Okamura K (2005) A trial to restore defective human sperm centrosomal function. Hum Reprod 20:1933–1937

Navara CS, First NL, Schatten G (1996) Phenotypic variations among paternal centrosomes expressed within the zygote as disparate microtubule lengths and sperm aster organization: correlations between centrosome activity and developmental success. Proc Natl Acad Sci U S A 93:5384–5388

Palazzo RE, Vogel JM, Schnackenberg BJ, Hull DR, Wu X (2000) Centrosome maturation. Curr Top Dev Biol 49:449–470

Pelican KM, Wildt DE, Pukazhenthi B, Howard J (2006) Ovarian control for assisted reproduction in the domestic cat and wild felids. Theriogenology 66:37–48

Penfold LM, Jost L, Evenson DP, Wildt DE (2003) Normospermic versus teratospermic domestic cat sperm chromatin integrity evaluated by flow cytometry and intracytoplasmic sperm injection. Biol Reprod 69:1730–1735

Pukazhenthi BS, Neubauer K, Jewgenow K, Howard J, Wildt DE (2006) The impact and potential etiology of teratospermia in the domestic cat and its wild relatives. Theriogenology 66:112–121

Ringleb J, Waurich R, Wibbelt G, Streich WJ, Jewgenow K (2011) Prolonged storage of epididymal spermatozoa does not affect their capacity to fertilise in vitro-matured domestic cat (*Felis catus*) oocytes when using ICSI. Reprod Fertil Dev 23:818–825

Sánchez-Partida LG, Simerly CR, Ramalho-Santos J (2008) Freeze-dried primate sperm retains early reproductive potential after intracytoplasmic sperm injection. Fertil Steril 89:742–745

Sato N, Oura C (1984) The fine structure of the neck region of cat spermatozoa. Okajimas Folia Anat Jpn 61:267–285

Schatten H, Sun QY (2011) New insights into the role of centrosomes in mammalian fertilisation and implications for ART. Reproduction [Epub ahead of print]

Schmehl ML, Graham EF (1989) Ultrastructure of the domestic tom cat (*Felis domestica*) and tiger (*Panthera tigris altaica*) spermatozoa. Theriogenology 31:861–874

Sun QY, Lai L, Park KW, Kuhholzer B, Prather RS, Schatten H (2001) Dynamic events are differently mediated by microfilaments, microtubules, and mitogen-activated protein kinase during porcine oocyte maturation and fertilization in vitro. Biol Reprod 64:879–889

Tachibana M, Terada Y, Ogonuki N, Ugajin T, Ogura A, Murakami T, Yaegashi N, Okamura K (2009) Functional assessment of centrosomes of spermatozoa and spermatids microinjected into rabbit oocytes. Mol Reprod Dev 76:270–277

Terada Y, Nakamura SI, Hewitson L, Simerly C, Horiuchi T, Murakami T, Okamura K, Schatten G (2002) Human sperm aster formation after intracytoplasmic sperm injection with rabbit and bovine eggs. Fertil Steril 77:1283–1284

Terada Y, Schatten G, Hasegawa H, Yaegashi N (2010) Essential roles of the sperm centrosome in human fertilization: developing the therapy for fertilization failure due to sperm centrosomal dysfunction. Tohoku J Exp Med 220:247–258

Ugajin T, Terada Y, Hasegawa H, Nabeshima H, Suzuki K, Yaegashi N (2010) The shape of the sperm midpiece in intracytoplasmic morphologically selected sperm injection relates sperm centrosomal function. J Assist Reprod Genet 27:75–81

Van Blerkom J, Davis P (1995) Evolution of the sperm aster after microinjection of isolated human sperm centrosomes into meiotically mature human oocytes. Hum Reprod 10:2179–2182

Wildt DE, Comizzoli P, Pukazhenthi B, Songsasen N (2010) Lessons from biodiversity—the value of nontraditional species to advance reproductive science, conservation, and human health. Mol Reprod Dev 77:397–409

Xu YN, Cui XS, Sun SC, Jin YX, Kim NH (2011) Cross species fertilization and development investigated by cat sperm injection into mouse oocytes. J Exp Zool A Ecol Genet Physiol 315:349–357

Chapter 4
Nuclear–Centrosome Relationships During Fertilization, Cell Division, Embryo Development, and in Somatic Cell Nuclear Transfer Embryos

Heide Schatten and Qing-Yuan Sun

Abstract Nuclear–centrosome relationships are critical for synchronized cell cycle progression. During fertilization and syngamy, centrosome–nuclear relationships are important for pronuclear migration and to allow synchronized maturation of the sperm nucleus into the male pronucleus and of the sperm's centriole–centrosome complex into a division-competent centrosome that is able to separate chromosomes precisely into the dividing daughter cells. Abnormalities in nuclear–centrosome interactions are among the underlying causes for male and female factor infertility, for developmental disorders, and disease. Centrosome–nuclear abnormalities are also encountered in somatic cell nuclear transfer (SCNT) embryos when centrosome reprogramming is defective. The present review is focused on 1) The sperm centriole–centrosome complex and associations with the sperm nucleus before fertilization; 2) Centrosome–nuclear relationships during pronuclear/zygote stage, cell division, and embryo development; and 3) Centrosome–nuclear interactions and centrosome reprogramming abnormalities in SCNT embryos.

H. Schatten (✉)
Department of Veterinary Pathobiology, University of Missouri-Columbia,
1600 E Rollins Street, Columbia MO 65211, USA
e-mail: SchattenH@missouri.edu

Q.-Y. Sun (✉)
State Key Laboratory of Reproductive Biology, Institute of Zoology,
Chinese Academy of Sciences, Beichen West Road,
Chaoyang 100101, Beijing, China
e-mail: sunqy@ioz.ac.cnsunqy@yahoo.com

4.1 Introduction

Structural and functional relationships between the nucleus and centrosomes are critically important for successful fertilization, accurate cell division, and proper embryo development. In somatic interphase cells, centrosomes are closely associated with the nuclear surface and structural centrosome–nuclear relationships have also been shown for *C. elegans* embryos (Meyerzon et al. 2009) while studies to explore the structural and functional centrosome–nuclear relationships in mammalian embryo cells are still only at the beginning.

Is has been well shown in somatic cell systems that precise nuclear–centrosome synchrony is essential for mitosis and cytokinesis when centrosomes, microtubules, and chromosomes need to interact precisely to fulfill mitotic checkpoint licensing and coordinate chromosome separation for accurate cell division (reviewed in Schatten 2008 and Chaps. 8, 9, 10, and 11 of this book). In the developing embryo, centrosomes and the nucleus have to coordinate both symmetric and asymmetric cell divisions to distribute cellular components and cell fate determinants to the dividing daughter cells which is important for cell differentiation. As our previous reviews have addressed the role of centrosomes in oocyte maturation (Schatten and Sun 2011b), fertilization (Schatten and Sun 2009a, 2009b, 2010, 2011a, 2011c), and male factor infertility (Schatten et al. 2011) we will focus the present review on 1) The sperm centriole–centrosome complex and associations with the sperm nucleus before fertilization; 2) Centrosome–nuclear–synchrony during pronuclear/zygote stage, cell division, and embryo development; and 3) Centrosome–nuclear interactions and reprogramming abnormalities in somatic cell nuclear transfer (SCNT) embryos.

4.2 The Sperm Centriole–Centrosome Complex and Associations with the Sperm Nucleus Before Fertilization

In non-rodent mammalian systems, the sperm contains one proximal and one distal centriole that are organized perpendicular to each other (Fig. 4.1) and located within the sperm's connecting piece between the midpiece and the sperm's nucleus. Only the proximal centriole is associated with the sperm nucleus and surrounded by a small amount of centrosomal proteins including γ-tubulin and centriole-associated centrin as well as other newly discovered centrosomal components (Goto et al. 2010) for which functions remain to be determined (reviewed in Schatten and Sun 2011a, 2011b, 2011c). The distal centriole is associated with the sperm tail and its functions primarily include organization and assembly of sperm tail microtubules. The close association and functional relationships of the proximal centriole with the sperm nucleus has been assessed on structural levels before fertilization and it becomes clearly apparent after fertilization when the

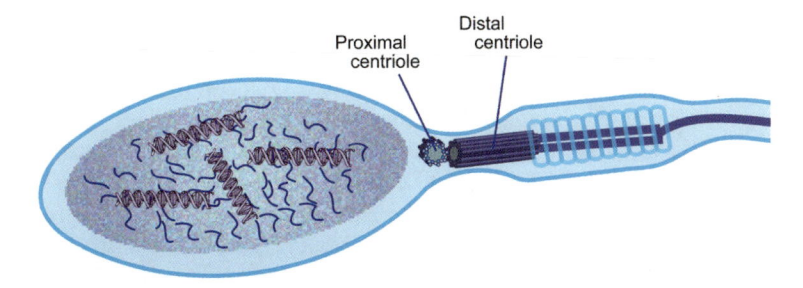

Fig. 4.1 *Schematic representation of non-rodent sperm*: showing sperm head with DNA and nuclear matrix proteins and the centriole complex. The sperm's centriole complex consists of two perpendicularly oriented centrioles. The centriole close to the nucleus is termed the proximal centriole and contains sparse centrosomal material that will become important for microtubule nucleation of sperm aster, zygote aster, and mitotic apparatus in the fertilized oocyte while the centriole associated with the sperm tail (distal centriole; basal body) will not participate in the embryo's microtubule formations and will be subjected to degeneration along with the sperm tail after fertilization

sperm head decondenses and the centriole–centrosome complex matures within the zygote to become a division-competent centrosome that forms the mitotic spindle poles and separates chromosomes equally to the dividing daughter cells, as detailed in Sect. 3 of this chapter.

Excellent electron micrographs are available on the centriole complex in human sperm that have been presented in several original papers and review articles (Chemes 2000; Chemes and Rawe 2003; Mitchell et al. 2006; Rawe et al. 2002; Rawe and Chemes 2009; Sathananthan et al. 1991, 1996, 2001; Sathananthan 1997, Sathananthan 2009) displaying clear structural associations between the sperm nucleus and the proximal centriole. Examples are presented in Chaps. 2 and 5 of this book (by Hector E. Chemes and A. Henry Sathananthan, respectively), providing remarkable detail of normal structure and structural abnormalities in the proximal centriole-sperm nucleus associations that impact fertilization. Structural abnormalities and dysfunctions associated with male factor infertility have been documented and it has also been shown that in some cases of infertility centriole–nuclear detachment sites are impaired (Liska et al. 2009; Kierszenbaum et al. 2011). While morphological abnormalities have clearly been identified as underlying causes for sperm centriole–centrosomal dysfunctions, molecular abnormalities have been determined by using a variety of molecular methods including immunoblotting techniques (Bohring and Krause 2003; reviewed in Schatten and Sun 2009a, 2009b). Several studies have been performed on human sperm that correlate decreased γ-tubulin and decreased centrin to lower fertilizability in humans (Hinduja et al. 2008; 2010) and correlated below-normal quantities to decreased sperm aster formation and developmental capacity, as also discussed by Comizzoli and Wildt in Chap. 3 of this book.

While recent research has focused on the sperm's centrosome complex as causes of infertility, studies related to nuclear components within the sperm nucleus have been sparse which may in part be related to the density of the sperm nucleus that has made experimental and analytical approaches difficult (reviewed in Johnson et al. 2011). The mature sperm nucleus is distinguished from other nuclei by its extreme chromatin condensation state which is achieved during spermiogenesis when the majority of histones are replaced by protamines, small basic proteins that are bound to sperm DNA and become cross-linked through the formation of disulfide bridges when spermatozoa transit through the epididymis (reviewed in Delbés et al. 2010). Proper chromatin compaction is important for male factor fertility in which accurate protamine, histone, and nuclear matrix component functions are essential. Only recently has it been possible to dissect structural aspects within the sperm nucleus and it has clearly been shown that nuclear matrix components are present in sperm (reviewed in Johnson et al. 2011) which opens up speculations that the nuclear mitotic apparatus (NuMA) protein may exist within the sperm nucleus. The multifunctional protein NuMA has been best studied in somatic cells (reviewed in Sun and Schatten 2006) and it has been shown that it plays important roles as nuclear matrix protein in interphase and as centrosome-associated protein during meiosis and mitosis (reviewed in Sun and Schatten 2006). We do not yet know when NuMA functions become important for sperm nuclear functions and for nuclear–centrosome relationships but we know for certain that it plays an important role in the decondensing sperm pronucleus after fertilization (reviewed in Sun and Schatten 2006; Alvarez-Sedo et al. 2011; Schatten et al. 2012) which will be addressed in Sect. 2 of this chapter. To better understand yet unexplained causes of male factor infertility it will be important to determine new methods to better analyze nuclear components in sperm that play a role in nuclear–centrosome synchrony after fertilization.

4.3 Centrosome–Nuclear Relationships During Pronuclear/ Zygote Stage, Cell Division, and Embryo Development

A schematic representation of the nuclear and centriole–centrosome cycle within the first embryonic cell cycle is shown in Fig. 4.2a–d and described in the figure legend. Significant changes in sperm chromatin structure begin immediately after fertilization when protamines are replaced by histones and several epigenetic modifications take place (Chao et al. 2012). Sperm chromatin becomes decondensed and the sperm nucleus matures into the male pronucleus while the sperm-derived centriole–centrosome complex matures by recruiting and accumulating centrosomal components from the sperm-activated oocyte (reviewed in Schatten and Sun 2010, 2011a, 2011c).

Intimate structural and functional relationships between nuclear and centrosome proteins are important to fulfill cell cycle-specific functions including microtubule organization, chromosome alignment, and chromosome separation during the embryo's first cell cycle (reviewed in Schatten 2008; Schatten and Sun

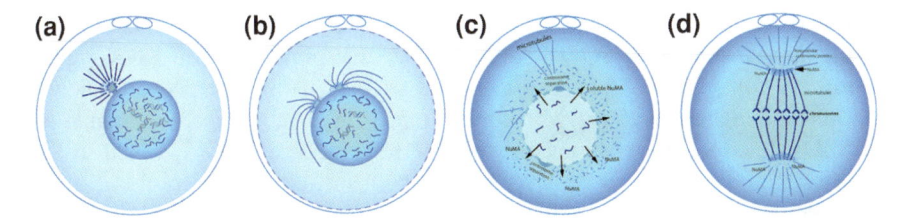

Fig. 4.2 a–d: schematic representation of the nuclear and centriole–centrosome cycle within the first embryonic cell cycle. **a** Sperm aster formation from the sperm-derived proximal centriole–centrosome complex; DNA and the nuclear matrix protein NuMA are localized in the nucleus; **b** duplication of centrioles at pronuclear stage; **c** duplicated centriole–centrosome complex separates and migrates around the zygote nucleus, relocating to opposite poles to form the centers of the mitotic spindle poles; NuMA becomes a centrosome-associated protein and participates in the formation of the spindle poles by forming a crescent around the centrosome area facing chromosomes; **d** mitosis of the first cell cycle

2011a, 2011c); synchronized centrosome and nuclear maturation within the fertilized oocyte is critical for all subsequent cell divisions within the developing embryo. Nuclear–centrosome cell cycle synchronization has been studied in detail for the S phase in somatic cells which correlates to the pronuclear stage in the zygote embryo cell and to the S phases of all cells in the developing embryo when synchronized DNA and centrosome duplication takes place. In somatic cells, critical regulatory processes have been identified including activation of CDK2-cyclin E (Okuda et al. 2000; Tokuyama et al. 2001; Ferguson and Maller 2010) and other cell cycle-specific proteins (reviewed in Chap. 8 of this book by Fisk and in Chap 11 of this book by Boutros). In addition, centrosome-related protein degradation becomes important, as detailed in Chap. 8 of this book by Fisk and in Chap. 9 of this book by Posser and Fry.

In normal cell cycles synchronization is tightly controlled through cell cycle checkpoints, coordinated signal transduction cascades, and several other regulatory processes that, while well-studied in somatic cells, remain only partly explored in embryo cells during preimplantation development. We do know that the maturation promoting factor (MPF) and mitogen-activated protein kinase (MAPK) are important for nuclear maturation and both are important for regulation of several cell cycle events during oocyte maturation, in the fertilized embryo cell, and in the embryo's first cell cycle (reviewed in Fan and Sun 2004; Snook et al. 2011) but we do not know details on cell cycle checkpoints and regulatory processes that drive centrosome maturation and dispersion of nuclear proteins into the cytoplasm for subsequent specific functions during cell cycle progression of the first and subsequent cell cycles in embryonic cells. It is clear that synchronized signaling of nuclear and centrosome dynamics is important to ensure accurate participation in the mitotic process during first and subsequent cell divisions. Studies have begun to investigate the regulation of NuMA as one of the essential nuclear and centrosome-associated proteins important for successful fertilization and embryonic cell divisions.

NuMA is a multifunctional 236 kDa protein that in interphase is a component of the nuclear matrix, a proteinaceous network that plays a role in DNA organization. NuMA's specific roles in the nucleus are only partly understood (reviewed in Sun and Schatten 2006) while significantly more studies are available on NuMA's functions as centrosome-associated protein in mitosis (reviewed in Sun and Schatten 2006). We do not yet know whether NuMA plays a role in sperm DNA organization and nuclear decompaction in the zygote embryo but we know that the nuclear matrix is important for DNA replication in the zygote (reviewed by Yamauchi et al. 2011; Johnson et al. 2011). Research on the sperm's nuclear matrix has accelerated in recent years, and it has been proposed that non-genetic male factor infertility problems may be related to nuclear matrix instability (reviewed in Johnson et al. 2011), contributing to transgenerational non-genetic instability. It has further been shown that chronic exposure of sperm to low doses of specific toxins is correlated with an altered nuclear matrix protein profile and includes abnormal chromatin condensation (Codrington et al. 2007a, 2007b) which affects fertilization and embryo development and may also affect nuclear matrix-centrosome interactions and nuclear–centrosome synchronization. This and other examples indicate the effects of non-genetic components on the nuclear matrix that may affect synergistic interactions with centrosomes and may have implications in male factor infertility.

NuMA is an important link in synchronizing nuclear and centrosome maturation events after fertilization; studies in somatic cells have shown that NuMA requires precise regulation including regulation by cyclin B to move out of the nucleus into the cytoplasm during prophase and associate with mitotic centrosomes to stabilize centrosome–microtubule interactions for the formation of the mitotic apparatus. NuMA is not associated with the interphase centrosome; it strictly serves as nuclear protein in interphase and becomes a centrosome-associated protein only in mitosis (reviewed in Sun and Schatten 2006). NuMA is highly insoluble in the nucleus but at the time of nuclear envelope breakdown NuMA becomes hyperphosphorylated by $p34^{cdc2}$ which allows dispersion of NuMA into the cytoplasm and subsequent translocation to the spindle poles in a dynein-mediated process. NuMA remains associated with the spindle poles until anaphase; it dissociates from spindle poles after dephosphorylation as a result of Cdk1 inactivation and loss of cyclin B that occurs by proteasome-mediated degradation (Gehmlich et al. 2004).

Most of the studies on NuMA have been performed in somatic cells while only a few detailed studies are available for embryonic cells. Our recent studies in human oocytes showed a requirement for dynein to mediate NuMA translocation to the spindle poles during first mitosis (Alvarez Sedó et al. 2011; Schatten et al. 2012). In the MI and MII meiotic spindles NuMA was localized to the meiotic spindle poles and displayed abnormalities in aged oocytes. It also displayed abnormalities in fertilized oocytes in which sperm decondensation failed which coincided with abnormal NuMA immunofluorescence staining patterns and suggests that NuMA abnormalities are associated with fertilization failures. Dispersed NuMA fluorescence staining patterns were seen in male and female pronuclei in normal fertilization (Liu et al. 2006; Alvarez-Sedó et al. 2011; Schatten et al. 2012).

Detailed studies on NuMA before and after fertilization will be important to determine whether NuMA dysfunctions play a role in male-factor infertility, embryo abnormalities, and whether NuMA dysfunctions at this early stage of an embryo's development will result in adulthood diseases, as misregulation of NuMA can result in the formation of multipolar mitoses which are hallmark features for cancer development and progression (Kammerer et al. 2005; reviewed in Sun and Schatten 2006). In this context it is also worth noting that NuMA becomes extensively modified after herpes simplex virus (HSV) infection which induces solubilization and relocalization of NuMA (Yamauchi et al. 2008); it may affect subsequent NuMA dynamics that may play a role in mitotic abnormalities underlying diseases such as cancer.

Close attachment between the centrosome and the nucleus is important for coordinated pronuclear migration and apposition of male and female pronuclei and formation of the zygote aster that evolves into the mitotic apparatus to separate the parental genomes equally to the dividing daughter cells (reviewed in Schatten and Sun 2009a, 2009b, 2010, 2011a, 2011c). The structural associations of centrosomes with the nucleus have been described in somatic cells (Meyer et al. 2011) and studies in *C. elegans* have explored the mechanisms of centrosome–nuclear relationships but only sparse information is available on such structural and functional relationships in mammalian embryo cells. In *C. elegans*, ZYG-12, SUN-1, and LIS-1 interact with dynein to contribute to the attachment of centrosomes to the nucleus in early development and it was proposed that recruitment of dynein to the cytoplasmic surface of the nuclear envelope is critical for the attachment of centrosomes (Malone et al. 2003; Meyerzon et al. 2009); two genes, zyg-12, and sun-1 were shown to be essential for centrosome attachment and embryonic development in this system (Fridkin et al. 2004; Malone et al. 2003). It was further proposed that the inner nuclear membrane and nuclear lamina proteins are involved in centrosome–nucleus attachment (Askjaer et al. 2002; Galy et al. 2006). In *C. elegans*, Meyerzon et al. (2009) described a novel role for nuclear lamina proteins in centrosome attachment to the nuclear envelope. The mechanisms described for nuclear–centrosomal attachment in *C. elegans* have been explored for the first few cell divisions in early embryogenesis but may be different during later development, as ZYG-12 is not required for later stages of embryogenesis. These studies in *C. elegans* provide important steps toward understanding embryonic centrosome–nuclear attachment mechanisms while we still know only little about such molecular mechanisms for mammalian embryos.

In fibroblast cells, it was shown that lamin and the integral inner nuclear membrane protein emerin are involved in centrosome attachment, as fibroblasts from emerin-defective human patients or lamin A/C mutant mice displayed centrosome detachment phenotypes (Lee et al. 2007; Salpingidou et al. 2007). Emerin is present at both the inner and outer nuclear membrane and it was determined that emerin interacts with tubulin which led to the proposed model that emerin on the outer nuclear envelope directly interacts with microtubules to attach the centrosome to the nuclear envelope (Salpingidou et al. 2007).

Another nuclear protein of interest regarding centrosome–nuclear relationships is the multifunctional structural protein 4.1R that localizes within nuclei, at the nuclear envelope, and in the cytoplasm. It has recently been shown that 4.1R, the nuclear envelope protein emerin and the intermediate filament protein lamin A/C co-immunoprecipitate in human cells and that its depletion affects the distribution of NuMA as well as other nuclear proteins (Meyer et al. 2011). These studies extend on the previous findings in different cell systems reporting that emerin couples centrosomes to the nuclear envelope (Markiewicz et al. 2006; Salpingidou et al. 2007). Meyer et al. showed that the functional effects of 4.1R deficiency included disruption of its association with emerin and A-type lamins and an increase in nucleus–centrosome distances, affecting centrosome–nuclear envelope association (Meyer et al. 2011). Previous reports had shown that 4.1 binds NuMA and contributes to the organization of the nucleoskeleton and nuclear membrane proteins; it plays a role in the attachment of centrosomes to the nucleus in *C. elegans* (Meyerzon et al. 2009; Simon and Wilson 2011). While we do not yet have a complete understanding of the processes and molecular interactions involved in nucleus–centrosome attachments, the role of nuclear proteins in centrosome attachment to the nucleus is beginning to emerge although our current still fragmented knowledge has been derived from different cell systems and different mechanisms may be employed in different cell systems. Because protein 4.1R is also integral to mitotic spindle and centrosome assembly and structure (Krauss et al. 2004, 2008) it may further allow us to generate new insights into nuclear–mitotic centrosome relationships as an important step toward understanding cell cycle-specific nuclear–centrosome synchronization.

As mentioned above, the synchronized distribution of centrosomal and genetic material to the dividing daughter cells is facilitated by the tight association of centrosomes with the nucleus; this interaction is important for all cells in the developing embryo and the close association also allows synchronized distribution of critical cellular components to the dividing daughter cells. The role of centrosomes during embryo development includes gathering and distribution of cellular components including critical cell fate determinants to the dividing daughter cells. Such cell fate determinants differ in different systems (reviewed in Knoblich 2010) but substantial evidence exists that they are transported along microtubules during interphase and distributed to the differentiating blastomeres during subsequent cell divisions. It includes cytoplasmic factors such as transcripts of developmental genes, eventually resulting in different gene activity; the inheritance of mRNA localized at centrosomes has been described for *Ilyanassa* (Lambert and Nagy 2002; Kingsly et al. 2007).

Transport along microtubules to centrosomes is also utilized for protein degradation which has been described for embryonic stem cell divisions, allowing asymmetric inheritance of proteins destined for different degradation in the dividing daughter cells (reviewed in Chap. 8 of this book by Fisk).

4.4 Centrosome–Nuclear Interactions and Reprogramming Abnormalities in Somatic Cell Nuclear Transfer Embryos

The tight structural connections of the centrosome with the nucleus are also apparent in nuclear isolations in which the centrosome is typically co-isolated along with the nucleus unless specific centrosome–nucleus separation methods are employed. This close connection becomes important for SCNT embryos in which an oocyte is enucleated and genomic material is replaced with a somatic cell nucleus that also contributes the nucleus-associated centriole–centrosome complex, thereby providing the centrosomal core structure that is normally contributed by sperm during fertilization. Like the somatic cell nucleus, the somatic cell's centriole–centrosome complex needs to be reprogrammed in SCNT to fulfill functions that are normally carried out by the blended sperm–oocyte centrosomal complex (reviewed in Schatten et al. 2009a, 2009b) which includes formation of the mitotic apparatus during cell division. Reprogramming of the somatic cell's centrosome complex depends on regulation by the enucleated oocyte to provide components that are important for embryonic centrosome cell cycles. While reconstructed SCNT eggs provide an ideal analysis system for centrosome regulation very few studies have been performed so far on centrosome regulation in the SCNT embryo system. However, studying the complexities between requirements for embryonic cells compared to somatic cells may bring about further insights into centrosome biology and correlations between nuclear–centrosome interactions.

Live births resulting from SCNT reconstructed embryos have been obtained for most animal species; however, in most cases the success rate is limited to 1–5 % which indicates incompatibilities of the oocyte to reprogram the somatic cell nucleus and its associated centrosomal complex. Indeed, our studies in porcine SCNT reconstructed eggs revealed that 39.4 % of reconstructed eggs displayed centrosomal abnormalities during the first cell cycle (Zhong et al. 2007) as determined by γ-tubulin and/or centrin-2 and correlated microtubule staining patterns. These centrosomal mitotic abnormalities may result in developmental abnormalities or contribute to cellular pathologies that may be manifested as adulthood diseases later in life. Reprogramming of somatic cell centrosomes begins shortly after SCNT within the first embryonic cell cycle which spans ca. 24 h and is a very short time for centrosomal reprogramming considering that somatic cell centrosomes are different from reproductive cell centrosomes, perhaps containing different centrosomal compositions and different capabilities to perform cell cycle-specific functions that are precisely provided by regulatory factors in the somatic cell cytoplasm for somatic cell cycles; centrosome functions may require different regulation in embryonic cells.

Centrosomal abnormalities may account for abnormal cell divisions during different stages of development and may be part of the cellular incompatibilities that allows only 1–5 % of reconstructed embryos to develop to full term resulting in healthy offspring. Details and thoughts on the possible reasons underlying centrosomal incompatibilities are provided in our more detailed previous review on SCNT (Schatten and Sun 2009a).

The case may be made that in the 10 times smaller somatic cell the requirements for somatic (donor) cell centrosomes are likely to be different compared to those in the huge reconstructed egg of about 100 μm. We know from somatic cell studies that microtubule lengths and numbers are regulated by changes in γ-tubulin recruitment to the centrosome (reviewed in Schatten 2008). As the sperm contributes only a small amount of γ-tubulin to the regular fertilization process most of the γ-tubulin components come from the oocyte after fertilization. We do not yet have detailed information on the recruitment of γ-tubulin from the oocyte to the somatic cell centrosome complex which may differ from recruitment to the sperm centrosome during normal fertilization resulting in abnormal microtubule formations in SCNT reconstructed eggs.

While we do not yet know the underlying causes for centrosomal abnormalities in SCNT reconstructed embryos, our knowledge of nuclear reprogramming has increased in recent years (reviewed in Prather 2000; Sun and Schatten 2007; Schatten and Sun 2009a) and we now have indications that nuclear matrix dysfunctions may play a role in the low success rate following SCNT, as nuclear matrix dysfunctions impacts DNA replication (reviewed by Yamauchi 2011) and it may also impact centrosome functions. As NuMA is part of the nuclear matrix, NuMA-derived spindle abnormalities have been reported for cancer cells (Kammerer et al. 2005); NuMA dysfunctions may be among the reasons for the multipolar spindle pole formations that we find in SCNT porcine embryo cells (Zhong et al. 2007).

To study the contributions of spindle pole centrosomal components in SCNT eggs, Zhong et al. (2005) used intraspecies and interspecies SCNT reconstructed eggs to determine specific centrosomal components that are contributed by the donor cell centrosome complex and the donor cell nucleus. This study used mouse MII oocytes as recipients, mouse fibroblasts, rat fibroblasts, or porcine granulosa cells as donors to produce intraspecies and interspecies nuclear transfer embryos. Specifically, to study NuMA dynamics in SCNT reconstructed eggs, a specific NuMA antibody was employed that did not recognize NuMA protein of mouse oocytes but recognized NuMA in porcine granulosa cells, thereby being able to distinguish NuMA contributed by the oocyte and donor. The results clearly showed that NuMA was localized to the donor cell nucleus and was translocated out of the nucleus into the cytoplasm followed by translocation to the mitotic spindle poles where donor cell NuMA participated in spindle pole formation during first mitosis in SCNT eggs. Further analysis of NuMA translocation out of the nucleus in porcine SCNT eggs (Liu et al. 2006) revealed that NuMA was contributed by fetal fibroblast donor cells to reconstructed porcine eggs and it took about 6 h after nuclear transfer before NuMA could be visualized with immunofluorescence microscopy, indicating a lag period for NuMA reprogramming by the reconstructed egg which is significantly longer than NuMA detection in decondensing sperm nuclei that takes place within minutes after fertilization (Liu et al. 2006). This study concluded that cytoplasmic factors in the recipient porcine oocyte were able to remodel the donor cell's NuMA although it took 6 h for the remodeling to take place.

While we do not yet know the exact molecular composition of the zygotic centrosome and we also do not yet know how it compares to the interphase somatic cell centrosome and to the centrosome in reconstructed eggs we know that precise regulation is important for centrosome functions. Phosphorylation plays a significant role in centrosome functions and we know from cancer cell centrosomes that abnormal increases in phosphorylation results in increased microtubule nucleation with consequences for abnormal spindle formation (Lingle et al. 1998). In somatic cells, NuMA is in part regulated by cyclin B (reviewed in Sun and Schatten 2006) but the regulation of NuMA is still unknown for mammalian embryonic cells. It is possible that the NuMA-related abnormalities that we found in human oocytes (Alvarez-Sedó et al. 2011; Schatten et al. 2012) may have been the result of inaccurate regulation by the fertilized ooplasm. It is also possible that NuMA may be part of the sperm's nuclear matrix that may play a role in nuclear matrix instability and dysfunctions (reviewed by Johnson et al. 2011).

Taken together, while we have started to analyze nuclear–centrosome relationships and centrosome-nuclear reprogramming more detailed studies are needed to determine how the somatic cell centrosome becomes remodeled to fulfill the functions of the embryo's blended centrosome that contains precise amounts and compositions of centrosomal proteins that are precisely regulated by the fertilized oocyte to serve embryo-specific functions including symmetric and asymmetric cell divisions during embryo differentiation and development.

4.5 Conclusion and Future Perspectives

Numerous lines of evidence have established the importance of centrosome–nuclear interactions and synchronized cell cycle progression to ensure accurate fertilization, zygote formation, and cell divisions during embryogenesis. However, most of our knowledge of centrosome–nuclear regulation comes from somatic cells and more research is needed to study abnormalities underlying male and female factor infertility problems and developmental abnormalities related to dysfunctions in centrosome–nuclear interactions.

References

Alvarez Sedó CA, Schatten H, Combelles C, Rawe VY (2011) The nuclear mitotic apparatus protein NuMA: localization and dynamics in human oocytes, fertilization and early embryos. Mol Hum Reprod 17(6):392–398

Askjaer P et al (2002) Ran GTPase cycle and importins alpha and beta are essential for spindle formation and nuclear envelope assembly in living Caenorhabditis elegans embryos. Mol Biol Cell 13:4355–4370

Bohring C, Krause W (2003) Immune infertility: towards a better understanding of sperm (auto)-immunity, the value of proteomic analysis. Hum Reprod 18(5):915–924

Chao, S-B, Guo, L, Ou, X-H, Luo, S-M, Wang, Z-B, Schatten, H, Gao, G-L, Sun, Q-L (2012) Heated spermatozoa: effects on embryonic development and epigenetics. Hum Reprod (first published online February 7, 2012). doi:10.1093/humrep/des005

Chemes HE (2000) Phenotypes of sperm pathology: genetic and acquired forms in infertile men. J Androl 21(6):799–808

Chemes HE, Rawe VY (2003) Sperm pathology: a step beyond descriptive morphology. Origin, characterization and fertility potential of abnormal sperm phenotypes in infertile men. Hum Reprod Update 9(5):405–428

Codrington AM, Hales BF, Robaire B (2007a) Chronic cyclophosphamide exposure alters the profile of rat sperm nuclear matrix proteins. Biol Reprod 77:303–311

Codrington AM, Hales BF, Robaire B (2007b) Exposure of male rats to cyclophosphamide alters the chromatin structure and basic proteome in spermatozoa. Human Reprod 22:1431–1442

Delbés G, Hales BF, Robaire B (2010) Toxicants and human sperm chromatin integrity. Mol Hum Reprod 16(1):14–22

Fan H-Y, Sun Q-Y (2004) Involvement of mitogen-activated protein kinase cascade during oocyte maturation and fertilization in mammals. Biol Reprod 70:535–547

Ferguson RL, Maller JL (2010) Centrosomal localization of cyclin E-Cdk2 is required for initiation of DNA synthesis. Curr Biol 20:856–860

Fridkin A et al (2004) Matefin, a Caenorhabditis elegans germ line-specific SUN domain nuclear membrane protein, is essential for early embryonic and germ cell development. Proc Natl Acad Sci U S A 101:6987–6992

Galy V et al (2006) MEL-28, a novel nuclear-envelope and kinetochore protein essential for zygotic nuclear-envelope assembly in C. elegans. Curr Biol 16:1748–1756

Gehmlich K, Haren L, Merdes A (2004) Cyclin B degradation leads to NuMA release from dynein/dynactin and from spindle poles. EMBO Rep 5:97–103

Goto M, O-Brien DA, Eddy EM (2010) Speriolin is a novel human and mouse sperm centrosome protein. Human Reprod 25(8):1884–1894

Hinduja I, Zaveri K, Baliga N (2008) Human sperm centrin levels and outcome of intracytoplasmic sperm injection (ICSI)–a pilot study. Indian J Med Res 128:606–610

Hinduja I, Zaveri K, Baliga N (2010) Correlation of human sperm centrosomal proteins with fertility. J Hum Reprod Sci 3(2):95–101

Johnson GD, Lalancette C, Linnemann AK, Leduc F, Boissonneault G, Krawetz SA (2011) The sperm nucleus: chromatin, RNA, and the nuclear matrix. Reproduction 141:21–36

Kammerer S, Roth RB, Hoyal CR, Reneland R, Marnellos G, Kiechle M, Schwarz-Boeger U, Griffiths LR, Ebner F, Rehbock J, Cantor CR, Nelson MR, Brown A (2005) Association of the NuMA region on chromosome 11q13 with breast cancer susceptibility. Proc Natl Acad Sci U S A 102(6):2004–2009

Kierszenbaum AL, Rivkin E, Tres LL, Yoder BK, Haycraft CJ, Bornens M, Rios RM (2011) GMAP210 and IFT88 are present in the spermatid Golgi apparatus and participate in the development of the acrosome-acroplaxome complex, head-tail coupling apparatus and tail. Dev Dyn 240:723–736

Kingsley EP, Chan XY, Duan Y, Lambert JD (2007) Widespread RNA segregation in a spiralian embryo. Evol Dev 9(6):527–539

Knoblich JA (2010) Asymmetric cell division: recent developments and their implications for tumour biology. Nat Rev 11:849–860

Krauss SW, Lee G, Chasis JA, Mohandas N, Heald R (2004) Two protein 4.1 domains essential for mitotic spindle and aster microtubule dynamics and organization in vitro. J Biol Chem 279:27591–27598

Krauss SW, Spence JR, Bahmanyar S, Barth AI, Go MM, Czerwinski D, Meyer AJ (2008) Downregulation of protein 4.1R, a mature centriole protein, disrupts centrosomes, alters cell cycle progression, and perturbs mitotic spindles and anaphase. Mol Cell Biol 28:2283–2294

Lambert JD, Nagy LM (2002) Asymmetric inheritance of centrosomally localized mRNAs during embryonic cleavages. Nature 420(6916):682–686

Lee JS et al (2007) Nuclear lamin A/C deficiency induces defects in cell mechanics, polarization, and migration. Biophys J 93:2542–2552

Lingle L, Lutz WH, Ingle JN, Maihle NJ, Salisbury JL (1998) Centrosome hypertrophy in human breast tumors: implications for genomic stability and cell polarity. Proc Natl Acad Sci U S A 95:2950–2955

Liska F, Gosele C, Rivkin E, Tres L, Cardoso MC, Domaing P, Krejcí E, Snajdr P, Lee-Kirsch MA, de Rooij DG, Kren V, Krenová D, Kierszenbaum AL, Hubner N (2009) Rat hd mutation reveals an essential role of centrobin in spermatid head shaping and assembly of the head-tail coupling apparatus. Biol Reprod 81:1196–1205

Liu ZH, Schatten H, Hao YH, Lai L, Wax D, Samuel M, Zhong ZS, Sun QY, Prather RS (2006) The nuclear mitotic apparatus (NuMA) protein is contributed by the donor cell nucleus in cloned porcine embryos. Front Biosci 11:1945–1957

Malone CJ et al (2003) The C. elegans hook protein, ZYG-12, mediates the essential attachment between the centrosome and nucleus. Cell 115:825–836

Markiewicz E, Tilgner K, Barker N, van de Wetering M, Clevers H, Dorobek M, Hausmanowa-Petrusewicz I, Ramaekers FC, Broers JL, Blankesteijn WM et al (2006) The inner nuclear membrane protein emerin regulates beta-catenin activity by restricting its accumulation in the nucleus. EMBO J 25:3275–3285

Meyer AJ, Almendrala DK, Go MM, Krauss SW (2011) Structural protein 4.1R is integrally involved in nuclear envelope protein localization, centrosome–nucleus association and transcriptional signaling. J Cell Sci 124:1433–1444. doi:10.1242/jcs.077883

Meyerzon M, Gao Z, Liu J, Wu J-C, Malone CJ, Starr DA (2009) Centrosome attachment to the C. elegans male pronucleus is dependent on the surface area of the nuclear envelope. Dev Biol 327:433–446

Mitchell V, Rives N, Albert M, Peers MC, Selva J, Clavier B, Escudier E, Escalier D (2006) Outcome of ICSI with ejaculated spermatozoa in a series of men with distinct ultrastructural flagellar abnormalities. Hum Reprod 21(8):2065–2074

Okuda M, Horn HF, Tarapore P, Tokuyama Y, Smulian AG, Chan PK, Knudsen ES, Hofmann IA, Snyder JD, Bove KE et al (2000) Nucleophosmin/B23 is a target of CDK2/cyclin E in centrosome duplication. Cell 103:127–140

Prather RS (2000) Perspectives: cloning. Pigs is pigs. Science 289:1886–1887

Rawe VY, Terada Y, Nakamura S, Chillik CF, Olmedo SB, Chemes HE (2002) A pathology of the sperm centriole responsible for defective sperm aster formation, sygamy and cleavage. Hum Reprod 17:2344–2349

Rawe, VY, Chemes, H (2009) Exploring the cytoskeleton during intracytoplasmic sperm injection in humans. In: Carroll DJ (ed.) Microinjection: methods and applications, vol 518. Humana Press

Salpingidou G, Smertenko A, Hausmanowa-Petrucewicz I, Hussey PJ, Hutchison CJ (2007) A novel role for the nuclear membrane protein emerin in association of the centrosome to the outer nuclear membrane. J Cell Biol 178:897–904

Sathananthan AH (1997) Mitosis in the human embryo. The vital role of the sperm centrosome (centriole)−review. Histol Histopathol 12:827–856

Sathananthan AH (2009) Human centriole: origin and how it impacts fertilization, embryogenesis, infertility and cloning. Indian J Med Res 129:348–350

Sathananthan AH, Kola I, Osborne J, Trounson A, Ng SC, Bongso A, Ratnam SS (1991) Centrioles in the beginning of human development. Proc Nat Acad Sci U S A 88:4806–4810

Sathananthan AH, Ratnam SS, Ng SC, Tarin JJ, Gianoroli L, Trounson A (1996) The sperm centriole: its inheritance, replication and perpetuation in early human embryos. Hum Reprod 11:345–356

Sathananthan AH, Ratnasooriya WD, de Silva PK, Menezes J (2001) Characterization of human gamete centrosomes for assisted reproduction. Ital J Anat Embryol 106(2 suppl 2):61–73

Schatten H (2008) The mammalian centrosome and its functional significance. Histochem Cell Biol 129:667–686

Schatten H, Sun Q-Y (2009a) The functional significance of centrosomes in mammalian meiosis, fertilization, development, nuclear transfer, and stem cell differentiation. Environ Mol Mutagen 50(8):620–636

Schatten H, Sun Q-Y (2009b) The role of centrosomes in mammalian fertilization and its significance for ICSI. Mol Hum Reprod 15(9):531–538

Schatten H, Sun Q-Y (2010) The role of centrosomes in fertilization, cell division and establishment of asymmetry during embryo development. Semin Cell Dev Biol 21:174–184

Schatten H, Rawe VY, Sun Q-Y (2011) The sperm centrosome: its role and significance in nature and human assisted reproduction. J Reprod Stem Cell Biotechnol 2(2):121–127

Schatten H, Sun QY (2011a) The significant role of centrosomes in stem cell division and differentiation. Microsc Microanal 17(4):506–512

Schatten H, Sun Q-Y (2011b) Centrosome dynamics during meiotic spindle formation in oocyte maturation. Mol Reprod Develop 78:757–768

Schatten H, Sun QY (2011c) New insights into the role of centrosomes in mammalian fertilisation and implications for ART. Reproduction 142:793–801

Schatten, H, Rawe, VY, Sun, Q-Y (2012) Cytoskeletal architecture of human oocytes with focus on centrosomes and their significant role in fertilization. In: Agarwal, A, Varghese, A, Nagy, ZP (eds.) Practical manual of in vitro fertilization: advanced methods and novel devices, Humana Press (Springer Science + Business Media)

Simon, DN, Wilson, KL (2011) The nucleoskeleton as a genome-associated dynamic network of networks. Nat Rev Mol Cell Biol (In press)

Snook RR, Hosken DJ, Karr TL (2011) The biology and evolution of polyspermy: insights from cellular and functional studies of sperm and centrosomal behavior in the fertilized egg. Reproduction 142:779–792

Sun Q-Y, Schatten H (2006) Multiple roles of NuMA in vertebrate cells: review of an intriguing multifunctional protein. Front Biosci 11:1137–1146

Sun Q-Y, Schatten H (2007) Centrosome inheritance after fertilization and nuclear transfer in mammals. In: Sutovsky, P (ed.) Somatic cell nuclear transfer. Landes Bioscience Adv Exp Med Biol 591:58–71

Tokuyama Y, Horn HF, Kawamura K, Tarapore P, Fukasawa K (2001) Specific phosphorylation of nucleophosmin on Thr(199) by cyclin-dependent kinase 2-cyclin E and its role in centrosome duplication. J Biol Chem 276:21529–21537

Yamauchi, Y, Kiriyama, K, Kimura, H, Nishiyama Y (2008) Herpes simplex virus induces extensive modification and dynamic relocalisation of the nuclear mitotic apparatus (NuMA) protein in interphase cells. J Cell Sci 15121(pt 12):2087–2096 (Epub 2008 May 27)

Yamauchi Y, Shaman JA, Ward WS (2011) Non-genetic contributions of the sperm nucleus to embryonic development. Asian J Androl 13(1):31–35. doi:10.1038/aja.2010.75

Zhong Z-S, Zhang G, Meng X-Q, Zhang Y-L, Chen D-Y, Schatten H, Sun Q-Y (2005) Function of donor cell centrosome in intraspecies and interspecies nuclear transfer embryos. Exp Cell Res 306:35–46

Zhong Z, Spate L, Hao Y, Li R, Lai L, Katayama M, Sun QY, Prather RS, Schatten H (2007) Remodeling of centrosomes in intraspecies and interspecies nuclear transfer porcine embryos. Cell Cycle 6(12):1510–1521

Chapter 5
Human Centrosomal Dynamics During Gametogenesis, Fertilization, and Embryogenesis and Its Impact on Fertility: Ultrastructural Analysis

A. Henry Sathananthan

Abstract A review of the human centrosome at the beginning of life is presented, based on our research over the past 25 years at Monash on gametes, fertilization, and embryos in conjunction with our pioneering work on human in vitro fertilization (IVF) and assisted reproduction. The digital images are mainly based on fine structure presentations from published work, conference, and Web presentations.

5.1 Introduction

The centrosome has been an enigma in cell biology for over 120 years, discovered by Theodore Boveri (1900). Some cells have centrioles within the centrosome complex but others do not. Oocytes, some rodents, some flies, and plants do not have centrioles. We now know that centrioles originate from the sperm cell at fertilization in humans (Sathananthan et al. 1991, 1996, 2006; Sathananthan 1991, 1997, 1998; Schatten 1994, 2008; Simerly et al. 1995; Sutovsky and Schatten 1999; Palermo et al. 1997) and in most animals from round worms to primates, who obey Boveri's rule; unfortunately, Boveri's brilliant work (Fig. 5.1) and its impact on infertility was not recognized for over a 100 years (Sathananthan et al. 2006). The role of the centrosome in mitosis of the embryo, organizing the first mitotic spindle and initiating the process of human development is now well established (Sathananthan et al. 1996; Sathananthan 1997, 1998). It is also involved in organizing the cytoskeleton, establishes cell polarity, and plays a role

---

A. H. Sathananthan (✉)
Monash Immunology and Stem Cell Laboratories, Melbourne, Australia
e-mail: henry.sathananthan@monash.eduhenry.sathananthan@med.monash.edu.au

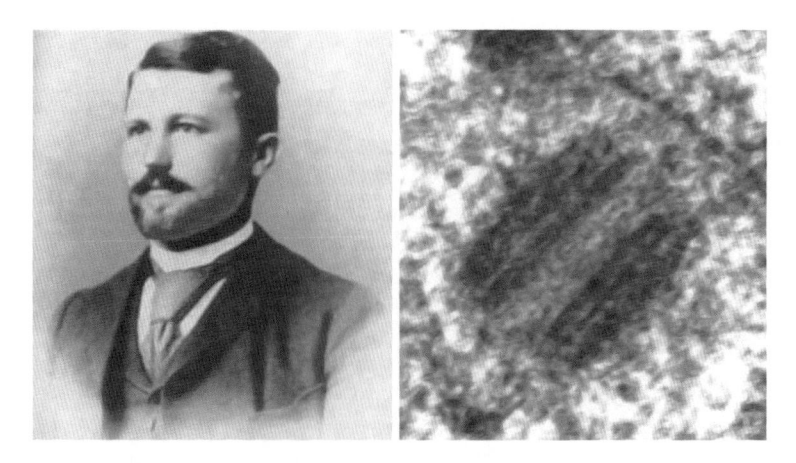

Fig. 5.1 Theodore Boveri (1887, 1901)—The father of the centrosome and his predictions, based on his research on Ascaris and the sea urchin. *Right* Ascaris centrosome in a spermatocyte TEM x200,000. Note its remarkable resemblance to the human centrosome (Sathananthan et al. 2006a, b)

in cytokinesis during the cell cycle (Fig. 5.2). The centrosomal cycle is closely integrated with the chromosomal cycle in embryonic and somatic cells (Mazia 1987). It plays a significant role in the cell cycle and cell division in most cells. Like chromosomes, centrioles are self-replicating organelles which duplicate during interphase, when they are located close to the nucleus. The process of egg activation by sperm and initiation of embryonic cleavage was little understood, until we discovered the important role of the centriole in humans in 1991(Sathananthan et al. 1991; Sathananthan 1991). We first detected a centriole in a human embryo in 1986 (Chen and Sathananthan 1986) and this prompted us to investigate this organelle in the following years. Pioneering work on centrosomes originated in Australia in the early 1990s.

5.2 Ultrastructure

Transmission electron microscopy (TEM) is still the best way to study centrioles, which are minute organelles (~ 2 μm in diameter), barrel-shaped, and presenting a unique '9 + 0' organization of microtubules (MTs) resembling a pin-wheel. Centrioles are typically surrounded by pericentriolar material (PCM) that nucleates MTs in somatic cells (Figs. 5.3, 5.4). The centriole and PCM complex (centrosome) becomes functional in the oocyte only after fertilization, when the sperm centrosome forms a sperm aster and the duplicated zygote centrosome forms the zygote aster (Sathananthan et al. 1996; Schatten 1994; Simerly et al. 1995). Centrosomes are oftentimes indirectly localized using fluorescent microscopy

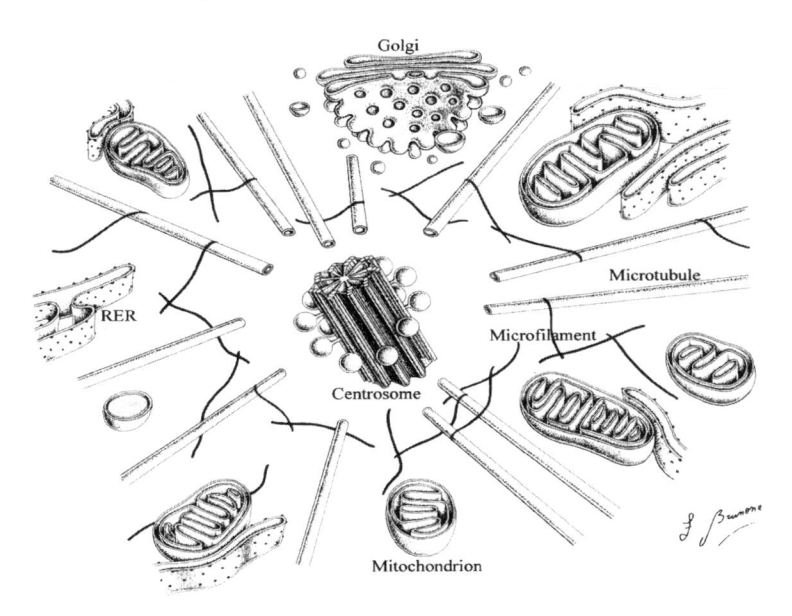

Fig. 5.2 The centrosome or cell center organizes the cell cytoskeleton and mitotic spindles composed of MTs. It also establishes the polarity of the cell. Note organelles located between MTs, which act as guide rails. Courtesy Late Professor P. Motta, Rome

(FM) to label MTs surrounding functional centrosomes (Fig. 5.5) or with antibodies specific to proteins associated with centrioles and PCM, such as alpha tubulin or centrin (Schatten 1994, 2008; Simerly et al. 1995; Sutovsky and Schatten 1999). Centrioles have been traced in all stages of preimplantation embryogenesis by TEM and even in embryonic stem cells; the sperm centriole is undoubtedly the precursor of centrioles in all fetal and adult somatic cells (Sathananthan et al. 1996; Sathananthan 1997). The maternal centrosome is greatly reduced in human oocytes, unlike in mouse oocytes that contain dominant maternal centrosomal material without centrioles. Hence, mice can easily develop parthenogenetically. Only one dominant, functional centrosome, paternal or maternal, is required to ensure normal development. The inheritance of two centrioles, as in dispermic fertilization, leads to abnormal development of the ensuing triploid embryos that may develop to term in humans.

5.3 Methodology

Our studies over three decades are based on the examination of human gametogenesis, gametes, and embryos at fertilization and during preimplantation development, where over 300 normal bipronuclear and abnormal tripronuclear

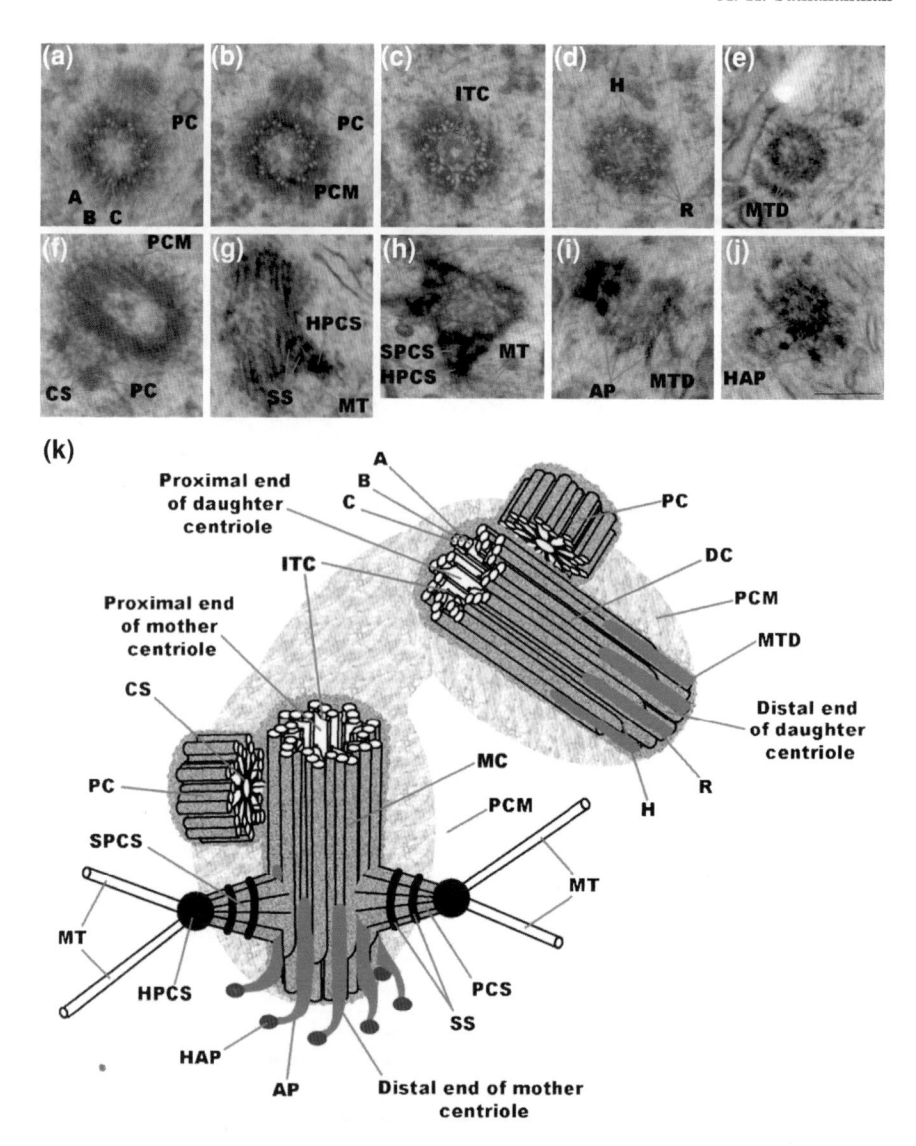

Fig. 5.3 Centrosome structure in an animal cell at the end of G1-phase, beginning of S-phase. From "a" to "j", electron microscopy images (*top line*—serial sections of daughter centriole, *second line*—selected sections of mother centriole, scale bar 200 nm) illustrating different centrosome structures shown on the centrosome scheme—"k". MC—mother (mature) centriole; DC—daughter centriole; PC—procentriole; PCM—pericentriolar material (pericentriolar matrix); A—microtubule of triplet; B—microtubule of triplet; C—microtubule of triplet; H—hook of C microtubule; MTD—A-B microtubule duplex (in distal part of centriolar cylinder); ITC—internal triplets connections system; CS—cartwheel structure (axis with spokes); PCS—pericentriolar satellite (=sub-distal appendage); HPCS—head of pericentriolar satellite; SPCS—stem of pericentriolar satellite (connected to three triplets in this case); SS—striated structure of pericentriolar satellite stem; MT—microtubule; AP—appendage (=distal appendage); HAP—head of appendage; R—rib. (Courtesy Uzbekov and Prigent 2007)

Fig. 5.4 A mammalian centrosome is composed of two centrioles surrounded by a meshwork of a proteins embedded in matrix called the pericentriolar material (PCM). Gamma-tubulin and the gamma-tubulin ring complex that nucleate microtubules along with associated proteins are embedded in the PCM. Highlighted in this diagram are two centrosomal complexes, the microtubule nucleating complex and the microtubule anchoring complex (Schatten 2008)

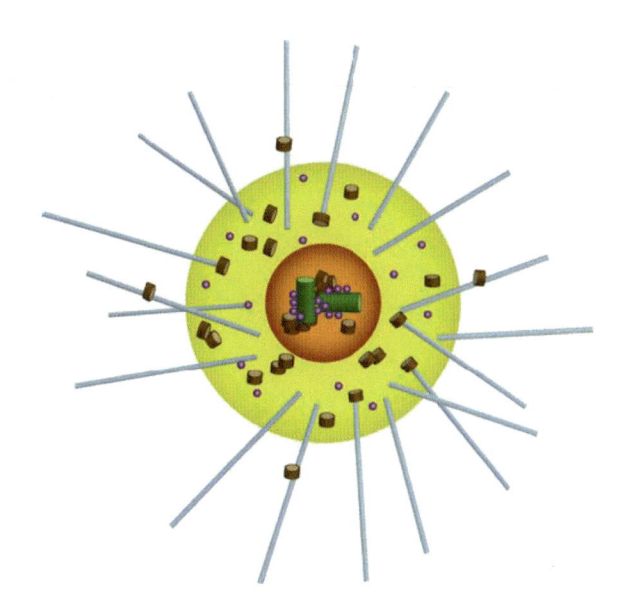

cleavage-stage embryos were examined by routine TEM. Centrosomes were also examined in several mature and immature oocytes during IVF, and 25 sperm samples (fertile and infertile) to characterize maternal and paternal centrosomes. Early cleavage-stage embryos were examined by serial sectioning up to the hatching blastocyst stage. Our investigation also extended to oogonia in fetal ovaries, testis and testicular biopsies, and embryonic stem cells. TEM is the best method currently available to reveal the structure of centrioles within centrosomes. Most other studies use confocal fluorescent microscopy (FM) to reveal chromosomes and microtubules (MTs) by immunocytochemistry (Schatten 1994; Simerly et al. 1995) or labeling of centrosomes with specific antibodies for proteins-like alpha tubulin and centrin associated with centrioles. These images elegantly demonstrate whole asters, spindles, and chromosomes, while TEM demonstrates the fine structure of the centrosome in serial sections. These two techniques are complimentary in the visualization of nearly all components of meiotic and mitotic figures in oocytes and early embryos. Locating centrosomes in oocytes and embryonic cells is both laborious and time-consuming in serial sections. They are usually found close to the nucleus during interphase and located at spindle poles in mitosis. It is, however, easier to locate the sperm centriole which is found below the basal plate in the neck of the sperm cell (Fig. 5.6). An optimal method to analyse sperm centrioles is by cutting a pellet of sperm which allows visualization of centrioles in all planes of sectioning. Our atlas (Sathananthan 1996) and website www.sathembryoart.com show some of these images.

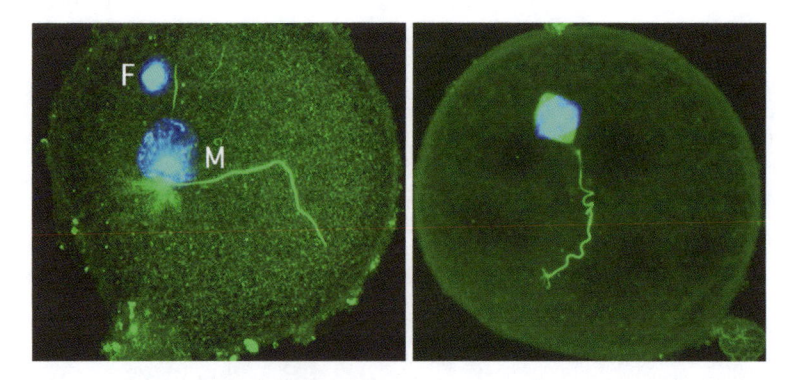

Fig. 5.5 Rhesus fertilization after ICSI (fluorescent microscopy). The sperm aster (*left*) is associated with the centrosome associated with the male pronucleus (M). The female pronucleus (F) is attracted to the male but has not decondensed yet. The first mitotic bipolar spindle (*right*) shows chromosomes at metaphase. Note distinct sperm tail in both fluorescent images. The monkey is similar to the human in centrosomal inheritance and dynamics, like most animals are (Courtesy Dr. Laura Hewitson, USA)

5.4 Gamete Centrosomes

Only one functional centrosome is required in either gamete for normal fertilization and embryo development. It is either the paternal (as in humans and most other animals) or maternal (as in rodents such as mice). Therefore, there is a reduction or inactivation of the centrosome in one of the gametes—sperm or egg. Those with dominant paternal centrosomes obey Boveri's rule of centrosomal inheritance postulated in 1890 (Boveri 1900), while rodents present the exception to the rule where the maternal centrosome is dominant (Schatten 1994). Humans and most other animals (including sea urchins, roundworms, and farm animals) follow Boveri's rule with respect to centriole inheritance and dynamics during early development (Sathananthan et al. 1991, 1996; Sathananthan 1997)

5.5 The Paternal Centrosome (Centriole)

In humans, the sperm centrosome also shows partial reduction during spermiogenesis, when round spermatids metamorphose into motile sperm cells (Sathananthan et al. 1996, 2001; de Kretser and Kerr 1994; Holstein and Roosen-Runge 1981; Manandhar et al. 2000). The round spermatid contains two centrioles in the centrosome complex, like most other somatic cells, while the mature sperm cell has only one centriole—the proximal (PC) concealed in a 'black box' in the sperm neck (connecting piece) beneath the basal plate (Figs. 5.7, 5.8, 5.9, 5.10, 5.11, 5.12, 5.13, 5.14). The distal centriole (DC) is committed to the formation of the flagellar axoneme in the elongated spermatid and is reduced to a vestige in mature

Fig. 5.6 Mature sperm: ultrastructure—(Sathananthan 1996) *A* acrosome (*black*), *AC* acrosome cap, *AX* axoneme, *BP* basal plate, *CD* cytoplasmic droplet, *ES* equatorial segment Acrosome, *FS* fibrous sheath, *IAM* inner acrosome membrane, *M* mitochondria, *MP* midpiece, *NE* nuclear envelope, *OAM* outer acrosome membrane, *PA* post acrosomal segment (fusogenic), *PC* proximal centriole, *PP* principal piece (tail), *R* ribs, *TS* transverse sections

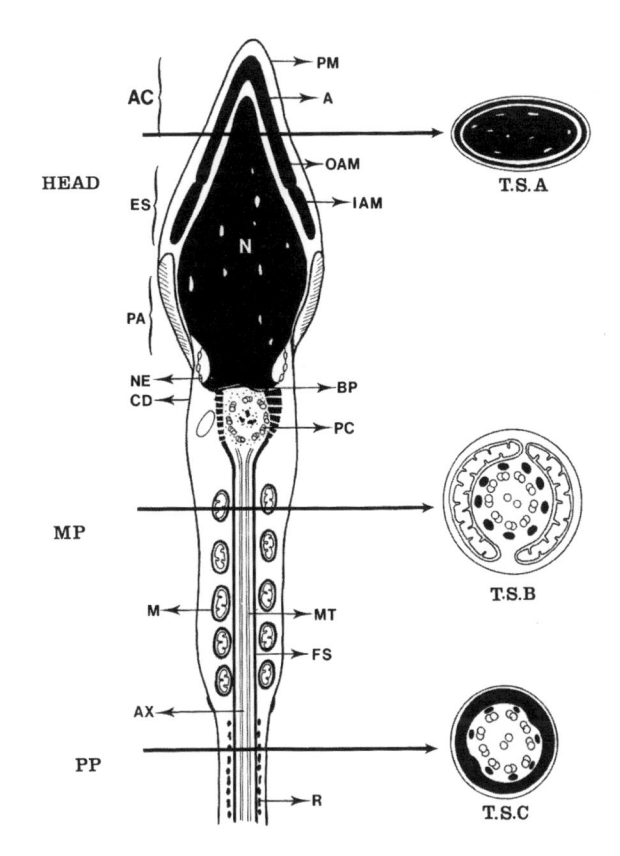

sperm and is apparently non-functional. The PC retains its typical '9 + 0' organization of triplets of MTs presenting the classical pinwheel structure; it is inherited by the embryo, while the DC is represented by a few disorganized peripheral MTs with a central doublet extending from the midpiece (MP) axoneme to the lower limit of the PC (Figs. 5.7, 5.8, 5.9), where it terminates in an accumulation of dense PCM (Sathananthan et al. 1996, 2001; Sathananthan 1997; de Kretser and Kerr 1994; Holstein and Roosen-Runge 1981). For this reason it is argued that the DC cannot function as a true centriole (Zamboni 1992; Zamboni and Stefanini 1970). The DC is more disorganized cranially and less so caudally toward the flagellar axoneme in the MP, as revealed in serial transverse sections (Sathananthan 1996; Manandhar et al. 2000). Apart from the centriole, there are other structures associated with both the PC and DC. The PC is hidden in a vault or 'black box' composed of the capitulum situated immediately beneath the basal plate and is flanked laterally by nine segmented columns that merge with the nine outer dense fibers surrounding the MP axoneme and the MT of the DC that are closely associated with the dense fibers (Figs. 5.6, 5.7). For an excellent review of sperm neck structure see Zamboni (1992). The sequence of events in spermiogenesis and oogenesis is also available on the Web www.sathembryoart.com.

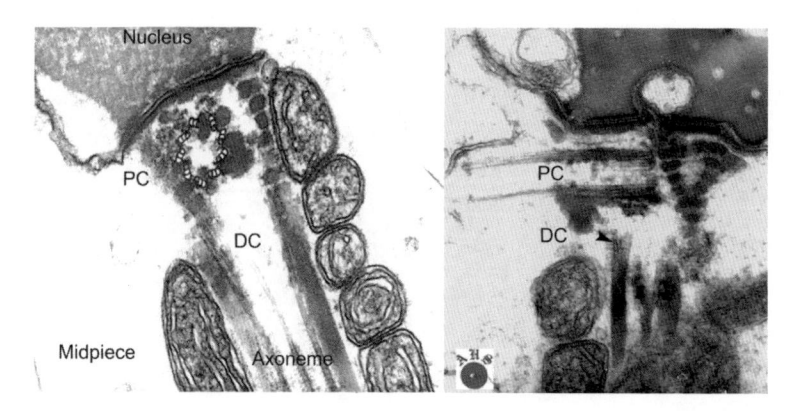

Fig. 5.7 Sperm centrosome TS and LS (TEM). The proximal centriole (PC) is hidden in a 'black box' beneath the basal plate and flanked by nine segmented columns. It shows the typical '9 + 0' structure of microtubule (MT) triplets. Note disorganized MTs of distal centriole (DC) and centriolar adjunct of PC (*right*) x65,000 (Sathananthan et al. 1996)

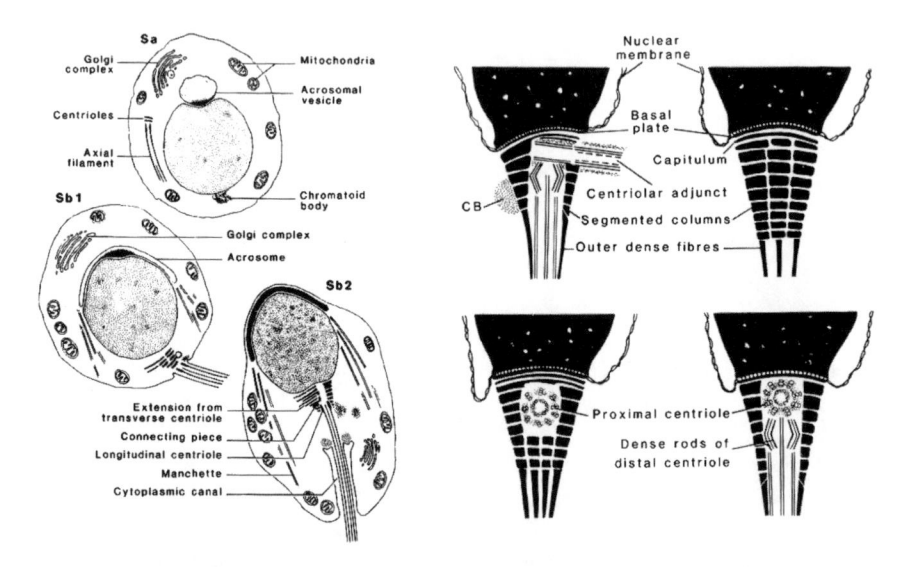

Fig. 5.8 Spermiogenesis and centrioles: The sperm neck region contains the proximal and distal centrioles. The latter gives rise to the axoneme and becomes a vestige in mature sperm, while the former persists as the functional centriole that is passed on to the embryo. Both are hidden in a "black box" composed of segmented columns, capitulum, and basal plate (Courtesy Prof. David deKretser, 1996)

The segmented columns show a complex structure in cross-sections of the neck region, where the centriolar adjunct (CA) extends laterally from the PC (Figs. 5.7, 5.13). The CA is a tuft of MTs that shows a disorganized triplet structure and may represent an attempt to form an axoneme by the PC (de Kretser and Kerr 1994).

5 Human Centrosomal Dynamics   81

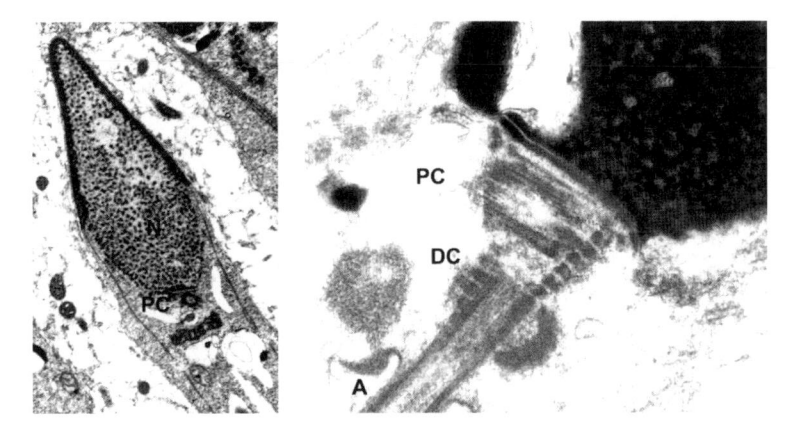

Fig. 5.9 Elongating spermatids: centrioles.These spermatids (early and late) show nuclear condensation and centrioles. The proximal centrioles (PC) are closer to the nucleus, while the distal centriole (DC) is organizing the axoneme (A). The two centrioles form the paternal centrosome x17,000, x85,000 (www.sathembryoart.com)

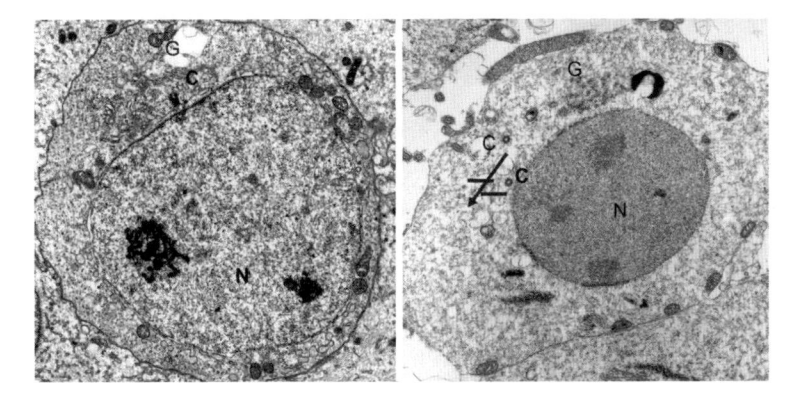

Fig. 5.10 Centrioles are usually associated with nuclei and Golgi complexes in early stages of spermatogenesis, directing secretion and polarizing the cells. In spermatids they migrate toward the basal pole away from the Golgi before spermiogenesis. *C* centrioles, *G* golgi, *N* nucleus. x10,200, x85,000 www.sathembryoart.com

The PC is associated with finely-granular PCM along its length (Fig. 5.7), which resembles the MT—nucleating PCM in zygote centrosomes (Sathananthan et al. 1996; Sathananthan 1997, 1998). This material is also found within the PC and gamma-tubulin has been reported in both locations in somatic cell centrioles (Figs. 5.3, 5.4).

Fig. 5.11 Round spermatids showing centrioles, (C). They duplicate (*left*) and migrate to the pole opposite the acrosome, prior to spermiogenesis (*right*)
A acrosome cap *N* nucleus.
x3,400, x85,000
(www.sathembryoart.com)

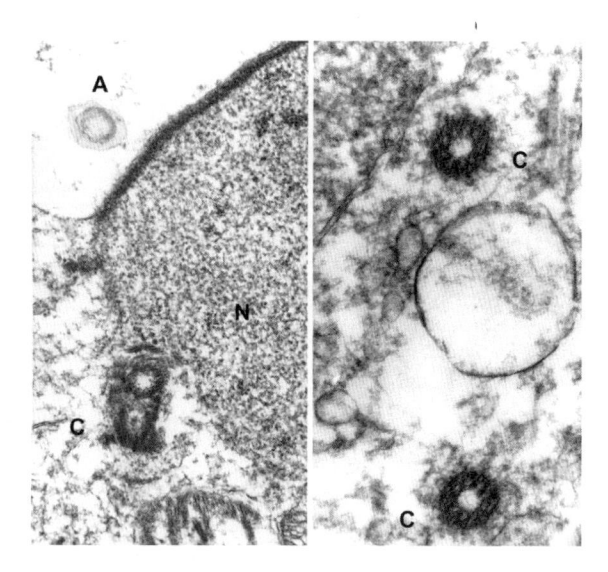

5.6 The Maternal Centrosome

Unlike the sperm centrosome, the centrosome in the mature human oocyte is both reduced and inactivated to become nonfunctional in the meiotically arrested oocyte (Figs. 5.15, 5.16). The metaphase II spindle, at either pole, neither contains centrioles nor dense, granulovesicular, centrosomal material that clearly nucleate MT in mouse oocytes (Fig. 5.17), which do contain a functional maternal centrosome. Centrioles are generally absent in mammalian oocytes and are also not found in mouse sperm (Sathananthan 1996; Manandhar et al. 2000). Although the mature human oocyte has no visible centrosome, a functional centrosome with two typical centrioles is found in fetal oogonia (Fig. 5.15), which conforms to those of other somatic cells (Sathananthan et al. 2000, 2006). These are, as usual, juxta-nuclear in position, have PCM which nucleate MTs and seem to organize the oogonial cytoskeleton (a true centrosome). Therefore, reduction and loss of the maternal centrosome has to occur either during oogenesis or in the final stages of oocyte maturation when the first polar body (PB1) is abstricted, as in starfish oocytes (Sluder and Rieder 1993). We have examined several maturing oocytes by TEM and have not yet located a centriole in PB1, nor in immature germinal vesicle oocytes. It seems possible that the human follows the starfish pattern of maternal centrosomal reduction since both follow Boveri's rule of paternal centrosomal dominance and inheritance. The sequence of events of centriologenesis during spermiogenesis and oogenesis are also available on the Web www.sathembryoart.com

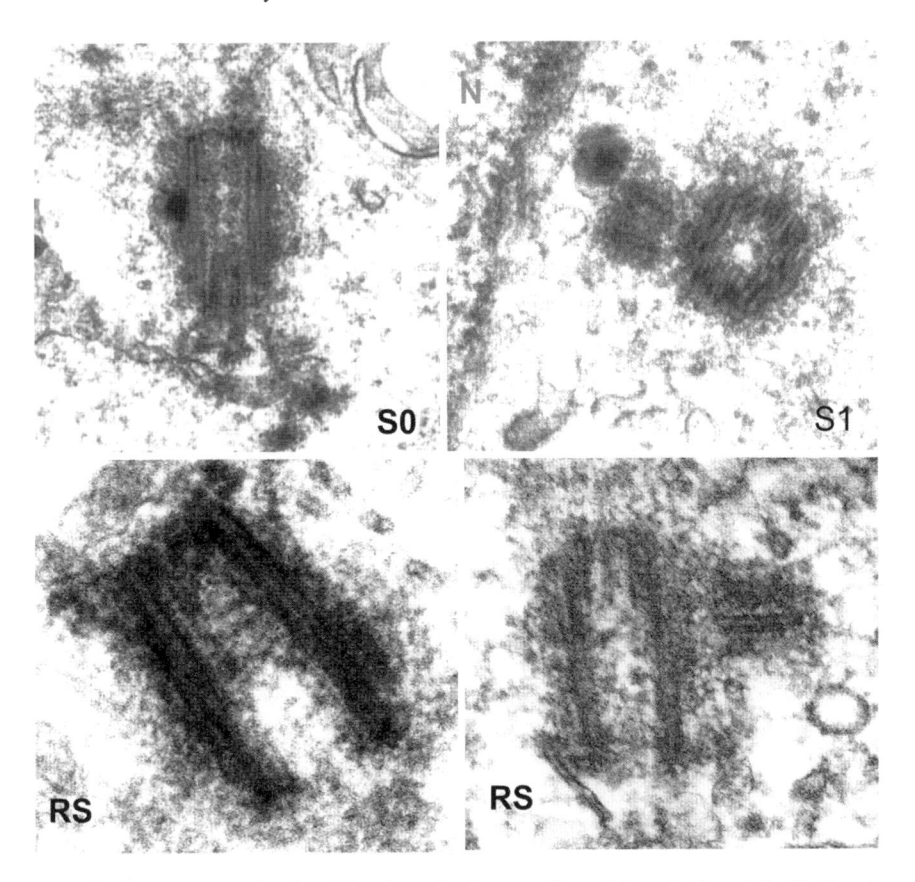

Fig. 5.12 Spermatogenesis: Centriolar dynamics images of centrioles, single and duplicating, in spermatogenic cells at interphase. Note dense pericentriolar material. *RS* round spermatids, *S0* spermatogonium, *S1* primary spermatocyte x85,000, x102,000 www.sathembryoart.com

5.7 The Zygote Centrosome

The zygote centrosome appears to be a blending of both paternal and maternal components as revealed by fluorescent microscopy (FM) and TEM and more recently by molecular studies (Sathananthan et al. 1996; Schatten 1994, 2008; Simerly et al. 1995; Sutovsky and Schatten 1999). There appears to be a progressive addition of granular PCM around the duplicated sperm centriole from the pronuclear stage to syngamy, soon after fertilization (Sathananthan et al. 1996; Sathananthan 1997, 1998). This PCM nucleates MTs of the sperm monoaster after sperm fusion and incorporation (Figs. 5.18, 5.19), which duplicates to form the two poles of the first bipolar mitotic spindle (Figs. 5.20, 5.25). Two centrioles compose the heart of the zygote centrosome soon after its release from the 'black box'. The sperm centriole is duplicated, while acquiring increasingly dense PCM

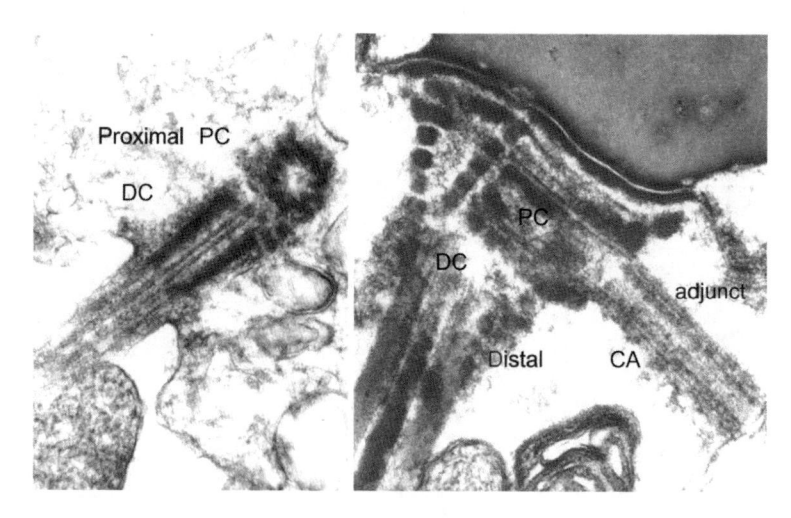

Fig. 5.13 Spermatid centrosomes TEM: The spermatid has two functional centrioles (*left*), while mature sperm (*right*) have only one centriole, the proximal centriole (PC) (centrosomal reduction). The distal is a vestige. Note centriolar adjunct (CA) on the right with microtubules x85,000 (www.sathembryoart.com)

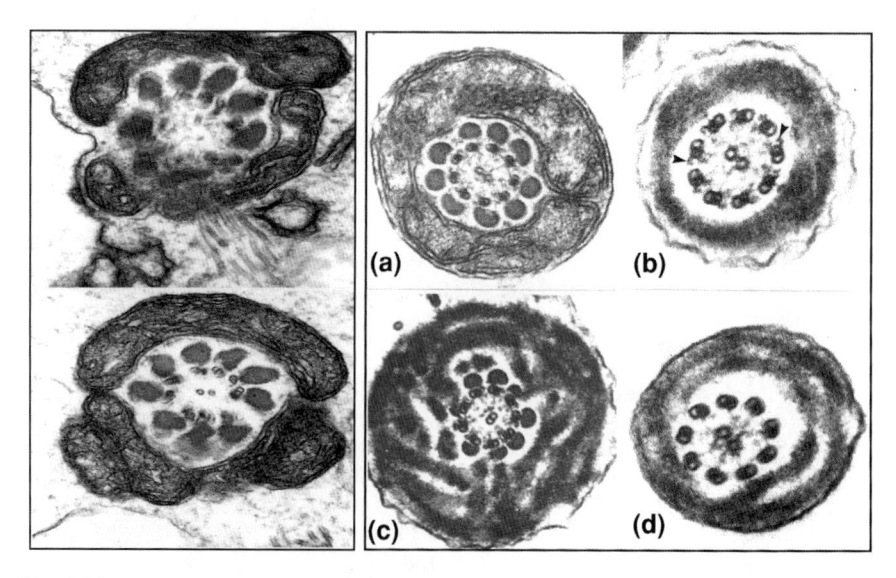

Fig. 5.14 Disorganized distal centriole in midpiece and tail axonemes in transverse sections. The centriole has disorganized microtubules (MTs) surrounded by nine dense fibers and mitochondria. Note MT doublet within midpiece showing axoneme, dense fibers, and mitochondria (*left*). Tails show the fibrous sheaths as well and the 9 + 2 organization of MTs (**a–d**) x87,500, x70,000 (www.sathembryoart.com)

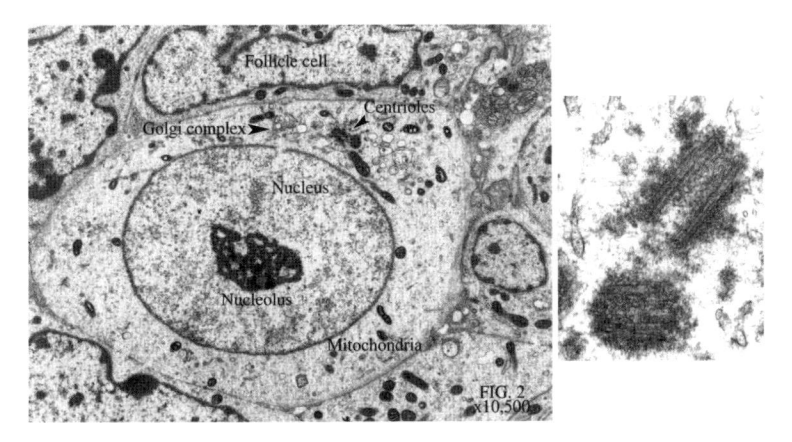

Fig. 5.15 Functional centrosome in a fetal oogonium. The centrosome has two typical centrioles with PCM (*right*), like in somatic cells. It is lost or reduced in the oocyte during oogenesis but its fate is unknown. Is it abstricted into the first polar body during maturation? x10,500, x87,500 (Sathananthan et al. 2000)

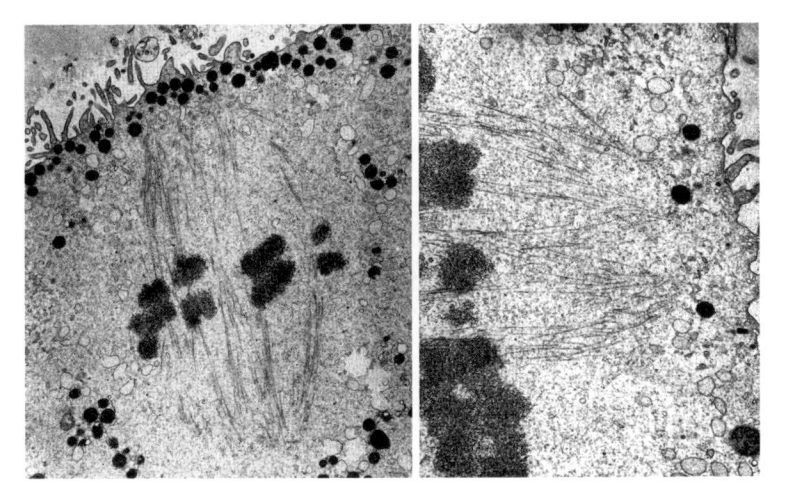

Fig. 5.16 Mature oocyte at metaphase II (TEM): the centrosome is absent or reduced at the spindle poles. Microtubules terminate in the egg cortex, abruptly. Note chromosomes at the equator and cortical granules in the cortex. x8,500, x15,400 (Sathananthan 1998)

that nucleates MTs, thus becoming a functional centrosome. This is the forerunner of all centrosomes in embryonic, fetal, and adult somatic cells (Sathananthan et al. 1996). An extensive MT network is formed by the sperm aster (Fig. 5.5), which reorganizes the whole cytoskeleton of the oocyte soon after sperm incorporation, as revealed by FM (Schatten 1994; Simerly et al. 1995). This aster is initially directed toward the female pronucleus (FPN) and may be involved in drawing the FPN toward the male pronucleus (MPN) to take up a more centralized position in

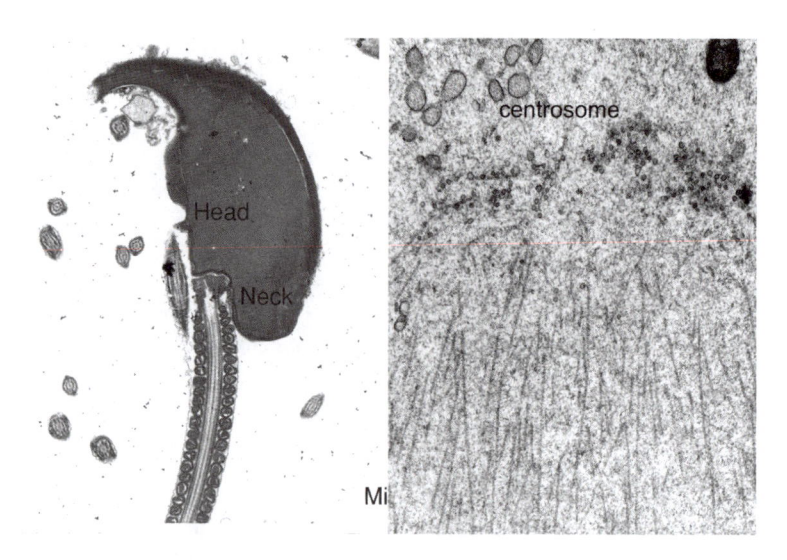

Fig. 5.17 Mouse centrosomes. The sperm has no centrosome (centriole) in its neck, since the maternal centrosome is dominant (*right*). Note functional centrosome composed of fine granulovesicular material in the mouse oocyte, nucleating MTs. Note differences in sperm structure compared to that of humans x7,000, x50,000

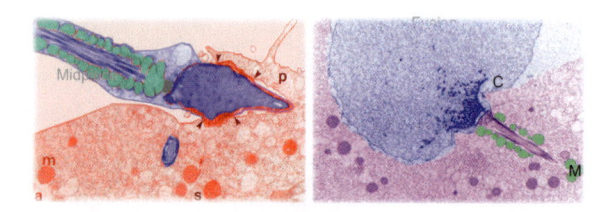

Fig. 5.18 Mouse centrosomes. The sperm has no centrosome (centriole) in its neck, since the maternal centrosome is dominant (*right*). Note functional centrosome composed of fine granulovesicular material in the mouse oocyte, nucleating MTs. Note differences in sperm structure compared to that of humans x7,000, x50,000

the ooplasm at the bipronuclear stage of fertilization (Sathananthan et al. 1996; Sathananthan 1996, 1997). Further, the sperm centriole is already associated with some PCM and this could act as a template to attract maternal centrosomal material (such as gamma–tubulin). Thus both paternal and maternal centrosomes are actively engaged to produce a composite zygote centrosome. This activation could be mediated by sperm cytosolic factors and the Ca^{++} transient that occurs soon after sperm-egg fusion at the onset of fertilization (Schatten 1994; Simerly et al. 1995; Sutovsky and Schatten 1999). A Ca^{++} activated protein, associated with the sperm centrosome, might be also involved in its release and activation.

5 Human Centrosomal Dynamics

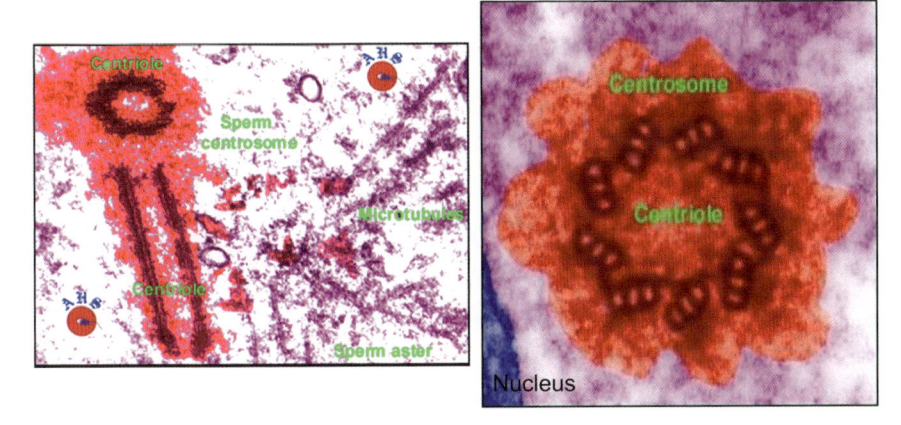

Fig. 5.19 Centrioles in monospermic embryos (TEM): Zygote centrioles associated with a sperm aster soon after fertilization. It is now functional with duplicated centrioles and PCM (*red*) nucleating MTs (*left*). A classic image of a juxta-nuclear centrosome in an eight-cell blastomere (*right*). The centriole is located in the heart of the centrosome. The zygote centrosome is replicated and perpetuated in embryos, fetal, and adult cells. Every cell has two centrioles. Colored by computer x50,000, x170,000 (Sathananthan et al. 1966)

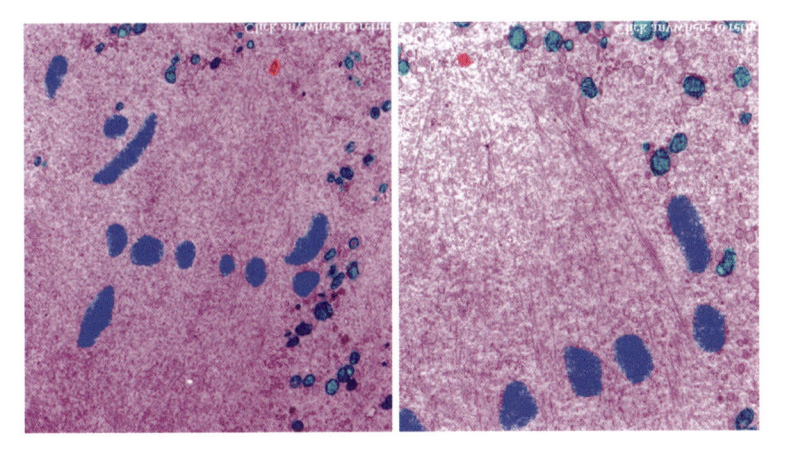

Fig. 5.20 Bipolar spindle at syngamy in a dispermic ovum showing centrioles at either pole (*red*). Note chromosomes (*blue*) at the equator and few outside the spindle zone, composed of MTs. Mitochondria are *green*. Computer coloured. x12,750 (Sathananthan and Edwards 1995)

5.8 Centrioles in Embryos

Embryos cleave by repeated mitoses involving centrosomes. Descendants of the sperm centrosome were found at every stage of preimplantation development and were traced from fertilization through the first four cleavage cell cycles to the morula and hatching blastocyst stages (Sathananthan et al. 1996; Sathananthan

1997). Single or double centrioles (diplosomes) were associated with nuclei at interphase (Figs. 5.21, 5.22), when they were often replicating or occupying pivotal positions on spindle poles during mitoses (Fig. 5.20). Sperm remnants were sometimes associated with centrioles and were found at most cleavage stages. Centrioles were also seen in trophoblast, embryoblast, and endoderm cells in blastocysts. PCM was associated with most centrioles and nucleated spindle MTs indicating their functionality. We were fortunate to detect three centrioles in an ICM cell of a blastocyst (Fig. 5.23).

5.9 Centriolar Duplication

According to Boveri's classical theory (Boveri 1900), the centrosome is a self-duplicating organelle. The replication of centrioles has been reported in detail in our previous publications (Sathananthan et al. 1996; Sathananthan 1997). It occurs during interphase, beginning at the pronuclear stage prior to syngamy and later at embryonic interphases. They duplicate only once during each cell cycle and the daughter centriole arises as a perpendicular outgrowth of the parent (Mazia 1987; Uzbekov and Prigent 2007) it is seen to grow progressively and acquires PCM as it migrates toward the poles of the mitotic spindle. Such immature centrioles have less PCM and do not nucleate MTs. There is also dense material within each centriole, as was seen within the sperm centriole. However, we have also observed a centriole at a spindle pole at syngamy, which appeared as an annular condensation of dense material devoid of MTs. This, called a procentriole, elongates by accretion of tubulin material to its free end and the pin-wheel arrangement of MT triplets gradually appears in this dense ring. Classically, in cell division, two diplosomes (two doublet centrioles) are formed from an original pair of centrioles and take up positions at opposite poles of the mitotic spindle (Uzbekov and Prigent 2007; Alieva and Uzbekov 2008). We have observed two centrioles, sometimes one, at a spindle pole during syngamy and at later cleavage (Figs. 5.21, 5.22) and in two instances a single centriole at the opposite pole. It is obvious that due to technical difficulties we are not detecting one of the pair of centrioles at each pole. We have detected three centrioles in an ICM cell of a blastocyst (Fig. 5.23). It is very likely that centriolar duplication occurs in early embryonic cells that is very much like that observed in somatic cells, and diplosomes so formed migrate to opposite poles of the spindle. These centrioles are very likely the ancestors of those seen in fetal and adult somatic cells and can be demonstrated in embryonic stem cells, as well. As expected, the behavior of centrosomes during the formation of the bipolar mitotic figure is similar to that in cell cycles of somatic cells, as expected (Fig. 5.25).

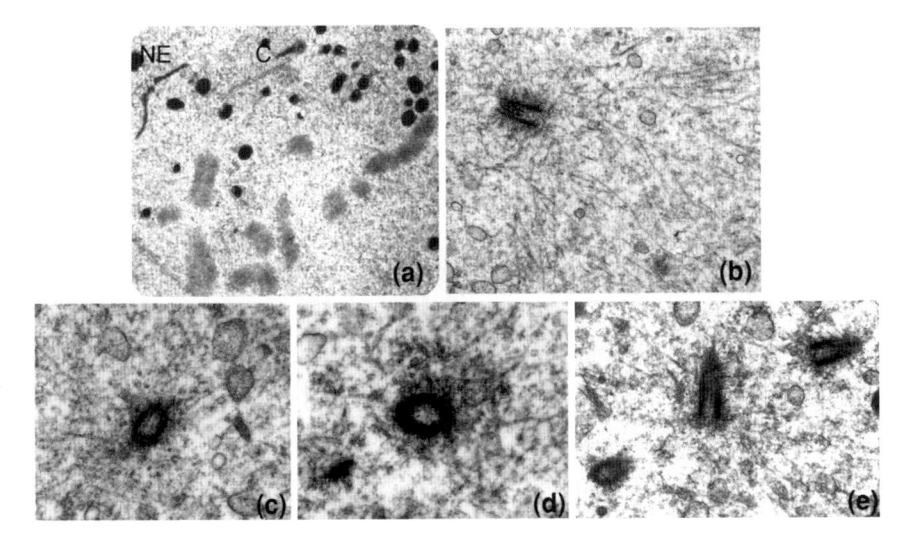

Fig. 5.21 Dispermic embryos **a** Sperm aster of a 1-cell embryo showing pronuclear envelope (NE) breakdown and duplicated centrioles (C). **b** Bipolar spindle at syngamy in a 1-cell embryo showing a centriole at one pole. **c–e** Anaphase spindle in an 8-cell blastomere. **c** Centriole at one pole **d** centriole at opposite pole. **e** Three supernumerary centrioles on the side of spindle—Is this centrosomal amplification, a prelude to cancer? x8,500, x20,400, x51,000, x20,400 (Sathananthan et al. 1999)

5.10 Molecular Aspects

Several attempts have been made to unravel the molecular structure of somatic centrosomes (Uzbekov and Prigent 2007; Alieva and Uzbekov 2008; Sathananthan et al. 2002; Fouquet et al. 1998; Fuller et al. 1995; Anderson 1999). Biochemically they are complex structures associated with a diversity of different proteins. The sperm centrosome has specific regulatory proteins, such as centrin, pericentrin, gamma tubulin, associated with disulfide bonds, sulfhydryl, and phosphates among other molecules. The zygote centrosome containing duplicated centrioles and maternal gamma-tubulin is further added to the PCM in the ooplasm to compose a functional centrosome (Fig. 5.24) when the centrosome forms the sperm monoaster. Both centrioles (mother and daughter) then duplicate again and move to opposite poles establishing a bipolar spindle at syngamy, the first mitosis of the human embryo (Fig. 5.25). This ensures the cleavage of two equal blastomeres completing the first cell cycle. Centrin is a universal, centrosomal protein playing a key role in centriolar duplication; it occurs in pericentriolar material (PCM) in fibers linking centrioles to one another. Clearly, more research is needed to characterize the proteins associated with centrosomes, especially in gametes and embryos.

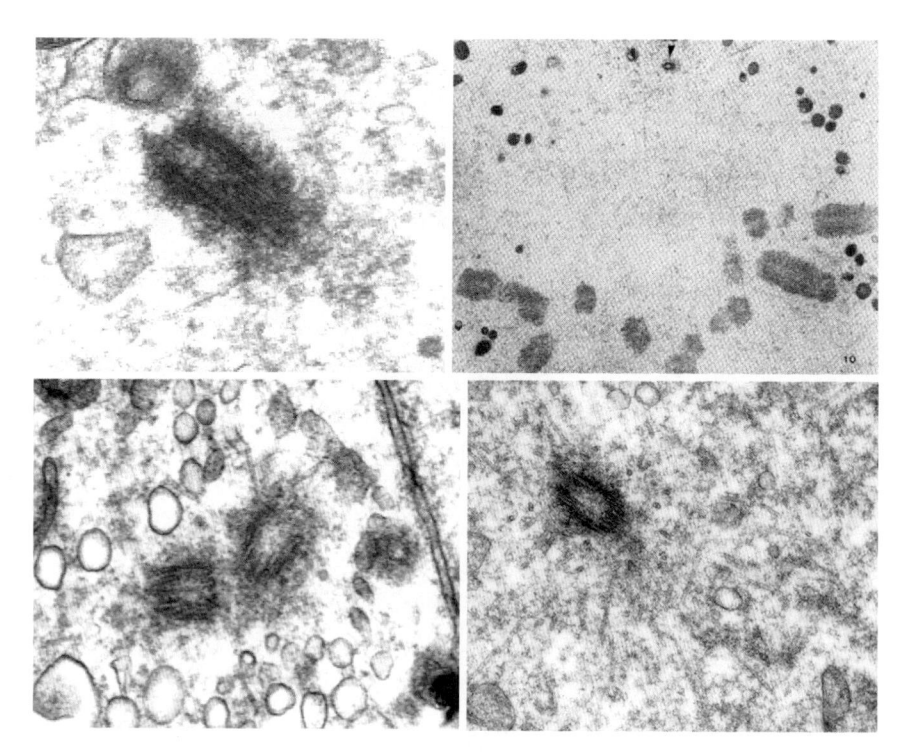

Fig. 5.22 Centrioles in a normal 2PN embryo **a** two-cell blastomere. **b** Centriole located pivotally in a dividing four-cell blastomere. **c** Double centriole in a three-cell blastomere, located next to its nucleus **d** Centriole in a eight-cell blastomere x25,000, x8,500 x20,000 (Sathananthan et al. 1996)

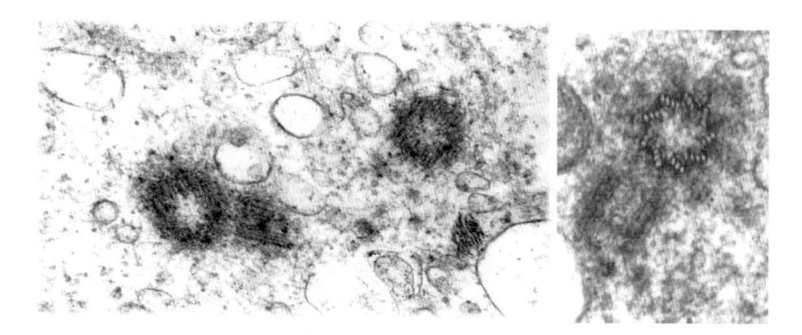

Fig. 5.23 Centriolar duplication in ICM cells of blastocysts. When centrioles duplicate at the onset of mitosis, the daughter centriole grows from the side of the mother, as is usual. Three of the four centrioles are shown on the *left* and two on the *right* during duplication. Note abundance of PCM around typical centrioles (*right*) x85,000 (www.sathembryoart.com)

5 Human Centrosomal Dynamics

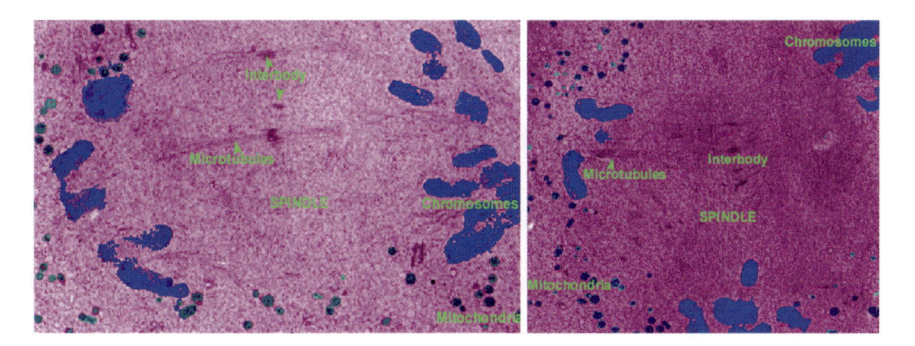

Fig. 5.24 Abnormal centrioles in sperm of infertile men with OATS. The functional centriole is absent or abnormal in the 'black box'. Note peculiar material in adjacent cytoplasmic droplet x 85,000 (Sathananthan 1996)

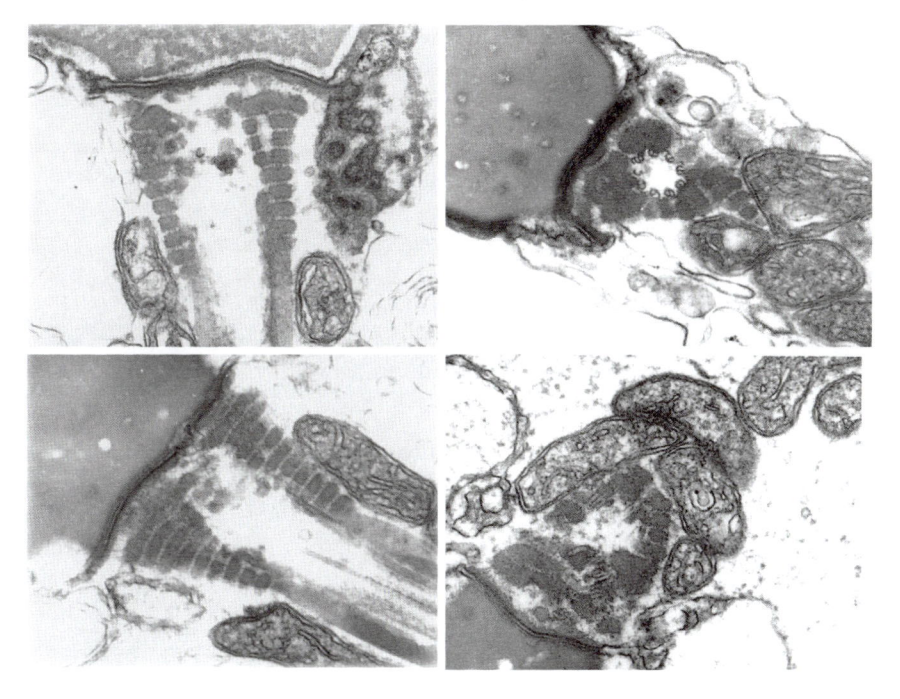

Fig. 5.25 Centrosome: molecular dissection. The sperm centrosome is unravelled, expanded, and everted in the zygote. The binding of Ca^{++} ions to centrin during activation is thought to release the sperm centrosome (Courtesy Schatten 1994)

5.11 Sperm Centrosomes and Infertility

In 1991, we postulated a hypothesis that if a defective male centrosome is inherited by a human oocyte, it might lead to abnormal cleavage and compromise embryonic development, a new dimension in the assessment of infertility (Sathananthan et al. 1991, 1996, 2001; Sathananthan 1991, 1997). Such centrioles are present in poorly motile or immotile sperm, commonly used in ICSI. Since the distal centriole gave rise to the sperm axoneme during spermiogenesis, it is logical to suppose that the proximal centriole that is inherited by the embryo could cause aberrant mitosis, as both are derived from the same mother centriole in spermatids. Our hypothesis was based on defects observed in the proximal centriole in the neck of sperm from motile, poorly motile, and immotile sperm samples (Sathananthan 1996; Sathananthan et al. 2001). Our TEM studies also indicate that the majority of sperm (80–90 %) from oligoasthenospermic men present centrosomal defects involving the centriole, PCM, and even the 'black box' surrounding the centrosomal complex. It is easier to detect centriolar MT defects than to assess aberrations of the PCM associated with the centriole. Both components are important in the overall assessment of the sperm centrosome. Centriolar aberrations may also be found in a few normal sperm and conversely normal centrioles could be located in a few poor quality sperm with 0–10 % motility. Of course, defects in the axoneme of sperm tails have been extensively documented.

The main aberrations of the sperm PC organization are

- disorganization of the MT triplet structure
- loss of MTs within triplets (doublets and singlets)
- loss of MT triplets (half and partial centrioles)
- total loss of the centriole
- displacement of the centriole from its vault within the 'black-box'
- aberrant distribution of PCM—difficult to assess
- disorganization of the 'black box' components
- Intrusion of mitochondria into the neck region
- Intrusion of vesicular elements and dense bodies within the centriole
- double centrioles with single or double axonemes

Some of these aberrations are portrayed in Fig. 5.26 and in our atlas (Sathananthan 1996).

Normal centriolar configuration is essential to ensure duplication and the functionality of the centrosome to organize embryonic spindles. There is now increasing clinical evidence to support our hypothesis that poor sperm correlates with poor embryos (Sathananthan 2009; Asch et al. 1995; Vandervorst et al. 1997; Chatzimeletiou et al. 2008; Van Blerkom 1996; Hinduja et al. 2010; Hewitson et al. 1997; Navara et al. 1995). A variety of chromosomal aberrations have been documented in early human embryos, such as aneuploidy, polyploidy, and mosaicism (Munne 2006) that could partly be attributed to centrosomal dysfunction. Centrosomal defects can lead to failure in fertilization, and cause embryonic arrest through the formation of abnormal spindles and the accumulation of chromosomally abnormal cells that derive from them (Fig. 5.27). These are some of the causes of early embryo loss in assisted reproduction causing infertility.

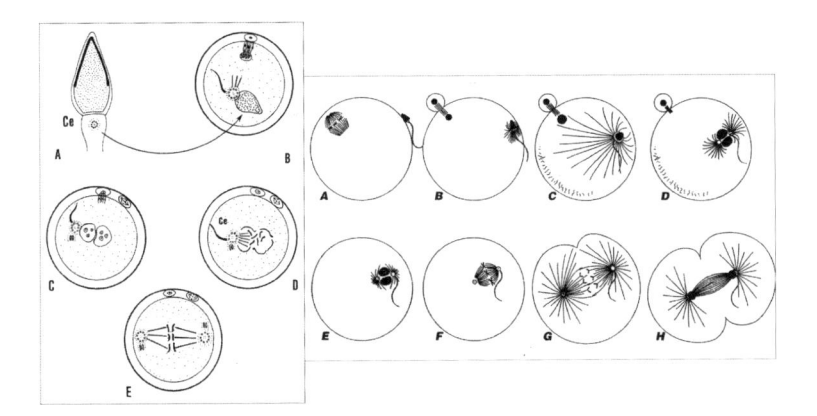

Fig. 5.26 Sperm centrosome inheritance. The centriole (**a–e**, *left*) and aster (**a–h**, *right*) is inherited at fertilization, duplicated and moves to either pole of a bipolar spindle to establish the first mitotic spindle, at syngamy (bipolarization). Note sperm tail attached to centrosome. The diagrams also show sperm incorporation, bipronuclear association, and nuclear disorganization, 2nd polar body extrusion at fertilization and the first cleavage. Sathananthan et al. (1996) TEM; Courtesy Simerly et al. (1995) FM

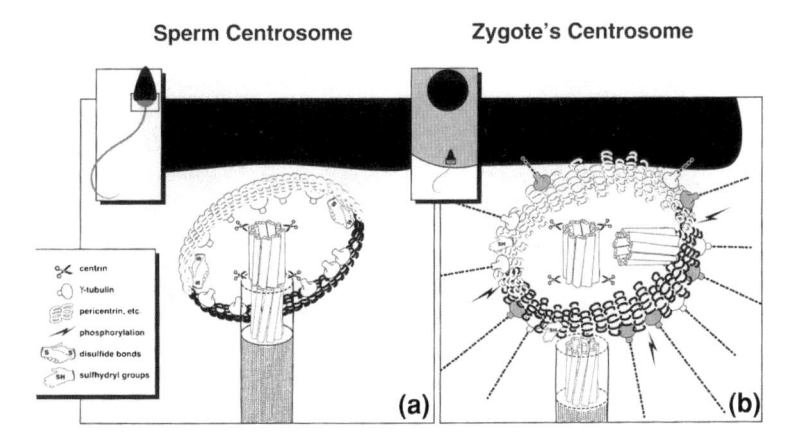

Fig. 5.27 Abnormal multinucleated IVF embryos A. One-cell (fragmented) B. two-cell (micronucleated) and C,D. three-cell embryos (multinucleated). Such defects reflect chromosomal aberrations that could be caused by centrosomal dysfunction. x6,000, x4,000 www.sathembryoart.com

5.12 Centrosomes and Cancer

Boveri predicted the possible role of the centrosome in cancer (Munne 2006) based on his studies on abnormal early development in the sea urchin following dispermic fertilization, where he proposed an explanation for the origin of aneuploidy in cancer through asymmetric mitoses and/or multipolar mitoses due to an

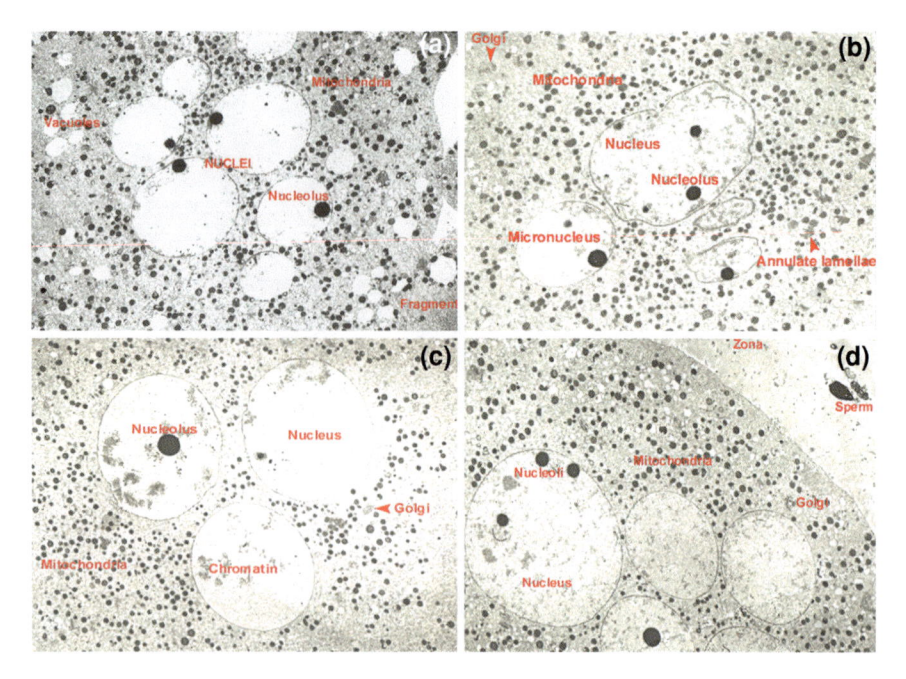

Fig. 5.28 Bipolar anaphase spindle in a monospermic ovum with chromosomes (*blue*) at either pole, which will divide into two cells (*left*). Tripolar anaphase spindle in a dispermic ovum with chromosomes at each pole, which will cleave into three cells (*right*). Mitochondria (*green*) are excluded from the spindle zone. Computer colored x10,000, x8,000 (Sathananthan and Edwards 1995; Sathananthan et al. 1999)

abnormal increase of centrosomes. Our research on human dispermic fertilization clearly shows three pronuclei (two male and one female) and the formation of two sperm asters and tripolar spindles (Fig. 5.28) and often bipolar spindles, as well (Sathananthan and Edwards 1995; Sathananthan et al. 1999). Boveri's and recent work on fruit flies (Boveri 1914; Salisbury 2005; Basto et al. 2008) show that centrosome amplification can initiate tumor formation. Many cells with extra centrosomes initially form multipolar spindles, which later become bipolar. Centrosomal amplification was also evident in human dispermic embryos (Fig. 5.21) which may be considered a prelude to cancer in humans? Aneuploidy is often encountered in human IVF embryos, as well as polyploidy (Munne 2006). This is an important area of research that needs to be pursued. Fortunately, we do not transfer dispermic embryos in IVF, but aneuploidy is a problem and may originate from either sperm or egg, which is now well documented. However, IVF children and young adults have a very slight risk of cancer according to a Reuters Health survey, conducted by Swedish pediatricians

5.13 Centrosomes in Embryonic Stem Cells

We have documented centrosomes in embryonic stem cells (ESCs), establishing their role in cell division and cell polarity (Sathananthan et al. 2002). These ESCs are pluripotent, diploid, self-reproducing cells, derived from the inner cell mass (ICM) of the blastocyst; they are capable of differentiating into all three germ layers, thus to any cell type of the human body. They are also capable of spontaneous differentiation and in essence are somatic cells with typical centrosomes, though with a much more simplified structure than ICM cells of blastocysts. This we attributed to cell de-differentiation.

5.14 Problems with Cloning

One of the challenges of modern reproductive technology is therapeutic cloning and reproductive cloning in mammals, including primates. Cloning or somatic cell nuclear transfer (SCNT) into enucleated oocytes to produce blastocysts, ESC and offspring has not been very successful, even in mice or monkeys (Sathananthan 2009; Simerly et al. 2004). There are evidently problems with centrosomal compatibility, cell cycle asynchrony, nuclear reprogramming, and genomic imprinting after SCNT, where a diploid somatic cell is electro-fused with or injected into a haploid enucleated oocyte and activated to develop. The oocyte is at metaphase II of meiosis, while the somatic cell is at interphase of mitosis with two centrioles, closely associated with its nucleus. Further, the oocyte centrosome is absent or inactive, while the somatic centrosome is functional and forms the bipolar spindle to initiate mitosis in the cloned cell. Abnormal MT formations were reported in monkey SCNT constructs but not in the bovine constructs (Simerly et al. 2004).

5.15 Conclusion and Future Directions

It is essential that the fundamental concepts of human fertilization and embryogenesis be clearly understood and applied to the new technologies of assisted reproduction. We need to explore further centrosomal dysfunction in embryos derived from poor quality sperm used in ICSI and follow these up during cleavage, implantation, and in offspring. We need to identify sperm with possible centrosomal defects in poor sperm samples with little or no motility, using specific antibodies for proteins associated with sperm centrosomes, before sperm injection by ICSI. Gamete aneuploidies and chromosomal defects, including DNA fragmentation, need to be closely monitored in assisted reproductive technologies (ART), since they will contribute to abnormal embryo development. It must be also remembered that the centrosome and chromosomal cycles are closely intertwined in each cleavage cycle. We can now identify nuclear defects in live sperm using high powered lenses, but the centrosome is beyond resolution.

If we can correlate centrosomal dysfunction to sperm motility, we might find a non-invasive method of sperm selection, particularly for ICSI in cases of severe male infertility.

References

Alieva IB, Uzbekov RE (2008) The centrosome is a polyfunctional multiprotein cell complex. Biochem 73:626–643
Anderson SS (1999) Molecular characteristics of the centrosome. J Int Rev Cytol 187:51–109
Asch R, Simerly C, Ord T et al (1995) The stages at which fertilization arrests in humans: defective sperm centrosomes and sperm asters as causes of human infertility. Hum Reprod 10:1897–1906
Basto R, Brunk K, Vinadogrove T, Peel N, Franz A, Hhodjakov A, Raff JW (2008) Centrosome amplification can initiate tumorigenesis in flies. Cell 133:1032–1042
Boveri T (1900) Ueber die Natur der Centrosomen. IV. Zellen – Studien: Heft 4. Gustav Fischer, Jena
Boveri T (1914) Zur Frage der Entstehumg Maligner Tumoren. Fischer Verlag, Jena
Chatzimeletiou K, Morrison EE, Prapas N, Prapas Y, Handyside AH (2008) The centrosome and early embryogenesis: clinical insights. Reprod Biomed Online 16:485–491
Chen C, Sathananthan AH (1986) Early penetration of human sperm through the vestments of human eggs in vitro. Arch Androl 16:183–197
de Kretser DM, Kerr JB (1994) The cytology of the testis. In: Knobil E, Neill JD (eds) The physiology of reproduction, 2nd edn. Raven Press, New York, pp 1177–1290
Fouquet JP, Kann ML, Combeau C et al (1998) Gamma-tubulin during the differentiation of spermatozoa in various mammals and man. Mol Hum Reprod 4:1122–1129
Fuller SD, Gowen BE, Reinsch S et al (1995) The core of the mammalian centriole contains gamma-tubulin. Curr Biol 5:1384–1393
Hewitson L, Simerly C, Schatten G (1997) Inheritance defects of the sperm centrosome in humans and its possible role in male infertility. Int J Androl 20(3):35–43
Hinduja I, Baliga NB, Zaveri K (2010) Correlation of human sperm centrosomal proteins with fertility. J Hum Reprod Sci 3:95–101
Holstein AF, Roosen-Runge EC (1981) Atlas of human spermatogenesis. Grosse Verlag, Berlin
Manandhar G, Simerly C, Schatten G (2000) Highly degenerated distal centrioles in rhesus and human spermatozoa. Hum Reprod 15:256–263
Mazia D (1987) The chromosome cycle and the centrosome cycle in the mitotic cycle. Int Rev Cytol 100:49–92
Munne S (2006) Chromosome abnormalities and their relationship to morphology and development of human embryos. Reprod BioMed Online 12:324–353
Navara C, Simerly C, Zoran S, Schatten G (1995) The sperm centrosome during fertilization in mammals: implications for fertility and reproduction. Reprod Fertil Dev 7:747–754
Palermo GD, Liliana T, Colombero LT, Rosenwaks Z (1997) The human sperm centrosome is responsible for normal syngamy and early embryonic development. Rev Reprod 2:19–27
Salisbury JL (2005) On the origin of centrosome amplification and chromosomal instability in cancer. Micros Microanal 11(2):1002–1003
Sathananthan AH (1991) Inheritance of paternal centrioles and male infertility. In: Abstracts (No. 1629) of theXIIIth world congress of obstetrics and gynaecology, Singapore
Sathananthan AH (ed) (1996) Visual atlas of human sperm structure and function for assisted reproductive technology. National University of Singapore, Serono p 279

Sathananthan AH (1997) Mitosis in the human embryo: the vital role of the sperm centrosome (centriole)—review. Histol Histopathol 12:827–856

Sathananthan AH (1998) Paternal centrosomal dynamics in early human development and infertility. J Assist Reprod Genet 15:129–139

Sathananthan AH (2009) Human centriole: origin and how it impacts fertilization, embryogenesis, infertility and cloning. Indian J Med Res 129:348–350

Sathananthan H, Edwards RG (1995) From sperm binding to syngamy—computer emhanced pictures of human fertilization. Hum Reprod Update 1:1

Sathananthan AH, Kola I, Osborne J, Trounson A, Ng SC, Bongso A, Ratnam SS (1991) Centrioles in the beginning of human development. Proc Nat Acad Sci. U S A 88:4806–4810

Sathananthan AH, Ratnam SS, Ng SC, Tarin JJ, Gianoroli L, Trounson A (1996) The sperm centriole: its inheritance, replication and perpetuation in early human embryos. Hum Reprod 11:345–356

Sathananthan AH, Tarin JJ, Gianaroli L, Ng SC et al (1999) Development of the human dispermic embryo. Hum Reprod Update 5:553–560

Sathananthan AH, Selvaraj K, Trounson A (2000) Fine structure of human oogonia in the foetal ovary. Mol Cell Endocrin 161:3–8

Sathananthan AH, Ratnasooriya WD, de Silva PKRA, Menezes J (2001) Characterization of human gamete centrosomes for assisted reproduction. It J Anat Embryol. 106:61–73

Sathananthan AH, Pera M, Trounson A (2002) A The fine structure of human embryonic stem cells. Reprod Biomed Online 4:56–61

Sathananthan AH, Ratnasooriya WD, Silva AD, Randeniya P (2006a) Rediscovering Boveri's centrosome in Ascaris (1888): Its impact on human fertlity and development. Reprod BioMed Online 12:254–257

Sathananthan AH, Selvaraj K, Girijashankar ML, Ganesh V, Selvaraj P, Trounson AO (2006b) From oogonia to mature oocytes: inactivation of the maternal centrosome in humans. Micros Res Tech 69:396–407

Schatten G (1994) The centrosome and its mode of inheritance: the reduction of the centrosome during gametogenesis and its restoration during fertilization. Dev Biol 165:299–335

Schatten H (2008) The mammalian centrosome and its functional significance. Histochem Cell Biol 129:667–686

Simerly C, Wu G, Zoran S et al (1995) The paternal inheritance of the centrosome, the cell's microtubule-organizing center, in humans and the implications for infertility. Nature Med. 1:47–53

Simerly C, Navara C, Hyun SH, Lee BC, Kang SK, Capuano S, Gosman G et al (2004) Embryogenesis and blastocyst development after somatic cell nuclear transfer in nonhuman primates: overcoming defects caused by meiotic spindle extraction. Dev Biol. 276:237–252

Sluder G, Rieder CL (1993) Centriole number and the reproductive capacity of spindle poles. J Cell Biol 100:887–896

Sutovsky P, Schatten G (1999) Paternal contributions to the mammalian zygote: fertilization after sperm—egg fusion. Int Rev Cytol 195:1–65

Tassin AM, Bornens M (1999) Centrosome structure and microtubule nucleation in animal cells. Biol Cell 91:343–354

Uzbekov RE, Prigent C (2007) Clockwise or anticlockwise? Turning the centriole triplets in the right direction! FEBS Lett 581:1251–1254

Van Blerkom J (1996) Sperm centrosome dysfunction: a possible new cause of male factor infertility in the human. Human Mol Reprod 2:349–354

Vandervorst M, Tournaye H, Camus M et al (1997) Patients with absolutely immotile spermatozoa and intracytoplasmic sperm injection. Hum Reprod 12:2429–2433

Zamboni L (1992) Sperm structure and its relevance to infertility. Arch Pathol Lab Med 116:325–344

Zamboni L, Stefanini M (1970) The fine structure of the neck of mammalian spermatozoa. Anat Rec 169:155–172

Chapter 6
Asymmetric Centrosome Behavior in Stem Cell Divisions

Therese M. Roth, Yukiko M. Yamashita and Jun Cheng

Abstract Stem cells are well known for their self-renewal ability and differentiation potential. It is critical to regulate stem cell self-renewal and differentiation, both during fast growth in development and tissue homeostasis in adulthood. One way to maintain tissue homeostasis is through asymmetric stem cell division, in which centrosomes play an important role in establishing mitotic spindles by acting as a microtubule organization center (MTOC). In this chapter, the asymmetric behavior of centrosomes during stem cell division will be discussed based on their structural, behavioral, and developmental asymmetry.

6.1 Structural and Functional Asymmetries of the Centrosome

Centrosomes were discovered more than 100 years ago as a cytoplasmic organelle that is often located at the center of the cell. They are known to contribute to various cellular processes, particularly cell migration and cell division, through their involvement in the organization of microtubules (MTs). The centrosome is

---

T. M. Roth · Y. M. Yamashita
Life Sciences Institute, Center for Stem Cell Biology,
Ann Arbor, MI, USA

T. M. Roth · Y. M. Yamashita
Department of Cell and Developmental Biology, University of Michigan,
Ann Arbor, MI, USA

J. Cheng (✉)
Department of Bioengineering, University of Illinois, Chicago, IL, USA
e-mail: juncheng@uic.edu

composed of a pair of centrioles and their surrounding pericentriolar material (PCM) (Bettencourt-Dias and Glover 2007). The centriole duplication cycle, as shown by electron microscopy, implies that there is a fundamental intrinsic asymmetry in the age of centrioles. The centrosome cycle includes four main phases: disengagement of the two centrioles, nucleation of the daughter centrioles, elongation of the daughter centrioles, and separation of the new centrosomes. Therefore, within one centrosome, there will always be one older centriole, which was assembled at least one cell cycle prior to, and one younger centriole that was assembled during, the current cell cycle. Furthermore, the "age" of the centrosome can be denoted by the "age" of older centrioles within the centrosome: "the mother centrosome" contains a centriole that was assembled at least two cell cycles prior and a centriole that was assembled in the current cell cycle, while "the daughter centrosome" contains a centriole that was assembled in one cell cycle prior and a centriole that was assembled in the current cell cycle. The age of the centrioles, and by extension the age of the centrosome, affect their maturity and function (Yamashita 2009a). In addition to the asymmetric composition of mother/daughter centrosomes, there are cellular effectors that concentrate on the centrosome in a cell cycle-dependent manner (Bornens 2002).

The centrosomes display functional and behavioral asymmetry, primarily due to structural asymmetry in the mother centrioles. Prior to centrosome duplication, the mother centriole has distal and surrounding subdistal appendages along its proximo-distal axis, while similar appendages are not found in the daughter centriole (Sluder 2005). Therefore, after centrosome duplication the mother centriole of the mother centrosome, which contains the appendages, is more mature compared to the mother centriole of the daughter centrosome throughout most of the cell cycle (Vorobjev and Chentsov Yu 1982). As the distal and subdistal appendages are major MT-anchoring sites, the mother centrosome has a greater capability to initiate and organize cytoplasmic MTs (Bornens 2002). Furthermore, vertebrate primary cilia can only grow from the mature mother centriole in the centrosome, and after cell division the cell that inherited the oldest centriole (i.e. the cell that inherited the mother centrosome) usually grows a primary cilium first (Anderson and Stearns 2009). Proteins that specifically localize to the appendages have been identified. For example, ε-tubulin and ODF2 asymmetrically localize to the mother centriole, while centrobin asymmetrically localizes to the daughter centrosome (Chang et al. 2003; Chang and Stearns 2000; Nakagawa et al. 2001; Zou et al. 2005) (Fig. 6.1).

A group of proteins are known to play important roles in centriole-duplication and PCM recruitment during centrosome biogenesis. In *C. elegans*, the kinase ZYG-1 (a functional ortholog of protein kinase PLK4) is involved in centrosome formation through phosphorylating SAS-6 (Kitagawa et al. 2009). It has been shown that SAS-4 and SAS-6 are required for centriole duplication in *C. elegans* (Dammermann et al. 2004; Kirkham et al. 2003), *Drosophila melanogaster* (Basto et al. 2006) and humans (Leidel and Gonczy, 2005), and polo kinase and centrosomin (Cnn) are localized to the centrosome and important for PCM recruitment (Yamashita 2009b). Cnn is asymmetrically incorporated into the PCM and the rate of Cnn incorporation is responsible for regulating the size of the centrosome (Conduit et al. 2010). Therefore,

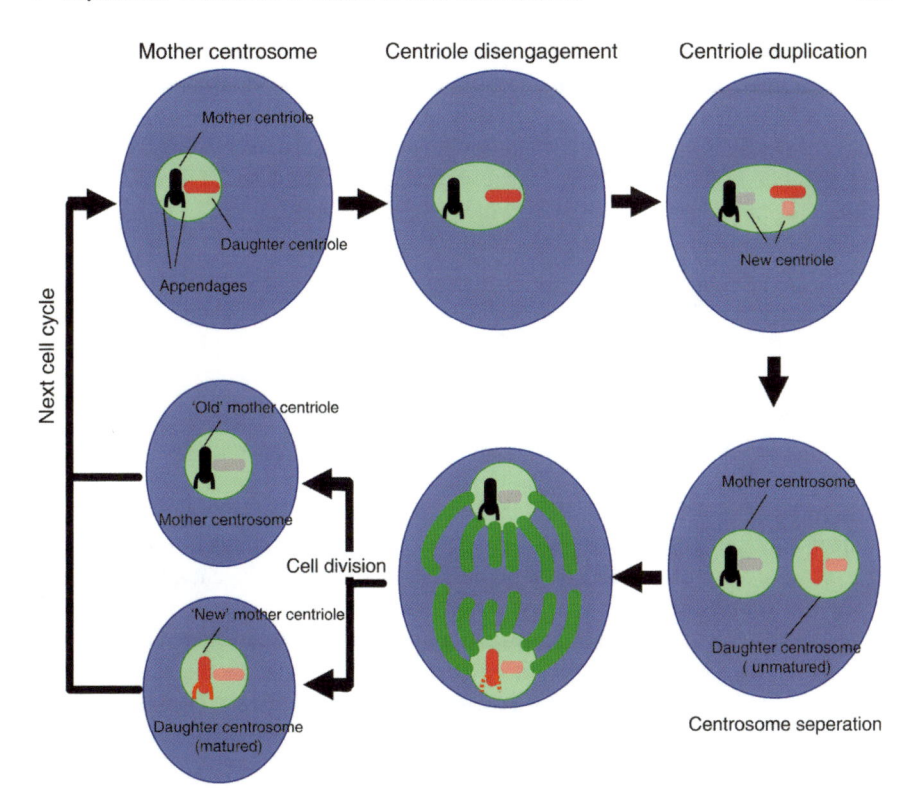

Fig. 6.1 *Centrosome duplication cycle: the intrinsic centrosome asymmetry.* Prior to centrosome duplication, a cell has a single centrosome (named mother centrosome) containing one "older" centriole (i.e., the mother centriole) and one "newer" centriole (i.e., the daughter centriole). The mother centriole (*black*) has distal and subdistal appendages (*black branches*), while the daughter centriole (*red*) does not. Preparing for centrosome duplication, two centrioles disengage from each other and then each centriole begins duplicating with new centrioles (*gray* and *purple*) nucleating by their sides. At a certain time point, the two pairs of centrioles separate from each other, forming two centrosomes. The mother centrosome contains the original mother centriole (*black*), and the daughter centrosome contains the original daughter centriole (*red*). The daughter centrosome is in her unmatured state at this time point, and will take time (e.g., by the end of current cell cycle or even next cell cycle) to become fully matured. After cell division, the mother centrosome is inherited by one daughter cell. In the other daughter cell, the original daughter centriole in the daughter centrosome matures as the "new" mother centriole with developed appendages (*red branches*), and the matured daughter centrosome can be denoted as the mother centrosome upon entry into the next cell cycle

the mother centrosome typically organizes a greater amount of Cnn than the daughter centrosome, since mother centrosomes are more mature and begin incorporating Cnn earlier. Furthermore, centrosomes behave asymmetrically during abscission, the last stage of cytokinesis (Gromley et al. 2005; Kuo et al. 2011). It has been shown that in Hela cells the mother centriole moves very close to the midbody right before abscission (Piel et al. 2001), and secretory vesicles carrying components required for abscission move to the site of abscission from the cell containing the daughter centrosome, leading to the inheritance of the midbody ring by the cell with the mother centrosome (Gromley et al. 2005; Kuo et al. 2011). The cell containing the mother centrosome was found to have multiple midbody derivatives through numerous successive cell divisions (Kuo et al. 2011). Multiple pathways have been discovered to alleviate the overaccumulation of midbody derivatives: they could be either digested by autophagy or released from cells (Kuo et al. 2011; Pohl and Jentsch 2009).

6.2 Stem Cells and Asymmetric Stem Cell Division

The balance between stem cell self-renewal and differentiation is critical for sustaining tissue growth during early development, repairing damaged tissue after injury, and maintaining tissue homeostasis during adulthood (Morrison and Kimble 2006; Morrison et al. 1997; Watt and Hogan 2000). Failure to regulate the number and function of stem cells has been speculated to lead to tumorigenesis due to the over-proliferation of stem cells or tissue aging/degeneration due to a decline in stem cell functions (Brunet and Rando 2007; Clarke and Fuller 2006; Clevers 2005; Kirkwood 2005; Rando 2006). Asymmetric stem cell division creates one daughter cell that retains a stem cell identify and a daughter cell that undergoes differentiation, thereby precisely balancing self-renewal and differentiation. There are two major regulatory mechanisms that ensure asymmetric stem cell division: extrinsic and intrinsic fate determinants.

6.2.1 Asymmetric Stem Cell Division Regulated by Extrinsic Fate Determinants

Many stem cells reside in a microenvironment, or niche, that provides extrinsic fate determinants that specify stem cell identity (Morrison and Spradling 2008). These extrinsic fate determinants, locally secreted within the stem cell niche, promote the ability of stem cells to self-renew and/or repress their differentiation. The *Drosophila* male germline stem cell (GSC) niche is among the best studied of these systems, where stem cells undergo asymmetric stem cell divisions in the context of the stem cell niche. The male GSC niche is composed of hub cells and cyst stem cells (CySCs, also known as cyst progenitor cells). The hub cells are located at the apical tip of the testis, to which the GSCs and the CySCs attach via adherens junctions (Hardy et al. 1979; Raymond et al. 2009). Hub cells secret the ligand Unpaired (Upd), which specifies the stem cell identity through activating JAK-STAT signaling pathways within GSCs and CySCs (Kiger et al. 2001; Leatherman and Dinardo 2008, 2010; Tulina and Matunis 2001). GSCs divide asymmetrically, producing one GSC and one gonialblast, and CySCs also divide asymmetrically, producing one CySC and one cyst cell (Cheng et al. 2011; Yamashita et al. 2003) (Fig. 6.2a).

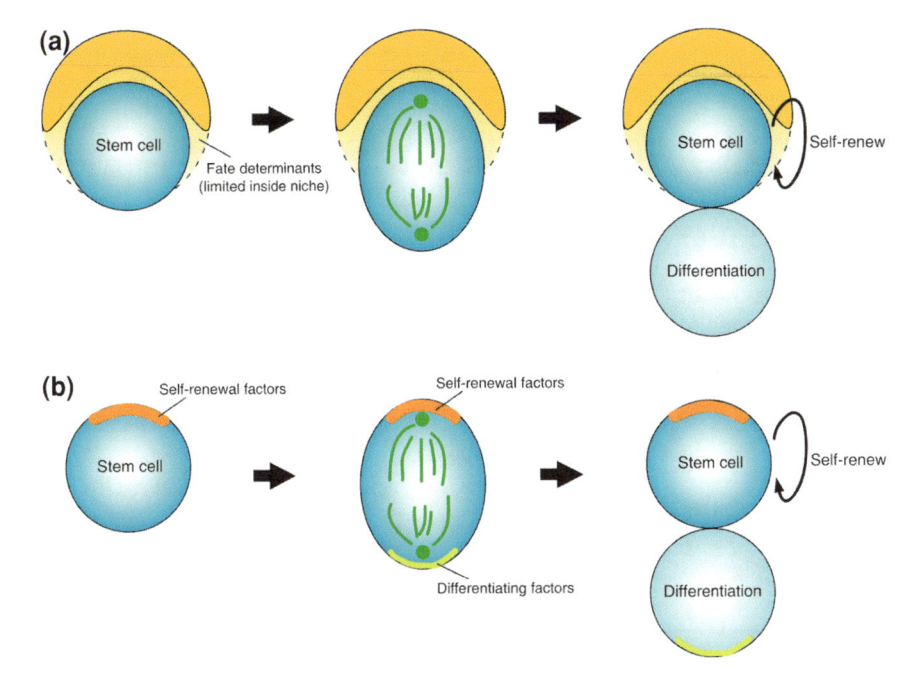

Fig. 6.2 *Asymmetric stem cell division regulated by extrinsic/intrinsic fate determinants.* **a** Asymmetric stem cell division regulated by extrinsic fate determinants. The extrinsic fate determinants (*brown*) are short ranged and localized within the stem cell niche, promoting self-renewal and/or repressing differentiation. Stem cells (*blue*) are located inside the niche, receiving fate determinants and maintaining stem cell function. Upon stem cell division, the mitotic spindle is stereotypically positioned so that the two daughters will be placed either inside or outside the stem cell niche, becoming either stem cell or differentiating cell. **b** Asymmetric stem cell division regulated by intrinsic fate determinants. The polarity of the stem cell is established by intrinsic fate determinants (*brown*) preferentially localized on the apical end of the cell. Upon cell division, self-renewal and differentiating factors are segregated so that self-renewal factors are concentrated at the apical end and differentiating factors are localized at the basal end. After cell division, one daughter cell receiving self-renewal factors maintains self-renewal ability while the other daughter cell acquiring differentiating factors differentiates

6.2.2 Asymmetric Stem Cell Division Regulated by Intrinsic Fate Determinants

Asymmetric stem cell division can also be achieved through the asymmetric segregation of intrinsic fate determinants (Knoblich 2008; Yamashita et al. 2010). Before/during cell division, intrinsic fate determinants are asymmetrically localized in the stem cell, and then unequally segregated into two daughter cells after the completion of mitosis, leading to the asymmetric fates of the two daughter cells (Fig. 6.2b). *Drosophila* neural stem cells, also known as neuroblasts (NB), provide one of the best-studied examples of the regulation of asymmetric stem cell division by intrinsic fate determinants (Knoblich 2008; Prehoda 2009). In NBs, apical-basal polarity is established by different sets of molecules. The apical cortex of neuroblasts is bounded by Par and Pins complexes. The Par complex consists of Bazooka (PAR3), Par6, and atypical protein kinase C (aPKC), while the Pins complex consists of Partner of Inscuteable (Pins), Discs large (Dlg), and Gαi. Proteins bound to the basal cortex of NBs include Miranda, Prospero, Brain tumor (Brat), and Numb. The orientation of the mitotic spindle is directed along the apical/basal axis, generating an apical daughter cell and a basal daughter cell that receive different fate determinants and asymmetric fates: the one receiving apical determinants remains a NB and the other, which receives basal determinants, becomes the ganglion mother cell (GMC), respectively. It was found that the ratio of apical to basal determinants also plays an important role in determining the distinct fates of the daughter cells (Cabernard and Doe 2009). In experiments where spindle orientation was disrupted, apical determinants were partitioned between two daughter cells, while the basal determinants were segregated asymmetrically into only one daughter cell, resulting in two NBs. Recently, Doe and colleagues found that the position of the cleavage furrow, which leads to the segregation of determinants, is not solely dependent upon the mitotic spindle, but is also correlated with the polarity proteins, particularly the Pins complex at the apical tip (Cabernard et al. 2010).

6.3 Asymmetric Behavior of Centrosomes in Stem Cells

In recent years, due to the centrosome's important role in organizing the mitotic spindle and directing mitotic spindle orientation in dividing cells, it has been increasingly recognized that it plays critical roles in cell divisions. Recent progresses in studying centrosome behavior during asymmetric stem cell divisions make it clear that the structural asymmetry of mother and daughter centrosomes is an important underlying mechanism by which stem cells coordinate their fate and cellular asymmetry. A few examples of asymmetric centrosome behaviors during stem cell divisions will be discussed below.

In *Drosophila* male GSCs, the mother centrosome stereotypically remains at the apical side of the stem cell near the hub, while the daughter centrosome migrates to the basal end of the cell, setting up a centrosome orientation perpendicular to the hub (Yamashita et al. 2003; Yamashita et al. 2007). This stereotypical centrosome orientation ensures asymmetric GSC division; one daughter cell retains stem cell identity containing the mother centrosome, while the other daughter cell inheriting the daughter centrosome undergoes differentiation (Fig. 6.3a). This stereotypical centrosome orientation and asymmetric centrosome inheritance requires centrosomin (Cnn). Cnn, an integral centrosome component, is important in PCM recruitment and normal astral microtubule function (Dobbelaere et al. 2008; Megraw et al. 2001). If cnn is mutated, most PCM components cannot properly localize to the centrosomes (Conduit et al. 2010). In cnn mutant flies, male GSCs display abnormal centrosome orientation and show an almost random mother/daughter centrosome inheritance

(Yamashita et al. 2007). Electron micrographs showed that there are indeed more astral microtubules around the mother centrosome than the daughter centrosome in wild type GSCs, presumably explaining the asymmetrical positioning and motility of the mother/daughter centrosomes. Furthermore, centrosome separation in GSCs occurs unusually early, suggesting that GSCs could take advantage of the difference in motility and/or stability of mother versus daughter centrosomes to position the daughter centrosome at the basal side of the cell, thus ensuring asymmetric stem cell division, and thus their asymmetric inheritance. Similarly, in the mammalian neocortex, the mother centrosome is preferentially inherited by the neural progenitor cell while the daughter centrosome is largely associated with the differentiating neuron (Wang et al. 2009). Strikingly, this centrosome asymmetry is closely related to the stem cell fate and maintenance. In the event of RNAi knockdown of ninein, a component of subdistal appendages in the mature centrosome, this asymmetric centrosome inheritance is disrupted, resulting in the depletion of the neural progenitors (Fig. 6.3b).

Another interesting example is found in *Drosophila* NBs. NBs undergo asymmetric stem cell division, generating two daughter cells with unequal sizes. During mitosis, spindle orientation is critical and must be in line with the polarity axis of the NB to ensure an asymmetric division (Rebollo et al. 2007; Rusan and Peifer 2007); pulling forces acting on the astral MTs work to determine the position of the mitotic spindle and are responsible for the differences in NB daughter cell sizes (Neumuller and Knoblich 2009). However, contrary to the centrosome inheritance in fly GSCs and mammalian neural progenitors, the daughter centrosome is consistently retained by the stem cell while the mother centrosome is inherited by the GMC (Conduit and Raff 2010; Januschke et al. 2011). To achieve this, Cnn is downregulated in the mother centrosome and the MTOC activity of the mother centrosome is reduced, while the level of Cnn is maintained in the daughter centrosome and the daughter centrosome functions as MTOC every cell cycle (Conduit and Raff 2010). Instead of relying on the passive maintenance of the MTOC, fly NBs utilize this elaborate mechanism of actively regulating the MTOC of each centrosome every cell cycle to achieve asymmetric centrosome segregation and inheritance, implying that NBs do have a reason to do so, although such a reason has yet to be elucidated (Fig. 6.3c).

The asymmetric centrosome segregation and inheritance during asymmetric stem cell divisions implies that stem cells do have reason(s) to inherit one particular centrosome over the other, provoking further studies to understand the biological meaning of this phenomenon. One possible explanation is that the mother/daughter centrosomes asymmetrically harbor fate determinants, so that the daughter cell inheriting the "stemness"-promoting centrosome remains as the stem cell, while the other daughter cell receiving the "differentiation"-initiating centrosome differentiates. For example, in mollusc embryos, certain mRNAs asymmetrically associated with the centrosome are asymmetrically distributed during division, producing two daughter cells with asymmetric fates (Lambert and Nagy 2002).

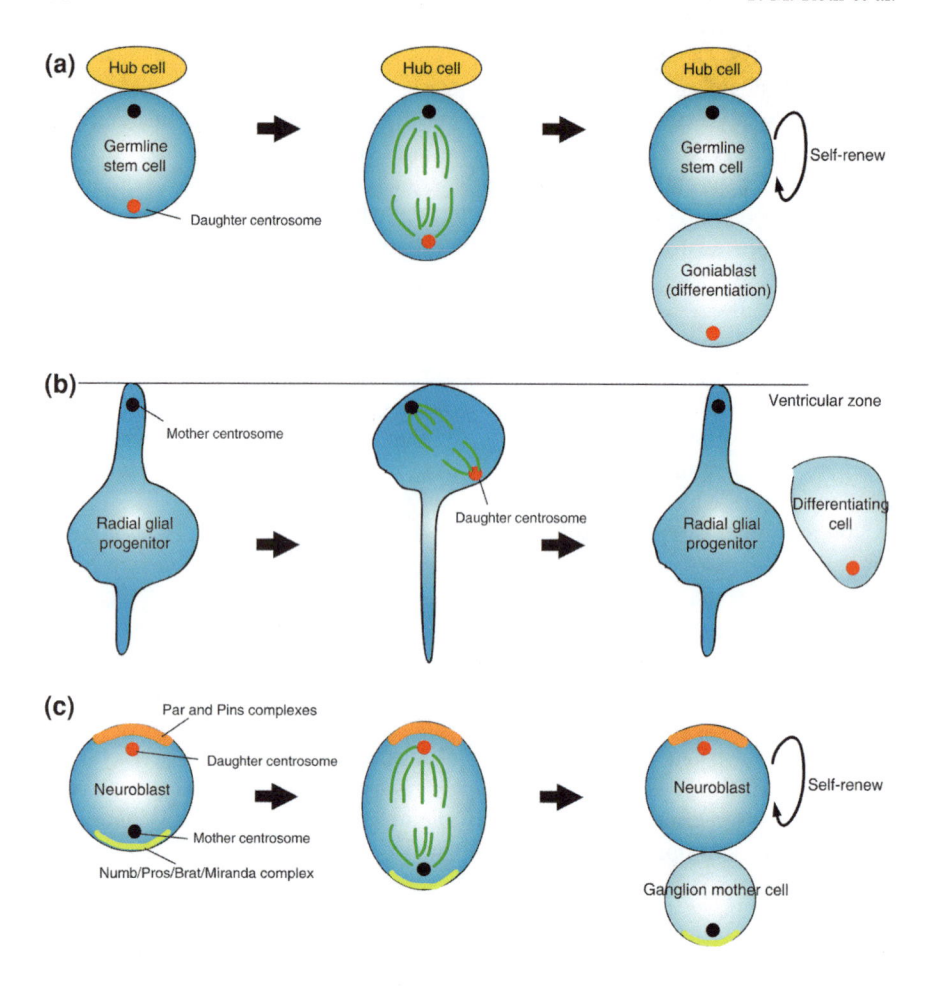

6.4 The Loss of Asymmetry of Centrosomes Relates to Centrosome Dysfunction and Tumorigenesis

As described above, the structural asymmetry between the two centrosomes is important in many cellular processes during cell divisions. The intriguing questions are, when cells contain extra centrosomes, whether this asymmetry is still maintained, and if not, how the extra centrosomes affect the cellular processes. Basto and colleagues answered these questions through driving centriole overduplication in fruit flies by overexpressing SAK, a protein kinase critical for initiating centriole duplication (Basto et al. 2008). Although the adult animals with SAK overexpression are viable and fertile, their development is significantly delayed and the mitosis period is longer. Cells with extra centrosomes initially form multipolar mitotic spindles, but later on almost all multipolar spindles become bipolar by metaphase, leading to normal partitioning of the chromosomes into two daughter cells. This

◀ **Fig. 6.3** *Asymmetric centrosome segregation and inheritance during asymmetric stem cell division.* **a** *Drosophila* germline stem cells divide asymmetrically and the mother centrosome is inherited by the stem cell daughter. At the center of the stem cell niche, hub cells produce external fate determinants within the niche. The mother centrosome (*black*) locates close to the stem cell-hub interface while the daughter centrosome (*red*) is mostly found to be at the basal side. During mitosis, germline stem cells form stable and stereotypical oriented mitotic spindles with mother/daughter centrosome asymmetrically positioned. After cell division, one daughter cell receiving the mother centrosome maintains contact with the hub cell and retains self-renewal ability. The other daughter cell receiving the daughter centrosome is displaced away from the hub and starts differentiation. **b** Radial glial progenitors in the ventricular zone of the mouse neocortex divide asymmetrically, and the self-renewing daughter inherits the mother centrosome (*black*). The centrosome in radial glial progenitor locates very close to the ventricular zone. Upon cell division, a mitotic spindle with asymmetric centrosomes forms at the surface of the ventricular zone. Therefore, the self-renewing radial glial progenitor inherits the mother centrosome, while the daughter centrosome is received by the differentiating cell. **c** *Drosophila* neuroblasts divide asymmetrically, but the daughter centrosome is received by the self-renewing daughter while the mother centrosome is retained by the differentiating daughter. Intrinsic self-renewal fate determinants, such as Par and Pins complexes, form a crescent (*brown*) and localize at the apical end; and differentiation-promoting proteins, such as Numb/Pros/Brat/Miranda complex, form the basal crescent (*yellow*). Upon cell division, the daughter centrosome (*red*) locates near the apical crescent and the mother centrosome (*black*) positions near the basal crescent. Therefore, this spindle positioning leads to asymmetric segregation of centrosomes and intrinsic fate determinants, producing one self-renewing neuroblast in a larger size with the daughter centrosome and one differentiating ganglion mother cell in a smaller size with the mother centrosome

process, formation of a bipolar spindle in a cell containing more than two centrosomes, requires a functional spindle assembly checkpoint, which inhibits the anaphase-promoting complex, and thus delays mitotic exit. This study showed that centrosome overamplification does not necessarily lead to multipolar cell division, leading to genomic instability, as has been suggested as a possible underlying mechanism that explains frequent centrosome overamplification in cancer cells. Interestingly, however, an unusually high percentage of NBs undergo symmetric division upon overamplification of centrosomes (Basto et al. 2008).

Almost a century ago, abnormalities in centrosome number were hypothesized to result in chromosomal alternations due to a failure in equal partitioning of the genome, triggering cancer development. This seemed to fit well with the observation that many cancer cells often show centrosome overamplification. Multiple lines of evidence have since shown that genetic alterations and instability are involved in tumor development and cancer, but the causative relationship between centrosome abnormalities and genetic alterations in cancer development remained elusive. A recent study provided the first critical evidence that genomic instability might not be the reason for tumor formation upon centrosome dysfunction (Castellanos et al. 2008). The authors systematically examined the tumor-forming ability of cells derived from fly brain upon transplantation to a host fly abdomen. Cells defective in centrosome function (such as mutants of dsas-4, polo or aurA) resulted in tumor formation, but without significant changes in the cellular DNA content. Instead, these cells with defective centrosomes showed a failure in asymmetric NB division. Critically, cells defective in genome stability (such as mutants of atm, a gene required for the DNA damage checkpoint) did not cause any tumor formation upon transplantation. These results led the authors to propose that the defective genomic stability upon centrosome dysfunction is not the cause of tumor formation; instead, centrosome dysfunction causes tumors through disturbing asymmetric stem cell divisions (Castellanos et al. 2008).

Because of its duplication mechanism during centriole biogenesis, the centrosome displays inherent asymmetry in cellular processes. Now we know several examples in which centrosome asymmetry is integrated into a higher level of asymmetry, i.e., fate, during stem cell divisions. While it is tempting to speculate that centrosome asymmetry is a universal mechanism for cells to divide asymmetrically, such assumptions and underlying mechanisms remain to be elucidated.

Acknowledgment The authors wish to acknowledge financial support from Chicago Biomedical Consortium from The Searle Funds at The Chicago Community Trust to J.C., Postdoctoral Fellowship from the Training Program in Organogenesis (T-32-HD007505) to T.M.R., and MacArthur Foundation to Y.M.Y.

References

Anderson CT, Stearns T (2009) Centriole age underlies asynchronous primary cilium growth in mammalian cells. Curr Biol 19:1498–1502

Basto R, Brunk K, Vinadogrova T, Peel N, Franz A, Khodjakov A, Raff JW (2008) Centrosome amplification can initiate tumorigenesis in flies. Cell 133:1032–1042

Basto R, Lau J, Vinogradova T, Gardiol A, Woods CG, Khodjakov A, Raff JW (2006) Flies without centrioles. Cell 125:1375–1386

Bettencourt-Dias M, Glover DM (2007) Centrosome biogenesis and function: centrosomics brings new understanding. Nat Rev Mol Cell Biol 8:451–463

Bornens M (2002) Centrosome composition and microtubule anchoring mechanisms. Curr Opin Cell Biol 14:25–34

Brunet A, Rando TA (2007) Ageing: from stem to stern. Nature 449:288–291

Cabernard C, Doe CQ (2009) Apical/basal spindle orientation is required for neuroblast homeostasis and neuronal differentiation in Drosophila. Dev Cell 17:134–141

Cabernard C, Prehoda KE, Doe CQ (2010) A spindle-independent cleavage furrow positioning pathway. Nature 467:91–94

Castellanos E, Dominguez P, Gonzalez C (2008) Centrosome dysfunction in Drosophila neural stem cells causes tumors that are not due to genome instability. Curr Biol 18:1209–1214

Chang P, Stearns T (2000) Delta-tubulin and epsilon-tubulin: two new human centrosomal tubulins reveal new aspects of centrosome structure and function. Nat Cell Biol 2:30–35

Chang P, Giddings TH Jr, Winey M, Stearns T (2003) Epsilon-tubulin is required for centriole duplication and microtubule organization. Nat Cell Biol 5:71–76

Cheng J, Tiyaboonchai A, Yamashita YM, Hunt AJ (2011) Asymmetric division of cyst stem cells in Drosophila testis is ensured by anaphase spindle repositioning. Development 138:831–837

Clarke MF, Fuller M (2006) Stem cells and cancer: two faces of eve. Cell 124:1111–1115

Clevers H (2005) Stem cells, asymmetric division and cancer. Nat Genet 37:1027–1028

Conduit PT, Raff JW (2010) Cnn dynamics drive centrosome size asymmetry to ensure daughter centriole retention in Drosophila neuroblasts. Curr Biol 20:2187–2192

Conduit PT, Brunk K, Dobbelaere J, Dix CI, Lucas EP, Raff JW (2010) Centrioles regulate centrosome size by controlling the rate of Cnn incorporation into the PCM. Curr Biol 20:2178–2186

Dammermann A, Muller-Reichert T, Pelletier L, Habermann B, Desai A, Oegema K (2004) Centriole assembly requires both centriolar and pericentriolar material proteins. Dev Cell 7:815–829

Dobbelaere J, Josue F, Suijkerbuijk S, Baum B, Tapon N, Raff J (2008) A genome-wide RNAi screen to dissect centriole duplication and centrosome maturation in Drosophila. PLoS Biol 6:e224

Gromley A, Yeaman C, Rosa J, Redick S, Chen CT, Mirabelle S, Guha M, Sillibourne J, Doxsey SJ (2005) Centriolin anchoring of exocyst and SNARE complexes at the midbody is required for secretory-vesicle-mediated abscission. Cell 123:75–87

Hardy RW, Tokuyasu KT, Lindsley DL, Garavito M (1979) The germinal proliferation center in the testis of Drosophila melanogaster. J Ultrastruct Res 69:180–190

Januschke J, Llamazares S, Reina J, Gonzalez C (2011) Drosophila neuroblasts retain the daughter centrosome. Nat commun 2:243

Kiger AA, Jones DL, Schulz C, Rogers MB, Fuller MT (2001) Stem cell self-renewal specified by JAK-STAT activation in response to a support cell cue. Science 294:2542–2545

Kirkham M, Muller-Reichert T, Oegema K, Grill S, Hyman AA (2003) SAS-4 is a C. elegans centriolar protein that controls centrosome size. Cell 112:575–587

Kirkwood TB (2005) Understanding the odd science of aging. Cell 120:437–447

Kitagawa D, Busso C, Fluckiger I, Gonczy P (2009) Phosphorylation of SAS-6 by ZYG-1 is critical for centriole formation in C. elegans embryos. Dev Cell 17:900–907

Knoblich JA (2008) Mechanisms of asymmetric stem cell division. Cell 132:583–597

Kuo TC, Chen CT, Baron D, Onder TT, Loewer S, Almeida S, Weismann CM, Xu P, Houghton JM, Gao FB et al (2011) Midbody accumulation through evasion of autophagy contributes to cellular reprogramming and tumorigenicity. Nat Cell Biol 13:1214–1223

Lambert JD, Nagy LM (2002) Asymmetric inheritance of centrosomally localized mRNAs during embryonic cleavages. Nature 420:682–686

Leatherman JL, Dinardo S (2008) Zfh-1 controls somatic stem cell self-renewal in the Drosophila testis and nonautonomously influences germline stem cell self-renewal. Cell Stem Cell 3:44–54

Leatherman JL, Dinardo S (2010) Germline self-renewal requires cyst stem cells and stat regulates niche adhesion in Drosophila testes. Nat Cell Biol 12:806–811

Leidel S, Gonczy P (2005) Centrosome duplication centrosome duplication and nematodes: recent insights from an old relationship. Dev Cell 9:317–325

Megraw TL, Kao LR, Kaufman TC (2001) Zygotic development without functional mitotic centrosomes. Curr Biol 11:116–120

Morrison SJ, Kimble J (2006) Asymmetric and symmetric stem-cell divisions in development and cancer. Nature 441:1068–1074

Morrison SJ, Spradling AC (2008) Stem cells and niches: mechanisms that promote stem cell maintenance throughout life. Cell 132:598–611

Morrison SJ, Shah NM, Anderson DJ (1997) Regulatory mechanisms in stem cell biology. Cell 88:287–298

Nakagawa Y, Yamane Y, Okanoue T, Tsukita S (2001) Outer dense fiber 2 is a widespread centrosome scaffold component preferentially associated with mother centrioles: its identification from isolated centrosomes. Mol Biol Cell 12:1687–1697

Neumuller RA, Knoblich JA (2009) Dividing cellular asymmetry: asymmetric cell division and its implications for stem cells and cancer. Genes Dev 23:2675–2699

Piel M, Nordberg J, Euteneuer U, Bornens M (2001) Centrosome-dependent exit of cytokinesis in animal cells. Science 291:1550–1553

Pohl C, Jentsch S (2009) Midbody ring disposal by autophagy is a post-abscission event of cytokinesis. Nat Cell Biol 11:65–70

Prehoda KE (2009) Polarization of Drosophila neuroblasts during asymmetric division. Cold Spring Harb Perspect Biol 1:a001388

Rando TA (2006) Stem cells, ageing and the quest for immortality. Nature 441:1080–1086

Raymond K, Deugnier MA, Faraldo MM, Glukhova MA (2009) Adhesion within the stem cell niches. Curr Opin Cell Biol 21:623–629

Rebollo E, Sampaio P, Januschke J, Llamazares S, Varmark H, Gonzalez C (2007) Functionally unequal centrosomes drive spindle orientation in asymmetrically dividing Drosophila neural stem cells. Dev Cell 12:467–474

Rusan NM, Peifer M (2007) A role for a novel centrosome cycle in asymmetric cell division. J Cell Biol 177:13–20

Sluder G (2005) Two-way traffic: centrosomes and the cell cycle. Nat Rev Mol Cell Biol 6:743–748

Tulina N, Matunis E (2001) Control of stem cell self-renewal in Drosophila spermatogenesis by JAK-STAT signaling. Science 294:2546–2549

Vorobjev IA, Chentsov yu S (1982) Centrioles in the cell cycle I. Epithelial cells. J Cell Biol 93:938–949

Wang X, Tsai JW, Imai JH, Lian WN, Vallee RB, Shi SH (2009) Asymmetric centrosome inheritance maintains neural progenitors in the neocortex. Nature 461:947–955

Watt FM, Hogan BL (2000) Out of Eden: stem cells and their niches. Science 287:1427–1430

Yamashita YM (2009a) The centrosome and asymmetric cell division. Prion 3:84–88

Yamashita YM (2009b) Regulation of asymmetric stem cell division: spindle orientation and the centrosome. Front Biosci 14:3003–3011

Yamashita YM, Jones DL, Fuller MT (2003) Orientation of asymmetric stem cell division by the APC tumor suppressor and centrosome. Science 301:1547–1550

Yamashita YM, Mahowald AP, Perlin JR, Fuller MT (2007) Asymmetric inheritance of mother versus daughter centrosome in stem cell division. Science 315:518–521

Yamashita YM, Yuan HB, Cheng J, Hunt AJ (2010) Polarity in stem stem cell division: asymmetric stem cell division in tissue homeostasis. Cold Spring Harb Perspect Biol 2:a001313

Zou C, Li J, Bai Y, Gunning WT, Wazer DE, Band V, Gao Q (2005) Centrobin: a novel daughter centriole-associated protein that is required for centriole duplication. J Cell Biol 171: 437–445c

Part II
Cell and Molecular Biology of Centrosomes

Chapter 7
Functional Associations Between the Golgi Apparatus and the Centrosome in Mammalian Cells

Breanne Karanikolas and Christine Sütterlin

Abstract The pericentrosomal positioning of the mammalian Golgi apparatus has been observed for many years, but, until recently, its functional significance remained unclear. Several studies have now demonstrated that there are associations between the Golgi and the centrosome that are critical for the establishment of cell polarity, the organization of the centrosome, and proper cell cycle progression. In this chapter, we will review the major factors that control the positioning of the mammalian Golgi apparatus next to the centrosome. We will also discuss the functional associations between the Golgi and the centrosome during interphase, when there is physical proximity between these two organelles, and during mitosis, when the physical Golgi-centrosome proximity is temporarily lost.

7.1 Introduction

The region next to the centrosome is a major site of membrane trafficking in mammalian cells (De Matteis and Luini 2008; Wilson et al. 2011). The most prominent organelle in this area is the Golgi apparatus, which is composed of stacks of 6–8 flattened membrane cisternae. Individual Golgi stacks are laterally connected to form the so-called Golgi ribbon, in which regions with tight cisternae packing ('compact zones') are separated by connecting tubular elements ('non-compact zones'). Newly synthesized proteins are delivered via anterograde transport from the Endoplasmic Reticulum (ER) to the *cis* face of the Golgi

---

B. Karanikolas · C. Sütterlin (✉)
Department of Developmental and Cell Biology, University of California,
Irvine, CA, USA
e-mail: suetterc@uci.edu

apparatus, where protein sorting occurs. Proteins and lipids destined for the endosomal/lysosomal system, the cell surface, or the extracellular space transit the cisternae of the Golgi stacks in a directional manner and are glycosylated in a stepwise process. In contrast, proteins that have "escaped" from the ER are recycled back to the ER by retrograde transport. In the most *trans* cisternae, the Trans Golgi Network (TGN), cargo molecules are packaged into transport carriers for delivery to their final destinations.

In addition to the Golgi apparatus, elements of the ER-Golgi Intermediate Compartment (ERGIC) have been detected in the pericentrosomal region (Marie et al. 2009). As the name implies, the ERGIC is a dynamic organelle at the interface between ER exit sites and the *cis* Golgi, which functions as a major site of post-ER protein sorting. Interestingly, the pericentrosomal domain of the ERGIC appears to be independent of the Golgi, maintaining its close association with the centrosome under conditions when Golgi membranes are dispersed (Marie et al. 2009).

The Endocytic Recycling Compartment (ERC) is also found in the pericentrosomal region (Lin et al. 2002). This organelle, which is morphologically and functionally distinct from the early endosome, is a collection of tubular endosomes concentrated near the centrosome. It is involved in the vesicle-mediated transport of proteins and lipids that are endocytosed and then targeted either to the lysosome or recycled back to the plasma membrane. For example, some cell surface proteins, such as transferrin and low-density lipoprotein receptors, use this pathway: they are first endocytosed, then separated from their respective ligands, and finally recycled back to the cell surface via the ERC.

The placement of multiple distinct trafficking organelles next to the centrosome indicates that this particular localization may be advantageous for a mammalian cell. One obvious benefit is the convergence of trafficking compartments with the microtubule network, which facilitates efficient vesicle-mediated transport to the cell center and the cell periphery. In addition, the close proximity of the Golgi, the ERGIC, and the ERC facilitates the rapid exchange of cargo among these compartments. For example, when TGN resident proteins, such as TGN38, escape from the Golgi to the cell surface, they are recycled back to the TGN via the ERC (Ghosh et al. 1998). This trafficking route is also used by bacterial toxins (Mallard et al. 1998). For instance, Shiga toxin is taken up by endocytosis, but avoids lysosomal destruction by trafficking from the ERC to the TGN, and finally to the ER, where it is released into the cytoplasm to inhibit ribosomal activity. Interestingly, under conditions when normal recycling pathways are blocked, transferrin receptor travels from the ERC to the pericentrosomal ERGIC and then back to the plasma membrane (Marie et al. 2009). Thus, the pericentrosomal region of the cell appears to serve as a trafficking "hub" that promotes efficient cargo transfer between organelles.

While seen in some eukaryotes, including mammals, frogs, fish, and the unicellular amoeba *Dictyostelium* (Distel et al. 2010; Rehberg et al. 2005; Reilein et al. 2003; Thyberg and Moskalewski 1999), a pericentrosomal Golgi ribbon is not universal (Wilson et al. 2011). In fact, Golgi membranes of many lower eukaryotes are organized as isolated, unconnected membrane stacks, or individual cisternae distributed throughout the cytoplasm. These Golgi membranes are fully functional for protein transport. In the yeast *Saccharomyces cerevisiae*, for example, individual Golgi cisternae are dispersed throughout the cytosol (Preuss et al. 1992; Rambourg et al. 2001). In the fission yeast *Schizosaccharomyces pombe*, Golgi mini-stacks exist, but they are distributed throughout the cytoplasm and do not form a single connected Golgi ribbon (Chappell and Warren 1989). Interestingly, studies in *Drosophila* have shown that Golgi organization can be cell type-specific (Stanley et al. 1997). In pre- and post-cellularized *Drosophila* embryos, as well as in S2 tissue culture cells, Golgi membranes appear as dispersed punctate structures that correspond to isolated Golgi mini-stacks (Kondylis and Rabouille 2009; Stanley et al. 1997). These mini-stacks localize close to the transitional ER, forming a unit for translation and transport that is also seen in *Toxoplasma*, *Plasmodium* and plants (Kondylis and Rabouille 2009). However, the *Drosophila* Golgi has also been observed as unstacked cisternae during distinct stages of development, or as a ribbon-like structure in spermatids, indicating that Golgi membranes can be organized differently within the same organism (Kondylis and Rabouille 2009).

In this chapter, we will focus on the spatial and functional relationship between the Golgi and the centrosome in mammalian cells. We will first discuss the mechanism by which Golgi membranes are positioned in the pericentrosomal region. We will then review studies on the functional connections between the Golgi and the centrosome in interphase, including the role of the pericentrosomal Golgi in directional protein secretion and cell polarization. Finally, we will discuss the link between the Golgi and the centrosome during mitosis, the stage of the cell cycle when the physical Golgi–centrosome connection is temporarily lost due to extensive Golgi fragmentation and dispersal.

7.2 Mechanisms of Golgi Positioning in Interphase Mammalian Cells

Although not essential for the typical function of the Golgi in protein and lipid transport, the positioning of the Golgi apparatus in mammalian cells next to the centrosome is actively maintained (Table 7.1). The preservation of this specific localization involves both the microtubule and actin cytoskeletal networks and their associated motor proteins and regulators (Brownhill et al. 2009). These cytoskeletal networks are linked to the Golgi by binding to either structural Golgi proteins, such as Hook3, or by directly associating with specialized phospholipids in the Golgi membranes, such as PtdIns(4,5)P_2 (Godi et al. 1998; Walenta et al. 2001). Both microtubule- and actin-associated motor proteins have been detected on the Golgi. Because microtubule motors actively contribute to the positioning of the Golgi apparatus, they will be discussed in detail. In contrast, Golgi-localized actin-associated motor proteins will not be covered because they function predominantly in the movement of vesicles to and from the Golgi, and do not seem to control Golgi localization.

Table 7.1 Factors that control the positioning of the Golgi ribbon next to the centrosome

Protein class Examples	Role in Golgi positioning	Golgi phenotype (when activity of the regulatory factor is altered)	References
Microtubules and associated proteins			
Centrosome-nucleated microtubules	Radial array with plus ends extending in all directions toward the cell periphery Critical for C-stage of Golgi reassembly	Dispersed Golgi mini-stacks	Cole et al. (1996)
Golgi-nucleated microtubules	Asymmetric array with plus ends extending toward the leading edge of the cell Critical for G-stage of Golgi reassembly	Golgi membrane clusters in periphery, no ribbon	Efimov et al. (2007), Miller et al. (2009)
Dynein	Responsible for directional movement of Golgi toward the centrosome (toward the minus ends of microtubules)	Dispersed Golgi mini-stacks	Corthesy-Theulaz et al. (1992), Harada et al. (1998)
Kinesin	Providing opposing force to dynein activity Encourages lateral spreading of Golgi membranes Minus-end directed kinesin activity has similar role as dynein	Collapsed Golgi in cell center Dispersed Golgi mini-stacks	Feiguin et al. (1994) Xu et al. (2002)

(continued)

Table 7.1 (continued)

Protein class Examples	Role in Golgi positioning	Golgi phenotype (when activity of the regulatory factor is altered)	References
Actin and associated proteins			
Actin	Similar role to kinesin in opposing dynein activity	Collapsed Golgi in cell center	di Campli et al. (1999)
	Creates tension by anchoring at the plasma membrane to encourage lateral spreading of Golgi membranes		
Spectrin and ankyrin	Creates flexible meshwork in close association with Golgi membranes	Inhibition in early anterograde transport	Godi et al. (1998)
	May create membrane microdomains necessary to maintain Golgi morphology		
Golgi-associated proteins			
Hook3	Links microtubules to Golgi	Fragmented Golgi	Walenta et al. (2001)
GMAP210	Not known	Dispersed Golgi mini-stacks	Yadav et al. (2009)
Golgin-160	Not known	Dispersed Golgi mini-stacks	Yadav et al. (2009)
Golgin-84	Not known	Dispersed Golgi mini-stacks	Diao et al. (2003)
Centrosome-associated proteins			
TBCCD1		Fragmented Golgi, mispositioned centrosome	Goncalves et al. (2010)

A typical mammalian cell contains two distinct populations of microtubules, which both originate in the pericentrosomal region. About half of cellular microtubules are nucleated and anchored at the centrosome, from which they extend radially toward the plasma membrane (Efimov et al. 2007). The other half are nucleated at the Golgi, from which they form an asymmetrical array, with their plus ends extending predominantly toward the leading edge of the cell (Chabin-Brion et al. 2001; Efimov et al. 2007; Vinogradova et al. 2009). Golgi-nucleated microtubules associate with CLASP proteins, which are recruited by the peripheral TGN protein GCC185 (Efimov et al. 2007). The binding of CLASP proteins to these microtubules increases their overall stability, which, together with their enhanced acetylation and tyrosination, explains why they are more difficult to depolymerize (Chabin-Brion et al. 2001; Efimov et al. 2007; Rivero et al. 2009; Thyberg and Moskalewski 1999).

Microtubules play a critical role in the organization and positioning of the Golgi apparatus (Thyberg and Moskalewski 1999). Their depolymerization by treatment with the compound nocodazole leads to the loss of the pericentrosomal Golgi ribbon, which is converted into mini-stacks at ER exit sites (Cole et al. 1996; Rogalski and Singer 1984). These mini-stacks are fully functional for protein processing and secretion, and are reminiscent of the ER-Golgi units of *Drosophila*. Upon nocodazole removal, microtubules repolymerize and promote the reassembly of the Golgi ribbon through active transport of Golgi mini-stacks toward the cell center. Interestingly, Golgi reassembly after treatment with nocodaozle occurs in two distinct steps that are each dependent on a different microtubule population (Miller et al. 2009). In the first step (Golgi- or G-phase), Golgi mini-stacks spread along microtubules and fuse into larger structures in the cell periphery. This step is dependent on Golgi-nucleated microtubules and does not occur in CLASP-depleted cells, in which this microtubule subset is absent. In the second step (centrosome- or C-phase), peripheral Golgi clusters are transferred from the cell periphery to their normal position next to the centrosome. This translocation of Golgi membranes to the cell center requires centrosome-nucleated microtubules and still takes place in CLASP-depleted cells, in which the fusion of mini-stacks in the cell periphery is prevented (Miller et al. 2009). In these cells, Golgi mini-stacks are positioned normally adjacent to the centrosome, but they are unable to form an interconnected ribbon. Thus, Golgi-nucleated microtubules are responsible for the integrity and morphology of the Golgi ribbon, whereas centrosome-nucleated microtubules determine the localization of Golgi membranes next to the centrosome (Miller et al. 2009).

Microtubules serve as tracks for the movement of Golgi membranes, but the actual locomotive force for the G- and C-stages of Golgi assembly is provided by the microtubule motor proteins dynein and kinesin (Miller et al. 2009). Microtubule motors are mechano-chemical enzymes that transport cargo along microtubule tracks, with dyneins moving toward microtubule minus ends, and kinesins generally moving toward plus ends. Dyneins associate with the Golgi apparatus and are important for Golgi organization and positioning (Allan et al. 2002; Thyberg and Moskalewski 1999). Disrupting their function by depleting dynein or ATP from the cytosol blocked the directional movement of the Golgi toward the centrosome, producing a Golgi fragmentation phenotype (Corthesy-Theulaz et al.

1992). In addition, knock-out mice for cytoplasmic dynein 1 (CD1), a TGN-associated dynein (Fath et al. 1994), developed into the blastocyst stage before embryonic death and reabsorption. Cells recovered from this CD1 knockout blastocyst had fragmented Golgi membranes in the form of peripheral mini-stacks (Harada et al. 1998). Another dynein family member, cytoplasmic dynein 2 (CD2), also localized to the Golgi apparatus, although its distribution was more ubiquitous than that of CD1 (Vaisberg et al. 1996). Blocking CD2 function by microinjection of a specific monoclonal antibody resulted in the conversion of the Golgi ribbon into peripheral mini-stacks. Similarly, interfering with the function of the minus-end directed kinesin family member KIFC3 caused Golgi fragmentation (Xu et al. 2002). These experiments provide strong support for a role of minus-end directed microtubule motors in Golgi organization and position.

The motor protein kinesin, which, in general, moves toward microtubule plus-ends, are also important for the maintenance of Golgi organization. Indeed, kinesins have been proposed to provide an opposing force to dyneins. Similar to what has been observed for dyneins, various kinesin family members localize to the Golgi (Allan et al. 2002; Gyoeva et al. 2000; Lippincott-Schwartz et al. 1995; Thyberg and Moskalewski 1999). Their RNAi-mediated knockdown resulted in the collapse of the Golgi into a circular body in the cell center, with reduction in the overall size, number, and spreading of Golgi cisternae (Feiguin et al. 1994). Thus, a balance between dynein and kinesin activity may be the critical determinant for the pericentrosomal organization of the Golgi ribbon.

In addition, there are reports implicating the actin cytoskeleton in the maintenance of a pericentrosomal Golgi apparatus. Disruption of actin filaments, specifically of branched actin structures, resulted in a collapse of Golgi morphology (di Campli et al. 1999). Under these conditions, Golgi cisternae remained in a stacked conformation, but they were swollen and condensed around the nucleus (Valderrama et al. 1998). This phenotype was reminiscent of Golgi morphology in cells with disrupted kinesin activity, suggesting that actin filaments, just like kinesins, may be antagonistic to dynein activity. It is possible that actin filaments oppose dynein activity by attaching to the plasma membrane, generating tension and encouraging lateral spreading of Golgi membranes.

Actin can also control Golgi organization through its interacting proteins spectrin and ankyrin, which have both been detected on the Golgi (Beck 2005; Beck et al. 1994, 1997; Devarajan et al. 1996; Fath et al. 1994; Stankewich et al. 1998). Spectrin is a cytoskeletal protein that is known to control membrane organization, stability, and shape by linking membranes to motor proteins or other cytoskeletal elements. Spectrin binds to membranes via the adaptor protein ankyrin, which itself associates with integral membrane proteins or membrane phospholipids. Membrane-bound spectrin-ankyrin complexes are cross-linked by short actin filaments, creating a flexible meshwork across the surface of a membrane (De Matteis and Morrow 2000; Godi et al. 1998). At the plasma membrane, the spectrin network has been found to promote the formation of specialized membrane domains by preventing the free diffusion of integral membrane proteins (Holleran and Holzbaur 1998). Spectrin and ankyrin may have an analogous role on the Golgi and maintain the integrity of this organelle by restricting access of Golgi resident proteins to budding transport vesicles (Holleran and Holzbaur, 1998). Furthermore, Golgi-associated spectrin may serve as a scaffold for the recruitment of signaling proteins from the cytosol. The actin-related protein Arp1 is a good example for such a spectrin-binding protein (Holleran et al. 2001). Spectrin-dependent recruitment of Arp1, a central component of the dynactin complex, to the Golgi may allow this protein to control dynein activity and contribute to the regulation of Golgi positioning.

Golgi and centrosome-associated proteins also control the pericentrosomal positioning of the Golgi. Loss of either Golgin-160, GMAP210 or Golgin-84, all putative structural Golgi proteins, resulted in the dispersal of Golgi membranes into ER-associated mini-stacks, producing a phenotype that is reminiscent of nocodazole-treated cells (Diao et al. 2003; Yadav et al. 2009). Dispersal of Golgi membranes was also observed in cells depleted of the centrosomal protein TBCCD1 (Goncalves et al. 2010). Interestingly, in these cells, the centrosome was mislocalized away from the nucleus. In conclusion, diverse groups of proteins are involved in, and required for, the proper positioning of the Golgi apparatus next to the centrosome (Table 7.1), suggesting that sustaining this specific localization may be important for cell homeostasis.

7.3 A Role for the Pericentrosomal Golgi Apparatus in Cell Polarity

Cell polarity is a feature of eukaryotic cells that allows them to carry out their specialized functions. For example, neurons depend on their polarization to transmit electrical signals from one cell to the next, whereas epithelial cells use their polarized organization to protect the body from its environment. The Golgi apparatus and the centrosome both have independent roles in the establishment of cell polarity. Golgi membranes control the sorting of proteins, which is important for their delivery to the leading edge of a cell (Bergmann et al. 1983). This so-called directional transport is a prerequisite for cell polarization, and was recently shown to involve Golgi-nucleated microtubules (Bergmann et al. 1983; Rivero et al. 2009). When Rivero and colleagues selectively disrupted this subset of microtubules, there were defects in cell polarization and migration (Rivero et al. 2009). Golgi membranes also recruit the Ste20-like kinase YSK1, a protein kinase required for Golgi organization and cell polarization (Preisinger et al. 2004). A pathway that involves the interaction between STK25, the mouse homolog of YSK1, and the Golgi protein GM130 is also required for Golgi organization and polarization in cultured neurons and in vivo (Matsuki et al. 2010).

Like the Golgi, the centrosome has long been anticipated to play an important role in cell polarization. In many migrating cells, including fibroblasts and macrophages, the centrosome localizes between the nucleus and the leading edge (Kupfer et al. 1982; Nemere et al. 1985), suggesting that the position of the centrosome may determine the direction of cell polarization. This idea is further supported by a recent study in Ptk2 cells, in which laser ablation of the centrosome caused a block in cell migration and cell polarization (Wakida et al. 2010). However, a leading role of the centrosome does not appear to be universal. In a study comparing centrosome position in migrating CHO and Ptk cells, the centrosome was localized toward the front of the nucleus in CHO, but not in Ptk cells (Yvon et al. 2002). In addition, there are conflicting results on the role of the centrosome in migrating neurons. Several studies have attributed a leading role to the centrosome in directing migration (Higginbotham and Gleeson 2007; Tsai and Gleeson 2005), but recent results from Zebrafish neurons showed that there was no correlation between centrosome positioning and cell migration (Distel et al. 2010). Indeed, it was found that the centrosome of migrating THN neurons often trailed the nucleus. Thus, additional experiments are needed to determine the exact role of the centrosome itself in cell polarization.

Interestingly, several studies support an additional role for the Golgi-centrosome relationship in the establishment of cell polarity. First, Bisel and colleagues reported that the Golgi and the centrosome move together toward the leading edge of migrating cells (Bisel et al. 2008). This coordinated movement of both organelles required the reorganization of Golgi membranes, which was mediated by ERK1-dependent phosphorylation of the peripheral Golgi protein GRASP65. In a second study, a role for the pericentrosomal Golgi ribbon in cell polarization and migration was identified (Yadav et al. 2009). By depleting the structural Golgi proteins Golgin-160 or GMAP210, Yadav and colleagues found that the Golgi ribbon was converted into dispersed mini-stacks. In these cells, normal protein transport to the cell surface occurred, but there was a specific block in directional transport toward the leading edge. As a consequence, cells did not polarize and migrate, indicating that a pericentrosomal Golgi ribbon is important for cell polarization. A third study asked directly whether it is Golgi organization or Golgi position that is important for cell polarization (Hurtado et al. 2011). Hurtado and colleagues expressed a specific domain of the Golgi scaffolding protein AKAP450, which resulted in the separation of a functional, interconnected Golgi ribbon from the centrosome. Intriguingly, these cells were unable to migrate in a wound-healing assay, indicating that the Golgi–centrosome proximity is necessary for directional protein transport and cell polarization. This study demonstrates for the first time that the physical proximity between the Golgi and the centrosome in interphase mammalian cells is important for cell polarization and is therefore of great functional significance.

7.4 Other Functional Interactions Between the Golgi and the Centrosome in Interphase

In addition to this emerging role for the Golgi–centrosome interaction in cell polarization, there are at least two other molecular associations between these two organelles during interphase. First, there are cellular functions that are performed by the Golgi and the centrosome. The most prominent example is the nucleation of microtubules that we discussed previously. Interestingly, microtubule nucleation at the Golgi and the centrosome depends on the same protein, the large scaffolding protein AKAP450, which associates with both organelles and recruits γ-tubulin as the microtubule nucleation factor (Rivero et al. 2009). Similarly, the γ-tubulin binding protein Cdk5RAP2 has been detected on both organelles and may have a complementary role in microtubule nucleation (Wang et al. 2010). There are several other proteins, including myomegalin, Golgin-97, FTCD (58 K), and CAP350 that associate with both organelles (Hagiwara et al. 2006; Hoppeler-Lebel et al. 2007; Takatsuki et al. 2002; Verde et al. 2001), but the functional significance of their specific localizations is not known.

Another example for a common function of the Golgi and the centrosome is their role as signaling platforms. For instance, several proteins controlling cell cycle progression have been detected at the centrosome. These include the kinase complex Cdk1-Cyclin B, which is required for entry into mitosis (Jackman et al. 2003) and centriolin, which controls cytokinesis and entry into S-phase (Gromley et al. 2003). In addition, signaling molecules, such as the protein kinases PKA and NRD, proteasomal subunits, and cytoskeletal elements, such as actin and Arp2/3, associate with the centrosome (Diviani et al. 2000; Hubert et al. 2011; Wigley et al. 1999). Similarly, Golgi membranes host a large number of signaling molecules. For instance, components of the ras and src pathways have been detected on the Golgi apparatus (Bard et al. 2002; Chiu et al. 2002; Wilson et al. 2011). In addition, cell cycle regulators, such as the mitotic cyclin B2 and the cytokinesis regulator Nir2, localize to the Golgi (Jackman et al. 1995; Litvak et al. 2004). It is clear that many of these Golgi- and centrosome-associated signaling molecules fulfill functions that are unrelated to the primary roles of the Golgi or the centrosome. Therefore, similar to the enrichment of membrane trafficking organelles in the pericentrosomal region, it may be beneficial for a cell to place signaling components in close proximity in its center.

A second association between the Golgi apparatus and the centrosome concerns the regulatory crosstalk between these two organelles. For instance, there is a signaling pathway by which the Golgi apparatus influences, and even controls, the proper organization and function of the centrosome (Kodani and Sütterlin 2008). This pathway involves the *cis* Golgi protein GM130, which forms a complex with the small GTPase Cdc42 and its specific guanine nucleotide exchange factor Tuba at the Golgi (Kodani et al. 2009; Kodani and Sütterlin 2008). Interfering with GM130, Tuba or Cdc42 causes the formation of a disorganized and non-functional centrosome, suggesting that each of these three proteins is required for the maintenance of normal centrosome morphology. It is not known at this point whether additional signaling molecules are involved in this pathway, and how the signal is transduced from one organelle to the other.

The mechanism(s) that support the functional interactions between the Golgi and the centrosome are only beginning to be understood. As discussed above, cell polarity appears to depend on the physical Golgi–centrosome proximity. However, for other common functions, such as protein localization and signaling, the importance of proximity between these organelles has not been tested. It will be interesting to address these questions in the future by expressing the AKAP450 domain that successfully disrupts Golgi–centrosome vicinity while leaving Golgi organization and functionality intact (Hurtado et al. 2011).

7.5 Functional Interactions Between the Golgi and the Centrosome in Mitosis

At the onset of mitosis, the Golgi apparatus of mammalian cells undergoes extensive reorganization. During this process, Golgi membranes lose their association with the centrosome, and are fragmented and dispersed throughout the cytoplasm. This fragmentation process is initiated in G2 with the disconnection of the non-compact zones of the Golgi ribbon and the generation of isolated, pericentrosomal mini-stacks (Colanzi et al. 2007; Feinstein and Linstedt 2007). This step depends on the recruitment of the mitotic kinase Aurora A to the centrosome, which, when blocked, prevents Golgi fragmentation and entry into mitosis (Persico et al. 2010). Next, in prophase, the isolated Golgi mini-stacks are converted into tubular vesicular elements, called Golgi "blobs". This step is mediated by the MAP kinase pathway components Raf1, MEK1, and Erk1c (Acharya et al. 1998; Colanzi et al. 2003b; Shaul and Seger 2006). Finally, Golgi "blobs" are broken down into the so-called Golgi "haze" by a mechanism that involves the protein kinases Plk1 and Cdc2 (Colanzi et al. 2003a; Lowe et al. 1998; Wei and Seemann 2009; Sütterlin et al. 2001). Upon completion of mitosis, Golgi fragments reassemble into the ribbon through the G- and C-stage steps that we have discussed previously for experimentally induced Golgi fragmentation (Miller et al. 2009). During this entire multi-step Golgi disassembly process, Golgi membranes remain separate and distinct from the ER (Jokitalo et al. 2001; Pecot and Malhotra 2004).

A number of studies support the existence of a functional link between the Golgi and the centrosome during mitosis. For example, mitotic Golgi reorganization, during which the physical Golgi–centrosome connection is lost, was found to be necessary for the regulation of cell cycle progression. When Golgi fragmentation was prevented by microinjection of GRASP65-related reagents, cells arrested in G2 and did not enter mitosis (Sütterlin et al. 2002). Preisinger and colleagues obtained similar results when they overexpressed GRASP65 (Preisinger et al. 2005). The inhibitory effect of GRASP65 on mitotic entry was only seen with wild-type GRASP65, and not with a non-phosphorylatable mutant, indicating that excess GRASP65 may titrate out the activity of a kinase important for Golgi fragmentation. Similarly, inhibiting the membrane fission protein CtBP3/BARS prevented mitotic Golgi fragmentation and blocked cells from entering mitosis (Hidalgo Carcedo et al. 2004).

There are several possible explanations for the link between Golgi fragmentation and mitotic entry. First, the conversion of the Golgi ribbon into smaller fragments may facilitate the equal partitioning of this single-copy organelle into the two daughter cells. Second, fragmentation may promote the release of mitotic signaling components that are normally sequestered on Golgi membranes. For example, ACBD3, a critical regulator of numb signaling, is released from the Golgi during mitotic fragmentation to promote asymmetric cell division (Zhou et al. 2007). Third, an intact Golgi ribbon may cause steric hindrance during centrosome maturation and restrict centrosome movement necessary for mitotic spindle formation.

In addition, there is support for an association between mitotic Golgi membranes and the mitotic spindle. In a careful live imaging study, Shima and colleagues detected the enrichment of mitotic Golgi fragments at spindle poles, indicating that the mitotic spindle may facilitate the ordered inheritance of Golgi fragments into daughter cells (Shima et al. 1998). Spindle poles were also found to contain factors that are important for Golgi ribbon formation (Wei and Seemann 2009). In this study, cells were induced to divide asymmetrically, with both spindle poles segregating into only one of the two daughter cells. Under these conditions, the pericentrosomal Golgi ribbon reformed only in the spindle pole-containing daughter cell, and not in the daughter cell that lacked a spindle pole. This result supports the notion that Golgi ribbon determinants associate with spindle poles for their inheritance into the daughter cells; however, the nature of these ribbon determinants is not known. Finally, Golgi proteins appear to control the formation of the mitotic spindle. Three functionally diverse, Golgi-associated proteins have been identified as having a role in mitotic spindle formation. These proteins include the poly-ADP ribosylase Tankyrase (Chang et al. 2005), the peripheral Golgi protein GRASP65 (Sütterlin et al. 2005) and the phosphoinositide phosphatase Sac1 (Liu et al. 2008). RNAi-mediated depletion of each of these proteins resulted in multi-polar spindles and defects in cell cycle progression. Furthermore, GM130 was found to be required for meiotic spindle formation during mouse oocyte maturation (Zhang et al. 2011). However, the mechanisms by which these diverse proteins regulate spindle formation are not understood.

7.6 Conclusion and Perspective

While a unique physical association between the mammalian Golgi apparatus and the centrosome has been observed over many years, recent studies have found that these two organelles are also linked functionally (Table 7.2). Such interactions occur primarily during interphase, when the Golgi and the centrosome are in close vicinity. During this stage of the cell cycle, the Golgi–centrosome interaction is important for cellular processes, such as the establishment of cell polarity, the nucleation of microtubules, and the control of centrosome structure and function. Other functional Golgi–centrosome interactions are independent of organelle proximity and occur during mitosis, when the physical proximity is disrupted. These include the regulation of mitotic entry and post-mitotic reassembly of the Golgi ribbon.

It has been a challenging task to determine which characteristic of the pericentrosomal Golgi apparatus is important for the regulation of processes such as

Table 7.2 Reported functional associations between the Golgi and the centrosome in mammalian cells

Regulatory functions of Golgi-centrosome interactions	Dependent on organelle proximity?	References
Interphase: adjacent Golgi and centrosome		
Cell polarization	Yes	Hurtado et al. (2011), Yadav et al. (2009)
Signaling Platforms	Not known	Doxsey et al. (2005), Wilson et al. (2011)
Centrosome organization	Not known	Kodani et al. (2009), Kodani and Sütterlin (2008); Wilson et al. (2011)
Mitosis: separated Golgi and centrosome		
Mitotic entry	Proximity has to be disrupted	Hidalgo Carcedo et al. (2004), Preisinger et al. (2005), Sütterlin et al. (2002)
Golgi ribbon formation	Not known	Wei and Seemann (2009)
Spindle formation	Not known	Chang et al. (2005), Liu et al. (2008), Sütterlin et al. (2005), Zhang et al. (2011)

cell polarization. Either the ribbon structure, or the position, or both, could be contributing regulatory elements. To distinguish between these possibilities, one would want to separate the two organelles without losing their structural and functional integrity—i.e. to position an intact, interconnected Golgi ribbon away from the centrosome. One can think of at least three different ways to achieve this goal. First, a genetic screen could be performed, in which mutants that display a physical separation of the Golgi and the centrosome are selected. However, Golgi–centrosome proximity has predominantly been observed and studied in mammalian cells, and such a genetic approach would therefore be highly complicated. Second, small molecules could be screened for the specific phenotype of disrupting Golgi–centrosome proximity. Natural compounds, such as Brefeldin A, Norrisolide, and Ilimaquinone have been widely used to understand different aspects of Golgi regulation (Guizzunti et al. 2006; Lippincott-Schwartz et al. 1989; Takizawa et al. 1993). However, to date, there is no compound that separates the intact Golgi ribbon from the centrosome. We have recently identified a natural compound with a completely novel Golgi disrupting activity (Schnermann et al. 2010). This molecule, by the name of MacFarlandin E, converts the Golgi ribbon into small fragments, but does not disperse them. The identification of a compound with this remarkable and novel Golgi modifying activity gives hope that there may be a compound that can specifically mislocalize the Golgi ribbon without altering its overall organization and functionality. Third, interfering with Golgi-localized proteins by RNAi-mediated depletion or by overexpression of dominant negative forms could disrupt the Golgi–centrosome proximity. A recent study has successfully used this approach. By overexpressing a fragment of the Golgi-associated scaffolding protein AKAP450, Hurtado and colleagues managed to move the Golgi ribbon away from the centrosome. Under these conditions, cell polarity and cell migration was significantly reduced, demonstrating for the first time that the pericentrosomal position of the Golgi ribbon is critical for the regulation of cell polarity (Hurtado et al. 2011). This system can now be used to test the effects of a physical separation of the Golgi and the centrosome on cellular processes that are controlled by the Golgi–centrosome interaction.

It is important to extend these observations from mammalian cells to organisms that do not have a pericentrosomal Golgi apparatus, and that may therefore lack functional Golgi–centrosome interactions. Obviously, such organisms must have developed alternative mechanisms to control important processes such as polarization. A straightforward means to achieve directional protein transport in the absence of a pericentrosomal Golgi is to deliver mRNA to specific ER-Golgi subunits for localized protein translation, sorting, and transport. This system is utilized by yeast, in which mRNAs encoding for membrane proteins are transported along the acto-myosin network to the bud tip, prior to local translation and delivery to the plasma membrane (Takizawa et al. 2000). Similarly, in *Drosophila* embryos, the mRNA of the developmental protein Gurken is positioned so that its translation and modification is restricted to a subset of ER-Golgi subunits (Herpers and Rabouille 2004). Alternatively, it is possible that the specific pericentrosomal positioning of the Golgi apparatus is a reflection of increased evolutionary complexity in higher organisms, providing an additional level of regulation required for cellular processes specific to these organisms.

References

Acharya U, Mallabiabarrena A, Acharya JK, Malhotra V (1998) Signaling via mitogen-activated protein kinase kinase (MEK1) is required for Golgi fragmentation during mitosis. Cell 92:183–192

Allan VJ, Thompson HM, McNiven MA (2002) Motoring around the Golgi. Nat Cell Biol 4:E236–E242

Bard F, Patel U, Levy JB, Jurdic P, Horne WC, Baron R (2002) Molecular complexes that contain both c-Cbl and c-Src associate with Golgi membranes. Eur J Cell Biol 81:26–35

Beck KA (2005) Spectrins and the Golgi. Biochim Biophys Acta 1744:374–382

Beck KA, Buchanan JA, Malhotra V, Nelson WJ (1994) Golgi spectrin: identification of an erythroid beta-spectrin homolog associated with the Golgi complex. J Cell Biol 127:707–723

Beck KA, Buchanan JA, Nelson WJ (1997) Golgi membrane skeleton: identification, localization and oligomerization of a 195 kDa ankyrin isoform associated with the Golgi complex. J Cell Sci 110(Pt 10):1239–1249

Bergmann JE, Kupfer A, Singer SJ (1983) Membrane insertion at the leading edge of motile fibroblasts. Proc Natl Acad Sci U S A 80:1367–1371

Bisel B, Wang Y, Wei JH, Xiang Y, Tang D, Miron-Mendoza M, Yoshimura S, Nakamura N, Seemann J (2008) ERK regulates Golgi and centrosome orientation towards the leading edge through GRASP65. J Cell Biol 182:837–843

Brownhill K, Wood L, Allan V (2009) Molecular motors and the Golgi complex: staying put and moving through. Semin Cell Dev Biol 20:784–792

Chabin-Brion K, Marceiller J, Perez F, Settegrana C, Drechou A, Durand G, Pous C (2001) The Golgi complex is a microtubule-organizing organelle. Mol Biol Cell 12:2047–2060

Chang P, Coughlin M, Mitchison TJ (2005) Tankyrase-1 polymerization of poly(ADP-ribose) is required for spindle structure and function. Nat Cell Biol 7:1133–1139

Chappell TG, Warren G (1989) A galactosyltransferase from the fission yeast Schizosaccharomyces pombe. J Cell Biol 109:2693–2702

Chiu VK, Bivona T, Hach A, Sajous JB, Silletti J, Wiener H, Johnson RL 2nd, Cox AD, Philips MR (2002) Ras signalling on the endoplasmic reticulum and the Golgi. Nat Cell Biol 4:343–350

Colanzi A, Suetterlin C, Malhotra V (2003a) Cell-cycle-specific Golgi fragmentation: how and why? Curr Opin Cell Biol 15:462–467

Colanzi A, Sütterlin C, Malhotra V (2003b) RAF1-activated MEK1 is found on the Golgi apparatus in late prophase and is required for Golgi complex fragmentation in mitosis. J Cell Biol 161:27–32

Colanzi A, Hidalgo Carcedo C, Persico A, Cericola C, Turacchio G, Bonazzi M, Luini A, Corda D (2007) The Golgi mitotic checkpoint is controlled by BARS-dependent fission of the Golgi ribbon into separate stacks in G2. EMBO J 26:2465–2476

Cole NB, Sciaky N, Marotta A, Song J, Lippincott-Schwartz J (1996) Golgi dispersal during microtubule disruption: regeneration of Golgi stacks at peripheral endoplasmic reticulum exit sites. Mol Biol Cell 7:631–650

Corthesy-Theulaz I, Pauloin A, Pfeffer SR (1992) Cytoplasmic dynein participates in the centrosomal localization of the Golgi complex. J Cell Biol 118:1333–1345

De Matteis MA, Luini A (2008) Exiting the Golgi complex. Nat Rev Mol Cell Biol 9:273–284

De Matteis MA, Morrow JS (2000) Spectrin tethers and mesh in the biosynthetic pathway. J Cell Sci 113(Pt 13):2331–2343

Devarajan P, Stabach PR, Mann AS, Ardito T, Kashgarian M, Morrow JS (1996) Identification of a small cytoplasmic ankyrin (AnkG119) in the kidney and muscle that binds beta I sigma spectrin and associates with the Golgi apparatus. J Cell Biol 133:819–830

di Campli A, Valderrama F, Babia T, De Matteis MA, Luini A, Egea G (1999) Morphological changes in the Golgi complex correlate with actin cytoskeleton rearrangements. Cell Motil Cytoskelet 43:334–348

Diao A, Rahman D, Pappin DJ, Lucocq J, Lowe M (2003) The coiled-coil membrane protein golgin-84 is a novel rab effector required for Golgi ribbon formation. J Cell Biol 160:201–212

Distel M, Hocking JC, Volkmann K, Koster RW (2010) The centrosome neither persistently leads migration nor determines the site of axonogenesis in migrating neurons in vivo. J Cell Biol 191:875–890

Diviani D, Langeberg LK, Doxsey SJ, Scott JD (2000) Pericentrin anchors protein kinase A at the centrosome through a newly identified RII-binding domain. Curr Biol: CB 10:417–420

Efimov A, Kharitonov A, Efimova N, Loncarek J, Miller PM, Andreyeva N, Gleeson P, Galjart N, Maia AR, McLeod IX, Yates JR 3rd, Maiato H, Khodjakov A, Akhmanova A, Kaverina I (2007) Asymmetric CLASP-dependent nucleation of noncentrosomal microtubules at the trans-Golgi network. Dev Cell 12:917–930

Fath KR, Trimbur GM, Burgess DR (1994) Molecular motors are differentially distributed on Golgi membranes from polarized epithelial cells. J Cell Biol 126:661–675

Feiguin F, Ferreira A, Kosik KS, Caceres A (1994) Kinesin-mediated organelle translocation revealed by specific cellular manipulations. J Cell Biol 127:1021–1039

Feinstein TN, Linstedt AD (2007) Mitogen-activated protein kinase kinase 1-dependent Golgi unlinking occurs in G2 phase and promotes the G2/M cell cycle transition. Mol Biol Cell 18:594–604

Ghosh RN, Mallet WG, Soe TT, McGraw TE, Maxfield FR (1998) An endocytosed TGN38 chimeric protein is delivered to the TGN after trafficking through the endocytic recycling compartment in CHO cells. J Cell Biol 142:923–936

Godi A, Santone I, Pertile P, Devarajan P, Stabach PR, Morrow JS, Di Tullio G, Polishchuk R, Petrucci TC, Luini A, De Matteis MA (1998) ADP ribosylation factor regulates spectrin binding to the Golgi complex. Proc Natl Acad Sci U S A 95:8607–8612

Goncalves J, Nolasco S, Nascimento R, Lopez Fanarraga M, Zabala JC, Soares H (2010) TBCCD1, a new centrosomal protein, is required for centrosome and Golgi apparatus positioning. EMBO Rep 11:194–200

Gromley A, Jurczyk A, Sillibourne J, Halilovic E, Mogensen M, Groisman I, Blomberg M, Doxsey S (2003) A novel human protein of the maternal centriole is required for the final stages of cytokinesis and entry into S phase. J Cell Biol 161:535–545

Guizzunti G, Brady TP, Malhotra V, Theodorakis EA (2006) Chemical analysis of norrisolide-induced Golgi vesiculation. J Am Chem Soc 128:4190–4191

Gyoeva FK, Bybikova EM, Minin AA (2000) An isoform of kinesin light chain specific for the Golgi complex. J Cell Sci 113(Pt 11):2047–2054

Hagiwara H, Tajika Y, Matsuzaki T, Suzuki T, Aoki T, Takata K (2006) Localization of Golgi 58 K protein (formiminotransferase cyclodeaminase) to the centrosome. Histochem Cell Biol 126:251–259

Harada A, Takei Y, Kanai Y, Tanaka Y, Nonaka S, Hirokawa N (1998) Golgi vesiculation and lysosome dispersion in cells lacking cytoplasmic dynein. J Cell Biol 141:51–59

Herpers B, Rabouille C (2004) mRNA localization and ER-based protein sorting mechanisms dictate the use of transitional endoplasmic reticulum-golgi units involved in gurken transport in Drosophila oocytes. Mol Biol Cell 15:5306–5317

Hidalgo Carcedo C, Bonazzi M, Spano S, Turacchio G, Colanzi A, Luini A, Corda D (2004) Mitotic Golgi partitioning is driven by the membrane-fissioning protein CtBP3/BARS. Science 305:93–96

Higginbotham HR, Gleeson JG (2007) The centrosome in neuronal development. Trends Neurosci 30:276–283

Holleran EA, Holzbaur EL (1998) Speculating about spectrin: new insights into the Golgi-associated cytoskeleton. Trends Cell Biol 8:26–29

Holleran EA, Ligon LA, Tokito M, Stankewich MC, Morrow JS, Holzbaur EL (2001) Beta III spectrin binds to the Arp1 subunit of dynactin. J Biol Chem 276:36598–36605

Hoppeler-Lebel A, Celati C, Bellett G, Mogensen MM, Klein-Hitpass L, Bornens M, Tassin AM (2007) Centrosomal CAP350 protein stabilises microtubules associated with the Golgi complex. J Cell Sci 120:3299–3308

Hubert T, Vandekerckhove J, Gettemans J (2011) Actin and Arp2/3 localize at the centrosome of interphase cells. Biochem Biophys Res Commun 404:153–158

Hurtado L, Caballero C, Gavilan MP, Cardenas J, Bornens M, Rios RM (2011) Disconnecting the Golgi ribbon from the centrosome prevents directional cell migration and ciliogenesis. J Cell Biol 193:917–933

Jackman M, Firth M, Pines J (1995) Human cyclins B1 and B2 are localized to strikingly different structures: B1 to microtubules, B2 primarily to the Golgi apparatus. EMBO J 14:1646–1654

Jackman M, Lindon C, Nigg EA, Pines J (2003) Active cyclin B1-Cdk1 first appears on centrosomes in prophase. Nat Cell Biol 5:143–148

Jokitalo E, Cabrera-Poch N, Warren G, Shima DT (2001) Golgi clusters and vesicles mediate mitotic inheritance independently of the endoplasmic reticulum. J Cell Biol 154:317–330

Kodani A, Sütterlin C (2008) The Golgi protein GM130 regulates centrosome morphology and function. Mol Biol Cell 19:745–753

Kodani A, Kristensen I, Huang L, Sütterlin C (2009) GM130-dependent control of Cdc42 activity at the Golgi regulates centrosome organization. Mol Biol Cell 20:1192–1200

Kondylis V, Rabouille C (2009) The Golgi apparatus: lessons from Drosophila. FEBS Lett 583:3827–3838

Kupfer A, Louvard D, Singer SJ (1982) Polarization of the Golgi apparatus and the microtubule-organizing center in cultured fibroblasts at the edge of an experimental wound. Proc Nat Acad Sci U S A 79:2603–2607

Lin SX, Gundersen GG, Maxfield FR (2002) Export from pericentriolar endocytic recycling compartment to cell surface depends on stable, detyrosinated (glu) microtubules and kinesin. Mol Biol Cell 13:96–109

Lippincott-Schwartz J, Yuan LC, Bonifacino JS, Klausner RD (1989) Rapid redistribution of Golgi proteins into the ER in cells treated with brefeldin A: evidence for membrane cycling from Golgi to ER. Cell 56:801–813

Lippincott-Schwartz J, Cole NB, Marotta A, Conrad PA, Bloom GS (1995) Kinesin is the motor for microtubule-mediated Golgi-to-ER membrane traffic. J Cell Biol 128:293–306

Litvak V, Argov R, Dahan N, Ramachandran S, Amarilio R, Shainskaya A, Lev S (2004) Mitotic phosphorylation of the peripheral Golgi protein Nir2 by Cdk1 provides a docking mechanism for Plk1 and affects cytokinesis completion. Mol Cell 14:319–330

Liu Y, Boukhelifa M, Tribble E, Morin-Kensicki E, Uetrecht A, Bear JE, Bankaitis VA (2008) The Sac1 phosphoinositide phosphatase regulates Golgi membrane morphology and mitotic spindle organization in mammals. Mol Biol Cell 19:3080–3096

Lowe M, Rabouille C, Nakamura N, Watson R, Jackman M, Jamsa E, Rahman D, Pappin DJ, Warren G (1998) Cdc2 kinase directly phosphorylates the cis-Golgi matrix protein GM130 and is required for Golgi fragmentation in mitosis. Cell 94:783–793

Mallard F, Antony C, Tenza D, Salamero J, Goud B, Johannes L (1998) Direct pathway from early/recycling endosomes to the Golgi apparatus revealed through the study of shiga toxin B-fragment transport. J Cell Biol 143:973–990

Marie M, Dale HA, Sannerud R, Saraste J (2009) The function of the intermediate compartment in pre-Golgi trafficking involves its stable connection with the centrosome. Mol Biol Cell 20:4458–4470

Matsuki T, Matthews RT, Cooper JA, van der Brug MP, Cookson MR, Hardy JA, Olson EC, Howell BW (2010) Reelin and stk25 have opposing roles in neuronal polarization and dendritic Golgi deployment. Cell 143:826–836

Miller PM, Folkmann AW, Maia AR, Efimova N, Efimov A, Kaverina I (2009) Golgi-derived CLASP-dependent microtubules control Golgi organization and polarized trafficking in motile cells. Nat Cell Biol 11:1069–1080

Nemere I, Kupfer A, Singer SJ (1985) Reorientation of the Golgi apparatus and the microtubule-organizing center inside macrophages subjected to a chemotactic gradient. Cell Motil 5:17–29

Pecot MY, Malhotra V (2004) Golgi membranes remain segregated from the endoplasmic reticulum during mitosis in mammalian cells. Cell 116:99–107

Persico A, Cervigni RI, Barretta ML, Corda D, Colanzi A (2010) Golgi partitioning controls mitotic entry through Aurora-A kinase. Mol Biol Cell 21:3708–3721

Preisinger C, Short B, De Corte V, Bruyneel E, Haas A, Kopajtich R, Gettemans J, Barr FA (2004). YSK1 is activated by the Golgi matrix protein GM130 and plays a role in cell migration through its substrate 14-3-3zeta. J Cell Biol 164:1009–1020

Preisinger C, Korner R, Wind M, Lehmann WD, Kopajtich R, Barr FA (2005) Plk1 docking to GRASP65 phosphorylated by Cdk1 suggests a mechanism for Golgi checkpoint signalling. EMBO J 24:753–765

Preuss D, Mulholland J, Franzusoff A, Segev N, Botstein D (1992) Characterization of the Saccharomyces Golgi complex through the cell cycle by immunoelectron microscopy. Mol Biol Cell 3:789–803

Rambourg A, Jackson CL, Clermont Y (2001) Three dimensional configuration of the secretory pathway and segregation of secretion granules in the yeast Saccharomyces cerevisiae. J Cell Sci 114:2231–2239

Rehberg M, Kleylein-Sohn J, Faix J, Ho TH, Schulz I, Graf R (2005) Dictyostelium LIS1 is a centrosomal protein required for microtubule/cell cortex interactions, nucleus/centrosome linkage, and actin dynamics. Mol Biol Cell 16:2759–2771

Reilein AR, Serpinskaya AS, Karcher RL, Dujardin DL, Vallee RB, Gelfand VI (2003) Differential regulation of dynein-driven melanosome movement. Biochem Biophys Res Commun 309:652–658

Rivero S, Cardenas J, Bornens M, Rios RM (2009) Microtubule nucleation at the cis-side of the Golgi apparatus requires AKAP450 and GM130. EMBO J 28:1016–1028

Rogalski AA, Singer SJ (1984) Associations of elements of the Golgi apparatus with microtubules. J Cell Biol 99:1092–1100

Schnermann MJ, Beaudry CM, Egorova AV, Polishchuk RS, Sütterlin C, Overman LE (2010) Golgi-modifying properties of macfarlandin E and the synthesis and evaluation of its 2,7-dioxabicyclo[3.2.1]octan-3-one core. Proc Natl Acad Sci U S A 107:6158–6163

Shaul YD, Seger R (2006) ERK1c regulates Golgi fragmentation during mitosis. J Cell Biol 172:885–897

Shima DT, Cabrera-Poch N, Pepperkok R, Warren G (1998) An ordered inheritance strategy for the Golgi apparatus: visualization of mitotic disassembly reveals a role for the mitotic spindle. J Cell Biol 141:955–966

Stankewich MC, Tse WT, Peters LL, Ch'ng Y, John KM, Stabach PR, Devarajan P, Morrow JS, Lux SE (1998) A widely expressed betaIII spectrin associated with Golgi and cytoplasmic vesicles. Proc Natl Acad Sci U S A 95:14158–14163

Stanley H, Botas J, Malhotra V (1997) The mechanism of Golgi segregation during mitosis is cell type-specific. Proc Nat Acad Sci U S A 94:14467–14470

Sütterlin C, Lin CY, Feng Y, Ferris DK, Erikson RL, Malhotra V (2001) Polo-like kinase is required for the fragmentation of pericentriolar Golgi stacks during mitosis. Proc Natl Acad Sci U S A 98:9128–9132

Sütterlin C, Hsu P, Mallabiabarrena A, Malhotra V (2002) Fragmentation and dispersal of the pericentriolar Golgi complex is required for entry into mitosis in mammalian cells. Cell 109:359–369

Sütterlin C, Polishchuk R, Pecot M, Malhotra V (2005) The Golgi-associated protein GRASP65 regulates spindle dynamics and is essential for cell division. Mol Biol Cell 16:3211–3222

Takatsuki A, Nakamura M, Kono Y (2002) Possible implication of Golgi-nucleating function for the centrosome. Biochem Biophys Res Commun 291:494–500

Takizawa PA, Yucel JK, Veit B, Faulkner DJ, Deerinck T, Soto G, Ellisman M, Malhotra V (1993) Complete vesiculation of Golgi membranes and inhibition of protein transport by a novel sea sponge metabolite, ilimaquinone. Cell 73:1079–1090

Takizawa PA, DeRisi JL, Wilhelm JE, Vale RD (2000) Plasma membrane compartmentalization in yeast by messenger RNA transport and a septin diffusion barrier. Science 290:341–344

Thyberg J, Moskalewski S (1999) Role of microtubules in the organization of the Golgi complex. Exp Cell Res 246:263–279

Tsai LH, Gleeson JG (2005) Nucleokinesis in neuronal migration. Neuron 46:383–388

Vaisberg EA, Grissom PM, McIntosh JR (1996) Mammalian cells express three distinct dynein heavy chains that are localized to different cytoplasmic organelles. J Cell Biol 133:831–842

Valderrama F, Babia T, Ayala I, Kok JW, Renau-Piqueras J, Egea G (1998) Actin microfilaments are essential for the cytological positioning and morphology of the Golgi complex. Eur J Cell Biol 76:9–17

Verde I, Pahlke G, Salanova M, Zhang G, Wang S, Coletti D, Onuffer J, Jin SL, Conti M (2001) Myomegalin is a novel protein of the golgi/centrosome that interacts with a cyclic nucleotide phosphodiesterase. J Biol Chem 276:11189–11198

Vinogradova T, Miller PM, Kaverina I (2009) Microtubule network asymmetry in motile cells: role of Golgi-derived array. Cell Cycle 8:2168–2174

Wakida NM, Botvinick EL, Lin J, Berns MW (2010) An intact centrosome is required for the maintenance of polarization during directional cell migration. PLoS ONE 5:e15462

Walenta JH, Didier AJ, Liu X, Kramer H (2001) The Golgi-associated hook3 protein is a member of a novel family of microtubule-binding proteins. J Cell Biol 152:923–934

Wang Z, Wu T, Shi L, Zhang L, Zheng W, Qu JY, Niu R, Qi RZ (2010) Conserved motif of CDK5RAP2 mediates its localization to centrosomes and the Golgi complex. J Biol Chem 285:22658–22665

Wei JH, Seemann J (2009) The mitotic spindle mediates inheritance of the Golgi ribbon structure. J Cell Biol 184:391–397

Wigley WC, Fabunmi RP, Lee MG, Marino CR, Muallem S, DeMartino GN, Thomas PJ (1999) Dynamic association of proteasomal machinery with the centrosome. J Cell Biol 145:481–490

Wilson C, Venditti R, Rega LR, Colanzi A, D'Angelo G, De Matteis MA (2011) The Golgi apparatus: an organelle with multiple complex functions. Biochem J 433:1–9

Xu Y, Takeda S, Nakata T, Noda Y, Tanaka Y, Hirokawa N (2002) Role of KIFC3 motor protein in Golgi positioning and integration. J Cell Biol 158:293–303

Yadav S, Puri S, Linstedt AD (2009) A primary role for Golgi positioning in directed secretion, cell polarity, and wound healing. Mol Biol Cell 20:1728–1736

Yvon AM, Walker JW, Danowski B, Fagerstrom C, Khodjakov A, Wadsworth P (2002) Centrosome reorientation in wound-edge cells is cell type specific. Mol Biol Cell 13:1871–1880

Zhang CH, Wang ZB, Quan S, Huang X, Tong JS, Ma JY, Guo L, Wei YC, Ouyang YC, Hou Y, Xing FQ, Sun QY (2011) GM130, a cis-Golgi protein, regulates meiotic spindle assembly and asymmetric division in mouse oocyte. Cell Cycle 10:1861–1870

Zhou Y, Atkins JB, Rompani SB, Bancescu DL, Petersen PH, Tang H, Zou K, Stewart SB, Zhong W (2007) The mammalian Golgi regulates numb signaling in asymmetric cell division by releasing ACBD3 during mitosis. Cell 129:163–178

Chapter 8
Many Pathways to Destruction: The Role of the Centrosome in, and Its Control by Regulated Proteolysis

Harold A. Fisk

Abstract Centrosome duplication must be precisely regulated to ensure the production of a bipolar mitotic spindle. As with other cell cycle events, irreversible protein destruction is critical for the fidelity of centrosome duplication, and the failure to properly destroy any of several critical centrosome regulators leads to the production of excess centrosomes that interfere with bipolar spindle assembly. Many pathways that regulate the degradation of these critical regulators are found at centrosomes, and in some cases the destruction occurs at the centrosome. This chapter discusses the various degradation machineries found at centrosomes, and some of the many aspects of centrosome biology that are controlled by protein degradation.

8.1 Introduction

Regulated protein destruction is a key event in many cell cycle transitions, in part because of its irreversibility. As well-known examples, the transition from metaphase to anaphase is controlled by the proteasome-dependent destruction of securin, the molecule responsible for preventing premature activation of the cysteine protease separase that destroys sister chromatid cohesion, and the exit from mitosis is triggered by the proteasome-dependent destruction of cyclin B and other proteins. When all goes well, regulated protein destruction contributes to a coordinated cell cycle that culminates in the assembly of a bipolar mitotic spindle that partitions a

H. A. Fisk (✉)
Department of Molecular Genetics, The Ohio State University,
484 West 12th Avenue, Columbus, OH 43210, USA
e-mail: fisk.13@osu.edu

single copy of the duplicated genome and one centrosome into each of two daughter cells. Like the genome the single centrosome must be duplicated, and to ensure formation of a bipolar mitotic spindle must not be duplicated again until the following cell cycle. Just as regulated proteolysis prevents the re-initiation of DNA replication, failing to properly destroy critical centrosome regulators leads to the production of excess centrosomes. Many aspects of centrosome duplication and function are controlled by protein degradation, and multiple protein degradation pathways are found at centrosomes, and in some cases the relevant destruction event occurs explicitly at centrosomes. Here, we discuss the various degradation machineries found at centrosomes, and the many ways in which centrosome biology is controlled by protein degradation.

8.2 The Centrosome

The centrosome is best known as a microtubule organizing center (MTOC). Centrosomes consist of a pair of centrioles, which are cylinders composed of nine radially symmetric triplet microtubules [reviewed in (Adams and Kilmartin 2000; Tsou and Stearns 2006)]. The central centriole pair is surrounded by a pericentriolar matrix (PCM) that is responsible for microtubule nucleation. As MTOCs, centrosomes function as spindle poles during mitosis, and in order to ensure faithful chromosome segregation centrosome number must be strictly controlled to ensure the assembly of a bipolar mitotic spindle (Fisk et al. 2002). Centrosome number is controlled through the tightly regulated process of centrosome duplication wherein a single new centriole is assembled at a site adjacent to each existing centriole [reviewed in (Azimzadeh and Marshall 2010; Pike and Fisk 2011)]. Centrosome duplication has many analogies with DNA replication, both of which are semi-conservative processes initiated by Cdk2. A vast amount of work in the past handful of years has identified a large number of centriolar and centrosomal proteins whose activities are coordinated to ensure that only two new centrioles are assembled in any given cell cycle [reviewed in (Pike and Fisk 2011)]. A proteomic analysis of the human centrosome identified hundreds of proteins (Andersen et al. 2003), while comparative genomics identified a set of genes likely to encode proteins that specifically function at basal bodies, cilia, and flagella (Li et al. 2004), and a combination of the two approaches using green algae identified a core set of centriolar proteins (Keller et al. 2005). Genome wide RNAi studies in *C. elegans* identified a core set of genes required for the assembly of new centrioles (Delattre et al. 2006; Pelletier et al. 2006), and live cell imaging in embryos led to the assignment of these proteins into an ordered centriole assembly pathway that is conserved in organisms as distinct as ciliated protazoa and humans (Delattre et al. 2006; Pelletier et al. 2006; Dammermann et al. 2004; Leidel and Gonczy 2005; O'Connell 2002; Kemp et al. 2004): Cdk2 regulates the targeting of Spd-2, the first protein known to arrive at the site of centriole assembly. The Zyg1 protein kinase is recruited to this site, which leads to recruitment of a Sas-5/Sas-6 complex that forms a hollow cylinder that recruits Sas-4, which determines centriole length and leads to assembly of centriolar microtubules. Many of these proteins have mammalian orthologs that regulate centriole assembly: Cdk2 triggers centrosome duplication, and both cyclin A and cyclin E direct Cdk2 to relevant centriolar substrates including NPM/B23 (Okuda et al. 2000), CP110 (Chen et al. 2002), and Mps1 (Kasbek et al. 2007) [reviewed in (Hinchcliffe and Sluder 2002)]. Cep192, the human ortholog of SPD-2 regulates centriole assembly and binds to Plk4 (Franck et al. 2010), the distant relative and presumptive functional counterpart to ZYG-1. Plk4 then leads to recruitment of hSas6, the human ortholog of Sas-6 (Strnad et al. 2007; Kleylein-Sohn et al. 2007), and CPAP/CENP-J, the human ortholog of Sas-4, (Tang et al. 2009; Schmidt et al. 2009; Kohlmaier et al. 2009), regulates centriole length. However, there are important differences between worms and humans. Notably, hSas6 forms the hub of a cartwheel structure that serves as a template around which a centriole is assembled, rather than the hollow tube seen in *C. elegans*, and vertebrate centrioles contain triplet microtubules rather than singlets. Moreover, in humans many additional proteins not found in worms regulate centriole assembly. For example, recruitment of Plk4 to the site of centriole assembly additionally requires the human ortholog of *D. melanogaster* asterless, Cep152 (Cizmecioglu et al. 2010; Hatch et al. 2010; Dzhindzhev et al. 2010), and centriole assembly and elongation involves additional factors such as Cep135 (Kleylein-Sohn et al. 2007; Kim et al. 2008), γ-tubulin, and CP110 (Kleylein-Sohn et al. 2007; Spektor et al. 2007), as well as δ- and ε- tubulins (Chang and Stearns 2000; Chang and Giddings 2003), Mps1 (Kasbek et al. 2007; Yang et al. 2010; Fisk et al. 2003), Centrin 2 (Yang et al. 2010; Salisbury et al. 2002), hPoc5 (Azimzadeh et al. 2009), and Cep76 (Tsang et al. 2009), among others.

While a combined proteomics and comparative genomics study found just 18 core centriole proteins in the unicellular green algae *C. reinhardtii* (Keller et al. 2005), proteomics analysis estimated at least 150 proteins present at the human centrosome (Andersen et al. 2003). One interesting feature of the centrosomal proteome is that many components of the proteasome were found at centrosomes. Indeed, several centriole assembly factors whose overexpression leads to the production of excess centrosomes have been shown to be controlled by proteasome-dependent degradation, including Mps1 (Kasbek et al. 2007, 2010; Fisk and Winey 2001), Plk4 (Guderian et al. 2010; Cunha-Ferreira et al. 2009), and Sas6 (Strnad et al. 2007), and failure to properly control degradation of these proteins causes the production of excess centrioles within a single cell cycle, also known as centrosome re-duplication. Moreover, many E3 ubiquitin ligases have been implicated in the control of centrosome function, as have ubiquitin-independent proteasome degradation and non-proteasomal proteolysis. In this chapter we will review the evidence linking these pathways to centrosome function.

8.3 The Ubiquitin Proteasome System

The proteasome is the major site of degradation for misfolded or damaged proteins, as well as for cell cycle regulators whose destruction drives cell cycle transitions. Initially described as macropain, ingestin, or the multicatyltic proteinase, the term proteasome was coined in 1988 (Arrigo et al. 1988). The functional proteasome within the cell is generally thought to be the 26S particle, and consists of a catalytic 20S core particle and either of two regulatory particles [reviewed in (Murata et al. 2009; Navon and Ciechanover 2009; Ravid and Hochstrasser 2008; Tomko 2011). The 20S particle is a roughly 700 kDa hollow cylinder formed by twenty eight subunits, two copies each of seven α and seven β subunits. The α and β subunits are assembled into four rings, with the peptidase activity coming from three of the β subunits. The α subunits form the outer two rings of the 20S particle and the β subunits form the inner two rings, such that in the assembled 20S particle access to the proteolytic core is limited by the narrow opening afforded by the α subunits. Thus, insertion into the proteasome of proteins destined for degradation is regulated by either the 19S or 11S regulatory particles [reviewed in (Murata et al. 2009; Tomko 2011)].

The 11S particle regulates proteasome activity by making it more selective for small peptides, and is thought to be involved in substrate processing during antigen presentation. In contrast, the 19S particle contains many ATPase activities, displays chaperone activity in vitro, binds to proteins that have been covalently modified with ubiquitin, and is thought to unfold proteins to allow their degradation by the proteasome [reviewed in (Tomko 2011)]. In general, proteins that are degraded by the proteasome are targeted for degradation through ubiquitylation, and thus the pathway leading up to proteasome degradation is often referred to as the ubiquitin proteasome system (UPS) [reviewed in (Ravid and Hochstrasser 2008)]. Ubiquitin is expressed as a precursor protein (either polyubiquitin, or a fusion with an unrelated protein such as UBCEP80 that is a fusion of ribosomal protein S27a and ubiquitin) that is proteolytically processed to generate free ubiquitin. Free ubiquitin is then transferred to one or more lysine residues on target proteins through an enzymatic cascade. In the only ATP-dependent step of substrate ubiquitylation, free ubiquitin is coupled to the active site cysteine of an E1, or ubiquitin-activating enzyme. Ubiquitin is then transferred from the E1 to the active site cysteine of an E2, or ubiquitin-conjugating enzyme. In the final step an E3 enzyme or ubiquitin ligase transfers ubiquitin to a lysine on the substrate protein [reviewed in (Ravid and Hochstrasser 2008; Rotin and Kumar 2009; Simpson-Lavy et al. 2010).

There are many types of E3 ligases that fall into two major classes, the HECT- (Homology to the E6-AP C-Terminus) and RING-type ligases [reviewed in (Rotin and Kumar 2009)]. In the case of the HECT-type ligases, ubiquitin is transferred to an active site cysteine in a HECT domain containing protein as an intermediate in the transfer of ubiquitin to substrate [reviewed in (Rotin and Kumar 2009)]. In contrast, RING-type ligases do not form this covalent intermediate with ubiquitin, but rather use a RING domain containing protein to bring an E2 into intimate association with substrates [reviewed in (Rotin and Kumar 2009; Skaar and Pagano 2009). RING-type ligases are multi subunit complexes that consist of a core scaffold that bridges a protein containing an E2-binding RING domain with a variety of targeting factors that bind to substrates. In the Skp/Cullin/F-box (SCF) E3 ligase, Cul1 acts as a scaffold for Skp1, which recruits both the RING domain protein Rbx1 and one of several F-box proteins that determines substrate specificity [reviewed in (Skaar and Pagano 2009)]. Because F-box proteins typically bind to phosphorylated proteins, substrate phosphorylation is generally the trigger for SCF-dependent degradation. In the anaphase promoting complex/cyclosome (APC/C) E3 ligase, the Cullin-like Apc2 protein serves as a scaffold for eleven other subunits, including the RING domain protein Apc11 that recruits the E2, and either Cdc20 or Cdh1 that determine substrate specificity (van Leuken et al. 2008). Phosphorylation can also regulate the binding of substrates to the APC/C, although unlike the SCF, phosphorylation more often prevents APC/C-dependent degradation (Simpson-Lavy et al. 2010).

8.4 Centrosomes as a Site of Action for the Proteasome

In 1993 the proteasome was seen to associate with spindle fibers (Amsterdam et al. 1993), leading to the suggestion that it regulated the cell cycle by degrading mitotic cyclins, and was later observed to associate with mitotic spindle poles (Wojcik et al. 1995). In 1996 it was recognized that treatment of cells with proteasome inhibitors promotes the formation of large cytoplasmic aggregates that contained both proteasome components and ubiquitylated proteins. These aggregates were located at a perinuclear region in close proximity with Golgi and were hypothesized to represent cellular proteolysis centers (Wojcik et al. 1996). Subsequently termed aggresomes (Johnston et al. 1998), these proteolysis centers were shown to be highly regulated structures containing proteasomes and chaperones that are assembled at the centrosome in response to misfolded or damaged proteins (Corboy et al. 2005) through dynein-mediated microtubule-dependent transport (Johnston et al. 2002). The association of proteasomes with MTOCs is evolutionarily conserved, as it was recently shown that aggresomes form and colocalize with the spindle pole body (SPB) when exon 1 of the human Huntington protein is expressed in budding yeast (Wang et al. 2009). Many cellular proteins destined for degradation have been found at centrosomes, including p53 and Hsp70 in cells expressing adenoviral E1A/E1B proteins (Brown et al. 1994), and misfolded nucleoprotein from the influenza virus (Anton et al. 1999). However, this is not unique to cell cycle regulators targeted for degradation in virally infected cells, and in 1998 it was found that misfolded cystic fibrosis transmembrane conductance regulator accumulated at centrosomes, together with the molecular chaperones thought to be responsible for its presentation to the proteasome (Johnston et al. 1998; Loo et al. 1998).

These initial investigations on aggresomes concentrated on the degradation of damaged or misfolded proteins, and contributed to a view of centrosomes as command centers where regulators mingle but do not reside (Doxsey 2001; Schatten 2008). However, with respect to proteasomal degradation the centrosome is more than simply a gathering point for critical regulators or a point of execution for degradation. In 1999 the 20S proteasome core, 19S and 11S regulatory subunits, ubiquitin, and several molecular chaperones were shown to concentrate at centrosomes and co-fractionate with γ-Tubulin independently of aggresome formation (Wigley et al. 1999), suggesting that all of the protein unfolding and proteolysis activities required for proteasome-dependent degradation are normally localized to centrosomes. Indeed, it was soon shown that the proteasomes present at centrosomes were fully active in protein degradation (Fabunmi et al. 2000). Moreover, during mitosis in embryonic stem cell divisions, proteins destined for degradation are transported along microtubules to the maternal centrosome, so that these proteins are asymmetrically inherited to produce one daughter cell that is relatively cleansed of damaged proteins (Fuentealba et al. 2008). As discussed below, the proteasome is the site of regulated destruction of many cell cycle regulators, as well as a site of accumulation for the major regulators of protein degradation. Because many aspects of centrosome biology are controlled by regulated proteolysis, it seems likely that the proteasome and its regulators represent bona fide resident centrosome proteins. Several components of the UPS were found in the centrosomal proteome (Andersen et al. 2003), including the ubiquitin precursor protein UBCEP80, the ubiquitin-activating enzyme UBE1, one beta subunit of the 20S particle, and seven subunits of the 19S particle. The centrosomal proteome study classified candidate proteins using a statistical method based on how closely their fractionation profile correlated to that of bona fide centrosome proteins (Andersen et al. 2003). The extraordinarily high correlation of UBE1 with the centrosome suggests that the core of the UPS is resident at centrosomes. Moreover, proteolysis appears to play a role in MTOC assembly and function that is conserved throughout eukaryotes. The budding yeast 19S proteasome cap subunit Rpt4p (McDonald and Byers 1997) and the ubiquitin-like Dsk2p have been shown to be required for the initiation of SPB duplication (Biggins et al. 1996), and modulating ER-dependent degradation can suppress the SPB duplication defects caused by mutations that result in misfolding of two transmembrane SPB proteins (McBratney and Winey 2002).

8.5 Proteasomal Degradation Pathways at the Centrosome

Classically, proteins to be degraded by the proteasome are first modified by ubiquitylation, and there is a huge diversity in the motifs that dictate how a protein is ubiquitylated [reviewed in (Ravid and Hochstrasser 2008)]. The E3 ligases responsible for the bulk of cell cycle degradation are the RING-type SCF and APC/C (Skaar and Pagano 2009), although both HECT-type ligases

(Rotin and Kumar 2009) and other classes of RING ligases such as the Forkhead and RING domain (FHA-RING) proteins are also involved in cell cycle control (Brooks and Heimsath 2008).

8.6 HECT-Type Ligases and the Centrosome

There is little data suggesting that HECT-type ligases regulate centrosome biology. However, E6-AP, the quintessential HECT E3 ligase, is recruited to aggresomes upon proteasome inhibition (Mishra et al. 2009), although no association of E6-AP with the centrosome has been reported under normal conditions. In contrast, the HECT-type ligase Smurf2 has been reported to localize to centrosomes (Osmundson et al. 2008). However, several spindle checkpoint proteins localize to centrosomes at the same time, and Smurf2 depletion inhibits the spindle assembly checkpoint, suggesting that the centrosomal localization may not be directly relevant to centrosome biology per se. Accordingly, it is not yet clear whether there is any centrosomal significance for the presence of HECT-type ligases at centrosomes.

8.7 APC/C and the Centrosome

Soon after its discovery in 1995, subunits of the APC/C were reported to localize to mitotic centrosomes in human cells (Kurasawa and Todokoro 1999; Tugendreich et al. 1995), and mutations in *makos*, which encodes the fruit fly Cdc27, an APC/C subunit, enhance centrosomal defects seen in *polo* mutants (Deak et al. 2003). In human cells Cdc14a, the phosphatase that regulates Cdh1, localizes predominantly to centrosomes (Bembenek and Yu 2001), and Cdc20 was shown to dynamically associate with mitotic centrosomes in a microtubule-independent manner (Kallio et al. 2002). Moreover, the Cdc20 binding protein Speriolin is required for centrosomal localization of Cdc20 during spermatogenesis (Goto and Eddy 2004), and the ability of the Cdc20 inhibitor RASSF1A to block Cdc20 activity requires its localization to centrosomes (Song et al. 2005), implicating centrosomes in the control of the APC/C.

Accordingly, it is perhaps not surprising that the first protein shown to be degraded specifically at centrosomes is the APC/C substrate cyclin B. The initial indication that cyclin B degradation was subject to spatial as well as temporal control came from cell fusion studies showing that spindles in fused cells exit mitosis independently despite being in a common cytoplasm, demonstrating local control over cyclin B degradation (Rieder et al. 1997). In support of this conclusion, cyclin B degradation was shown to be required for mitotic exit in syncytial fruit fly embryos (Su et al. 1998), despite previous observations that cyclin B is not completely degraded in early embryos (Edgar et al. 1994). Cyclin B degradation was subsequently shown to initiate specifically at centrosomes in both fruit fly embryos (Huang and Raff 1999; Raff et al. 2002) and human cells (Clute and Pines 1999). In flies it was shown that this spatial destruction of cyclin B is regulated by the distribution of Vihar, an E2 enzyme responsible for ubiquitylation of APC/C substrates (Mathe et al. 2004). Vihar localizes to centrosomes and is itself an APC/C substrate that disappears first from centrosomes. Interestingly, Vihar degradation leads to the auto-deactivation of APC/C at centrosomes, releasing the APC/C from centrosomes to initiate the subsequent waves of cyclin B degradation elsewhere along the spindle and in the cell (Mathe et al. 2004). The human ortholog of Vihar, UbcH10, also localizes to centrosomes and its overexpression leads to precocious degradation of cyclin B, centriole amplification, and aneuploidy (van Ree et al. 2010). UbcH10 is also elevated in human tumors, and its overexpression in mice promotes tumor formation (van Ree et al. 2010).

Several other centrosome-associated APC/C substrates have been described, including L2DTL, the human ortholog of fly lethal (2) denticleless (Pan et al. 2006), microtubule associated protein CKAP2 (Seki and Fang 2007), Ninein-like protein (Nlp) (Wang and Zhan 2007), hSas6 (Strnad et al. 2007), and Nek2 (Hames et al. 2001, 2005). Nlp regulates centrosome maturation, and is critical for chromosome segregation and cytokinesis. Nlp is an APC/C substrate whose levels peak at G2/M due to the combined activities of Cdc20 and Cdh1 (Wang and Zhan 2007). However, Plk1 phosphorylation prevents the interaction of Nlp with γ-Tubulin and leads to the release of Nlp from centrosomes, which is required for proper centrosome maturation and spindle assembly (Casenghi et al. 2003), suggesting that cytoplasmic Nlp may be the relevant APC/C substrate. Similarly, hSas6 is targeted for APC/C-dependent degradation by binding to Cdh1 at G2/M (Strnad et al. 2007), but because hSas6 can only be incorporated into centrioles at G1/S this degradation must largely occur in the cytoplasm. Regardless, increasing hSas6 levels causes centriole overproduction (Strnad et al. 2007), and the Cdh1-dependent degradation of hSas6 at G2/M ensures that its levels are low in the subsequent G1/S when procentrioles form. In contrast, the degradation of Nek2 has been shown to occur explicitly at centrosomes. Nek2A is targeted for destruction through its cyclin A-type D-box in early mitosis (Hames et al. 2001). Nek2 is targeted to centrosomes via centriolar satellites, but once at centrosomes is rapidly exchanged via APC/C-dependent degradation (Hames et al. 2005). That Nek2A degradation occurs at centrosomes was demonstrated using FRAP analysis, which showed that either proteasome inhibition or expression of non-degradable Nek2 mutants greatly reduced the rate of recovery of GFP-Nek2A at centrosomes (Hames et al. 2005); had Nek2A been released from centrosomes and degraded in the cytoplasm, the recovery would not have been expected to change in response to preventing Nek2A degradation.

In addition to APC/C-dependent regulation described above in proliferating cells, the APC/C has functions at centrosomes in differentiated cells as well. Cdc20 is enriched at the centrosomes in neurons, and its RNAi-mediated depletion impairs the formation of dendritic arbors (Kim et al. 2009). The relevant Cdc20 substrate was shown to be Id1, and the APC-dependent degradation of Id1 at centrosomes was shown to promote dendrite growth and elaboration. The centrosomal localization of Cdc20 was shown to be critical for this activity through an elegant series of experiments involving rescue from RNAi depletion using a Cdc20 mutant lacking residues responsible for its accumulation at centrosomes that was artificially targeted to centrosomes (Kim et al. 2009) via the pericentrin and AKAP450 centrosome targeting (PACT) domain (Gillingham and Munro 2000).

8.8 SCF and the Centrosome

Skp1 (Freed et al. 1999; Gstaiger et al. 1999) and Cul1 (Freed et al. 1999) have been shown to localize to the centrosome in mammalian cells, and the F-box protein βTrCP was found in the centrosomal proteome (Andersen et al. 2003). Moreover, several candidate F-box proteins were also found at centrosomes, suggesting the existence of multiple centrosomal SCF complexes (Freed et al. 1999). Interestingly, centrosomal targeting of Skp1 and Cul1 may be regulated by modification with NEDD8, a ubiquitin-like protein (Freed et al. 1999). Antibodies against Skp1 or Cul1 block centriole separation in an in vitro assay using *X. laevis* egg extracts, and proteasome inhibitors block centriole separation and centrosome duplication in *X. laevis* embryos, suggesting that SCF function is directly relevant to centrosome assembly (Freed et al. 1999). Cyclin E is critical for the initiation of centrosome duplication, and in addition to resulting in polyploidy and centrosome overduplication in mice, Skp2 deletion leads to the elevation of cyclin E and the Cdk inhibitor p27Kip1 (Nakayama et al. 2000). Similarly, in fruit flies, mutations in genes encoding either the F-box protein Slimb (the fly ortholog of βTrCP) (Wojcik et al. 2000) or SkpA (Murphy 2003) lead to centrosome overproduction. Because cyclin E has many functions in centrosome duplication, and has been shown to be degraded in an SCF-dependent manner (Dealy et al. 1999; Wang et al. 1999), failure to degrade cyclin E might explain excess centrosomes in mice lacking Skp2 and flies lacking Slimb or SkpA. In contrast, excess p27 would be predicted to suppress centrosome duplication by inhibiting Cdk2, suggesting that a Cdk inhibitor might be the relevant SCF substrate in the *X. laevis* in vitro centriole assembly assay (Freed et al. 1999). However, excess cyclin E could not rescue the centriole duplication defect caused by the inhibitory Skp1 or Cul1 antibodies (Freed et al. 1999), and excess centrosomes were still observed in *skpa cyc E* double mutant flies (Murphy 2003), suggesting that additional centrosomal SCF substrates whose degradation controls centrosome duplication remained to be identified.

While it seems likely that excess cyclin E contributes to centrosome defects in SCF-deficient cells, it was recently shown that Plk4 is regulated by SCF-dependent degradation. Overexpression of Plk4 leads to centriole overproduction (Kleylein-Sohn et al. 2007; Habedanck et al. 2005), demonstrating that the control of centrosome duplication requires tight regulation of the levels of Plk4. The Cul1 subunit of the SCF localizes to the maternal centriole, where it antagonizes

Plk4-dependent centriole overproduction (Korzeniewski et al. 2009), and Plk4 is regulated by the Slimb/βTrCP F-box protein in both flies (Rogers et al. 2009; Cunha-Ferreira et al. 2009) and humans (Guderian et al. 2010) [reviewed in (Sillibourne and Bornens 2010)]. SCF-dependent degradation is typically stimulated by substrate phosphorylation, and in the case of Plk4 it is an autophosphorylation event that promotes Slimb/βTrCP binding (Guderian et al. 2010). Accordingly, autoactivation of Plk4 ensures its own destruction, limiting the levels of active Plk4 as part of the regulation of the initiation of centriole assembly.

SCF has additionally been shown to control centrosome homeostasis through the cyclin F-dependent degradation of CP110. Despite being the protein for which the F-box was named, cyclin F remained an orphan protein until MudPIT analysis showed that it bound to Skp1 and Cul1 in an F-box independent manner, and to CP110 in an F-box dependent manner (D'Angiolella et al. 2010). Interestingly, the binding of CP110 to the F-box in cyclin F is not regulated by phosphorylation as is typical for other SCF ligases, but rather by an RxL cyclin-binding motif in CP110. Regardless, either the depletion of cyclin F or the expression of a CP110 mutant that fails to bind cyclin F leads to centrosome overproduction, punctuating the importance of SCF function in controlling centrosome homeostasis (D'Angiolella et al. 2010).

8.9 BRCA1 and the Centrosome

The BRCA1 breast cancer susceptibility gene encodes a RING domain protein that has E3 ligase activity when bound to its partner BARD1 [reviewed in (Parvin 2009; Deng 2006)]. In 1998 BRCA1 was seen to associate with mitotic centrosomes (Hsu and White 1998), and in 1999 it was shown that mice with a targeted deletion of BRCA1 exon 11 undergo centrosome amplification (Xu et al. 1999). It was later shown that BRCA1 binds to γ-Tubulin to regulate its recruitment to centrosomes (Hsu et al. 2001). BRCA1 binds to γ-Tubulin, and overexpression of the BRCA1 γ-Tubulin binding domain displaces endogenous BRCA1 from centrosomes and dominantly interferes with the centrosomal recruitment of γ-Tubulin (Hsu et al. 2001). Only hypophosphorylated BRCA1 was capable of binding to γ-Tubulin (Hsu et al. 2001), and it was subsequently shown that Aurora A phosphorylates BRCA1 and inhibits its centrosomal functions (Ouchi et al. 2004; Sankaran et al. 2007). In a search for relevant centrosomal targets of BRCA1 E3 ligase activity, BRCA1 was found to monoubiquitylate γ-Tubulin in vitro at lysine 48 (Starita et al. 2004). The ubiquitylation of γ-Tubulin by BRCA1 requires the BRCA1 RING domain (Starita et al. 2004) and its E3 ligase activity (Sankaran et al. 2006), as well as its partner BARD1 that is localized to centrosomes throughout the cell cycle (Sankaran et al. 2006). While BRCA1 appeared to ubiquitylate other proteins in isolated centrosomes, γ-Tubulin would appear to be largely responsible for the centrosomal functions of BRCA1, because expression of a K48R mutant γ-Tubulin that could not be ubiquitylated caused a dominant increase in centrosome amplification that phenocopied BRCA1 inactivation (Starita et al. 2004). However, BRCA1 no doubt has other centrosomal functions. For example, Nlp was recently identified in a yeast two-hybrid screen designed to identify novel BRCA1 interacting proteins (Jin et al. 2009). Nlp and BRCA1 physically interact, and BRCA1 is required for the centrosomal localization of Nlp (Jin et al. 2009). BRCA1 depletion leads to a loss of Nlp from centrosomes and to enhanced Nlp degradation, presumably because BRCA1 negatively regulates Plk1, which is elevated in BRCA1-depleted cells (Jin et al. 2009).

8.10 Ubiquitin-Independent Proteasomal Degradation at Centrosomes

Interestingly, although the 26S proteasome is typically thought to be the active proteasome species within the cell, the major cellular proteasome species is the 20S proteasome that makes up 1 % of total cellular protein [reviewed in (Zetter and Mangold 2005). Moreover, it has recently been estimated that as much as 20 % of cellular protein can be degraded independently of ubiquitylation (Baugh et al. 2009). While some disordered regions within proteins can allow direct access to the 20S proteasome (Baugh et al. 2009; Tsvetkov et al. 2010), such as is the case for p53 p21, c-fos and c-jun [reviewed in (Zetter and Mangold 2005; Tsvetkov et al. 2010; Jariel-Encontre et al. 2008)], ubiquitin-independent degradation can also be conferred by cellular or viral factors. For example the 20S-dependent degradation of p53 is regulated by NAD(P)H quinonie oxidoreductase (NQO1), which prevents its association with the 20S proteasome, while pRb family members and IκBα are targeted to the proteasome by the cytomegalovirus protein pp71 and the HTLV Tax protein, respectively, independently of ubiquitylation [reviewed in (Zetter and Mangold 2005; Jariel-Encontre et al. 2008)].

Relative to centrosomes, the most relevant ubiquitin-independent degradation pathway is mediated by ornithine decarboxylase antizyme (OAZ). Initially characterized as a regulator of polyamine biosynthesis, OAZ binds to ornithine decarboxylase (ODC), and the ODC/OAZ complex is targeted to the proteasome by binding to the same recognition site on the 19S regulatory subunit as ubiquitylated proteins; however, ODC degradation occurs completely in the absence of ubiquitylation [reviewed in (Zetter and Mangold 2005)]. OAZ has a handful of other substrates, including Smad1, SNIP1, Aurora A, and cyclin D1 [reviewed in (Jariel-Encontre et al. 2008)], many of which are also degraded via ubiquitin-dependent routes. It was recently demonstrated that both OAZ and its antagonist antizyme inhibitor (AZI) are found at centrosomes, and that OAZ activity suppresses centrosome amplification (Mangold et al. 2008). This could not be explained by the known OAZ substrates, and suggested that OAZ promoted the degradation of some protein whose proper degradation restricted centrosome duplication, and it was recently shown that OAZ regulates centrosome duplication by targeting Mps1 for proteasome-dependent degradation. Mps1 is a multifunctional protein kinase (Fisk 2011) that localizes to centrosomes (Kasbek et al. 2010; Fisk et al. 2003; Fisk and Winey 2001; Liu et al. 2003; Tyler et al. 2009) and participates in several aspects of centrosome duplication [reviewed in (Pike and Fisk 2011)] including procentriole assembly (Yang et al. 2010; Kasbek et al. 2010) and centriole maturation (Yang et al. 2010). These centrosomal activities of Mps1 are controlled by an Mps1 degradation signal (MDS) that promotes the proteasome-dependent degradation of Mps1 at centrosomes and whose function is suppressed by Cdk2 phosphorylation (Kasbek et al. 2007). While Cdk2 has been shown to antagonize the APC/C-dependent degradation of Cdc6 to limit DNA replication (Mailand and Diffley 2005), the MDS has no known APC/C or SCF targeting motifs (Kasbek et al. 2007). However, while Mps1 may be dispensable for centriole assembly (Pike and Fisk 2011), preventing its degradation at centrosomes causes centriole overproduction (Kanai et al. 2007; Kasbek 2009), and Mps1 thus fits the expectation of an OAZ substrate. Indeed, OAZ binds to Mps1 via the MDS, and targets Mps1 for proteasome-mediated degradation at centrosomes (Kasbek et al. 2010). Increasing OAZ activity promotes degradation of the centrosomal pool of Mps1 and causes a delay in centriole assembly, while decreasing OAZ activity increases the level of the centrosomal Mps1 pool and causes Mps1-dependent centrosome reduplication (Kasbek et al. 2010). This degradation appears to occur specifically at centrosomes, because Cdk2 has little affect on the whole cell levels of Mps1, and because the centrosomal levels of a version of Mps1 tethered to centrosomes via the PACT domain show proteasome- and Cdk2-dependence that reflect that of the endogenous protein (Kasbek et al. 2007). Interestingly, cytoplasmic Mps1 levels are controlled by a D-box in its N-terminus that can bind to both Cdc20 and Cdh1 and promote its APC/C-dependent degradation (Cui et al. 2010). This suggests that APC/C could cooperate with OAZ to control centrosome duplication by limiting the amount of cytoplasmic Mps1 available for delivery to centrosomes. However, as discussed above the APC/C is present at centrosomes, suggesting the possibility that multiple inputs might control the degradation of Mps1 at centrosomes.

8.11 Non-Proteasomal Degradation Pathways at the Centrosome

In addition to the proteasome-dependent degradation pathways discussed above, there are many additional proteolytic activities that have been localized to centrosomes and/or shown to regulate centrosome biology. One striking example is the protease separase, initially described for its role in destroying the cohesin ring at anaphase onset [reviewed in (Nasmyth 2005)]. In addition to its role in regulating sister chromatid cohesion, separase has been shown localize to centrosomes and regulate centriole disengagement both in vitro and in vivo (Tsou and Stearns 2006). It was further shown that centriole disengagement requires the catalytic activity of separase, but does not require sister chromatid disjunction (Tsou et al. 2009). Separase-dependent centriole disengagement during mitosis is required for timely centriole replication in the following cell cycle. Both cells lacking separase and cells expressing a catalytically inactive mutant show a variety of centrosome duplication defects that range from asynchronous centriole replication to a complete failure to replicate centrioles (Tsou et al. 2009). Some centriole disengagement and replication is observed in cells lacking separase activity, because Plk1 also contributes to centriole disengagement. Downregulation of either separase or Plk1 causes a similar phenotype, while their co-depletion completely blocks both centriole disengagement during mitosis and centriole replication in the subsequent cell cycle (Tsou et al. 2009). While it was not initially clear whether the centriolar substrate(s) of separase could be the same as those involved in sister chromatid cohesion, recent studies have shown that cohesin subunits localize to centrosomes and likely do regulate centriole engagement. Scc1/Rad21 localizes to centrosomes in a manner regulated by Aurora B and Plk1, and its presence at centrosomes prevents premature centriole disengagement (Nakamura et al. 2009). Interestingly, centrosomal localization of Rad21 requires its cleavage by separase (Gimenez-Abian et al. 2010), and while the nature of this requirement is not clear, it demonstrates that Scc1/Rad21 is a separase substrate that is relevant for the regulation of centriole engagement. Depleting any of the cohesin subunits Scc1/Rad21, Smc1, or Smc3 results in multipolar mitosis as a result of premature centriole disengagement, and the defects in centriole disengagement are direct rather than representing some indirect consequence of defects in sister chromatid cohesion (Beauchene et al. 2010; Diaz-Martinez et al. 2010). However, while the role of separase in centriole engagement may involve the same substrates as sister chromatid cohesion, it does not appear that cohesin regulates centriole disengagement in the same way that it regulates sister chromatid disjunction, because Scc1/Rad21 does not localize to centrosomes until after its cleavage by separase (Simmons Kovacs and Haase 2010).

There are many other examples of non-proteasomal protease activities that regulate centrosome biology, and inhibition of non-proteasomal proteases leads to aberrant centriole elongation (Korzeniewski et al. 2010). Although the main compound used in that study, $Z-L_3VS$, is a proteasome inhibitor, other proteasome inhibitors could not induce centriole elongation as potently. Moreover, the ability of $Z-L_3VS$ to induce centriole elongation correlated with its ability to inhibit cleavage of casein, which is not a proteasome substrate and was not greatly affected by other proteasome inhibitors. Although neither the specific proteases whose inhibition promotes centriole elongation nor any explicit protease substrates were identified, a directed RNAi screen identified several centriole assembly factors that enhanced or suppressed the effect of $Z-L_3VS$. This screen not only validated the recently described roles of CPAP/CENPJ and CP110 in centriole elongation (Tang et al. 2009; Kohlmaier et al. 2009), but also uncovered several additional proteins involved in the control of centriole length. Although it is not clear whether all of these proteins represent substrates of non-proteasomal proteolysis, two of these proteins, FOP and CAP350 were stabilized in $Z-L_3VS$ treated cells (Korzeniewski et al. 2010).

While the Z-L$_3$VS study did not identify specific proteases, caspase-2 (Narine et al. 2010), tripeptidyl-peptidase II [TPP II; (Stavropoulou et al. 2005)], and membrane type-1 matrix metalloproteinase [MT1-MMP; (Golubkov et al. 2005a)] have been reported at centrosomes. While no substrates were identified for caspase-2 or TPP II, TPP II is overexpressed in Burkitt's Lymphoma where it promotes centrosome amplification and mitotic infidelity (Stavropoulou et al. 2005). However, a relevant centrosomal substrate has been identified for MT1-MMP. Although best known for its pericellular activities, MT1-MMP was shown to traffic to the centrosome subsequent to its endocytosis (Golubkov et al. 2005a), where it directly cleaves centrosomal pericentrin (Golubkov et al. 2005b), leading to mitotic spindle abnormalities and aneuploidy (Golubkov et al. 2005a; Golubkov et al. 2005b). Overexpression of MT1-MMP leads to the accumulation of pericentrin cleavage fragments, multipolar spindles, and aneuploidy, and pericentrin cleavage fragments are observed in tumor biopsies (Golubkov et al. 2005b). MMPs have long been known to play a role in tumorigenesis through their cleavage of extracellular matrix proteins, which has clear implications for tumor metastasis. However, the ability of MT1-MMP to cleave pericentrin at centrosomes suggests that MMPs may additionally play a role in the establishment of aneuploidy, and it will be interesting to determine to what extent the centrosomal effects of MT1-MMP contribute to its ability to promote cellular transformation and tumorigenesis (Soulie et al. 2005).

8.12 Other Relevant Degradation Events

Recent studies suggest that proteasome activity may have a very broad role in centriole assembly and function. Protein phosphatase 2A (PP2A) was recently shown to control centrosome duplication in flies (Kotadia et al. 2008) and worms (Kitagawa et al. 2011; Schlaitz et al. 2007; Song et al. 2011), and in worms it does so at least in part through the proteasome-dependent control of the levels of Sas-5 (Song et al. 2011), although degradation of Sas-5 per se was not demonstrated. Interestingly, it now appears that the STIL protein is the human Sas-5 ortholog, based on its similarity to *D. melanogaster* Ana2 (Stevens et al. 2010). STIL/hSas5 was recently shown to regulate the function of CHFR (Castiel et al. 2011), a RING-type E3 ligase [reviewed in (Brooks and Heimsath 2008)] responsible for the regulation of the antephase checkpoint, a newly emerging checkpoint that prevents entry into mitosis in the absence of sufficient cellular energy and/or arrests mitosis in response to mitotic stress [reviewed in (Chin and Yeong 2010)]. CHFR was found to be upregulated in STIL depleted cells, leading to inhibition of proliferation and multiple centrosome defects, likely through the degradation of the CHFR substrate Plk1 (Castiel et al. 2011). While CHFR has not been reported to localize to centrosomes, the regulation of CHFR by STIL/hSas5 implicates centrosomes in the control of CHFR and the antephase checkpoint. It further suggests that centriole assembly factors may both be controlled by and control protein degradation pathways. In addition, proteasome inhibitors lead to an accumulation of ubiquitylated centrosomal proteins including γ-Tubulin, TUB-GCP4, NEDD1, Ninein, pericentrin, dynactin, and PCM-1 (Didier et al. 2008). This accumulation is independent of microtubules, suggesting that the centrosome is the normal site for degradation of these proteins. The accumulation of these non-degraded proteins reduces the ability of centrosomes to form microtubule asters, suggesting that proteasome activity regulates the turnover of centrosome proteins in order to regulate centrosome homeostasis, and may further suggest a potential role in centrosome maturation.

Moreover, noncentrosomal degradation events are also relevant to centrosome homeostasis. Emi1 is an APC/C inhibitor that keeps the APC/C inactive during S and G2, but must be inactivated for cells to enter mitosis. In mitosis Emi1 is phosphorylated by Cdk1, after which it can be targeted by βTrCP/Slimb for SCF-dependent degradation (Margottin-Goguet et al. 2003). Expression of nondegradable Emi1 causes delayed activation of the APC/C and prolonged stabilization of APC/C substrates, which leads to metaphase arrest followed by mitotic catastrophe and centrosome amplification (Margottin-Goguet et al. 2003). Although Emi does not directly regulate either a centrosomal degradation pathway or substrate, such non-centrosomal events can nonetheless have a major impact on centrosome homeostasis, and there are likely to be other non-centrosomal events that more directly impact centrosome biology.

8.13 Conclusions and Prospective

As discussed above, proteolysis controls centrosome duplication and function at many levels and through many pathways. While the role of proteolysis in MTOC assembly and function appears to be conserved throughout eukaryotes, there are clearly mysteries yet to be solved. For example, no Rpt4p substrate involved in SPB duplication in yeast has been identified, so the degree to which proteolysis controls the function of non-centriole containing MTOCs remains unclear, and if proteolysis plays a major role in SPB duplication and/or function that role remains to be identified. While this review is intended to be as inclusive as possible, there are certain to be other degradation pathways and events not discussed here that are relevant to centrosomes and/or SPBs. While some may be known but regrettably overlooked here, there are likely more that are not yet known. It seems likely that the story of centrosomal degradation has not completely unfolded, and that much about the control of centrosome biology by regulated protein destruction remains to be uncovered. In fact, the observation in worms that PP2A controls the levels of Sas-5 in a proteasome-dependent manner (Song et al. 2011) suggests that regulated degradation may be a common feature of centriole assembly factors, even if their degradation has not yet been studied. Interestingly, despite initial searches that suggested there was no human ortholog of Sas-5, the STIL protein is likely to be the human Sas5 ortholog based on its similarity to the fly Ana2 (Stevens et al. 2010), and whether this core centriole assembly factor is regulated by degradation across species is an open question. Given that phosphorylation is often the trigger for SCF-dependent degradation, it is tempting to speculate that PP2A removes a phosphate to antagonize SCF-dependent degradation. However, it has not yet been demonstrated that Sas-5 is degraded, and PP2A can also promote SCF-dependent degradation as it does for c-myc (Sears 2004). The role of PP2A in the control Sas-5 therefore can't be predicted, and it will be exciting to watch the story of regulated Sas-5 degradation develop. Moreover, at least three factors that can cause centriole overproduction in human cells, Mps1, Plk4, and hSas6, are regulated by degradation. Whether proteins such as Ana2 and Asterless that can cause centrosome re-duplication in flies (Stevens et al. 2010) are also controlled by regulated degradation is also not yet known. Given how critical it is for centrosome duplication to occur just once each cell cycle, it seems likely that regulated degradation will ultimately be shown to be a common feature of centriole assembly factors.

References

Adams IR, Kilmartin JV (2000) Spindle pole body duplication: a model for centrosome duplication? Trends Cell Biol 10:329–335
Amsterdam A, Pitzer F, Baumeister W (1993) Changes in intracellular localization of proteasomes in immortalized ovarian granulosa cells during mitosis associated with a role in cell cycle control. Proc Natl Acad Sci U S A 90:99–103
Andersen JS, Wilkinson CJ, Mayor T, Mortensen P, Nigg EA, Mann M (2003) Proteomic characterization of the human centrosome by protein correlation profiling. Nature 426:570–574
Anton LC, Schubert U, Bacik I, Princiotta MF, Wearsch PA, Gibbs J, Day PM, Realini C, Rechsteiner MC, Bennink JR, Yewdell JW (1999) Intracellular localization of proteasomal degradation of a viral antigen. J Cell Biol 146:113–124
Arrigo AP, Tanaka K, Goldberg AL, Welch WJ (1988) Identity of the 19S 'prosome' particle with the large multifunctional protease complex of mammalian cells (the proteasome). Nature 331:192–194
Azimzadeh J, Marshall WF (2010) Building the centriole. Curr Biol 20:R816–825
Azimzadeh J, Hergert P, Delouvee A, Euteneuer U, Formstecher E, Khodjakov A, Bornens M (2009) hPOC5 is a centrin-binding protein required for assembly of full-length centrioles. J Cell Biol 185:101–114
Baugh JM, Viktorova EG, Pilipenko EV (2009) Proteasomes can degrade a significant proportion of cellular proteins independent of ubiquitination. J Mol Biol 386:814–827
Beauchene NA, Diaz-Martinez LA, Furniss K, Hsu WS, Tsai HJ, Chamberlain C, Esponda P, Gimenez-Abian JF, Clarke DJ (2010) Rad21 is required for centrosome integrity in human cells independently of its role in chromosome cohesion. Cell Cycle 9:1774–1780
Bembenek J, Yu H (2001) Regulation of the anaphase-promoting complex by the dual specificity phosphatase human Cdc14a. J Biol Chem 276:48237–48242
Biggins S, Ivanovska I, Rose MD (1996) Yeast ubiquitin-like genes are involved in duplication of the microtubule organizing center. J Cell Biol 133:1331–1346
Brooks L 3rd, Heimsath EG Jr, Loring GL, Brenner C (2008) FHA-RING ubiquitin ligases in cell division cycle control. Cell Mol Life Sci 65:3458–3466
Brown CR, Doxsey SJ, White E, Welch WJ (1994) Both viral (adenovirus E1B) and cellular (hsp 70, p53) components interact with centrosomes. J Cell Physiol 160:47–60

Casenghi M, Meraldi P, Weinhart U, Duncan PI, Korner R, Nigg EA (2003) Polo-like kinase 1 regulates Nlp, a centrosome protein involved in microtubule nucleation. Dev Cell 5:113–125

Castiel A, Danieli MM, David A, Moshkovitz S, Aplan PD, Kirsch IR, Brandeis M, Kramer A, Izraeli S (2011) The Stil protein regulates centrosome integrity and mitosis through suppression of Chfr. J Cell Sci 124:532–539

Chang P, Giddings TH Jr, Winey M, Stearns T (2003) Epsilon-tubulin is required for centriole duplication and microtubule organization. Nat Cell Biol 5:71–76

Chang P, Stearns T (2000) δ-Tubulin and ϵ-tubulin: two new human centrosomal tubulins reveal new aspects of centrosome structure and function. Nat Cell Biol 2:30–35

Chen Z, Indjeian VB, McManus M, Wang L, Dynlacht BD (2002) CP110, a cell cycle-dependent CDK substrate, regulates centrosome duplication in human cells. Dev Cell 3:339–350

Chin CF, Yeong FM (2010) Safeguarding entry into mitosis: the antephase checkpoint. Mol Cell Biol 30:22–32

Cizmecioglu O, Arnold M, Bahtz R, Settele F, Ehret L, Haselmann-Weiss U, Antony C, Hoffmann I (2010) Cep152 acts as a scaffold for recruitment of Plk4 and CPAP to the centrosome. J Cell Biol 191:731–739

Clute P, Pines J (1999) Temporal and spatial control of cyclin B1 destruction in metaphase. Nat Cell Biol 1:82–87

Corboy MJ, Thomas PJ, Wigley WC (2005) Aggresome formation. Methods Mol Biol 301:305–327

Cui Y, Cheng X, Zhang C, Zhang Y, Li S, Wang C, Guadagno TM (2010) Degradation of the human mitotic checkpoint kinase Mps1 is cell cycle-regulated by APC-cCdc20 and APC-cCdh1 ubiquitin ligases. J Biol Chem 285:32988–32998

Cunha-Ferreira I, Rodrigues-Martins A, Bento I, Riparbelli M, Zhang W, Laue E, Callaini G, Glover DM, Bettencourt-Dias M (2009) The SCF/Slimb ubiquitin ligase limits centrosome amplification through degradation of SAK/PLK4. Curr Biol 19:43–49

Dammermann A, Muller-Reichert T, Pelletier L, Habermann B, Desai A, Oegema K (2004) Centriole assembly requires both centriolar and pericentriolar material proteins. Dev Cell 7:815–829

D'Angiolella V, Donato V, Vijayakumar S, Saraf A, Florens L, Washburn MP, Dynlacht B, Pagano M (2010) SCF(Cyclin F) controls centrosome homeostasis and mitotic fidelity through CP110 degradation. Nature 466:138–142

Deak P, Donaldson M, Glover DM (2003) Mutations in makos, a Drosophila gene encoding the Cdc27 subunit of the anaphase promoting complex, enhance centrosomal defects in polo and are suppressed by mutations in twins/aar, which encodes a regulatory subunit of PP2A. J Cell Sci 116:4147–4158

Dealy MJ, Nguyen KV, Lo J, Gstaiger M, Krek W, Elson D, Arbeit J, Kipreos ET, Johnson RS (1999) Loss of Cul1 results in early embryonic lethality and dysregulation of cyclin E. Nat Genet 23:245–248

Delattre M, Canard C, Gonczy P (2006) Sequential protein recruitment in *C. elegans* centriole formation. Curr Biol 16:1844–1849

Deng CX (2006) BRCA1: cell cycle checkpoint, genetic instability, DNA damage response and cancer evolution. Nucleic Acids Res 34:1416–1426

Diaz-Martinez LA, Beauchene NA, Furniss K, Esponda P, Gimenez-Abian JF, Clarke DJ (2010) Cohesin is needed for bipolar mitosis in human cells. Cell Cycle 9:1764–1773

Didier C, Merdes A, Gairin JE, Jabrane-Ferrat N (2008) Inhibition of proteasome activity impairs centrosome-dependent microtubule nucleation and organization. Mol Biol Cell 19:1220–1229

Doxsey SJ (2001) Centrosomes as command centres for cellular control. Nat Cell Biol 3:E105–108

Dzhindzhev NS, Yu QD, Weiskopf K, Tzolovsky G, Cunha-Ferreira I, Riparbelli M, Rodrigues-Martins A, Bettencourt-Dias M, Callaini G, Glover DM (2010) Asterless is a scaffold for the onset of centriole assembly. Nature 467:714–718

Edgar BA, Sprenger F, Duronio RJ, Leopold P, O'Farrell PH (1994) Distinct molecular mechanism regulate cell cycle timing at successive stages of Drosophila embryogenesis. Genes Dev 8:440–452

Fabunmi RP, Wigley WC, Thomas PJ, DeMartino GN (2000) Activity and regulation of the centrosome-associated proteasome. J Biol Chem 275:409–413

Fisk HA (2011) The Mip-ing link: Mip1 links Mps1 to the actin cytoskeleton. Cell Cycle 10:1026–1027

Fisk HA, Winey M (2001) The mouse mps1p-like kinase regulates centrosome duplication. Cell 106:95–104

Fisk HA, Mattison CP, Winey M (2002) Centrosomes and tumour suppressors. Curr Opin Cell Biol 14:700–705

Fisk HA, Mattison CP, Winey M (2003) Human Mps1 protein kinase is required for centrosome duplication and normal mitotic progression. Proc Natl Acad Sci U S A 100:14875–14880

Franck N, Montembault E, Rome P, Pascal A, Cremet JY, Giet R (2010) CDK11 is required for centriole duplication and Plk4 recruitment to mitotic centrosomes. PLoS ONE 6:e14600

Freed E, Lacey KR, Huie P, Lyapina SA, Deshaies RJ, Stearns T, Jackson PK (1999) Components of an SCF ubiquitin ligase localize to the centrosome and regulate the centrosome duplication cycle. Genes Dev 13:2242–2257

Fuentealba LC, Eivers E, Geissert D, Taelman V, De Robertis EM (2008) Asymmetric mitosis: unequal segregation of proteins destined for degradation. Proc Natl Acad Sci U S A 105:7732–7737

Gillingham AK, Munro S (2000) The PACT domain, a conserved centrosomal targeting motif in the coiled-coil proteins AKAP450 and pericentrin. EMBO Rep 1:524–529

Gimenez-Abian JF, Diaz-Martinez LA, Beauchene NA, Hsu WS, Tsai HJ, Clarke DJ (2010) Determinants of Rad21 localization at the centrosome in human cells. Cell Cycle 9:1759–1763

Golubkov VS, Boyd S, Savinov AY, Chekanov AV, Osterman AL, Remacle A, Rozanov DV, Doxsey SJ, Strongin AY (2005a) Membrane type-1 matrix metalloproteinase (MT1-MMP) exhibits an important intracellular cleavage function and causes chromosome instability. J Biol Chem 280:25079–25086

Golubkov VS, Chekanov AV, Doxsey SJ, Strongin AY (2005b) Centrosomal pericentrin is a direct cleavage target of membrane type-1 matrix metalloproteinase in humans but not in mice: potential implications for tumorigenesis. J Biol Chem 280:42237–42241

Goto M, Eddy EM (2004) Speriolin is a novel spermatogenic cell-specific centrosomal protein associated with the seventh WD motif of Cdc20. J Biol Chem 279:42128–42138

Gstaiger M, Marti A, Krek W (1999) Association of human SCF(SKP2) subunit p19(SKP1) with interphase centrosomes and mitotic spindle poles. Exp Cell Res 247:554–562

Guderian G, Westendorf J, Uldschmid A, Nigg EA (2010) Plk4 trans-autophosphorylation regulates centriole number by controlling betaTrCP-mediated degradation. J Cell Sci 123:2163–2169

Habedanck R, Stierhof YD, Wilkinson CJ, Nigg EA (2005) The Polo kinase Plk4 functions in centriole duplication. Nat Cell Biol 7:1140–1146

Hames RS, Wattam SL, Yamano H, Bacchieri R, Fry AM (2001) APC/C-mediated destruction of the centrosomal kinase Nek2A occurs in early mitosis and depends upon a cyclin A-type D-box. EMBO J 20:7117–7127

Hames RS, Crookes RE, Straatman KR, Merdes A, Hayes MJ, Faragher AJ, Fry AM (2005) Dynamic recruitment of Nek2 kinase to the centrosome involves microtubules, PCM-1, and localized proteasomal degradation. Mol Biol Cell 16:1711–1724

Hatch EM, Kulukian A, Holland AJ, Cleveland DW, Stearns T (2010) Cep152 interacts with Plk4 and is required for centriole duplication. J Cell Biol 191:721–729

Hinchcliffe EH, Sluder G (2002) Two for two: Cdk2 and its role in centrosome doubling. Oncogene 21:6154–6160

Hsu LC, White RL (1998) BRCA1 is associated with the centrosome during mitosis. Proc Natl Acad Sci U S A 95:12983–12988

Hsu LC, Doan TP, White RL (2001) Identification of a gamma-tubulin-binding domain in BRCA1. Cancer Res 61:7713–7718

Huang J, Raff JW (1999) The disappearance of cyclin B at the end of mitosis is regulated spatially in Drosophila cells. EMBO J 18:2184–2195

Jariel-Encontre I, Bossis G, Piechaczyk M (2008) Ubiquitin-independent degradation of proteins by the proteasome. Biochim Biophys Acta 1786:153–177

Jin S, Gao H, Mazzacurati L, Wang Y, Fan W, Chen Q, Yu W, Wang M, Zhu X, Zhang C, Zhan Q (2009) BRCA1 interaction of centrosomal protein Nlp is required for successful mitotic progression. J Biol Chem 284:22970–22977

Johnston JA, Ward CL, Kopito RR (1998) Aggresomes: a cellular response to misfolded proteins. J Cell Biol 143:1883–1898

Johnston JA, Illing ME, Kopito RR (2002) Cytoplasmic dynein/dynactin mediates the assembly of aggresomes. Cell Motil Cytoskeleton 53:26–38

Kallio MJ, Beardmore VA, Weinstein J, Gorbsky GJ (2002) Rapid microtubule-independent dynamics of Cdc20 at kinetochores and centrosomes in mammalian cells. J Cell Biol 158:841–847

Kanai M, Ma Z, Izumi H, Kim SH, Mattison CP, Winey M, Fukasawa K (2007) Physical and functional interaction between mortalin and Mps1 kinase. Genes Cells 12:797–810

Kasbek C, Yang C-H, Fisk HA (2009) Mps1 as a link between centrosomes and genetic instability. Environ Mol Mutagen 50:654–665

Kasbek C, Yang CH, Yusof AM, Chapman HM, Winey M, Fisk HA (2007) Preventing the degradation of mps1 at centrosomes is sufficient to cause centrosome reduplication in human cells. Mol Biol Cell 18:4457–4469

Kasbek C, Yang CH, Fisk HA (2010) Antizyme restrains centrosome amplification by regulating the accumulation of Mps1 at centrosomes. Mol Biol Cell 21:3879–3889

Keller LC, Romijn EP, Zamora I, Yates JR 3rd, Marshall WF (2005) Proteomic analysis of isolated chlamydomonas centrioles reveals orthologs of ciliary-disease genes. Curr Biol 15:1090–1098

Kemp CA, Kopish KR, Zipperlen P, Ahringer J, O'Connell KF (2004) Centrosome maturation and duplication in *C. elegans* require the coiled-coil protein SPD-2. Dev Cell 6:511–523

Kim K, Lee S, Chang J, Rhee K (2008) A novel function of CEP135 as a platform protein of C-NAP1 for its centriolar localization. Exp Cell Res 314:3692–3700

Kim AH, Puram SV, Bilimoria PM, Ikeuchi Y, Keough S, Wong M, Rowitch D, Bonni A (2009) A centrosomal Cdc20-APC pathway controls dendrite morphogenesis in postmitotic neurons. Cell 136:322–336

Kitagawa D, Flückiger I, Polanowska J, Keller D, Reboul J, Gönczy P (2011) PP2A phosphatase acts upon SAS-5 to ensure centriole formation in *C. elegans* embryos. Dev Cell 20:550–562

Kleylein-Sohn J, Westendorf J, Le Clech M, Habedanck R, Stierhof YD, Nigg EA (2007) Plk4-induced centriole biogenesis in human cells. Dev Cell 13:190–202

Kohlmaier G, Loncarek J, Meng X, McEwen BF, Mogensen MM, Spektor A, Dynlacht BD, Khodjakov A, Gönczy P (2009) Overly long centrioles and defective cell division upon excess of the SAS-4-related protein CPAP. Curr Biol 19:1012–1018

Korzeniewski N, Zheng L, Cuevas R, Parry J, Chatterjee P, Anderton B, Duensing A, Munger K, Duensing S (2009) Cullin 1 functions as a centrosomal suppressor of centriole multiplication by regulating polo-like kinase 4 protein levels. Cancer Res 69:6668–6675

Korzeniewski N, Cuevas R, Duensing A, Duensing S (2010) Daughter centriole elongation is controlled by proteolysis. Mol Biol Cell 21:3942–3951

Kotadia S, Kao LR, Comerford SA, Jones RT, Hammer RE, Megraw TL (2008) PP2A-dependent disruption of centrosome replication and cytoskeleton organization in Drosophila by SV40 small tumor antigen. Oncogene 27:6334–6346

Kurasawa Y, Todokoro K (1999) Identification of human APC10/Doc1 as a subunit of anaphase promoting complex. Oncogene 18:5131–5137

Leidel S, Gönczy P (2005) Centrosome duplication and nematodes: recent insights from an old relationship. Dev Cell 9:317–325

Li JB, Gerdes JM, Haycraft CJ, Fan Y, Teslovich TM, May-Simera H, Li H, Blacque OE, Li L, Leitch CC et al (2004) Comparative genomics identifies a flagellar and basal body proteome that includes the BBS5 human disease gene. Cell 117:541–552

Liu ST, Chan GK, Hittle JC, Fujii G, Lees E, Yen TJ (2003) Human MPS1 kinase is required for mitotic arrest induced by the loss of CENP-E from kinetochores. Mol Biol Cell 14:1638–1651

Loo MA, Jensen TJ, Cui L, Hou Y, Chang XB, Riordan JR (1998) Perturbation of Hsp90 interaction with nascent CFTR prevents its maturation and accelerates its degradation by the proteasome. EMBO J 17:6879–6887

Mailand N, Diffley JF (2005) CDKs promote DNA replication origin licensing in human cells by protecting Cdc6 from APC/C-dependent proteolysis. Cell 122:915–926

Mangold U, Hayakawa H, Coughlin M, Munger K, Zetter BR (2008) Antizyme, a mediator of ubiquitin-independent proteasomal degradation and its inhibitor localize to centrosomes and modulate centriole amplification. Oncogene 27:604–613

Margottin-Goguet F, Hsu JY, Loktev A, Hsieh HM, Reimann JD, Jackson PK (2003) Prophase destruction of Emi1 by the SCF(betaTrCP/Slimb) ubiquitin ligase activates the anaphase promoting complex to allow progression beyond prometaphase. Dev Cell 4:813–826

Mathe E, Kraft C, Giet R, Deak P, Peters JM, Glover DM (2004) The E2-C vihar is required for the correct spatiotemporal proteolysis of cyclin B and itself undergoes cyclical degradation. Curr Biol 14:1723–1733

McBratney S, Winey M (2002) Mutant membrane protein of the budding yeast spindle pole body is targeted to the endoplasmic reticulum degradation pathway. Genetics 162:567–578

McDonald HB, Byers B (1997) A proteasome cap subunit required for spindle pole body duplication in yeast. J Cell Biol 137:539–553

Mishra A, Godavarthi SK, Maheshwari M, Goswami A, Jana NR (2009) The ubiquitin ligase E6-AP is induced and recruited to aggresomes in response to proteasome inhibition and may be involved in the ubiquitination of Hsp70-bound misfolded proteins. J Biol Chem 284:10537–10545

Murata S, Yashiroda H, Tanaka K (2009) Molecular mechanisms of proteasome assembly. Nat Rev Mol Cell Biol 10:104–115

Murphy TD (2003) Drosophila skpA, a component of SCF ubiquitin ligases, regulates centrosome duplication independently of cyclin E accumulation. J Cell Sci 116:2321–2332

Nakamura A, Arai H, Fujita N (2009) Centrosomal Aki1 and cohesin function in separase-regulated centriole disengagement. J Cell Biol 187:607–614

Nakayama K, Nagahama H, Minamishima YA, Matsumoto M, Nakamichi I, Kitagawa K, Shirane M, Tsunematsu R, Tsukiyama T, Ishida N et al (2000) Targeted disruption of Skp2 results in accumulation of cyclin E and p27(Kip1), polyploidy and centrosome overduplication. EMBO J 19:2069–2081

Narine KA, Keuling AM, Gombos R, Tron VA, Andrew SE, Young LC (2010) Defining the DNA mismatch repair-dependent apoptotic pathway in primary cells: evidence for p53-independence and involvement of centrosomal caspase 2. DNA Repair (Amst) 9:161–168

Nasmyth K (2005) How do so few control so many? Cell 120:739–746

Navon A, Ciechanover A (2009) The 26 S proteasome: from basic mechanisms to drug targeting. J Biol Chem 284:33713–33718

O'Connell KF (2002) The ZYG-1 kinase, a mitotic and meiotic regulator of centriole replication. Oncogene 21:6201–6208

Okuda M, Horn HF, Tarapore P, Tokuyama Y, Smulian AG, Chan PK, Knudsen ES, Hofmann IA, Snyder JD, Bove KE, Fukasawa K (2000) Nucleophosmin/B23 is a target of CDK2/cyclin E in centrosome duplication. Cell 103:127–140

Osmundson EC, Ray D, Moore FE, Gao Q, Thomsen GH, Kiyokawa H (2008) The HECT E3 ligase Smurf2 is required for Mad2-dependent spindle assembly checkpoint. J Cell Biol 183:267–277

Ouchi M, Fujiuchi N, Sasai K, Katayama H, Minamishima YA, Ongusaha PP, Deng C, Sen S, Lee SW, Ouchi T (2004) BRCA1 phosphorylation by Aurora-A in the regulation of G2 to M transition. J Biol Chem 279:19643–19648

Pan HW, Chou HY, Liu SH, Peng SY, Liu CL, Hsu HC (2006) Role of L2DTL, cell cycle-regulated nuclear and centrosome protein, in aggressive hepatocellular carcinoma. Cell Cycle 5:2676–2687

Parvin JD (2009) The BRCA1-dependent ubiquitin ligase, gamma-tubulin, and centrosomes. Environ Mol Mutagen 50:649–653

Pelletier L, O'Toole E, Schwager A, Hyman AA, Muller-Reichert T (2006) Centriole assembly in *Caenorhabditis elegans*. Nature 444:619–623

Pike AN, Fisk HA (2011) Centriole assembly and the role of Mps1: defensible or dispensable? Cell Div 6:9

Raff JW, Jeffers K, Huang JY (2002) The roles of Fzy/Cdc20 and Fzr/Cdh1 in regulating the destruction of cyclin B in space and time. J Cell Biol 157:1139–1149

Ravid T, Hochstrasser M (2008) Diversity of degradation signals in the ubiquitin-proteasome system. Nat Rev Mol Cell Biol 9:679–690

Rieder CL, Khodjakov A, Paliulis LV, Fortier TM, Cole RW, Sluder G (1997) Mitosis in vertebrate somatic cells with two spindles: implications for the metaphase/anaphase transition checkpoint and cleavage. Proc Natl Acad Sci U S A 94:5107–5112

Rogers GC, Rusan NM, Roberts DM, Peifer M, Rogers SL (2009) The SCF Slimb ubiquitin ligase regulates Plk4/Sak levels to block centriole reduplication. J Cell Biol 184:225–239

Rotin D, Kumar S (2009) Physiological functions of the HECT family of ubiquitin ligases. Nat Rev Mol Cell Biol 10:398–409

Salisbury J, Suino K, Busby R, Springett M (2002) Centrin-2 is required for centriole duplication in mammalian cells. Curr Biol 12:1287

Sankaran S, Starita LM, Simons AM, Parvin JD (2006) Identification of domains of BRCA1 critical for the ubiquitin-dependent inhibition of centrosome function. Cancer Res 66:4100–4107

Sankaran S, Crone DE, Palazzo RE, Parvin JD (2007) Aurora-A kinase regulates breast cancer associated gene 1 inhibition of centrosome-dependent microtubule nucleation. Cancer Res 67:11186–11194

Schatten H (2008) The mammalian centrosome and its functional significance. Histochem Cell Biol 129:667–686

Schlaitz AL, Srayko M, Dammermann A, Quintin S, Wielsch N, MacLeod I, de Robillard Q, Zinke A, Yates JR 3rd, Muller-Reichert T et al (2007) The *C. elegans* RSA complex localizes protein phosphatase 2A to centrosomes and regulates mitotic spindle assembly. Cell 128:115–127

Schmidt TI, Kleylein-Sohn J, Westendorf J, Le Clech M, Lavoie SB, Stierhof YD, Nigg EA (2009) Control of centriole length by CPAP and CP110. Curr Biol 19:1005–1011

Sears RC (2004) The life cycle of C-myc: from synthesis to degradation. Cell Cycle 3:1133–1137

Seki A, Fang G (2007) CKAP2 is a spindle-associated protein degraded by APC/C-Cdh1 during mitotic exit. J Biol Chem 282:15103–15113

Sillibourne JE, Bornens M (2010) Polo-like kinase 4: the odd one out of the family. Cell Div 5:25

Simmons Kovacs LA, Haase SB (2010) Cohesin: it's not just for chromosomes anymore. Cell Cycle 9:1750–1753

Simpson-Lavy KJ, Oren YS, Feine O, Sajman J, Listovsky T, Brandeis M (2010) Fifteen years of APC/cyclosome: a short and impressive biography. Biochem Soc Trans 38:78–82

Skaar JR, Pagano M (2009) Control of cell growth by the SCF and APC/C ubiquitin ligases. Curr Opin Cell Biol 21:816–824

Song MS, Chang JS, Song SJ, Yang TH, Lee H, Lim DS (2005) The centrosomal protein RAS association domain family protein 1A (RASSF1A)-binding protein 1 regulates mitotic progression by recruiting RASSF1A to spindle poles. J Biol Chem 280:3920–3927

Song MH, Liu Y, Anderson DE, Jahng WJ, O'Connell KF (2011) Protein phosphatase 2A-SUR-6/B55 regulates centriole duplication in *C. elegans* by controlling the levels of centriole assembly factors. Dev Cell 20:563–571

Soulie P, Carrozzino F, Pepper MS, Strongin AY, Poupon MF, Montesano R (2005) Membrane-type-1 matrix metalloproteinase confers tumorigenicity on nonmalignant epithelial cells. Oncogene 24:1689–1697

Spektor A, Tsang WY, Khoo D, Dynlacht BD (2007) Cep97 and CP110 suppress a cilia assembly program. Cell 130:678–690

Starita LM, Machida Y, Sankaran S, Elias JE, Griffin K, Schlegel BP, Gygi SP, Parvin JD (2004) BRCA1-dependent ubiquitination of gamma-tubulin regulates centrosome number. Mol Cell Biol 24:8457–8466

Stavropoulou V, Xie J, Henriksson M, Tomkinson B, Imreh S, Masucci MG (2005) Mitotic infidelity and centrosome duplication errors in cells overexpressing tripeptidyl-peptidase II. Cancer Res 65:1361–1368

Stevens NR, Dobbelaere J, Brunk K, Franz A, Raff JW (2010) Drosophila Ana2 is a conserved centriole duplication factor. J Cell Biol 188:313–323

Strnad P, Leidel S, Vinogradova T, Euteneuer U, Khodjakov A, Gonczy P (2007) Regulated HsSAS-6 levels ensure formation of a single procentriole per centriole during the centrosome duplication cycle. Dev Cell 13:203–213

Su TT, Sprenger F, DiGregorio PJ, Campbell SD, O'Farrell PH (1998) Exit from mitosis in Drosophila syncytial embryos requires proteolysis and cyclin degradation, and is associated with localized dephosphorylation. Genes Dev 12:1495–1503

Tang CJ, Fu RH, Wu KS, Hsu WB, Tang TK (2009) CPAP is a cell-cycle regulated protein that controls centriole length. Nat Cell Biol 11:825–831

Tomko RJ Jr, Hochstrasser M (2011) Order of the proteasomal ATPases and eukaryotic proteasome assembly. Cell Biochem Biophys 60:13–20

Tsang WY, Spektor A, Vijayakumar S, Bista BR, Li J, Sanchez I, Duensing S, Dynlacht BD (2009) Cep76, a centrosomal protein that specifically restrains centriole reduplication. Dev Cell 16:649–660

Tsou MF, Stearns T (2006a) Controlling centrosome number: licenses and blocks. Curr Opin Cell Biol 18:74–78

Tsou MF, Stearns T (2006b) Mechanism limiting centrosome duplication to once per cell cycle. Nature 442:947–951

Tsou MF, Wang WJ, George KA, Uryu K, Stearns T, Jallepalli PV (2009) Polo kinase and separase regulate the mitotic licensing of centriole duplication in human cells. Dev Cell 17:344–354

Tsvetkov P, Reuven N, Shaul Y (2010) Ubiquitin-independent p53 proteasomal degradation. Cell Death Differ 17:103–108

Tugendreich S, Tomkiel J, Earnshaw W, Hieter P (1995) CDC27Hs colocalizes with CDC16Hs to the centrosome and mitotic spindle and is essential for the metaphase to anaphase transition. Cell 81:261–268

Tyler RK, Chu ML, Johnson H, McKenzie EA, Gaskell SJ, Eyers PA (2009) Phosphoregulation of human Mps1 kinase. Biochem J 417:173–181

van Leuken R, Clijsters L, Wolthuis R (2008) To cell cycle, swing the APC/C. Biochim Biophys Acta 1786:49–59

van Ree JH, Jeganathan KB, Malureanu L, van Deursen JM (2010) Overexpression of the E2 ubiquitin-conjugating enzyme UbcH10 causes chromosome missegregation and tumor formation. J Cell Biol 188:83–100

Wang Y, Zhan Q (2007) Cell cycle-dependent expression of centrosomal ninein-like protein in human cells is regulated by the anaphase-promoting complex. J Biol Chem 282:17712–17719

Wang Y, Penfold S, Tang X, Hattori N, Riley P, Harper JW, Cross JC, Tyers M (1999) Deletion of the Cul1 gene in mice causes arrest in early embryogenesis and accumulation of cyclin E. Curr Biol 9:1191–1194

Wang Y, Meriin AB, Zaarur N, Romanova NV, Chernoff YO, Costello CE, Sherman MY (2009) Abnormal proteins can form aggresome in yeast: aggresome-targeting signals and components of the machinery. FASEB J 23:451–463

Wigley WC, Fabunmi RP, Lee MG, Marino CR, Muallem S, DeMartino GN, Thomas PJ (1999) Dynamic association of proteasomal machinery with the centrosome. J Cell Biol 145:481–490

Wojcik C, Paweletz N, Schroeter D (1995) Localization of proteasomal antigens during different phases of the cell cycle in HeLa cells. Eur J Cell Biol 68:191–198

Wojcik C, Schroeter D, Wilk S, Lamprecht J, Paweletz N (1996) Ubiquitin-mediated proteolysis centers in HeLa cells: indication from studies of an inhibitor of the chymotrypsin-like activity of the proteasome. Eur J Cell Biol 71:311–318

Wojcik EJ, Glover DM, Hays TS (2000) The SCF ubiquitin ligase protein slimb regulates centrosome duplication in Drosophila. Curr Biol 10:1131–1134

Xu X, Weaver Z, Linke SP, Li C, Gotay J, Wang XW, Harris CC, Ried T, Deng CX (1999) Centrosome amplification and a defective G2-M cell cycle checkpoint induce genetic instability in BRCA1 exon 11 isoform-deficient cells. Mol Cell 3:389–395

Yang CH, Kasbek C, Majumder S, Mohd Yusof A, Fisk HA (2010) Mps1 phosphorylation sites regulate the function of Centrin 2 in centriole assembly. Mol Biol Cell 21:4361–4372

Zetter BR, Mangold U (2005) Ubiquitin-independent degradation and its implication in cancer. Future Oncol 1:567–570

Chapter 9
Regulation of the Centrosome Cycle by Protein Degradation

Suzanna L. Prosser and Andrew M. Fry

Abstract Irreversible protein destruction is a key regulatory mechanism controlling progression through the cell cycle. This is orchestrated by ubiquitin ligases, the two most prominent of which in the cell cycle are the anaphase promoting complex/cyclosome (APC/C) and the Skp1/Cullin/F-box (SCF) protein. Through targeting specific proteins for timely degradation, these complexes not only ensure accurate control of the cell cycle, but also ensure precise regulation of the centrosome duplication cycle. Disruption of the centrosome cycle can lead to formation of aberrant or supernumerary centrosomes that in turn contribute to cell division errors and genetic instability. Recent progress has revealed that protein degradation mechanisms are central to many aspects of centriole biogenesis. By strictly regulating the abundance of core centriole assembly proteins, the number of new centrioles formed within each cell cycle is tightly controlled. Moreover, protein destruction is equally important in ensuring that centrioles of the correct length are formed, while licensing of centriole duplication, which occurs during mitosis, is also controlled by protein degradation. The major ubiquitin-mediated degradation events that ensure fidelity of the centrosome cycle will be considered in this chapter.

9.1 Introduction: Protein Degradation in Cell Cycle Control

The cell cycle describes a sequential order of events that culminate in cell division. These events are driven by oscillations in the level of cyclin-dependent kinase (Cdk) activity (Nurse 2000). However, phosphorylation is a rapidly reversible form of

---

S. L. Prosser · A. M. Fry (✉)
Department of Biochemistry, University of Leicester,
Lancaster Road, Leicester LE1 9HN, UK
e-mail: amf5@le.ac.uk

regulation and does not alone provide a robust mechanism to ensure that cells do not go backwards in the cell cycle. Cells have therefore evolved a second control mechanism to provide directionality and ensure that events keep moving forwards, that is the irreversible destruction of proteins (Murray 2004). Cyclins are the archetypal example of proteins that accumulate and then are abruptly destroyed in a cell cycle-dependent manner and these determine the oscillations in Cdk activity (Glotzer et al. 1991). However, many other proteins are similarly destroyed at specific times when their continued presence would be detrimental to further cell cycle progression. Most cell cycle-dependent protein degradation is mediated by the 26S proteasome. However, recognition by the proteasome requires tagging of the substrate protein with polyubiquitin chains, a process that involves an E1 ubiquitin activating enzyme, an E2 ubiquitin conjugating enzyme, and an E3 ubiquitin ligase (Hershko and Ciechanover 1998). It is the E3 ligase that determines substrate specificity and there are two RING-family E3 ligases in particular that regulate cell cycle transitions, namely the anaphase-promoting complex/cyclosome (APC/C), and the Skp1/Cullin/F-box (SCF) protein (Skaar and Pagano 2009).

The APC/C is a large multisubunit complex, containing at least 15 different protein components (Peters 2006). Its catalytic core consists of a cullin subunit, Apc2, and a RING-H2 protein, Apc11. Together, these facilitate the transfer of ubiquitin from the E2 enzyme onto the substrate. However, substrate recognition by the APC/C also requires one of two APC/C co-activator proteins, Cdc20 or Cdh1, which directs the APC/C to specific substrates at defined points in the cell cycle. The substrates themselves have particular amino acid sequence motifs, such as a D-box or KEN-box, which enable the APC/C-co-activator complexes to identify them. The APC/C was originally discovered for its role in the degradation of Cyclin B that promotes the transition from metaphase to anaphase (Glotzer et al. 1991). However, it is also required at this time for the proteolysis of securin that triggers sister chromatid separation. In fact, the APC/C degrades many cell cycle control proteins, including kinases, such as the Aurora kinases, Polo-like kinases (Plks), and NIMA-related kinases (Neks), along with proteins involved in regulating mitotic spindle formation and DNA replication.

Importantly, the activity of the APC/C is cell cycle regulated with high activity from mitosis to late G1 and low activity from S-phase to late G2. Activity during early mitosis, from prophase to metaphase, is associated with APC/C^{Cdc20}, whereas activity from anaphase to late G1 is associated with APC/C^{Cdh1}. In addition to co-activator binding, APC/C activity is controlled by phosphorylation and inhibitor proteins, such as Emi1 or the mitotic checkpoint proteins, Mad2, and BubR1. The fact that both co-activators and inhibitors of the APC/C are themselves subject to cell cycle-dependent degradation adds a further layer of control to the system.

In contrast to the APC/C, the SCF is active throughout the cell cycle, but its ability to ubiquitylate its substrates depends on their post-translational modification, which only occurs at specific times in the cell cycle (Cardozo and Pagano 2004). Substrate modification allows recognition by an F-box protein, of which about 70 have been identified in humans, with Skp2, Fbxw7, and β-TrCP having the most well-defined roles in regulating the cell cycle. In addition to an F-box protein, the SCF also consists of a cullin subunit, Cul1, a RING-H2 protein, Hrt/

Rbx1, and a linker subunit, Skp1, that, as in the APC/C, catalyze the transfer of ubiquitin from the E2 enzyme to the substrate.

9.2 The Centrosome Cycle

Being a single copy organelle, the centrosome is duplicated just once during each cell cycle. Hence, cells enter mitosis with two centrosomes that give rise to two spindle poles. This process ensures both spindle bipolarity and inheritance of a single centrosome by each daughter cell. The centrosome cycle must therefore proceed in a timely fashion that is carefully coordinated with the cell cycle. Failure to coordinate these events could lead to the generation of supernumerary centrosomes and genetic instability (Nigg 2007; Tsou and Stearns 2006).

The centrosome cycle can be viewed as four discrete steps: centriole duplication, centrosome maturation, centrosome disjunction, and centriole disengagement. Centriole duplication commences in S-phase with the appearance of procentrioles lying perpendicular and in very close proximity to the proximal ends of the existing centrioles. This association establishes the tight orthogonal arrangement of the parental and progeny centrioles that is maintained through to late mitosis. The two procentrioles elongate in a proximal to distal direction, reaching full length in G2. In late G2, centrosome maturation (or enlargement) occurs through recruitment of additional pericentriolar material (PCM) in preparation for the increased microtubule nucleating activity required in mitosis. However, the duplicated centrosome, now containing four centrioles, still continues to act as a single microtubule organizing center due to the presence of a loose tether that links the proximal ends of the parental centrioles throughout interphase. As cells progress from G2 into mitosis, centrosome disjunction occurs and the tether is severed. The physical separation of the two centrosomes in space is driven by motor proteins and ultimately gives rise to the two poles of the mitotic spindle. The centrosome cycle is completed with centriole disengagement in late mitosis. Each centriole pair, residing at the spindle poles, loses its tight orthogonal attachment, thereby licensing centriole duplication in the next cell cycle. As cells enter G1, the pair of now disengaged centrioles become tethered once again through their proximal ends regenerating the single microtubule organizing center typical of an interphase cell.

Together, these events are subjected to strict spatial and numerical control to ensure that the centrosome cycle occurs once and only once per cell cycle (Nigg and Stearns 2011). In line with the classical cell cycle, the centrosome cycle is also subject to control by both reversible phosphorylation and irreversible protein degradation (Fig. 9.1). It has been known for more than a decade now that components of both the APC/C and SCF ubiquitin ligases localize to the centrosome putting them in position to directly regulate events of the centrosome cycle (Freed et al. 1999; Raff et al. 2002). Indeed, mechanistic details are now beginning to emerge about how the process of protein degradation controls different events of the centrosome cycle, as described below.

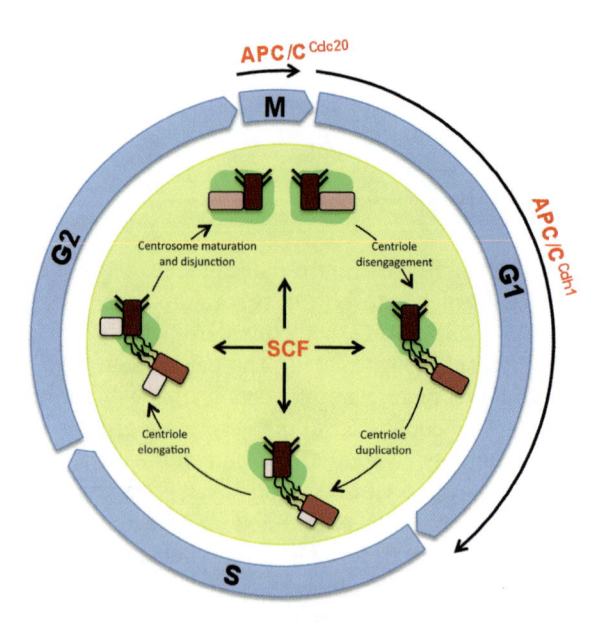

Fig. 9.1 *APC/C and SCF activity during the centrosome cycle.* The cell begins the centrosome cycle in G1 with one centrosome, consisting of a pair of centrioles loosely tethered at their proximal ends by a proteinaceous linker (*black strands*). The older, mother, centriole (*black cylinder*) is structurally distinct from the younger, daughter, centriole (*brown cylinder*) as it has additional appendages at its distal end; it is also associated with the bulk of the PCM (*green*). As the cell progresses into S-phase, centrosome duplication commences with the appearance of procentrioles (*pink cylinders*) at the proximal ends of the existing centrioles, a positioning that dictates the tight orthogonal arrangement of an engaged centriole pair, and which is maintained until late mitosis. The procentrioles elongate as the cell moves through G2, reaching full length at mitosis. At the G2/M transition, the linker between the original mother and daughter is lost, allowing the two centrosomes to move apart and form the poles of the bipolar mitotic spindle. At anaphase, centriole disengagement occurs, thereby licensing duplication in the subsequent cell cycle. Many key centrosomal proteins are regulated by protein degradation to ensure the fidelity of the centrosome cycle. The APC/C^{Cdc20} is active from early mitosis until anaphase, when the APC/C^{Cdh1} become active until late G1. The SCF is active throughout the cell cycle, being directed to its targets via their phosphorylation and subsequent recognition by specific F-box proteins

9.3 Regulation of Centriole Duplication by Protein Degradation

9.3.1 Regulation of Centriole Biogenesis

Early studies into the molecular control of centriole duplication revealed an essential requirement for Cdk2 (Hinchcliffe et al. 1999; Matsumoto et al. 1999; Lacey et al. 1999; Meraldi et al. 1999). This important discovery provided the first clue for how cells ensure that when they enter mitosis they will have duplicated both their DNA and centrosome. Cdk2 is activated at the G1/S transition through association with first, Cyclin E, and later, Cyclin A. Cyclins E and A are subjected to control by cell cycle-dependent degradation: Cyclin E by the SCF, in conjunction with the F-box protein, Fbxw7, and Cyclin A by the APC/C, in conjunction with first Cdc20, and later Cdh1. Cdk2 activity is also regulated by the inhibitor, p27, though this is degraded by the SCF, in conjunction with Skp2, at the onset of S-phase. Hence, cell cycle control of Cdk2 is mediated in large part by protein degradation and provides a neat mechanism to couple the onset of DNA replication and centriole duplication. However, while Cdk2 substrates that regulate DNA replication have been well characterized, validated substrates that promote, or possibly even inhibit, centrosome duplication remain frustratingly elusive.

Despite the uncertainty over how Cdk2 promotes centriole duplication, the molecular events of centriole assembly are now beginning to emerge (Bettencourt-Dias and Glover 2007; Loncarek and Khodjakov 2009; Strnad and Gonczy 2008). In brief, a core set of five proteins that are sufficient for centriole assembly was first identified in *C. elegans*; these include four structural proteins, SPD-2, SAS-4, SAS-5, and SAS-6, and a protein kinase, ZYG-1 (Pelletier et al. 2006; Leidel et al. 2005; Delattre et al. 2006; Delattre et al. 2004; Dammermann et al. 2004). SPD-2 and ZYG-1 act upstream of SAS-5 and SAS-6 to promote the assembly of a central tube. Singlet microtubules are then built on this tube under the action of SAS-4 to generate the simple centrioles typical of this organism. In higher eukaryotes, additional components are involved; however, these same five proteins have been highly conserved and also play crucial roles in the biogenesis of more complex centrioles. So, for example, SAS-6, together with the Bld10p/Cep135 protein, forms the cartwheel structure that assembles at the proximal end of a growing centriole and defines its 9-fold symmetry (Hiraki et al. 2007; Kitagawa et al. 2011; Nakazawa et al. 2007; van Breugel et al. 2011), SAS-5/Ana-2/STIL proteins are required for initiation of procentriole assembly (Stevens et al. 2010; Tang et al. 2011), while SAS-4/CPAP contributes to centriole elongation(Kohlmaier et al. 2009; Schmidt et al. 2009; Tang et al. 2009). Importantly, it is becoming clear that almost all these proteins are regulated at the level of protein degradation to ensure numerical and spatial control of centriole biogenesis.

Crucially, it would appear that many of these centriole assembly proteins are rate limiting with their expression level tightly regulated by protein degradation (Fig. 9.2). Perhaps the best studied of these so far is the polo-like kinase, Plk4, or SAK, which are the human and *Drosophila* orthologues of the *C. elegans* ZYG-1 kinase, respectively. Typical of these rate limiting proteins, Plk4 localizes to centrioles and experimental manipulation of Plk4 expression alters centriole numbers. So, downregulation of Plk4 leads to loss of centrioles over successive cell divisions, while overexpression leads to the formation of multiple procentrioles that form in a rosette around a single parent centriole (Bettencourt-Dias et al. 2005; Kleylein-Sohn et al. 2007; Habedanck et al. 2005). Plk4 levels also regulate centriole formation via the de novo assembly pathway, which occurs in the absence of a pre-existing centriole (Eckerdt et al. 2011; Peel et al. 2007; Rodrigues-Martins et al. 2007).

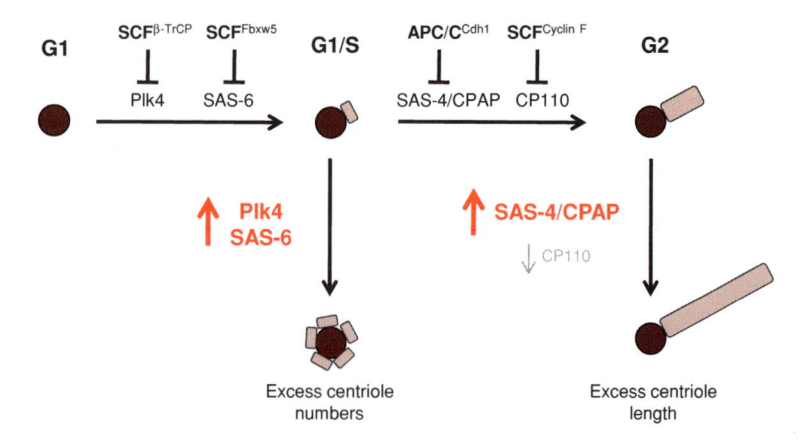

Fig. 9.2 *Control of centriole biogenesis and elongation by ubiquitin ligases.* Procentriole formation is dependent upon both Plk4 and SAS-6. Excess amounts of either protein leads to the formation of multiple procentrioles in a rosette arrangement around a single centriole. Their abundance is therefore strictly controlled to ensure only one procentriole forms per existing centriole. SCF$^{\beta\text{-TrCP}}$ targets Plk4 for destruction while SCF^{Fbxw5}, which is itself regulated by Plk4, targets SAS-6 for destruction. Procentriole elongation occurs as the cell progresses through S and G2 phases, and is regulated by the positive and negative regulators, SAS-4/CPAP and CP110, respectively. Excess SAS-4/CPAP leads to excessively long centrioles, while, conversely, loss of CP110 causes the same phenotype. However, the abundance of both proteins is strictly regulated, by the APC/C^{Cdh1} for SAS-4/CPAP, and SCF$^{Cyclin\ F}$ for CP110

Plk4 activity is regulated in large part then at the expression level and specifically by degradation, which is catalyzed by the SCF ubiquitin ligase, in conjunction with the β-TrCP (also called Slimb in *Drosophila*) F-box protein. β-TrCP was in fact demonstrated to have an important role in regulating centrosome duplication in *Drosophila* and mice well before the identification of Plk4 (Wojcik et al. 2000; Guardavaccaro et al. 2003), and has also been shown to localize to centrioles. It is now clear that β-TrCP interacts with Plk4 via a conserved β-TrCP binding motif on Plk4 and that mutation of this site, or depletion of β-TrCP, leads to Plk4 stabilization and centriole overduplication (Cunha-Ferreira et al. 2009; Rogers et al. 2009). As indicated above, the SCF recognizes its substrates following their post-translational modification, usually phosphorylation. In this case, Plk4 regulates its own degradation through autophosphorylation at a number of sites within a multiphosphodegron, within which lies the β-TrCP-binding site (Cunha-Ferreira et al. 2009; Rogers et al. 2009; Guderian et al. 2010; Holland et al. 2010). Plk4 stability is therefore under direct control of its own activity providing an important negative feedback mechanism that presumably helps to limit centriole duplication to once per cell cycle.

Interestingly, Plk4 activity does not peak at the time of centriole biogenesis in S-phase; instead, it remains at a relatively low level throughout interphase due to the stimulation of its own degradation. In contrast, Plk4 is stabilized in mitosis allowing its activity to rise at this point in the cell cycle. This stabilization can be explained by a peak in expression in mitosis of the protein phosphatase 2A (PP2A) regulatory subunit, Twins, which promotes dephosphorylation of the sites in Plk4 required for β-TrCP to bind (Brownlee et al. 2011). Plk4 does, though, also have a role in cytokinesis (Rosario et al. 2010); hence, its stabilization in mitosis may serve the purpose of cytokinesis, rather than centriole duplication. On the other hand, overexpression of Twins promotes centriole amplification in a manner analogous to Plk4 overexpression. This would suggest then that the activity of Plk4 in mitosis promotes centriole duplication in the ensuing S-phase.

A similar control to that described above exists in *C. elegans* where PP2A promotes centriole duplication through stabilizing the levels of not only ZYG-1, the Plk4 orthologue, but also SAS-5 (Song et al. 2011). This raises the prospect of course that SAS-5 levels are also controlled by phospho-dependent protein degradation, although at the current time the mechanism for this is unknown. Overexpression of the human SAS-5 orthologue, STIL, a microcephaly-related protein, promotes centriole overduplication while depletion inhibits centriole duplication (Tang et al. 2011). Furthermore, depletion of STIL blocks the centriole overduplication that occurs in response to Plk4 overexpression suggesting that STIL acts downstream of Plk4, in a manner that is analogous to the model in *C. elegans* that places SAS-5 downstream of ZYG-1.

Another key rate limiting step for centriole biogenesis is the expression level of SAS-6. As for Plk4 and SAS-5, the abundance of SAS-6 directly influences the rate of centriole formation with depletion from human cells resulting in centriole loss over successive cell cycles and overexpression inducing centriole overduplication (Leidel et al. 2005; Strnad et al. 2007). SAS-6 and SAS-5 play a mutually-dependent structural role in centriole assembly, and it would appear that strict control over the expression of both proteins is required to ensure that only one new procentriole forms per parental centriole. In this respect, it would seem odd that overexpression of one or the other is sufficient to drive centriole overduplication unless the proteins also mutually regulate the stability of each other. The levels of SAS-6 do oscillate during the cell cycle, with the protein accumulating from late G1 until it is degraded in mitosis. Consistent with this, SAS-6 is targeted for degradation by the APC/C^{Cdh1} through a KEN-box located at the C-terminal end of the protein (Strnad et al. 2007; Puklowski et al. 2011). While degradation of SAS-6 helps to limit the number of procentrioles seeded per parental centriole to one, maintaining low SAS-6 levels in G1 may also prevent centriole duplication commencing too early in the cell cycle (Strnad et al. 2007).

In addition to the APC/C^{Cdh1}, the SCF also targets SAS-6 for degradation via the Fbox protein, Fbxw5. Depletion of Fbxw5 from cells causes an increase in SAS-6 levels, centriole amplification, and multipolar spindles, while overexpression inhibits centriole formation and reduces the half-life of SAS-6 (Puklowski et al. 2011). Fbxw5 can bind SAS-6 and promote its ubiquitination, although a degron in SAS-6 has yet to be identified and it is unclear whether recognition by Fbxw5 depends on SAS-6 phosphorylation. Crucially, the levels of Fbxw5 are regulated by the APC/C, targeting it for degradation in mitosis, before the protein accumulates at the G1/S transition. However, with Fbxw5 present in S-phase, its binding to SAS-6 must be regulated to prevent complete degradation of SAS-6. Indeed, it appears that phosphorylation of Fbxw5 on Ser-151 by Plk4 somehow suppresses its ability to ubiquitinate SAS-6, preventing degradation. Thus, the low levels of Plk4 present in interphase may be sufficient to regulate the abundance of SAS-6 by controlling the ability of Fbxw5 to target SAS-6 for degradation. However, as Fbxw5 depletion only partially rescues the loss of centrioles caused by Plk4 depletion, Fbxw5 cannot be the only Plk4 target involved in centriole assembly (Puklowski et al. 2011). This leaves open the possibility that Plk4 may phosphorylate other centriole assembly factors, such as SAS-4/CPAP, SAS-5/STIL, Cep152, or even SAS-6 directly (Hatch et al. 2010; Kitagawa et al. 2009), to promote centriole biogenesis.

9.3.2 Regulation of Centriole Length

After procentrioles are formed in S-phase, they elongate in a proximal to distal direction, reaching full length in G2 or mitosis. It was originally thought that centriole length was restricted principally by kinetic constraints, with the centriolar structural components unable to stably form a longer structure than that already observed in the mother centriole. However, a number of studies have now shown that centriole length is also regulated through proteolysis (Kohlmaier et al. 2009; Schmidt et al. 2009; Tang et al. 2009; Korzeniewski et al. 2010; Vidwans et al. 2003). SAS-4/CPAP and CP110 localize to the distal ends of growing procentrioles and were first described as forming a cap under which the elongating centriolar microtubules were inserted (Kleylein-Sohn et al. 2007). They have since been shown to have potentially opposing roles as positive and negative regulators, respectively, of centriole length. Indeed, overexpression of SAS-4/CPAP induces excessive centriole elongation, while it is depletion of CP110 that has a similar consequence (Schmidt et al. 2009; Tang et al. 2009; Korzeniewski et al. 2010). Not surprisingly, then, the abundance of each protein is again carefully regulated by cell cycle-dependent degradation.

SAS-4/CPAP is absent in G1, with levels gradually increasing from early S-phase to mitosis. As cells exit mitosis, SAS-4/CPAP protein levels fall, concurrent with decreased abundance of SAS-6. As SAS-4/CPAP binds to Cdh1, but not Cdc20, and contains a functional KEN- and D-box, it is likely to be a target of APC/C^{Cdh1}. If SAS-4/CPAP levels are not controlled, for example, due to overexpression of wild-type or a non-degradable mutant, extended microtubule filaments are generated with properties that are highly reminiscent of elongated centriolar structures. In fact, both parental centrioles and procentrioles undergo elongation in the presence of excess SAS-4/CPAP, with no increase in overall centriole numbers (Schmidt et al. 2009). SAS-4/CPAP contains both tubulin dimer- and microtubule-binding domains, along with an ability to destabilize microtubules (Cormier et al. 2009; Hsu et al. 2008), suggesting that SAS-4/CPAP is able to regulate the assembly of centriolar microtubules. Indeed, mutation of the site via which SAS-4/CPAP binds tubulin dimers inhibits the formation of the elongated centrioles induced by overexpression of the wild-type protein (Tang et al. 2009).

CP110 caps the distal ends of centrioles (Kleylein-Sohn et al. 2007), where it participates in restricting the length of the growing centriole and/or regulates primary cilia generation. Depletion of CP110 leads to microtubule extensions growing from both parental centrioles and procentrioles similar to those seen upon SAS-4/CPAP overexpression (Schmidt et al. 2009; Tang et al. 2009). Additionally, loss of CP110, or a protein that interacts with it, CEP97, induces the aberrant formation of primary cilia in cycling cells, while in quiescent cells, enforced expression of CP110 suppresses primary cilium assembly (Spektor et al. 2007). Furthermore, depletion of the centriolar kinesin, Kif24, causes the loss of CP110 from mother centrioles, inducing cilia formation but not elongated centrioles per se (Kobayashi and Dynlacht 2011). Through an unbiased screen, CP110 was identified as an interacting partner and substrate of Cyclin F (D'Angiolella et al. 2010). Cyclin F, also called Fbxo1, is the founding member of the F-box protein family, being the first protein in which the F-box motif was described (Bai et al. 1996). CP110 and Cyclin F interact on centrioles in G2, with CP110 subsequently ubiquitylated by $SCF^{Cyclin\ F}$ and degraded (D'Angiolella et al. 2010). Depletion of Cyclin F, or expression of a stable CP110 mutant unable to bind Cyclin F, leads to both multipolar spindles and asymmetric, bipolar spindles with lagging chromosomes. If Cyclin F and CP110 were silenced together these phenotypes were prevented (D'Angiolella et al. 2010). Potentially, Cyclin F also regulates CP110 levels to limit centrosome duplication to once per cell cycle, as depletion of CP110 prevents Plk4-induced centriole amplification (Kleylein-Sohn et al. 2007). However, CP110 levels may be controlled in mitosis so that cells enter the following cell cycle with low CP110 abundance to allow primary cilium formation if the cell enters quiescence.

9.4 Regulation of Centrosomes in Mitosis by Protein Degradation

9.4.1 Centrosome Maturation and Disjunction

The onset of mitosis is accompanied by major changes in centrosome organization. First of all, the PCM increases dramatically in size in a process known as centrosome maturation. This includes a substantial increase in the amount of γ-tubulin and its associated partners leading to an overall increase in the microtubule nucleation capacity of the centrosome (Khodjakov and Rieder 1999). Both Plk1 and Aurora A are required for centrosome maturation and, although the precise pathways for this remain to be defined, it almost certainly involves regulation of multiple centrosomal components (Haren et al. 2009; Kettenbach et al. 2011). At around the same time, the filamentous intercentriolar linker that has held the two parental centrioles in close proximity throughout interphase disassembles in a process known as centrosome disjunction. This leaves the two pairs of duplicated centrioles as distinct and independent structures that can now be moved to opposite poles of the cell through the action of microtubule-based motor proteins, such as Eg5. Again, this process is phosphorylation-dependent, with the kinases, Nek2 and SIK2, directly phosphorylating centriolar linker proteins, and with these two kinases themselves the downstream targets of phosphoregulation involving Plk1 and Mst2, in the case of Nek2, and PKA, in the case of SIK2 (Ahmed et al. 2010; Mardin et al. 2011; Mardin et al. 2010; O'Regan et al. 2007).

To date, protein degradation has not been strongly implicated in regulating the onset of either centrosome maturation or disjunction. The remodeling of microtubule organization that occurs upon mitotic onset may be due in part to displacement of the ninein-like protein, Nlp, which anchors γ-tubulin ring complexes to the centrosome in interphase, but not mitosis (Casenghi et al. 2003). Displacement of Nlp occurs in response to Plk1, and possibly Nek2, phosphorylation (Casenghi et al. 2003; Rapley et al. 2005); however, Nlp is also targeted for degradation by the APC/C in a manner that may require phosphorylation by Cdk1 (Wang and Zhan 2007; Zhao et al. 2010). That said, it seems unlikely that the rapid change in centrosome size associated with maturation is in any way dependent on Nlp degradation. The centriole linker proteins, C-Nap1 and rootletin, are also displaced from the centrosome following phosphorylation, but in this case there is no evidence that these proteins are degraded (Bahe et al. 2005; Mayor et al. 2002). Indeed, as mitosis is a relatively short event and cells need to reestablish the interphase centrosome architecture from the beginning of G1, then it makes sense for the cell to retain what are often very large proteins. C-Nap1, for example, immediately loads back onto centrosomes at the end of mitosis presumably as a result of its dephosphorylation (Mayor et al. 2002).

In contrast, many of the regulators that trigger these two events are degraded at some point after mitotic entry and this may also be very important to allow rapid reassembly of the centrosome structure in G1. Nek2, Plk1, and Aurora A are all targeted by the APC/C for proteasomal-mediated degradation. Interestingly, Nek2 is degraded earlier than these other kinases in a manner that is independent of the spindle assembly checkpoint (Hayes et al. 2006). This requires Cdc20, although in the case of Nek2 degradation the role of Cdc20 is to promote its ubiquitylation rather than its recruitment to the APC/C (Kimata et al. 2008). Plk1 and Aurora A are degraded much later in mitosis presumably reflecting their requirement for additional mitotic processes, such as spindle organization, chromosome segregation, and cytokinesis. Their degradation is mediated by the APC/C in partnership with its other co-activator, Cdh1 (Floyd et al. 2008). The importance of Plk1 and Aurora A degradation for completion of cytokinesis and cell division is well established and almost certainly explains why interfering with APC/C^{Cdh1} function late in mitosis leads to multinucleated cells and centrosome amplification (Kim et al. 2011; Wang and Kung 2011).

9.4.2 Centriole Disengagement

The final step in the centrosome cycle occurs in late mitosis when, at each spindle pole, the centriole pair loses its tight orthogonal attachment. This process, originally called disorientation but now more commonly referred to as disengagement, is a key licensing step for the next round of centriole duplication that will take place in the subsequent cell cycle. While we know very little about the nature of the "glue" that binds procentrioles to their parents, pathways that regulate disengagement are beginning to emerge and, as for centriole duplication, there are key roles for protein degradation.

Intriguingly, much of what has been learned about the regulation of disengagement has come from the educated guess that the process of disengagement is not dissimilar to that of sister chromatid segregation that also occurs in late mitosis. The separation of sister chromatids results from the APC/C^{Cdc20}-mediated degradation of securin. This releases the cysteine protease, separase that in turn cleaves the Scc1/kleisin subunit of cohesin. Having found that APC/C activity is required for centriole disengagement in an in vitro disengagement assay established in *Xenopus* egg extracts, Tsou and Stearns hypothesized that separase may also play a key role in centriole disengagement (Tsou and Stearns 2006). Remarkably, they found that blocking separase activity in the egg extracts, either with an excess of cyclin B that can bind and inhibit separase or non-degradable securin, prevented centriole disengagement. A number of additional studies have since provided persuasive evidence that release of active separase is what triggers centriole disengagement in human cells (Thein et al. 2007; Tsou et al. 2009). Thus, activation of the APC/C and destruction of securin at the metaphase-anaphase transition provides an elegant mechanism of coupling the centrosome cycle with the cell division cycle. Moreover, it ensures that the two centrioles within a single spindle pole do not split apart until after the onset of chromosome segregation.

An obvious question is what separase cleaves to trigger centriole disengagement. To separate sister chromatids, it cleaves the cohesin ring that encircles the DNA sisters (Peters et al. 2008; Nasmyth 2011). It is not clear, though, how a ring could hold centriole pairs together. Although one can conceive of concatenated rings acting as a chain between the two centrioles, the most obvious solution, at least at first, was that separase has a distinct target at the centrosome. However, there is increasing evidence now for the existence of cohesin subunits at the centrosome and Stemmann and colleagues have elegantly demonstrated that ectopic cleavage of engineered cohesin rings can trigger centriole disengagement (Schockel et al. 2011). This highly provocative finding argues strongly that the cohesin ring may indeed represent both the sister chromatid and centriole pair "glue" explaining why separase triggers both events simultaneously at the onset of anaphase. However, the centrosomal protein, kendrin/pericentrin, has recently been proposed as an alternative separase target whose cleavage promotes centriole disengagement (Matsuo et al. 2012).

Further coupling of sister chromatid separation and centriole disengagement is afforded by the mitotic kinase, Plk1, which also plays a role in both processes. In the case of sister chromatids, Plk1 promotes the displacement of cohesin rings from chromosome arms in early mitosis by phosphorylating the Scc3/SA2 subunit of cohesin presumably leading to ring opening, while separase then cleaves the centromeric cohesin at the metaphase-anaphase transition (Peters et al. 2008; Nasmyth 2011). Regarding centriole disengagement, cells in which separase had been genetically deleted still completed disengagement, albeit with much delayed kinetics (Tsou et al. 2009). This delayed disengagement was absolutely dependent on Plk1. Moreover, it has now been found that arrest of cells in G2 through inhibition or depletion of Cdk1 leads to premature centriole disengagement and that loss of either separase or Plk1 can block this event (Loncarek et al. 2010; Steere et al. 2011; Prosser et al. 2012).

The role of the APC/C in promoting centriole disengagement through triggering degradation of securin and release of active separase is now clear. Indeed, the premature disengagement that occurs in G2 arrest relies on the untimely activation of the APC/C that occurs in response to Cdk1 inhibition (Prosser et al. 2012). Moreover, oscillation of APC/C activity during cell cycle arrest may explain the repeated rounds of centriole disengagement and reduplication that leads to centrosome amplification. Normally, during G2, the APC/C is held inactive by the inhibitor, Emi1 (Di Fiore and Pines 2007). However, Emi1 is itself subject to degradation triggered by the SCF$^{\beta\text{-TrCP}}$ ubiquitin ligase in response to phosphorylation of Emi1 by Plk1 (Hansen et al. 2004; Moshe et al. 2004). Thus, in the presence of sufficient Plk1 activity, Emi1 is degraded during prolonged arrest and the APC/C becomes active. In this sense, Plk1 acts upstream of the APC/C and separase in centriole disengagement. However, there is growing evidence that, as in the case of sister chromatid separation, Plk1 may also have an APC/C and separase independent role (Prosser et al. 2012), although whether this also involves direct phosphorylation of cohesin proteins is an intriguing question for the future.

It has been proposed that disengagement is a prerequisite for subsequent duplication because it frees up the site on the lateral wall of the parental centriole from which new procentrioles grow (Tsou and Stearns 2006). In other words, assuming that procentrioles emerge from a highly specific "birth-site" on the parental centriole and that there is only one such site per centriole, then while it is occupied another centriole cannot grow. While this is an elegant model, there is much to be learned about what defines the birth-site and why, when proteins like Plk4 or SAS-6 are overexpressed, centrioles can start growing from adjacent positions on the parental centriole.

9.5 Perspectives

As we begin to reveal the complex networks that regulate the centrosome duplication cycle, a few general paradigms are becoming clear. First, there is intimate coupling between the control of the DNA replication cycle and the centrosome cycle. This "cell cycle control" includes the use of many of the same regulatory enzymes, be they involved in phosphorylation or ubiquitination, and ensures that each centriole duplicates only once per cell cycle. Second, centriole biogenesis appears to be exquisitely sensitive to the expression level of centriole assembly factors. The extent of transcription, translation, and degradation of these proteins must be in perfect balance to ensure "copy number control", i.e., that only a single procentriole is nucleated by each parental centriole (Nigg 2007). Third, localization is critically important with the centrosome acting as the primary site of localization for both substrates for degradation and components of the ubiquitylation system, including E2 and E3 enzymes. This may also provide a mechanism for localized degradation that can be temporally, as well as spatially, separated from degradation throughout the rest of the cell (Mathe et al. 2004). Hence, the centrosome is not only under the control of degradation processes but can itself regulate such events.

Considerable advances have clearly been made in recent years into the molecular regulation of the centrosome duplication cycle and the role that protein degradation has in this process. However, much still remains to be learned; for example, what Plk4 phosphorylates in mitosis to regulate centriole assembly, what regulates the recognition of SAS-6 by Fbxw5, what controls expression of SAS-5, and what are the targets of separase and Plk1 that regulate centriole disengagement. With the current rapid rate of progress, one can expect answers to these and other related questions in the near future.

Acknowledgments We are grateful to all members of our laboratory for useful discussion and to The Wellcome Trust, the Biotechnology and Biological Sciences Research Council, the Association for International Cancer Research and Cancer Research, UK for supporting our research. AMF is a member of the Leicester Experimental Cancer Medicine Centre.

References

Ahmed AA et al (2010) SIK2 is a centrosome kinase required for bipolar mitotic spindle formation that provides a potential target for therapy in ovarian cancer. Cancer Cell 18:109–121

Bahe S, Stierhof YD, Wilkinson CJ, Leiss F, Nigg EA (2005) Rootletin forms centriole-associated filaments and functions in centrosome cohesion. J Cell Biol 171:27–33

Bai C et al (1996) SKP1 connects cell cycle regulators to the ubiquitin proteolysis machinery through a novel motif, the F-box. Cell 86:263–274

Bettencourt-Dias M, Glover DM (2007) Centrosome biogenesis and function: centrosomics brings new understanding. Nat Rev Mol Cell Biol 8:451–463

Bettencourt-Dias M et al (2005) SAK/PLK4 is required for centriole duplication and flagella development. Curr Biol 15:2199–2207

Brownlee CW, Klebba JE, Buster DW, Rogers GC (2011) The Protein Phosphatase 2A regulatory subunit Twins stabilizes Plk4 to induce centriole amplification. J Cell Biol 195:231–243

Cardozo T, Pagano M (2004) The SCF ubiquitin ligase: insights into a molecular machine. Nat Rev Mol Cell Biol 5:739–751

Casenghi M et al (2003) Polo-like kinase 1 regulates Nlp, a centrosome protein involved in microtubule nucleation. Dev Cell 5:113–125

Cormier A et al (2009) The PN2-3 domain of centrosomal P4.1-associated protein implements a novel mechanism for tubulin sequestration. J Biol Chem 284:6909–6917

Cunha-Ferreira I et al (2009) The SCF/Slimb ubiquitin ligase limits centrosome amplification through degradation of SAK/PLK4. Curr Biol 19:43–49

Dammermann A et al (2004) Centriole assembly requires both centriolar and pericentriolar material proteins. Dev Cell 7:815–829
D'Angiolella V et al (2010) SCF(Cyclin F) controls centrosome homeostasis and mitotic fidelity through CP110 degradation. Nature 466:138–142
Delattre M et al (2004) Centriolar SAS-5 is required for centrosome duplication in C. elegans. Nat Cell Biol 6:656–664
Delattre M, Canard C, Gonczy P (2006) Sequential protein recruitment in C. elegans centriole formation. Curr Biol 16:1844–1849
Di Fiore B, Pines J (2007) Emi1 is needed to couple DNA replication with mitosis but does not regulate activation of the mitotic APC/C. J Cell Biol 177:425–437
Eckerdt F, Yamamoto TM, Lewellyn AL, Maller JL (2011) Identification of a polo-like kinase 4-dependent pathway for de novo centriole formation. Curr Biol 21:428–432
Floyd S, Pines J, Lindon C (2008) APC/C Cdh1 targets aurora kinase to control reorganization of the mitotic spindle at anaphase. Curr Biol 18:1649–1658
Freed E et al (1999) Components of an SCF ubiquitin ligase localize to the centrosome and regulate the centrosome duplication cycle. Genes Dev 13:2242–2257
Glotzer M, Murray AW, Kirschner MW (1991) Cyclin is degraded by the ubiquitin pathway. Nature 349:132–138
Guardavaccaro D et al (2003) Control of meiotic and mitotic progression by the F box protein beta-Trcp1 in vivo. Dev Cell 4:799–812
Guderian G, Westendorf J, Uldschmid A, Nigg EA (2010) Plk4 transautophosphorylation regulates centriole number by controlling betaTrCP-mediated degradation. J Cell Sci 123:2163–2169
Habedanck R, Stierhof YD, Wilkinson CJ, Nigg EA (2005) The Polo kinase Plk4 functions in centriole duplication. Nat Cell Biol 7:1140–1146
Hansen DV, Loktev AV, Ban KH, Jackson PK (2004) Plk1 regulates activation of the anaphase promoting complex by phosphorylating and triggering SCFbetaTrCP-dependent destruction of the APC Inhibitor Emi1. Mol Biol Cell 15:5623–5634
Haren L, Stearns T, Luders J (2009) Plk1-dependent recruitment of gamma-tubulin complexes to mitotic centrosomes involves multiple PCM components. PLoS ONE 4:e5976
Hatch EM, Kulukian A, Holland AJ, Cleveland DW, Stearns T (2010) Cep152 interacts with Plk4 and is required for centriole duplication. J Cell Biol 191:721–729
Hayes MJ et al (2006) Early mitotic degradation of Nek2A depends on Cdc20-independent interaction with the APC/C. Nat Cell Biol 8:607–614
Hershko A, Ciechanover A (1998) The ubiquitin system. Annu Rev Biochem 67:425–479
Hinchcliffe, EH, Li, C, Thompson, EA, Maller, JL, Sluder, G (1999) Requirement of Cdk2-cyclin E activity for repeated centrosome reproduction in Xenopus egg extracts. Science 283:851–854
Hiraki M, Nakazawa Y, Kamiya R, Hirono M (2007) Bld10p constitutes the cartwheel spoke tip and stabilizes the 9-fold symmetry of the centriole. Curr Biol 17:1778–1783
Holland AJ, Lan W, Niessen S, Hoover H, Cleveland DW (2010) Polo-like kinase 4 kinase activity limits centrosome overduplication by autoregulating its own stability. J Cell Biol 188:191–198
Hsu WB et al (2008) Functional characterization of the microtubule-binding and -destabilizing domains of CPAP and d-SAS-4. Exp Cell Res 314:2591–2602
Kettenbach, AN et al (2011) Quantitative phosphoproteomics identifies substrates and functional modules of Aurora and Polo-like kinase activities in mitotic cells. Sci Signal 4, rs5
Khodjakov A, Rieder CL (1999) The sudden recruitment of gamma-tubulin to the centrosome at the onset of mitosis and its dynamic exchange throughout the cell cycle, do not require microtubules. J Cell Biol 146:585–596
Kim HS et al (2011) SIRT2 maintains genome integrity and suppresses tumorigenesis through regulating APC/C activity. Cancer Cell 20:487–499
Kimata Y, Baxter JE, Fry AM, Yamano H (2008) A role for the Fizzy/Cdc20 family of proteins in activation of the APC/C distinct from substrate recruitment. Mol Cell 32:576–583
Kitagawa D, Busso C, Fluckiger I, Gonczy P (2009) Phosphorylation of SAS-6 by ZYG-1 is critical for centriole formation in C. elegans embryos. Dev Cell 17:900–907
Kitagawa D et al (2011) Structural basis of the 9-fold symmetry of centrioles. Cell 144:364–375

Kleylein-Sohn J et al (2007) Plk4-induced centriole biogenesis in human cells. Dev Cell 13:190–202
Kobayashi T, Dynlacht BD (2011) Regulating the transition from centriole to basal body. J Cell Biol 193:435–444
Kohlmaier G et al (2009) Overly long centrioles and defective cell division upon excess of the SAS-4-related protein CPAP. Curr Biol 19:1012–1018
Korzeniewski N, Cuevas R, Duensing A, Duensing S (2010) Daughter centriole elongation is controlled by proteolysis. Mol Biol Cell 21:3942–3951
Lacey KR, Jackson PK, Stearns T (1999) Cyclin-dependent kinase control of centrosome duplication. Proc Nat Acad Sci U S A 96:2817–2822
Leidel S, Delattre M, Cerutti L, Baumer K, Gonczy P (2005) SAS-6 defines a protein family required for centrosome duplication in C. elegans and in human cells. Nat Cell Biol 7:115–125
Lindon C, Pines J (2004) Ordered proteolysis in anaphase inactivates Plk1 to contribute to proper mitotic exit in human cells. J Cell Biol 164:233–241
Loncarek J, Khodjakov A (2009) Ab ovo or de novo? Mechanisms of centriole duplication. Mol Cells 27:135–142
Loncarek J, Hergert P, Khodjakov A (2010) Centriole reduplication during prolonged interphase requires procentriole maturation governed by Plk1. Curr Biol 20:1277–1282
Mardin BR et al (2010) Components of the Hippo pathway cooperate with Nek2 kinase to regulate centrosome disjunction. Nat Cell Biol 12:1166–1176
Mardin BR, Agircan FG, Lange C, Schiebel E (2011) Plk1 controls the Nek2APP1gamma antagonism in centrosome disjunction. Curr Biol 21:1145–1151
Mathe E et al (2004) The E2-C vihar is required for the correct spatiotemporal proteolysis of cyclin B and itself undergoes cyclical degradation. Curr Biol 14:1723–1733
Matsuo K, Ohsumi K, Iwabuchi M, Kawamata T, Ono Y, Takahashi M (2012) Kendrin is a novel substrate for separase involved in the licensing of centriole duplication. Curr Biol 22 (in press)
Matsumoto Y, Hayashi K, Nishida E (1999) Cyclin-dependent kinase 2 (Cdk2) is required for centrosome duplication in mammalian cells. Curr Biol 9:429–432
Mayor T, Hacker U, Stierhof YD, Nigg EA (2002) The mechanism regulating the dissociation of the centrosomal protein C-Nap1 from mitotic spindle poles. J Cell Sci 115:3275–3284
Meraldi P, Lukas J, Fry AM, Bartek J, Nigg EA (1999) Centrosome duplication in mammalian somatic cells requires E2F and Cdk2-cyclin A. Nat Cell Biol 1:88–93
Moshe Y, Boulaire J, Pagano M, Hershko A (2004) Role of Polo-like kinase in the degradation of early mitotic inhibitor 1, a regulator of the anaphase promoting complex/cyclosome. Proc Natl Acad Sci U S A 101:7937–7942
Murray AW (2004) Recycling the cell cycle: cyclins revisited. Cell 116:221–234
Nakazawa Y, Hiraki M, Kamiya R, Hirono M (2007) SAS-6 is a cartwheel protein that establishes the 9-fold symmetry of the centriole. Curr Biol 17:2169–2174
Nasmyth K (2011) Cohesin: a catenase with separate entry and exit gates? Nat Cell Biol 13:1170–1177
Nigg EA (2007) Centrosome duplication: of rules and licenses. Trends Cell Biol 17:215–221
Nigg EA, Stearns T (2011) The centrosome cycle: centriole biogenesis, duplication and inherent asymmetries. Nat Cell Biol 13:1154–1160
Nurse P (2000) A long twentieth century of the cell cycle and beyond. Cell 100:71–78
O'Regan L, Blot J, Fry AM (2007) Mitotic regulation by NIMA-related kinases. Cell Div 2:25
Peel N, Stevens NR, Basto R, Raff JW (2007) Overexpressing centriole-replication proteins in vivo induces centriole overduplication and de novo formation. Curr Biol 17:834–843
Pelletier L, O'Toole E, Schwager A, Hyman AA, Muller-Reichert T (2006) Centriole assembly in Caenorhabditis elegans. Nature 444:619–623
Peters JM (2006) The anaphase promoting complex/cyclosome: a machine designed to destroy. Nat Rev Mol Cell Biol 7:644–656
Peters JM, Tedeschi A, Schmitz J (2008) The cohesin complex and its roles in chromosome biology. Genes Dev 22:3089–3114
Prosser SL, Samant M, Baxter JE, Morrison CG, Fry AM (2012) Oscillation of APC/C activity during cell cycle arrest promotes centrosome amplification (under revision)

Puklowski A et al (2011) The SCF-FBXW5 E3-ubiquitin ligase is regulated by PLK4 and targets HsSAS-6 to control centrosome duplication. Nat Cell Biol 13:1004–1009

Raff JW, Jeffers K, Huang JY (2002) The roles of Fzy/Cdc20 and Fzr/Cdh1 in regulating the destruction of cyclin B in space and time. J Cell Biol 157:1139–1149

Rapley J et al (2005) Coordinate regulation of the mother centriole component nlp by nek2 and plk1 protein kinases. Mol Cell Biol 25:1309–1324

Rodrigues-Martins A, Riparbelli M, Callaini G, Glover DM, Bettencourt-Dias M (2007) Revisiting the role of the mother centriole in centriole biogenesis. Science 316:1046–1050

Rogers GC, Rusan NM, Roberts DM, Peifer M, Rogers SL (2009) The SCF Slimb ubiquitin ligase regulates Plk4/Sak levels to block centriole reduplication. J Cell Biol 184:225–239

Rosario CO et al (2010) Plk4 is required for cytokinesis and maintenance of chromosomal stability. Proc Natl Acad Sci U S A 107:6888–6893

Schmidt TI et al (2009) Control of centriole length by CPAP and CP110. Curr Biol 19:1005–1011

Schockel L, Mockel M, Mayer B, Boos D, Stemmann O (2011) Cleavage of cohesin rings coordinates the separation of centrioles and chromatids. Nat Cell Biol 13:966–972

Skaar JR, Pagano M (2009) Control of cell growth by the SCF and APC/C ubiquitin ligases. Curr Opin Cell Biol 21:816–824

Song MH, Liu Y, Anderson DE, Jahng WJ, O'Connell KF (2011) Protein phosphatase 2A-SUR-6/B55 regulates centriole duplication in C. elegans by controlling the levels of centriole assembly factors. Dev Cell 20:563–571

Spektor A, Tsang WY, Khoo D, Dynlacht BD (2007) Cep97 and CP110 suppress a cilia assembly program. Cell 130:678–690

Steere N et al (2011) Centrosome amplification in CHO and DT40 cells by inactivation of cyclin-dependent kinases. Cytoskeleton (Hoboken) 68:446–458

Stevens NR, Dobbelaere J, Brunk K, Franz A, Raff JW (2010) Drosophila Ana2 is a conserved centriole duplication factor. J Cell Biol 188:313–323

Strnad P, Gonczy P (2008) Mechanisms of procentriole formation. Trends Cell Biol 18:389–396

Strnad P et al (2007) Regulated HsSAS-6 levels ensure formation of a single procentriole per centriole during the centrosome duplication cycle. Dev Cell 13:203–213

Tang CJ, Fu RH, Wu KS, Hsu WB, Tang TK (2009) CPAP is a cell-cycle regulated protein that controls centriole length. Nat Cell Biol 11:825–831

Tang CJ et al (2011) The human microcephaly protein STIL interacts with CPAP and is required for procentriole formation. EMBO J 30:4790–4804

Thein KH, Kleylein-Sohn J, Nigg EA, Gruneberg U (2007) Astrin is required for the maintenance of sister chromatid cohesion and centrosome integrity. J Cell Biol 178:345–354

Tsou MF, Stearns T (2006a) Controlling centrosome number: licenses and blocks. Curr Opin Cell Biol 18:74–78

Tsou MF, Stearns T (2006b) Mechanism limiting centrosome duplication to once per cell cycle. Nature 442:947–951

Tsou MF et al (2009) Polo kinase and separase regulate the mitotic licensing of centriole duplication in human cells. Dev Cell 17:344–354

van Breugel M et al (2011) Structures of SAS-6 suggest its organization in centrioles. Science 331:1196–1199

Vidwans SJ, Wong ML, O'Farrell PH (2003) Anomalous centriole configurations are detected in Drosophila wing disc cells upon Cdk1 inactivation. J Cell Sci 116:137–143

Wang, LY, Kung, HJ (2011) Male germ cell-associated kinase is overexpressed in prostate cancer cells and causes mitotic defects via deregulation of APC/C(CDH1). Oncogene

Wang Y, Zhan Q (2007) Cell cycle-dependent expression of centrosomal ninein-like protein in human cells is regulated by the anaphase-promoting complex. J Biol Chem 282:17712–17719

Wojcik EJ, Glover DM, Hays TS (2000) The SCF ubiquitin ligase protein slimb regulates centrosome duplication in Drosophila. Curr Biol 10:1131–1134

Zhao X, Jin S, Song Y, Zhan Q (2010) Cdc2/cyclin B1 regulates centrosomal Nlp proteolysis and subcellular localization. Cancer Biol Ther 10:945–952

Chapter 10
Molecular Links Between Centrosome Duplication and Other Cell Cycle-Associated Events

Kenji Fukasawa

Abstract The centrosome duplicates once in each cell cycle, and the duplication proceeds in coordination with other cell cycle events. Thus, centrosome duplication must crosstalk with other cell cycle events, including the growth signaling and DNA replication. Recent studies have identified several pathways that link the receptor tyrosine kinases (RTKs) activation and initiation of centrosome duplication. The molecular mechanisms of how those pathways link the RTK activation and initiation of centrosome duplication will be discussed in the first part of this chapter. Because centrosome duplication occurs at the time of S-phase entry, there is a mechanism that couples the initiation of centrosome duplication and DNA replication, and cyclin-dependent kinase 2 (CDK2)-cyclin E kinase complex is known to play a key role. In the second part of this chapter, through focusing of the p53-p21 pathway, the regulatory mechanisms underlying the coupling of initiation of centrosome duplication and DNA replication, and how loss of p53 leads to overduplication of centrosomes (centrosome amplification) will be discussed.

K. Fukasawa (✉)
Molecular Oncology Program, H. Lee Moffitt Cancer Center & Research Institute,
SRB-2, 12902 Magnolia Drive, Tampa, FL 33612-9416, USA
e-mail: kenji.fukasawa@moffitt.org

10.1 Link Between Activation of Growth Factor Receptors and Initiation of Centrosome Duplication: The Roles of the Rho-ROCK II Pathway and STAT Transcriptional Factors

When RTKs are activated by growth factor binding, they transmit the growth signals to a number of different pathways, and cells commence cell cycling processes. Because centrosome duplication is a cell cycle-dependent event, it is reasonable to predict that the activated RTKs also signal to initiation of centrosome duplication. It has been known that there is a close link between the activation of RTKs and initiation of centrosome duplication. For instance, in certain cell types, addition of epidermal growth factor (EGF) can rather rapidly induce physical separation of paired centrioles, which is an initial event of centrosome duplication (Sherline and Mascardo 1982). In the experimental system using Chinese hamster ovary cells that are cell cycle-arrested by exposure to DNA synthesis inhibitors, centrosomes continue to duplicate, resulting in generation of ≥ 3 centrosomes (centrosome amplification), but when serum is depleted from the media, centrosomes are no longer able to undergo duplication. Moreover, centrosomes resume duplication and reduplication in those arrested cells upon addition of serum or EGFs to the media (Balczon et al. 1995). It has also been known that oncogenic (constitutive) activation of RTKs such as the Met receptor leads to centrosome overduplication and amplification (Kanai et al. 2010; Nam et al. 2010; Fukasawa 2011).

What is the molecular pathway(s) that link the activation of RTKs occurring at cell membrane to the initiation of centrosome duplication occurring near the nuclear membrane? This question has recently been answered at least in part by the identification of ROCK II kinase as a key positive regulator of centrosome duplication (Ma et al. 2006) (Fig. 10.1). ROCK II is one of two members of the ROCK Ser/Thr kinase family. ROCK II is primed for activation by binding of GTP-bound Rho small GTPase (Rho-GTP): Rho binding disrupts the interaction between the kinase domain and autoinhibitory domain of ROCK II, freeing the kinase domain (Leung et al. 1996; Matsui et al. 1996). Rho cycles between an active GTP-bound state and inactive GDP-bound state, and many RTKs, when activated by the ligand binding, promote the exchange for Rho-bound GDP to GTP via activating the Rho guanine nucleotide exchange factors (Rho-GEFs) (Etienne-Manneville and Hall 2002). ROCK II was found to localize to centrosomes, and ectopic expression of the ROCK II mutant that lacks the negative regulatory domain (hence its activity is independent from Rho-binding) promotes initiation of centrosome duplication in a kinase activity and centrosome localization-dependent manners. Moreover, depletion of ROCK II results in significantly delayed initiation of centrosome duplication, indicating that ROCK II plays a critical role in the timely initiation of centrosome duplication. Of note, although the initiation of centrosome duplication is delayed in the ROCK II-depleted cell, they eventually duplicate because of the functional replacement by ROCK I, another member of the ROCK family, that shares ~ 65 % overall identity

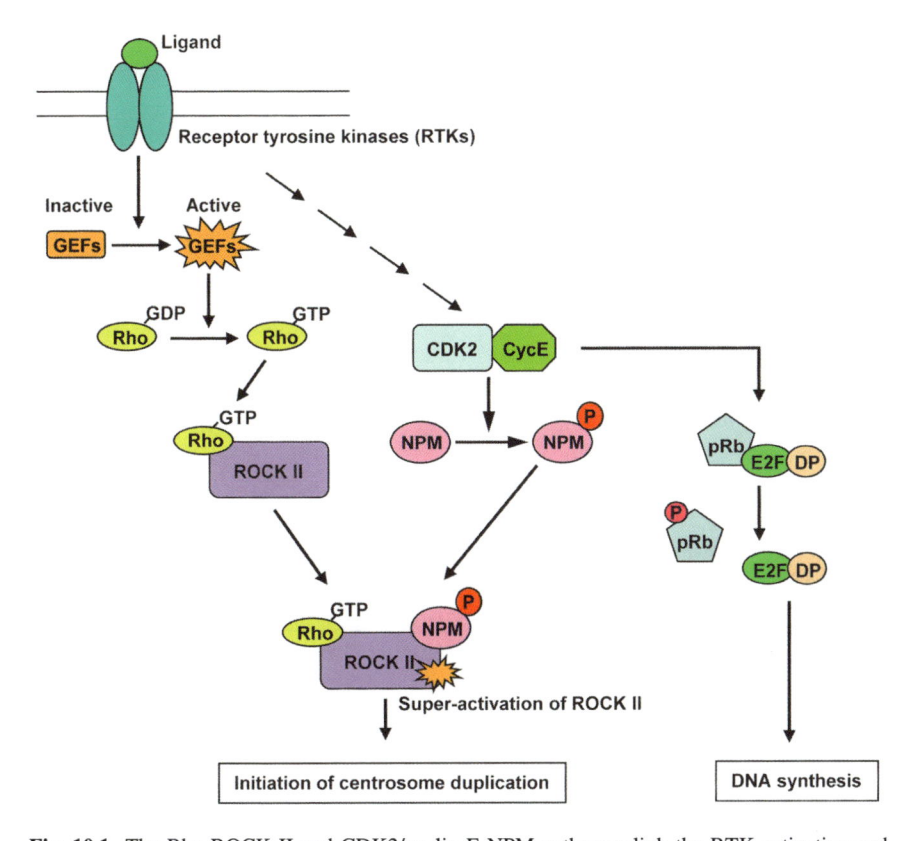

Fig. 10.1 The Rho-ROCK II and CDK2/cyclin E-NPM pathways link the RTK activation and initiation of centrosome duplication. The RTKs activated by the binding of growth factors activates Rho-GEFs, which in turn promotes exchange of Rho-GDP to Rho-GTP. Rho-GTP is then recruited to centrosomes, and binds to ROCK II at centrosomes. In late G1 phase, CDK2-cyclin E is activated, and phosphorylates NPM/B23 likely at centrosomes. NPM/B23 acquires a high binding affinity to ROCK II upon phosphorylation by CDK2-cyclin E, and binds to and superactivates ROCK II. The super activated ROCK II then triggers initiation of centrosome duplication. At the same time, CDK2-cyclin E targets proteins, including pRB which inhibits E2F transcriptional factor through direct binding. Upon phosphorylation by CDK2-cyclin E, pRB dissociates from E2F, resulting in activation of E2F and consequentially initiation of DNA replication

with ROCK II (Nakagawa et al. 1996). ROCK I also localize to centrosomes, and is implicated in proper positioning of centrosomes (Chevrier et al. 2002). ROCK I is dispensable for initiation of centrosome duplication as long as ROCK II is present. However, in the absence of ROCK II, ROCK I comes into replace the ROCK II function to promote centrosome duplication, but not as efficiently as ROCK II, resulting in the delay in the initiation of centrosome duplication. The reason behind the inefficient triggering of centrosome duplication by ROCK I is that unlike ROCK II, ROCK I cannot be super activated by nucleophosmin binding (discussed in details below) (Ma et al. 2006).

ROCK II is present at centrosomes throughout the cell cycle, and activated Rho (Rho-GTP) proteins are found at centrosomes much more than the inactive Rho (Rho-GDP) proteins (Kanai et al. 2010). Thus, Rho is likely recruited to centrosomes as Rho-GTP (after activation by Rho-GEFs), and binds to ROCK II at centrosomes. There are three major Rho isoforms, RhoA, B and C, and they share 85 % sequence identity (Etienne-Manneville and Hall 2002), yet each isoform is known to function in the specific cellular events (Wheeler and Ridley 2004). Although all isoforms are capable of binding and activating ROCK II in vitro, RhoA and RhoC, but not RhoB, are involved in the regulation of centrosome duplication. For instance, ectopic expression of constitutively active forms of RhoA (RhoA-V14) as well as RhoC (RhoC-V14) leads to promotion of centrosome duplication, while expression of RhoB-V14 has no effect on centrosome duplication (Kanai et al. 2010). The inability of RhoB to function in the regulation of centrosome duplication appears to be in part by its inability to localize to centrosomes. Although the primary target of RhoA and RhoC appears to be ROCK II for the regulation of centrosome duplication, both RhoA and RhoC are required for centrosome duplication. For instance, depletion of either RhoA or RhoC alone results in inhibition of centrosome duplication. Since expression of excess RhoA in the RhoC-depleted cells as well as expression of excess RhoC in the RhoA-depleted cells allow centrosome duplication, it is likely that RhoA and RhoC comprise the total amount of Rho proteins necessary for activating ROCK II (especially those present at centrosome) to promote centrosome duplication (Kanai et al. 2010). However, it remains as a possibility that RhoA and RhoC may also activate distinct targets in addition to ROCK II for promoting centrosome duplication.

Because activated RTKs signals to a number of pathways, the pathways other than the Rho-ROCK II pathway may also be involved in the promotion of centrosome duplication. For instance, activation of many RTKs leads to upregulation of STAT (signal transducer and activator) transcriptional factors. The activity of STAT3 has been shown to be essential for centrosome duplication by inducing the expression of some key centrosomal proteins such as PCM-1 and γ-tubulin (Metge et al. 2004). STAT3 induces expression of those proteins not by direct upregulation of the transcription of the respective genes, but does so indirectly likely by upregulating other transcriptional factor(s). STAT5, another member of the STAT family, has also been shown to promote centrosome duplication via inducing expression of Aurora-A (also known as STK15 and BTAK) (Hung et al. 2008), which is a positive regulator of centrosome duplication (Zhou et al. 1998). This study shows that the ligand-activated EGF receptor is translocated into the nucleus, and recruited to the AT-rich sequence sites of the Aurora-A promoter through interacting with STAT5.

In sum, RTKs activated by growth factor binding transmit the signal to centrosomes to duplicate through activation of the Rho-ROCK II pathway and transcriptional induction of the key centrosomal proteins and positive regulatory protein(s) essential for centrosome duplication. It should be noted here that other downstream pathways of RTKs may also function to link the RTK activation and initiation of centrosome duplication. For instance, the Ras-MAPK pathway and Akt (protein kinase B) pathway are both implicated in the regulation of centrosome duplication (Fukasawa and Vande Woude 1997; Zeng et al. 2010; Nam et al. 2010). Further studies should reveal the underlying mechanisms of the regulation of centrosome duplication by these pathways.

Overexpression and oncogenic mutation of RTKs are highly common in various types of cancers. It had been known that chromosomes become destabilized in cells transformed by oncogenically activated RTKs, but such a phenomenon had been belittled as an indirect consequence of the continuous firing of growth signals. However, the recent findings described above indicate that oncogenic activation of RTKs influences chromosome stability more directly than previously being thought via induction of centrosome amplification through upregulation of the Rho-ROCK II pathway and possibly other pathways as well as STAT-dependent transcription. Considering that chromosome instability plays a critical role in tumor progression, and centrosome amplification is one of the major causes of chromosome instability in cancer cells, induction of centrosome amplification and consequential destabilization of chromosomes should be appreciated as one of the key oncogenic activities of many RTKs.

10.2 Link Between Activation of Initiation of Centrosome Duplication and DNA Replication: The Roles of CDK2-Cyclin E and p53

As described in other chapters, CDK2-cyclin E plays a key role in the initiation of centrosome duplication (Hinchcliffe et al. 1999; Lacey et al. 1999; Tarapore et al. 2002). In normal cells, initiation of centrosome duplication occurs at the time of S-phase entry. Because CDK2-cyclin E is also a key triggering factor for DNA replication (Dulic et al. 1992; Koff et al. 1992), the coupling of initiation of centrosome duplication and DNA replication is at least in part achieved by the late G1-specific activation of CDK2-cyclin E resulting from the temporal increase of cyclin E expression. One of the targets of CDK2-cyclin E is nucleophosmin (NPM/B23). NPM/B23 is involved in the regulation of centrosome duplication both positively and negatively (Okuda et al. 2000; Grisendi et al. 2005; Ma et al. 2006), and CDK2-cyclin E-mediated phosphorylation of NPM/B23 on Thr199 residue simultaneously switches off the negative regulatory function and switches on the positive regulatory function. By the mediation of the Thr199-phosphorylated NPM/B23, the CDK2-cyclin E pathway and Rho-ROCK II pathway come together to trigger initiation of centrosome duplication (Fig. 10.1). ROCK II is not fully activated by the Rho binding: Rho binding results in only 1.5-fold increase in the kinase activity (Amano et al. 1996). NPM/B23 physically interacts with Rho-bound ROCK II (the NPM/B23-binding region of ROCK II is located near the kinase domain, and is masked by the negative regulatory domain in the nascent form of ROCK II, and thus NPM/B23 cannot bind to Rho-unbound ROCK II), and ROCK II is super activated (5–10-fold higher than unbound ROCK II) by the NPM/B23-binding (Ma et al. 2006). Although unphosphorylated NPM/B23 can bind to ROCK II, NPM/B23 acquires a significantly higher binding affinity to ROCK II upon Thr199 phosphorylation. Under a physiological condition where the protein concentrations are limited, especially the Rho-bound ROCK II at centrosomes, phosphorylation-dependent upregulation of the ROCK II-NPM/B23 interaction becomes essential. Indeed, most (if not all) of the ROCK II-bound NPM/B23 in cells are Thr199-phosphorylated. The superactivation of ROCK II by NPM/B23 binding is critical for the timely initiation of centrosome duplication, and is the primary downstream event of CDK2-cyclin E for the initiation of centrosome duplication. For instance, downregulation of the CDK2 activity either by expression of the dominant negative CDK2 or by depletion of cyclin E and cyclin A results in complete inhibition of centrosome duplication (Hanashiro et al. 2008), but introduction of the Rho-independent constitutively active ROCK II mutant can override the inhibition of centrosome duplication by inactivation of CDK2 in a NPM/B23 binding-dependent manner (Hanashiro et al. 2011). To sum up, in late G1, NPM/B23 acquires a high binding affinity to ROCK II by CDK2-cyclin E mediated phosphorylation, and binds to and superactivates the Rho-bound ROCK II, which in turn rapidly acts on the centrosomal target(s) to initiate centrosome duplication. At the same time, CDK2-cyclin E targets proteins like Rb to initiate DNA replication, and thus initiation of centrosome duplication and DNA replication occurs in a coordinated manner.

Because CDK2-cyclin E plays a key role in the initiation of centrosome duplication, the proteins that control the CDK2/cyclin E activity are also expected to be critically involved in the regulation of centrosome duplication. The p53 tumor suppressor protein and its transactivation target p21$^{\text{Waf1/Cip1}}$ (p21) CDK inhibitor are the well-known regulatory proteins of the CDK2 activity (Sherr and Roberts, 1999). The involvement of p53 in the regulation of centrosome duplication was initially recognized by the observations that cells and tissues from p53-deficient mice show a high frequency of centrosome amplification resulting from overduplication of centrosomes (Fukasawa et al. 1996, 1997). The subsequent studies have revealed how p53 participates in the regulation of centrosome duplication. p53 and p21 are known to present at a basal level in cycling cells, monitoring untimely activation of CDK2-cyclin E in early to mid G1 phase (Minella et al. 2002; Nevis et al. 2009). When cyclin E expression is induced at late G1, the concentration of active CDK2-cyclin E complexes rapidly increases to the level beyond the capacity of the p53-p21 monitoring system, leading to initiation of centrosome duplication as well as DNA replication. Indeed, overexpression of exogenously introduced cyclin E in cells harboring wild-type p53 (and thus, continual activation of CDK2-cyclin E beyond the capacity of the p53-p21 monitoring system) results in initiation of centrosome duplication in early G1 phase (Mussman et al. 2000). In the absence of p53, p21 cannot be transactivated, hence allowing fortuitous activation of CDK2-cyclin E in early and mid-G1. Because Rho-bound ROCK II are already available in early G1, CDK2-cyclin E prematurely triggers initiation of centrosome duplication through phosphorylation of NPM/B23 and consequential superactivation of ROCK II. However, because initiation of DNA replication requires many CDK2-cyclin E-independent events, CDK2-cyclin E can trigger DNA replication only after those events are completed, and thus the presence of active CDK2-cyclin E shortens the G1 duration for only few hours (Dulic et al. 1992; Koff et al. 1992). Thus, loss or mutational inactivation of p53 leads to uncoupling of initiation of centrosome duplication and DNA replication (Fig. 10.2). However, because uncoupling of initiation of centrosome duplication and DNA replication in cells lacking functional p53 depend on occurrence of "accidental" premature activation of CDK2-cyclin E, apparently the cells lacking functional p53 do not always experience the uncoupling of these two events, but in a long term, the majority of the p53-negative cells in a given population will experience uncoupling of centrosome duplication and DNA replication.

10.3 Loss of p53 and Centrosome Amplification

As described in other chapters, centrosome amplification leads to a high frequency of mitotic spindle defects and consequentially chromosome segregation errors. Centrosome amplification occurs frequently in various types of cancers, and is thought to be the major cause of chromosome instability in cancer cells (D'Assoro et al. 2002; Fukasawa 2005). Initially, induction of centrosome amplification by loss of p53 was identified in cells and tissues of p53-deficient mouse (Fukasawa et al. 1996, 1997). The mechanism of how loss of p53 leads to centrosome amplification was explored by the experimental system often referred to as "centrosome amplification (reduplication) assay", in which centrosomes undergo multiple rounds of duplication exposed to DNA synthesis inhibitors such as the DNA polymerase inhibitor (i.e., aphidicolin) and ribonucleotide reductase inhibitors (i.e., hydroxyurea (HU)), resulting in generation of amplified centrosomes. However, centrosome reduplication in the cell cycle-arrested cells occurs efficiently only when p53 is either inactivated or lost (Tarapore et al. 2001). In normal cells, p53 is upregulated in response to the physiological stress associated with the prolonged arrest by the ARF-mediated inhibition of Mdm2 (Sherr 2006) as well as DNA damages inflicted by the inhibitors by ATM/ATR- as well as Chk1/Chk2-mediated phosphorylation (Taylor and Stark 2001), leading to an increase in the intracellular level of p21, which in turn inhibits CDK2. Without the activity of CDK2, centrosome reduplication cannot be initiated. In contrast, in cells lacking functional p53, there will be no inhibitory mechanism for the CDK2 activity in response to the physiological and genotoxic stresses, and fortuitous activation of CDK2 leads to centrosome reduplication. This observation helped understanding the mechanism of how loss of p53 could lead to centrosome amplification (Fig. 10.3). Even under a normal growth condition/environment, cells are constantly subjected to internal as well as external stresses that temporarily halt cell

(a) Normal cells

- ⇨ CDK2-cyclin E-independent event
- ⇒ CDK2-cyclin E-dependent event

DNA replication

Centrosome duplication

| G1 | S |

Premature activation of CDK2/cyclin E

Late G1-specific activation of CDK2-cyclin E

p53→p21↑

(b) Cells with constitutively active CDk2/cyclin E

DNA replication

Centrosome duplication

| G1 | S |

Constitutively active CDK2-cyclin E at the level beyond the p53-p21 monitoring system

p53→p21↑

(c) Cells lacking functional p53

DNA replication

Centrosome duplication

| G1 | S |

Premature activation of CDK2-cyclin E

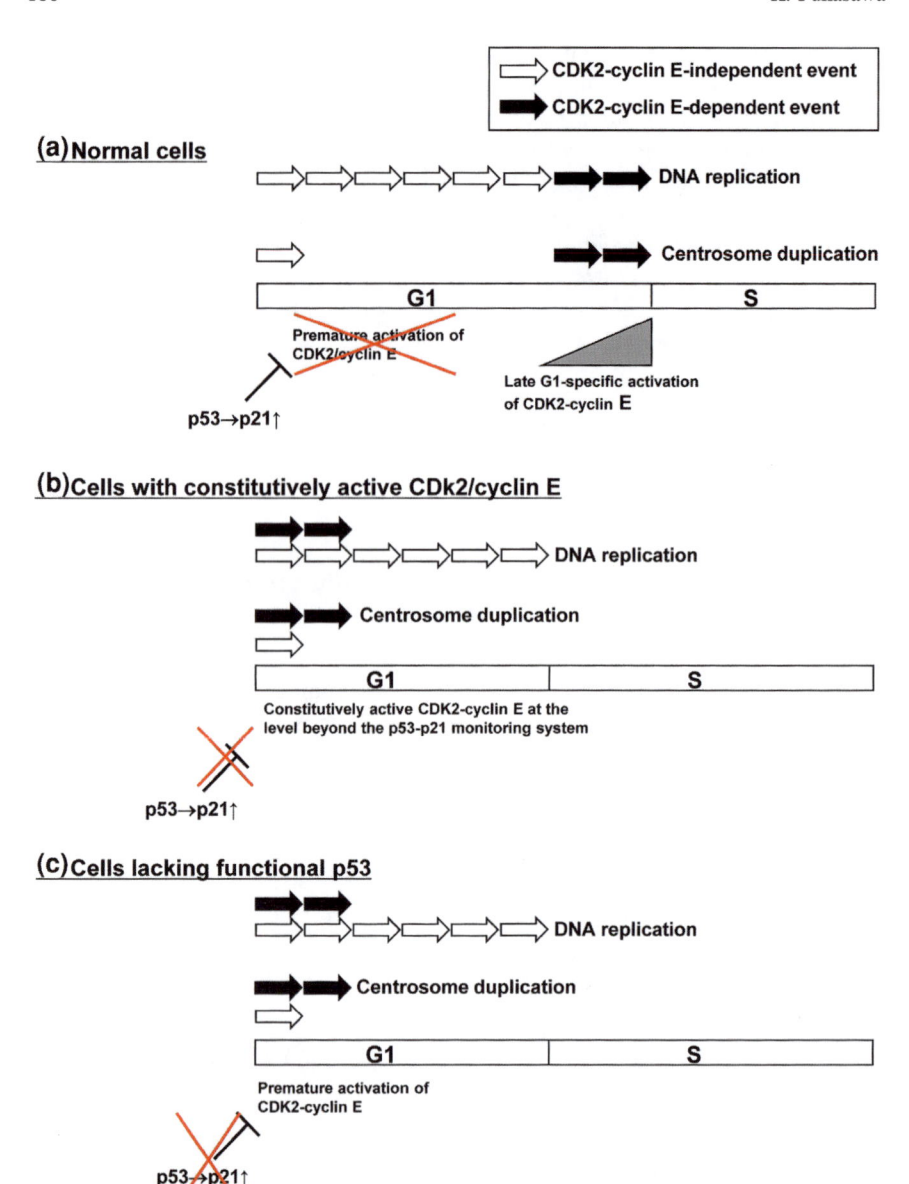

p53→p21↑

cycling irrespective of the p53 status (i.e., imbalance or deprivation of critical molecules such as dNTPs similar to the situation experimentally induced by HU treatment). In such cases, centrosomes reduplicate if cells lack functional p53, leading to centrosome amplification. Once the stress-causing problems are resolved, those cells resume cell cycling with amplified centrosomes.

◂**Fig. 10.2** The p53-p21 pathway monitors the premature activation of CDK2-cyclin E during G1 phase to ensure the coordinated initiation of centrosome duplication and DNA replication. CDK2-cyclin E is a triggering factor for both DNA replication and centrosome duplication. In normal cells, the basal levels of p53 and its transactivation target p21 monitor the premature activation of CDK2-cyclin E. In late G1, CDK2-cyclin E is activated by a temporal increase in cyclin E expression to the level beyond the p53-p21 monitoring capacity, and trigger initiation of both centrosome duplication and DNA replication (**a**). Initiation of DNA replication requires many CDK2-cyclin E-independent cellular events (*white arrows*) in addition to the CDK2-cyclin E-dependent events (*black arrows*). In contrast, initiation of centrosome duplication requires only few CDK2-cyclin E-independent cellular events (i.e., activation of ROCK II by Rho-binding) in addition to the CDK2-cyclin E-dependent events. Thus, constitutive activation of CDK2-cyclin E to the level beyond the p53-p21 monitoring capacity by cyclin E overexpression, centrosomes initiate duplication rapidly, while initiation of DNA replication occurs only after the completion of the CDK2-cyclin E-independent events. Thus, initiation of centrosome duplication and DNA replication is uncoupled (**b**). In the absence of p53, there would be no monitoring function to prevent premature activation of CDK2-cyclin E, and fortuitous activation of CDK2-cyclin E in early to mid G1 rapidly triggers centrosome duplication, but not DNA replication until the CDK2-cyclin E-independent events are completed. Thus, if p53 is lost or inactivated, cells experience uncoupling of initiation of centrosome duplication and DNA replication (**c**)

In mouse cells, loss or inactivation of p53 alone is sufficient to induce centrosome amplification at high frequencies (Fukasawa et al. 1996; Wang et al. 1998). However, it is not the case for human cells; inactivation/depletion of p53 in human primary fibroblasts by either expression of the dominant negative mutant p53 or small interfering RNA sequence targeting p53 does not efficiently induce centrosome amplification (Duensing et al. 2000; Bunz et al. 2002; Kawamura et al. 2004). Human cells are known to differ from mouse cells in the degrees of stringency in the regulation of cyclin E expression (Botz et al. 1996; Ekholm et al. 2001). In human cells, cyclin E expression is more strictly controlled than mouse cells, and occurs in a narrow window of late G1, and when cell cycle progression is halted, the activity of CDK2-cyclin E is tightly suppressed, and thus centrosome duplication remains blocked (Kawamura et al. 2004). In contrast, in mouse cells, cyclin E expression is relatively promiscuous, often showing the increased levels of cyclin E and active CDK2-cyclin E in the early-mid G1 (Mussman et al. 2000). In the absence of p53, the untimely activated CDK2-cyclin E is free from the p21-mediated inhibition, and triggers centrosome reduplication. In support of this scenario, loss of p53 together with cyclin E overexpression efficiently and synergistically induces centrosome amplification in human cells (Kawamura et al. 2004). These observations explain why the studies examining human cancer tissues have repeatedly shown conflicting results for the association between p53 mutation and chromosome instability (or centrosome amplification); while many studies detected a positive association between p53 mutation and chromosome instability/centrosome amplification, many failed to do so. The finding of the synergistic actions of p53 mutation and cyclin E overexpression for induction of centrosome amplification in human cells suggests that centrosome amplification (and consequential chromosome instability) by p53 mutation in tumors can be profoundly affected by the status of cyclin E expression and the activity of CDK2-

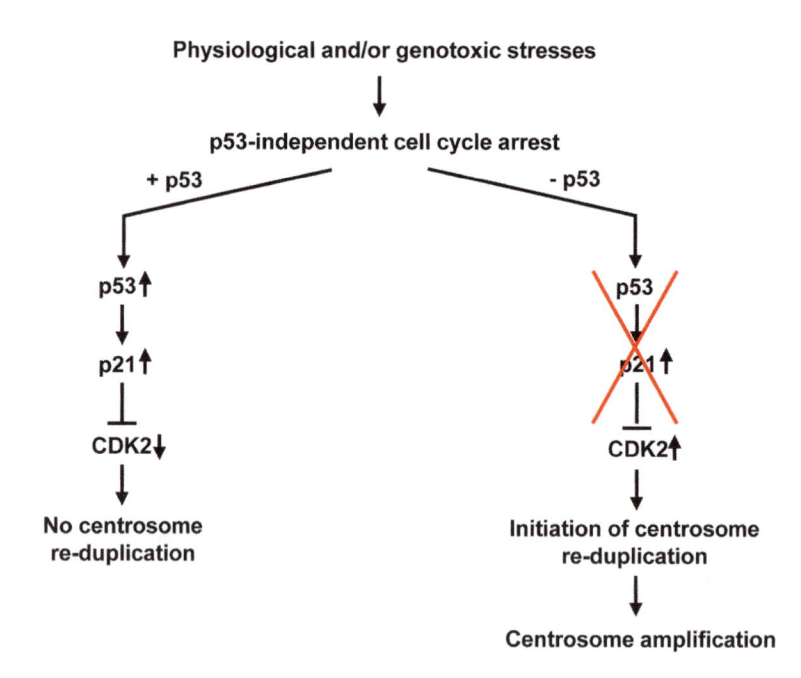

Fig. 10.3 Loss of p53 and centrosome amplification. Cells in any given populations and even under the optimal growth conditions are subjected to physiological and genotoxic stresses, resulting in the cell cycle arrest in a p53-independent manner. In cells with functional p53, p53 is upregulated by various mechanisms during the arrest. p53 then upregulates p21, which effectively inhibits CDK2. Without the activity of CDK2, newly duplicated centrosomes cannot undergo reduplication. In contrast, in cells lacking functional p53, p21 cannot be upregulated during the arrest, and if the active CDK2 is available, newly duplicated centrosomes undergo reduplication, resulting in centrosome amplification. Once the stress-causing problems are resolved, those cells will resume cell cycling with amplified centrosomes

cyclin E. The careful examination of bladder cancer specimens has revealed that this is indeed the case (Kawamura et al. 2004). This study shows that the occurrence of centrosome amplification parallels with increased frequencies of p53 mutation and cyclin E overexpression, and the multivariate analysis of the bladder cancer specimens in respect to status of p53, cyclin E expression and chromosome instability/centrosome amplification shows that there is a strong association between concomitant occurrence of p53 mutation and cyclin E overexpression and chromosome instability/centrosome amplification, but p53 mutation or cyclin E overexpression alone is not significantly associated with chromosome instability/centrosome amplification. Because cyclin E overexpression is frequent in many types of tumors (Keyomarsi and Herliczek 1997), the univariate analysis of p53 and chromosome instability/centrosome amplification tends to give a positive association. However, if cyclin E overexpression is a rare event in the tumor types under examination, the association between p53 mutation and chromosome instability/centrosome amplification will likely be weak. In support, it has been shown that upregulation of E2F activity, which can be equated to the activation of CDK2-cyclin E, synergistically induces chromosome instability (aneuploidy) with p53 mutation in lung carcinomas (Karakaidos et al. 2004). Moreover, considering that chromosome instability is the driving force for acquisition of more malignant phenotypes, it is also consistent with the earlier studies showing that concomitant occurrence of cyclin E overexpression and p53 mutation strongly correlates with poor prognosis of renal pelvis, ureter, and gastric carcinomas (Furihata et al. 1998; Sakaguchi et al. 1998).

10.4 Other CDK2 and p53 Modulators and Centrosome Amplification

Because of the involvement of p53 in the regulation of centrosome duplication, the proteins that control the stability of p53 is expected to participate in the regulation of centrosome duplication, and aberrant expression/activity of such proteins leads to centrosome amplification. For instance, Mdm2, an E3 ubiquitin ligase that promotes degradation of p53 (Haupt et al. 1997; Kubbutat et al. 1997), is frequently overexpressed in various types of cancers, especially in those retaining wild-type p53 (Momand and Zambetti 1997). When MDM2 is overexpressed in mouse cells harboring wild-type p53, the level of p53 decreases, leading to efficient induction of centrosome amplification (Carroll et al. 1999).

Besides the p53-p21 pathway, the activity of CDK2 is also controlled by other CDK inhibitors, including $p27^{Kip1}$ and p16(INK4a). Both $p27^{Kip1}$ and p16(INK4a) have been implicated in the regulation of centrosome duplication. For example, centrosome amplification associated with DNA damage requires downregulation of $p27^{Kip1}$ in certain cell types such as neuroblastoma cells (Sugihara et al. 2006). Loss of p16(INK4a) has also been shown to induce centrosome amplification (McDermott et al. 2006). In both cases, uncontrolled activation of CDK2-cyclin E was detected, which likely contributes to generation of amplified centrosomes.

10.5 Conclusion

It has been known that centrosome duplication occurs in coordination with other cell cycle-associated events. Thus, it is logical to predict that there are pathways that link centrosome duplication and the other cell cycle-associated events, including the RTK activation and DNA replication, and recent studies have started to identify those pathways. Regarding the molecular link between the RTK activation and centrosome duplication, the roles of the Rho-ROCK II pathway and STAT pathway were mainly discussed. However, the activated RTKs transmit the cell cycle signals to many downstream pathways, and other pathways may play equally important roles to link the growth stimulation and centrosome duplication, which remains to be determined in the future studies. As described in this chapter, the key cell cycle-associated events, including centrosome duplication, are linked by the common regulatory proteins and pathways, and that many of those regulatory proteins are oncogenic and tumor suppressor proteins frequently mutated in cancers (Fukasawa 2007). Mutational activation or inactivation of those regulatory proteins can lead to uncoupling of centrosome duplication from other cell cycle-associated events, which lays a ground for occurrence of centrosome amplification, and consequentially destabilization of chromosomes, and thus profoundly influences the tumor development.

Acknowledgments The preparation of this manuscript is in part supported by the grants from National Institute of Medicine and State of Florida.

References

Amano M, Ito M, Kimura K, Fukata Y, Chihara K, Nakano T, Matsuura Y, Kaibuchi K (1996) Phosphorylation and activation of myosin by Rho-associated kinase (Rho-kinase). J Biol Chem 271:20246–20249

Balczon R, Bao L, Zimmer WE, Brown K, Zinkowski RP, Brinkley BR (1995) Dissociation of centrosome replication events from cycles of DNA synthesis and mitotic division in hydroxyurea-arrested Chinese hamster ovary cells. J Cell Biol 130:105–115

Botz J, Zerfass-Thome K, Spitkovsky D, Delius H, Vogt B, Eilers M, Hatzigeorgiou A, Jansen-Durr P (1996) Cell cycle regulation of the murine cyclin E gene depends on an E2F binding site in the promoter. Mol Cell Biol 16:3401–3409

Bunz F, Fauth C, Speicher MR, Dutriaux A, Sedivy JM, Kinzler KW, Vogelstein B, Lengauer C (2002) Targeted inactivation of p53 in human cells does not result in aneuploidy. Cancer Res 62:1129–1133

Carroll PE, Okuda M, Horn HF, Biddinger P, Stambrook PJ, Gleich LL, Li YQ, Tarapore P, Fukasawa K (1999) Centrosome hyperamplification in human cancer: chromosome instability induced by p53 mutation and/or Mdm2 overexpression. Oncogene 18:1935–1944

Chevrier V, Piel M, Collomb N, Saoudi Y, Frank R, Paintrand M, Narumiya S, Bornens M, Job D (2002) The Rho-associated protein kinase p160ROCK is required for centrosome positioning. J Cell Biol 157:807–817

D'Assoro AB, Lingle WL, Salisbury JL (2002) Centrosome amplification and the development of cancer. Oncogene 21:6146–6153

Duensing S, Lee LY, Duensing A, Basile J, Piboonniyom S, Gonzalez S, Crum CP, Munger K (2000) The human papillomavirus type 16 E6 and E7 oncoproteins cooperate to induce mitotic defects and genomic instability by uncoupling centrosome duplication from the cell division cycle. Proc Natl Acad Sci U S A 97:10002–10007

Dulic V, Lees E, Reed SI (1992) Association of human cyclin E with a periodic G1-S phase protein kinase. Science 257:1958–1961

Ekholm SV, Zickert P, Reed SI, Zetterberg A (2001) Accumulation of cyclin E is not a prerequisite for passage through the restriction point. Mol Cell Biol 21:3256–3265

Etienne-Manneville S, Hall A (2002) Rho GTPases in cell biology. Nature 420:692–695

Fukasawa K, Choi T, Kuriyama R, Rulong S, Vande Woude GF (1996) Abnormal centrosome amplification in the absence of p53. Science 271:1744–1747

Fukasawa K, Wiener F, Vande Woude GF, and Mai S (1997) Genomic instability and apoptosis are frequent in p53 deficient mice. Oncogene 15:1295–1302

Fukasawa K, Vande Woude GF (1997) Synergy between the Mos/mitogen-activated protein kinase pathway and loss of p53 function in transformation and chromosome instability. Mol Cell Biol 17:506–518

Fukasawa K (2005) Centrosome amplification, chromosome instability and cancer development. Cancer Lett 230:6–19

Fukasawa K (2007) Oncogenes and tumour suppressors take on centrosomes. Nat Rev Cancer 7:911–924

Fukasawa K (2011) Aberrant activation of cell cycle regulators, centrosome amplification, and mitotic defects. Horm Cancer 2:104–112

Furihata M, Ohtsuki Y, Sonobe H, Shuin T, Yamamoto A, Terao N, Kuwahara M (1998) Prognostic significance of cyclin E and p53 protein overexpression in carcinoma of the renal pelvis and ureter. Br J Cancer 77:783–788

Grisendi S, Bernardi R, Rossi M, Cheng K, Khandker L, Manova K, Pandolfi PP (2005) Role of Nucleophosmin in embryonic development and tumorigenesis. Nature 437:147–153

Hanashiro K, Kanai M, Geng Y, Sicinski P, Fukasawa K (2008) Roles of cyclins A and E in induction of centrosome amplification in p53-compromised cells. Oncogene 27:5288–5302

Hanashiro K, Brancaccio M, Fukasawa K (2011) Activated ROCK II by-passes the requirement of the CDK2 activity for centrosome duplication and amplification. Oncogene 30:2188–2197

Haupt Y, Kazaz A, Oren M (1997) Mdm2 promotes the rapid degradation of p53. Nature 387:296–299

Hinchcliffe EH, Li C, Thompson EA, Maller JL, Sluder G (1999) Requirement of Cdk2-cyclin E activity for repeated centrosome reproduction in Xenopus egg extracts. Science 283:851–854

Hung LY, Tseng JT, Lee YC, Xia W, Wang YN, Wu ML, Chuang YH, Lai CH, Chang WC (2008) Nuclear epidermal growth factor receptor (EGFR) interacts with signal transducer and activator of transcription 5 (STAT5) in activating Aurora-A gene expression. Nucleic Acids Res 36:4337–4351

Kanai M, Crowe MS, Zheng Y, Vande Woude GF, Fukasawa K (2010) RhoA and RhoC are both required for the ROCK II-dependent promotion of centrosome duplication. Oncogene 29:6040–6050

Karakaidos P, Taraviras S, Vassiliou LV, Zacharatos P, Kastrinakis NG, Kougiou D, Kouloukoussa M, Nishitani H, Papavassiliou AG, Lygerou Z, Gorgoulis VG, Karakaidos P, Taraviras S, Vassiliou LV, Zacharatos P, Kastrinakis NG, Kougiou D, Kouloukoussa M, Nishitani H, Papavassiliou AG, Lygerou Z, Gorgoulis VG (2004) Overexpression of the replication licensing regulators hCdt1 and hCdc6 characterizes a subset of non-small-cell lung carcinomas: synergistic effect with mutant p53 on tumor growth and chromosomal instability–evidence of E2F-1 transcriptional control over hCdt1. Am J Pathol 165:1351–1365

Kawamura K, Izumi H, Ma Z, Ikeda R, Moriyama M, Tanaka T, Nojima T, Levin LS, Fujikawa-Yamamoto K, Suzuki K, Fukasawa K (2004) Induction of centrosome amplification and chromosome instability in human bladder cancer cells by p53 mutation and cyclin E overexpression. Cancer Res 64:4800–4809

Keyomarsi K, Herliczek TW (1997) The role of cyclin E in cell proliferation, development and cancer. Prog Cell Cycle Res 3:171–191

Koff A, Giordano A, Desai D, Yamashita K, Harper JW, Elledge S, Nishimoto T, Morgan DO, Franza BR, Roberts JM (1992) Formation and activation of a cyclin E-cdk2 complex during the G1 phase of the human cell cycle. Science 257:1689–1694

Kubbutat MH, Jones SN, Vousden K (1997) Regulation of p53 stability by Mdm2. Nature 387:299–303

Lacey KR, Jackson PK, Stearns T (1999) Cyclin-dependent kinase control of centrosome duplication. Proc Natl Acad Sci U S A 96:2817–2822

Leung T, Chen XQ, Manser E, Lim L (1996) The p160 RhoA-binding kinase ROK alpha is a member of a kinase family and is involved in the reorganization of the cytoskeleton. Mol Cell Biol 16:5313–5327

Matsui T, Amano M, Yamamoto T, Chihara K, Nakafuku M, Ito M, Nakano T, Okawa K, Iwamatsu A, Kaibuchi K (1996) Rho-associated kinase, a novel serine/threonine kinase, as a putative target for small GTP binding protein Rho. EMBO J 15:2208–2216

Ma Z, Kanai M, Kawamura K, Kaibuchi K, Ye K, Fukasawa K (2006) Interaction between ROCK II and nucleophosmin/B23 in the regulation of centrosome duplication. Mol Cell Biol 26:9016–9034

McDermott KM, Zhang J, Holst CR, Kozakiewicz BK, Singla V, Tlsty TD (2006) p16(INK4a) prevents centrosome dysfunction and genomic instability in primary cells. PLoS Biol 4:e51

Metge B, Ofori-Acquah S, Stevens T, Balczon R (2004) Stat3 activity is required for centrosome duplication in Chinese hamster ovary cells. J Biol Chem 279:41801–41806

Minella AC, Swanger J, Bryant E, Welcker M, Hwang H, Clurman BE (2002) p53 and p21 form an inducible barrier that protects cells against cyclin E-cdk2 deregulation. Curr Biol 12:1817–1827

Momand J, Zambetti G (1997) Mdm-2: "big brother" of p53. J Cell Biochem 64:343–352

Mussman JG, Horn HF, Carroll PE, Okuda M, Donehower LA, Fukasawa K (2000) Synergistic induction of centrosome hyperamplification by loss of p53 and cyclin E overexpression. Oncogene 19:1635–1946

Nakagawa O, Fujisawa K, Ishizaki T, Saito Y, Nakao K, Narumiya S (1996) ROCK-I and ROCK-II, two isoforms of Rho-associated coiled-coil forming protein serine/threonine kinase in mice. FEBS Lett 392:189–193

Nam HJ, Chae S, Jang SH, Cho H, Lee JH (2010) The PI3 K-Akt mediates oncogenic Met-induced centrosome amplification and chromosome instability. Carcinogenesis 31:1531–1540

Nevis KR, Cordeiro-Stone M, Cook JG (2009) Origin licensing and p53 status regulate Cdk2 activity during G(1). Cell Cycle 15:1952–1963

Okuda M, Horn HF, Tarapore P, Tokuyama Y, Smulian AG, Chan P-K, Knudsen ES, Hofmann IA, Snyder JD, Bove KE, Fukasawa K (2000) Nucleophosmin/B23 is a target of CDK2-cyclin E in centrosome duplication. Cell 103:127–140

Sakaguchi T, Watanabe A, Sawada H, Yamada Y, Yamashita J, Matsuda M, Nakajima M, Miwa T, Hirao T, Nakano H (1998) Prognostic value of cyclin E and p53 expression in gastric carcinoma. Cancer 82:1238–1243

Sherline P, Mascardo RN (1982) Epidermal growth factor induces rapid centrosomal separation in HeLa and 3T3 cells. J Cell Biol 93:507–512

Sherr CJ (2006) Divorcing ARF and p53: an unsettled case. Nat Rev Cancer 6:663–673

Sherr CJ, Roberts JM (1999) CDK inhibitors: positive and negative regulators of G1-phase progression. Genes Dev 13:1501–1512

Sugihara E, Kanai M, Saito S, Nitta T, Toyoshima H, Nakayama K, Nakayama KI, Fukasawa K, Schwab M, Saya H, Miwa M (2006) Suppression of centrosome amplification after DNA damage depends on p27 accumulation. Cancer Res 66:4020–4029

Tarapore P, Horn HF, Tokuyama Y, Fukasawa K (2001) Direct regulation of the centrosome duplication cycle by the p53-p21$^{Waf1/Cip1}$ pathway. Oncogene 20:3173–3184

Tarapore P, Okuda M, Fukasawa K (2002) A mammalian in vitro centriole duplication system: Evidence for involvement of CDK2/cyclin E and Nucleophosmin/B23 in centrosome duplication. Cell Cycle 1:75–81

Taylor WR, Stark GR (2001) Regulation of the G2/M transition by p53. Oncogene 20:1803–1815

Wang XJ, Greenhalgh DA, Jiang A, He D, Zhong L, Brinkley BR, Roop DR (1998) Analysis of centrosome abnormalities and angiogenesis in epidermal-targeted p53 172H mutant and p53-knockout mice after chemical carcinogenesis: evidence for a gain of function. Mol Carcinog 23:185–192

Wheeler AP, Ridley AJ (2004) Why three Rho proteins? RhoA, RhoB, RhoC, and cell motility. Exp Cell Res 301:43–49

Zeng X, Shaikh FY, Harrison MK, Adon AM, Trimboli AJ, Carroll KA, Sharma N, Timmers C, Chodosh LA, Leone G, Saavedra HI (2010) The Ras oncogene signals centrosome amplification in mammary epithelial cells through cyclin D1/Cdk4 and Nek2. Oncogene 29:5103–5112

Zhou H, Kuang J, Zhong L, Kuo WL, Gray JW, Sahin A, Brinkley BR, Sen S (1998) Tumour amplified kinase STK15/BTAK induces centrosome amplification, aneuploidy and transformation. Nat Genet 20:189–193

Chapter 11
Regulation of Centrosomes by Cyclin-Dependent Kinases

Rose Boutros

Abstract In eukaryotic cells, each cell division cycle involves two distinct replication cycles. These are the chromosomal DNA replication and the centrosome replication cycles. Progression through the cell division cycle is regulated by the activities of the cyclin–cyclin-dependent kinase complexes. These enzymes control both DNA replication and centrosome replication and ensure that the two cycles occur in synchrony.

11.1 CDK-Cyclins and Cell Cycle Progression

Cyclin-dependent kinases (CDKs) are a large family of serine/threonine protein kinases. The founding member, Cdc2 was identified in genetic screens from yeast as a mutant that caused cell division cycle defects (Russell and Nurse 1986). The human homologue (CDK1) was subsequently identified by its ability to rescue the yeast Cdc2 mutants (Lee and Nurse 1987). There are now 11 known genes that encode CDKs and nine genes that encode CDK-like proteins in mammalian cells (Malumbres and Barbacid 2005). The protein products of the best characterised CDKs control progression through the cell division cycle, in complex with their regulatory subunits, the cyclins (Fig. 11.1). Binding of a cyclin to a CDK induces a conformational change within the active site of the CDK and allows the kinase to become activated (Bourne et al. 1996; De Bondt et al. 1993). Thus, cell cycle

---

R. Boutros (✉)
Children's Medical Research Institute, The University of Sydney,
214 Hawkesbury Road, Westmead, NSW 2145, Australia
e-mail: rboutros@cmri.org.au

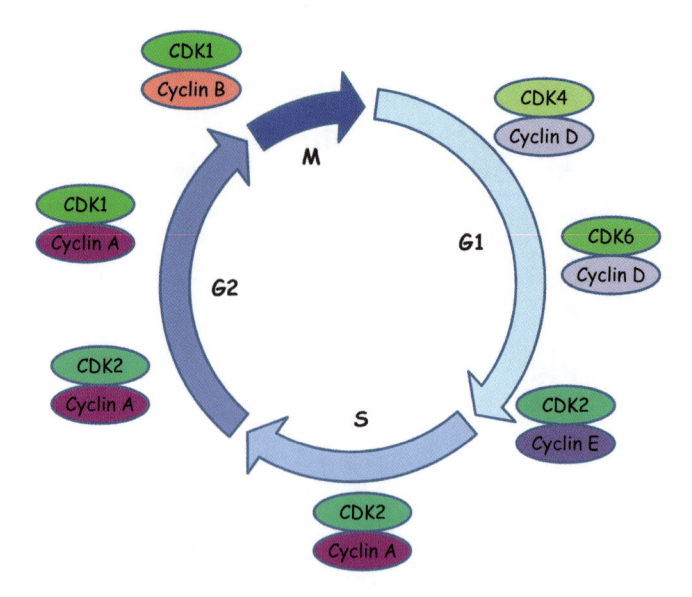

Fig. 11.1 Progression through the cell division cycle is controlled by the activities of specific CDK-cyclin complexes at each phase

progression is believed to be driven primarily by combinations of these CDKs and their cyclin partners.

Cyclins are a diverse family of proteins, all of which contain a stretch of 150 amino acids termed the 'cyclin box' (Malumbres and Barbacid 2005). They were first identified in marine invertebrates as proteins whose abundance oscillated during the cell cycle (Evans et al. 1983). There are at least 29 genes encoding cyclins in human cells, although not all have known CDK partners (Malumbres and Barbacid 2005). Those cyclins that do have known CDK partners and that regulate progression through the cell cycle fall into four major classes: D, E, A, and B type cyclins (Satyanarayana and Kaldis 2009). The D type cyclins bind CDK4 and CDK6 in G1 phase, the E type cyclins bind CDK2 at the G1–S phase transition, the A type cyclins bind CDK2 during S phase and CDK1 during G2 phase and the B type cyclins bind CDK1 during the G2–M transition and early mitosis (Satyanarayana and Kaldis 2009) (Fig. 11.1).

11.1.1 Regulation of CDK-Cyclin Activities

The catalytic activities of CDKs are tightly regulated in a strict spatio-temporal manner by a number of complementary mechanisms, including cyclin binding, changes in cyclin levels (determined by gene expression and proteolysis), protein phosphorylation and dephosphorylation, binding to CDK inhibitors, and subcellular localisation (Morgan 1995; King et al. 1996; Booher et al. 1989).

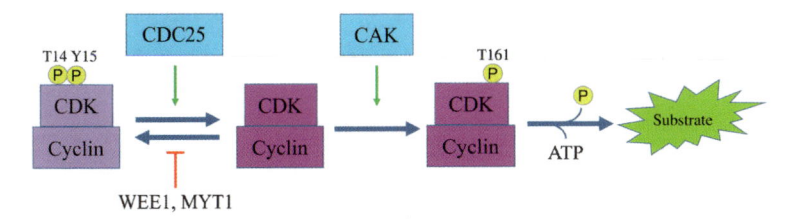

Fig. 11.2 Regulation of CDK-cyclin activity by phosphorylation. Phosphorylation of CDK on T14 and Y15 by the Wee1 and Myt1 kinases keeps the complex inactive. Dephosphorylation of these two residues by the CDC25 phosphatases activates the complex. Phosphorylation of one further residue, T161 within the activation loop of CDK is required for full activation of the complex. Once activated, CDK-cyclins can phosphorylate and activate their downstream substrates

Successive oscillations of cyclin levels throughout cell division control the overall activity of a given CDK-cyclin complex during each cell cycle phase. Cyclin protein levels are regulated by a balance between gene expression and protein degradation. Cyclin E for example, is expressed during a very narrow window at the G1-S transition. It becomes rapidly expressed in late G1 phase and then degraded very soon after forming a complex with CDK2, by ubiquitin-mediated proteolysis (Clurman et al. 1996; Won and Reed 1996). In contrast, Cyclin A is more stable throughout the cell cycle. It is expressed from early S phase and its protein levels continue to increase throughout S and G2 phase, during which the proteins complexes with CDK2 and CDK1, respectively. Cyclin A is then rapidly degraded during prometaphase (Hunt et al. 1992; Pines and Hunter 1990). Cyclins have also been reported to contribute to the substrate specificity of each CDK-cyclin complex (Peeper et al. 1993).

Phosphorylation of both CDK and cyclin subunits has been shown to regulate the level of activity of CDK-cyclin complexes. Phosphorylation of three critical residues (corresponding to T161, T14 and Y15 in mammalian cells) regulate CDK activity (Morgan 2007) (Fig. 11.2). T14 and Y15 phosphorylation by the Wee1 and Myt1 kinases keep the complex in an inactive form (Malumbres and Barbacid 2005). The opposing activities of the CDC25 protein phosphatases, which dephosphorylate CDK on these two residues, activate the complex (Fig. 11.2). Three CDC25 isoforms exist in mammalian cells (CDC25A, B and C), all of which co-operate to regulate the activities of the various CDK-cyclin complexes throughout cell division (Boutros et al. 2006, 2007), and all of which are found at the centrosome (Schmitt et al. 2006; Bonnet et al. 2008; Shreeram et al. 2008). Phosphorylation of T161 on CDK by CDK-activating kinase (CAK) is required for full activation of the kinase (Kaldis 1999) (Fig. 11.2).

CDK activities are also regulated by CDK inhibitors. These bind to and inactivate CDKs, either prior to the requirement of their activity or in response to cellular stress signals, such as DNA damage. Mammalian cells have a number of CDK inhibitors, including p21, p27 and p57, which inhibit CDK2 and p15, p16, p18 and p19 which inhibit CDK4 and CDK6 (Morgan 2007).

Some cyclins contain sequences that target them and their CDK partner to specific subcellular localisations. For example, cyclin B1 encodes a cytoplasmic retention sequence whose phosphorylation triggers the nuclear localisation of CDK1-cyclin B1. This is essential for phosphorylation of nuclear lamina A which triggers the breakdown of the nuclear envelope in prophase (Li et al. 1997). Recently, the A and E type cyclins have been found to encode centrosome localisation sequences that target these proteins to the centrosome (Matsumoto and Maller 2004; Pascreau et al. 2010).

11.2 CDK Control of the Centrosome

Each cell inherits a single centrosome at the end of cell division and a single copy of DNA. In order for accurate chromosome segregation during the next cell division, both the chromosomal DNA and the centrosome must replicate, once. Centrosome replication thus commences with DNA replication at the G1-S phase transition. CDK2-cyclin E in the nucleus initiates DNA replication through phosphorylation of the Retinoblastoma (Rb) protein and activation of the E2F transcription factor (Stevaux and Dyson 2002). Similarly, CDK2-cyclin E at the centrosome is believed to initiate centrosome replication by phosphorylation of its centrosome substrates, such as nucleophosmin (Hinchcliffe and Sluder 2002).

11.2.1 CDK2-Cyclin E/Cyclin A Control of Centrosome Replication

The first step in centriole replication is centriole disorientation—the loss of orthogonal association between the mother and daughter centrioles—in late G1 phase. This process was shown to be dependent on CDK2-cyclin E. Centriole disorientation was found to occur in late G1 phase in isolated mammalian cells that were incubated with control Xenopus egg extracts containing normal CDK2-cyclin E activity but not in extracts that had been treated with the CDK2-cyclin E inhibitors p21 or p27 (Lacey et al. 1999).

Centriole disorientation is followed in early S phase, by the appearance of small procentriole structures oriented at right angles to the original centrioles, that elongate during S phase. Under normal cell cycle conditions, only one procentriole forms perpendicular to each existing centriole (Tsou and Stearns 2006). However, if cells are arrested in S phase for prolonged periods, by inhibitors of DNA or protein synthesis for example, multiple procentrioles can form next to each existing centriole through repeated cycles of centrosome replication (Balczon et al. 1995). This experimental uncoupling of centrosome replication from DNA replication was exploited in a series of reports in 1999 that demonstrated that the formation of procentrioles in S phase is dependent on the activity of CDK2. The Sluder laboratory used Xenopus egg extracts and sperm nuclei to demonstrate that cells incubated with control egg extracts could form multiple centrosomes following S phase arrest with the DNA polymerase inhibitor aphidicolin. However, cells incubated with egg extracts containing recombinant Xic-1^{p27}, a CDK2–cyclin E inhibitor, did not undergo repeated rounds of centrosome replication (Hinchcliffe et al. 1999). The Stearns lab used S phase-arrested Xenopus embryos in which individual blastomeres were microinjected with p21 or p27. Compared to control non-injected blastomeres that underwent multiple rounds of centrosome doubling, repeated centrosome doubling was not observed in blastomeres from the same embryo that had been microinjected with the CDK2 inhibitors (Lacey et al. 1999). Similar observations were made in mammalian cells. Nishida and colleagues demonstrated that Chinese hamster ovary (CHO) cells that were arrested in S phase by hydroxyurea treatment had high levels of CDK2 and could undergo repeated centrosome replication, whilst cells arrested in G1 phase by mimosine treatment had low levels of CDK2 and could not undergo multiple rounds of centrosome replication (Matsumoto et al. 1999). In addition, blocking CDK2 activity in S phase-arrested cells by treatment with roscovitine or butyrolactone or expression of p21 significantly inhibited the formation of extra centrioles (Matsumoto et al. 1999). The Nigg laboratory used hydroxyurea-arrested CHO cells to demonstrate a role for the Retinoblastoma (Rb)-E2F transcription factor pathway in centrosome reduplication (Meraldi et al. 1999). They found that CDK2 in complex with cyclin A, rather than cyclin E, was necessary for centrosome reduplication in this system (Meraldi et al. 1999).

More recent evidence for a requirement for CDK2 in centrosome replication came from a study on p53−/− CHO cells, which formed multiple centrin foci following G1/S arrest with hydroxyurea. These centrin foci were found to represent procentrioles that could mature into functional centrosomes and were dependent on the presence of active CDK2 (Prosser et al. 2009). We have also found that overexpression of the CDK-cyclin activator CDC25B in G1/S-arrested U2OS cells results in multiple centrosomes formed within a single S phase (Boutros et al. 2007). We found that centrosome re-replication in these cells could be blocked by specific inhibition of CDK2 but not CDK1 activity (R Boutros, unpublished data).

CDK2 in complex with cyclin E and/or cyclin A therefore appears to be required for the initiation of centrosome replication. However, the exact mechanism for this is unclear. And the findings that CDK2 knockout mice are viable suggest that other kinases, most likely CDK1, can take on the role of CDK2 in both centrosome replication and DNA replication (Berthet et al. 2003; Ortega et al. 2003). Nonetheless, a number of kinase targets for CDK2–cyclins E/A in centrosome replication have been identified to date, which suggest that CDK2 exerts its effects on centrosome replication through the timely phosphorylation of its substrates. The polo kinase 4 (PLK4), also essential for centrosome replication, functions in co-operation with CDK2 (Habedanck et al. 2005).

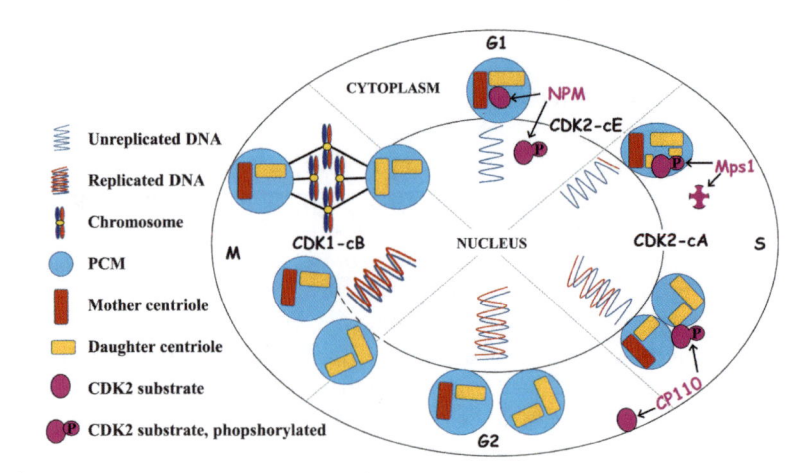

Fig. 11.3 The centrosome and DNA replication cycles. Both are initiated by the activity of CDK2-cyclin E at G1/S and maintained by CDK2-cyclin A in S/G2 phases. CDK1-cyclin B coordinates centrosome maturation with nuclear envelope breakdown and chromosome condensation at G2/M. Known centrosomal CDK2 substrates NPM, Mps1 and CP110 are shown (*pink*)

11.2.2 CDK2 Centrosomal Substrates

Once the importance of CDK2-cyclin E in centrosome replication was established, efforts turned to identifying its centrosomal targets. At least three centrosome proteins have been found to be directly phosphorylated by CDK2 in complex with cyclin E and/or cyclin A. These are Nucleophosmin (NPM) (Okuda et al. 2000), Monopolar spindle 1 (Mps1) (Fisk and Winey 2001), and Centrosome Protein of 110 kDa (CP110) (Chen et al. 2002) (Fig. 11.3).

11.2.2.1 NPM

To identify centrosomal CDK2-cyclin E targets, the Fukasawa laboratory performed an *in vitro* kinase reaction using centrosomes isolated from quiescent 3T3 mouse fibroblast cells, as substrate (Okuda et al. 2000). A single protein in the centrosome prep was found to be phosphorylated by CDK2-cyclin E and was identified as NPM (Okuda et al. 2000), also known as B23, a previously identified component of nucleolar granules (Yung et al. 1985). NPM was found to be recruited to the centrosomes during mitosis and remained at the unreplicated centrosome in early G1 phase, but was then lost following centrosome replication (Okuda et al. 2000). Microinjection of antibodies to NPM or overexpression of either a deletion mutant (NPMΔ186-239) or a non-phosphorylable mutant (NPM-T199A) of NPM blocked centrosome replication (Okuda et al. 2000; Tokuyama et al. 2001). The NPM-T199A mutant also remained associated with the centrosomes throughout the cell cycle and resulted in the formation of monopolar spindles in mitosis (Tokuyama et al. 2001).

NPM localises between the centriole pair of the mother centrosome in G1 phase to negatively regulate centrosome replication (Okuda et al. 2000; Grisendi et al. 2005). In late G1 phase, NPM is phosphorylated by CDK2–cyclin E on T199. This triggers its dissociation from the centrosome and its relocalisation to the nucleus and the start of centrosome replication (Tokuyama et al. 2001) (Fig. 11.3). NPM phosphorylation by CDK2–cyclin E therefore functions as a licensing factor for centrosome replication (Okuda et al. 2000). CDK2-cyclin A can also phosphorylate NPM on T199 *in vitro*, suggesting that NPM phosphorylation continues through S phase and may ensure that centrosome replication is not initiated a second time once cyclin E has been degraded (Tokuyama et al. 2001).

CDK1-cyclin B was shown to phosphorylate NPM on two alternative sites (T234 and T237) *in vitro* and it is possible that phosphorylation of these are involved in the recruitment of NPM to the centrosomes during mitosis, in preparation for the centrosome replication during the next cell division cycle (Tokuyama et al. 2001).

11.2.2.2 Mps1

The Mps protein kinases were first identified in yeast as temperature-sensitive mutants that were defective in the replication of the spindle pole, thus resulting in the formation of monopolar spindles during mitosis (Winey et al. 1991). Rather than arresting in metaphase, these mutants were found to continue cycling and segregate their DNA inappropriately, thus identifying a second role for Mps in the mitotic spindle assembly checkpoint (SAC) (Weiss and Winey 1996; Abrieu et al. 2001).

In addition to its mitotic roles, mammalian Mps1 was subsequently found to play a role in centrosome replication (Fisk and Winey 2001). Mouse Mps1 (mMps1) was found to localise to the centrosome in interphase as well as during mitosis. Overexpression of mMps1 caused centrosome re-replication in S phase arrested NIH3T3 cells and overexpression of a kinase-dead form (mMps1-KD) blocked centrosome replication (Fisk and Winey 2001). Initial functional analyses of human Mps1 (hMps1) did not support a direct role for this kinase in centrosome replication (Stucke et al. 2002). However, subsequent studies in human cells revealed that overexpression of hMps1 in S phase-arrested U2OS cells results in centrosome re-replication, while overexpression of a hMps1-KD blocked centrosome replication in a number of human cell lines (Fisk et al. 2003; Kanai et al. 2007). Recently, hMps1 overexpression was found to promote centrosome replication through phosphorylation of the structural centriole component centrin 2 (Yang et al. 2010). Mps1 phosphorylation of centrin 2 on three threonine residues (T45, T47, T118) stimulates the formation of new centrioles (Yang et al. 2010).

Mps1 itself is regulated by phosphorylation. Inhibition of CDK2 activity in S phase by treatment with chemical inhibitors of CDK2, resulted in loss of the centrosomal localisation of Mps1 and blocked centrosome re-replication induced by S phase arrest (Fisk and Winey 2001). Further examination of the role of CDK2 in Mps1 regulation at the centrosome revealed that CDK2 functions to promote the stability of Mps1 protein (Fisk and Winey 2001). A deletion mutant of Mps1 (Mps1Δ12/13, deletion of exons 12 and 13) was found to remain at the centrosome after CDK2 inhibition, suggesting that CDK2 phosphorylation of Mps1 within the region coded by exons 12–13 is responsible for regulation of its protein stability (Kasbek et al. 2007). Three phosphorylation sites were identified within this region (S436, T453, T468), which are regulated by the activities of both CDK2–cyclin E and CDK2–cyclin A kinases (Kasbek et al. 2007). A non-phosphorylable mutant of Mps1 (Mps1T468A) resulted in a loss in accumulation of Mps1 at the centrosome, suggesting that CDK2–cyclin A mediated phosphorylation of Mps1 at the centrosome protects the protein from proteasome-mediated degradation (Kasbek et al. 2007) (Fig. 11.3). Preventing the degradation of Mps1 at the centrosome, in the absence of CDK2 activity, was found to be sufficient to cause centrosome re-replication in S phase arrested cells. Thus, phosphorylation of Mps1 by CDK2 controls the level of Mps1 protein at the centrosome and restricts the number of centrosome replication cycles during each cell division cycle (Kasbek et al. 2007).

11.2.2.3 CP110

CP110 was identified as a CDK substrate during a screen of a human cDNA expression library with a dominant negative form of CDK2-cyclin E (Chen et al. 2002). It was found to be phosphorylated by CDK2-cyclin E, CDK2-cyclin A and CDK1-cyclin B *in vitro*. CP110 was subsequently found to be a centrosomal protein, which specifically co-localised with centrin to the centrioles (Chen et al. 2002). Similarly to NPM and Mps1, CP110 depletion was found to suppress centrosome re-replication in S phase arrested U2OS cells. Expression of a CP110 phosphorylation site mutant caused premature centrosome separation, which resulted in unscheduled mitotic entry and subsequent accumulation of polyploid cells (Chen et al. 2002). Phosphorylation of CP110 by CDK2 therefore suppresses premature centrosome separation, thereby regulating the timing of mitotic entry (Chen et al. 2002) (Fig. 11.3).

CP110 has also been shown to play a number of roles at the centrosome which are independent of phosphorylation by CDK-cyclins. CP110 contributes to the regulation of centriole elongation, by localising to the distal end of both the mother and daughter centrioles and functioning as a cap to limit centriole length (Schmidt et al. 2009). CP110 also plays a role in cytokinesis, through interactions with the proteins centrin 2 and calmodulin (Tsang et al. 2006).

11.2.3 CDK1-Cyclin B Control of Spindle Assembly and Mitosis

The G2-M transition is regulated by CDK1 in complex with cyclin B (Fig. 11.1). CDK1-cyclin B becomes activated in prophase by the CDC25 phosphatases (Gavet and Pines 2010), first at the centrosome and then in the nucleus (Jackman et al. 2003). Once activated, CDK1-cyclin B phosphorylates many mitotic substrates, resulting in massive architectural changes to the cell, such as centrosome maturation and separation, chromosome condensation and nuclear envelope breakdown. For example, CDK1-mediated phosphorylation of motor proteins, such as the kinesin-like protein Eg5, regulates centrosome separation and transformation into the bipolar spindle (Blangy et al. 1995). Nuclear CDK1-cyclin B phosphorylates other proteins, such as condensin, which results in chromosome condensation (Kimura et al. 1998; Kimura and Hirano 1997) and lamins, which cause nuclear membrane breakdown (Peter et al. 1990). However, CDK1 is not solely responsible for driving mitosis, as other kinases, such as the polo-like and aurora kinase families, also play key roles in regulating mitotic progression (Glover et al. 1998; Eyers and Maller 2003).

11.3 Conclusions

Centrosome replication is largely controlled by the activity of CDK2 in complex with cyclins E/A. Three CDK2 centrosome substrates have been identified to date and together, these are involved in all stages of centrosome replication—initiation, elongation and separation. In addition, phosphorylation of all three CDK substrates appears to control their local protein concentration at the centrosome. Dissociation of NPM and CP110 from the centrosome cause a decrease and protection of Mps1 from degradation causes an increase in centrosomal protein levels. Misregulation of any one of these can cause centrosome re-replication, resulting in abnormal centrosome numbers and abnormal mitotic spindles. Such defects are found in most human cancers and may contribute to tumourigenesis.

References

Russell P, Nurse P (1986) Cell 45:781–782
Lee MG, Nurse P (1987) Nature 327:31–35
Malumbres M, Barbacid M (2005) Trends Biochem Sci 30:630–641
Bourne Y, Watson MH, Hickey MJ, Holmes W, Rocque W, Reed SI, Tainer JA (1996) Cell 84:863–874
De Bondt HL, Rosenblatt J, Jancarik J, Jones HD, Morgan DO, Kim SH (1993) Nature 363:595–602
Evans T, Rosenthal ET, Youngblom J, Distel D, Hunt T (1983) Cell 33:389–396
Satyanarayana A, Kaldis P (2009) Oncogene 28:2925–2939
Morgan DO (1995) Nature 374:131–134
King RW, Deshaies RJ, Peters JM, Kirschner MW (1996) Science 274:1652–1659
Booher RN, Alfa CE, Hyams JS, Beach DH (1989) Cell 58:485–497
Clurman BE, Sheaff RJ, Thress K, Groudine M, Roberts JM (1996) Genes Dev 10:1979–1990
Won KA, Reed SI (1996) EMBO J 15:4182–4193
Hunt T, Luca FC, Ruderman JV (1992) J Cell Biol 116:707–724
Pines J, Hunter T (1990) Nature 346:760–763
Peeper DS, Parker LL, Ewen ME, Toebes M, Hall FL, Xu M, Zantema A, van der Eb AJ, Piwnica-Worms H (1993) EMBO J 12:1947–1954

Morgan DO (2007) The cell cycle: principles of control. Oxford University Press, Oxford
Boutros R, Dozier C, Ducommun B (2006) Curr Opin Cell Biol 18:185–191
Boutros R, Lobjois V, Ducommun B (2007a) Nat Rev Cancer 7:495–507
Schmitt E, Boutros R, Froment C, Monsarrat B, Ducommun B, Dozier C (2006) J Cell Sci 119:4269–4275
Bonnet J, Coopman P, Morris MC (2008) Cell Cycle 7:1991–1998
Shreeram S, Hee WK, Bulavin DV (2008) Mol Cell Biol 28:7442–7450
Kaldis P (1999) Cell Mol Life Sci 55:284–296
Li J, Meyer AN, Donoghue DJ (1997) Proc Natl Acad Sci U S A 94:502–507
Matsumoto Y, Maller JL (2004) Science 306:885–888
Pascreau G, Eckerdt F, Churchill ME, Maller JL (2010) Proc Natl Acad Sci U S A 107:2932–2937
Stevaux O, Dyson NJ (2002) Curr Opin Cell Biol 14:684–691
Hinchcliffe EH, Sluder G (2002) Oncogene 21:6154–6160
Lacey KR, Jackson PK, Stearns T (1999) Proc Natl Acad Sci U S A 96:2817–2822
Tsou MF, Stearns T (2006) Curr Opin Cell Biol 18:74–78
Balczon R, Bao L, Zimmer WE, Brown K, Zinkowski RP, Brinkley BR (1995) J Cell Biol 130:105–115
Hinchcliffe EH, Li C, Thompson EA, Maller JL, Sluder G (1999) Science 283:851–854
Matsumoto Y, Hayashi K, Nishida E (1999) Curr Biol 9:429–432
Meraldi P, Lukas J, Fry AM, Bartek J, Nigg EA (1999) Nat Cell Biol 1:88–93
Prosser SL, Straatman KR, Fry AM (2009) Mol Cell Biol 29:1760–1773
Boutros R, Lobjois V, Ducommun B (2007b) Cancer Res 67:11557–11564
Berthet C, Aleem E, Coppola V, Tessarollo L, Kaldis P (2003) Curr Biol 13:1775–1785
Ortega S, Prieto I, Odajima J, Martin A, Dubus P, Sotillo R, Barbero JL, Malumbres M, Barbacid M (2003) Nat Genet 35:25–31
Habedanck R, Stierhof YD, Wilkinson CJ, Nigg EA (2005) Nat Cell Biol 7:1140–1146
Okuda M, Horn HF, Tarapore P, Tokuyama Y, Smulian AG, Chan PK, Knudsen ES, Hofmann IA, Snyder JD, Bove KE, Fukasawa K (2000) Cell 103:127–140
Fisk HA, Winey M (2001) Cell 106:95–104
Chen Z, Indjeian VB, McManus M, Wang L, Dynlacht BD (2002) Dev Cell 3:339–350
Yung BY, Busch RK, Busch H, Mauger AB, Chan PK (1985) Biochem Pharmacol 34:4059–4063
Tokuyama Y, Horn HF, Kawamura K, Tarapore P, Fukasawa K (2001) J Biol Chem 276:21529–21537
Grisendi S, Bernardi R, Rossi M, Cheng K, Khandker L, Manova K, Pandolfi PP (2005) Nature 437:147–153
Winey M, Goetsch L, Baum P, Byers B (1991) J Cell Biol 114:745–754
Weiss E, Winey M (1996) J Cell Biol 132:111–123
Abrieu A, Magnaghi-Jaulin L, Kahana JA, Peter M, Castro A, Vigneron S, Lorca T, Cleveland DW, Labbe JC (2001) Cell 106:83–93
Stucke VM, Sillje HH, Arnaud L, Nigg EA (2002) EMBO J 21:1723–1732
Fisk HA, Mattison CP, Winey M (2003) Proc Natl Acad Sci U S A 100:14875–14880
Kanai M, Ma Z, Izumi H, Kim SH, Mattison CP, Winey M, Fukasawa K (2007) Genes Cells 12:797–810
Yang CH, Kasbek C, Majumder S, Yusof AM, Fisk HA (2010) Mol Biol Cell 21:4361–4372
Kasbek C, Yang CH, Yusof AM, Chapman HM, Winey M, Fisk HA (2007) Mol Biol Cell 18:4457–4469
Schmidt TI, Kleylein-Sohn J, Westendorf J, Le Clech M, Lavoie SB, Stierhof YD, Nigg EA (2009) Curr Biol 19:1005–1011
Tsang WY, Spektor A, Luciano DJ, Indjeian VB, Chen Z, Salisbury JL, Sanchez I, Dynlacht BD (2006) Mol Biol Cell 17:3423–3434
Gavet O, Pines J (2010) Dev Cell 18:533–543
Jackman M, Lindon C, Nigg EA, Pines J (2003) Nat Cell Biol 5:143–148

Blangy A, Lane HA, d'Herin P, Harper M, Kress M, Nigg EA (1995) Cell 83:1159–1169
Kimura K, Hirano M, Kobayashi R, Hirano T (1998) Science 282:487–490
Kimura K, Hirano T (1997) Cell 90:625–634
Peter M, Nakagawa J, Doree M, Labbe JC, Nigg EA (1990) Cell 61:591–602
Glover DM, Hagan IM, Tavares AA (1998) Genes Dev 12:3777–3787
Eyers PA, Maller JL (2003) Cell Cycle 2:287–289

Part III
Centrosome Abnormalities in Cancer

Chapter 12
Disruption of Centrosome Duplication Control and Induction of Mitotic Instability by the High-Risk Human Papillomavirus Oncoproteins E6 and E7

Nina Korzeniewski and Stefan Duensing

Abstract Centrosome abnormalities and genomic instability are hallmarks of major human malignancies and have been implicated in malignant progression as well as therapy resistance. Since the etiology of most cancers is complex and incompletely understood, it is vital to utilize tumors which are caused by limited oncogenic stimuli to explore causes and consequences of centrosome aberrations in cancer cells. High-risk HPV-associated neoplasms are suitable model systems since only two viral oncoproteins, E6 and E7, are consistently overexpressed in HPV-associated cancers, for example, of the uterine cervix. HPV-16 E6 and E7 have been instrumental in a number of ways to better understand centrosome aberrations in cancer. Using these two oncoproteins, it has been shown that centrosome overduplication and centrosome accumulation are fundamentally different processes but can co-exist in a tumor. In this chapter we highlight the importance of HPV oncoproteins as tools to dissect basic cellular processes in human cancer and to provide a basis for novel translational approaches to prevent and treat cancer.

12.1 Introduction

The concept that centrosome abnormalities, genomic instability, and cancer are intimately linked biological events was postulated over 100 years ago by Theodor Boveri (Boveri 2008). Experimentally proving Boveri's hypothesis has been a

N. Korzeniewski · S. Duensing (✉)
Section of Molecular Urooncology, Department of Urology,
University of Heidelberg, School of Medicine, Im Neuenheimer Feld 517,
69120 Heidelberg, Germany
e-mail: stefan.duensing@med.uni-heidelberg.de

daunting task but certain model systems such as high-risk human papillomavirus (HPV) oncoproteins have been particularly useful tools to understand the causes and consequences of centrosome aberrations on both genomic stability and carcinogenic progression (Duensing and Munger 2004).

Importantly, centrosome aberrations are observed in many non-HPV-associated carcinomas such as breast, prostate, and colon cancer. In non-HPV-associated lesions it is difficult to determine whether centrosome abnormalities are the driving force for malignant progression or occur as secondary events following the initial mutations promoting carcinogenesis. Studying HPV-associated malignant progression presents a unique opportunity because expression of the two high-risk HPV oncoproteins, HPV E6 and E7, by themselves has been shown to drive genomic instability and malignant progression while at the same time being the major transforming proteins in high-risk HPV-associated neoplasms (White et al. 1994; zur Hausen 1996).

Experiments analyzing high-risk HPV infection have demonstrated that centrosome amplification is present in precancerous high-risk HPV-containing cells and may potentially drive genomic alterations necessary for carcinogenesis (Duensing et al. 2001). Additionally, studies in HPV-associated anal tumors have demonstrated that centrosome overduplication correlates with the frequency of cell division errors (Duensing et al. 2008). In line with this, multipolar, specifically tripolar, mitoses are a hallmark of high-risk HPV-associated carcinomas (Duensing et al. 2000). Expression of the two viral oncoproteins drives high-risk HPV-associated malignant progression and allows a detailed dissection of the role of these drivers in promoting centrosome abnormalities, genomic instability and ultimately carcinogenesis.

Recently, prophylactic vaccines have been developed against HPV-6, -11, -16, and -18 and a small but growing proportion of the worldwide population are being vaccinated to prevent HPV infection (Schiller and Lowy 2009). However, the vaccines are currently still expensive and prevention of HPV-associated carcinoma is only effective in people with no prior exposure to high-risk HPV. Understanding the exact mechanisms by which high-risk HPV oncoproteins promote centrosome amplification and genomic instability will not only provide novel insights into basic biological processes but may also contribute to the development of alternative preventive and improved therapeutic options.

12.2 General Biology of HPVs

HPV infection is associated with squamous cell carcinomas (SCCs) of the anogenital, and a subset of oropharyngeal tract carcinomas. HPVs are small, circular double-stranded DNA viruses, approximately 8000 base pairs in length that infect cutaneous and mucosal epithelial tissues (Longworth and Laimins 2004). Over 200 types of HPV have been characterized and classified into two groups based on tissue tropism: cutaneous and mucosal. Infection with cutaneous HPVs, such as HPV-1 and HPV-2, typically promotes the formation of benign lesions which do not undergo malignant progression, such as skin and plantar warts. An exception to this are patients suffering from epidermodysplasia verruciformis (EV), a rare inherited autosomal recessive genetic disorder, which predisposes patients to the formation of cutaneous SCCs following infection with certain HPV types, such as HPV-5 and HPV-8 (Berkhout et al. 2000; Orth 1986).

Mucosal HPVs are further subdivided into two categories: low-risk and high-risk (de Villiers et al. 2004). Low-risk HPVs, such as HPV-6 and HPV-11, are associated with benign lesions of the oropharyngeal and anogenital tracts such as oral and laryngeal papillomas and anogenital mucosal condylomata acuminata (Schiller and Lowy 2006). High-risk HPVs such as HPV-16, -18, -31, -33, and -45 are associated with malignant progression and the development of SCCs. Epidemiological and biological studies have shown that HPV-16 and -18 are the most oncogenic types within the high-risk group accounting for 50 % and 20 %, respectively, of cervical cancers (Munoz et al. 2003). HPV-16 is also the most commonly found type in HPV-positive head and neck SCCs (Paz et al. 1997).

12.3 HPV Genome Organization and Life Cycle

Oncogenic HPV genomes contain eight open reading frames (ORFs), which are expressed as polycistronic mRNAs in a temporal manner under control of the non-coding long control region (LCR) (Longworth and Laimins 2004). The LCR contains the viral origin of DNA replication and important transcriptional control elements recognized by both cellular and virally encoded regulatory proteins. HPV early (E) transcripts control viral gene transcription and deregulate host targets to allow amplification of the viral genome in terminally growth-arrested cells. Early transcripts of high-risk HPV-16 are under control of a promoter, P_{97}, contained within the LCR (Longworth and Laimins 2004). HPV late (L) transcripts, L1 and L2, encode for the major and minor viral capsid proteins, respectively. Late transcripts of high-risk HPV-16 are under control of the late promoter P_{670}, residing within the E7 coding region, which only becomes active in terminally differentiating keratinocytes (Grassmann et al. 1996).

The HPV life cycle is intimately linked to the differentiation state of the infected host cell (Longworth and Laimins 2004). HPV infection is promoted by microabrasions that occur in the stratified squamous epithelial lining of the skin, anogenital, and oropharyngeal tracts. These microabrasions allow the virus to obtain access to cells within the basal, or basement, layer that supports the stratified epithelium. The basal layers contain stem cells that give rise to transiently amplifying cells, which represent the majority of cells within the cervical epithelium capable of cell division (Longworth and Laimins 2004). Upon infection of these cells, E1 and E2 viral gene expression is activated. The E1 protein functions as a DNA helicase and interacts with the E2 protein to bind to the viral origin of replication (Yang et al. 1993; Sedman and Stenlund 1995, 1998). E2 also acts as a transcriptional regulator both activating and tightly limiting the expression of the viral genes E6 and E7 during the early phase of virus infection (Longworth and Laimins 2004). During this stage of the HPV life cycle, the HPV genome is maintained as a circular episome at low copy number, approximately 50–100 extrachromosomal copies per infected cell (Stubenrauch and Laimins 1999). The HPV genome can persist for years, even decades, in the nuclei of these infected cells in a non-productive state (McLaughlin-Drubin and Munger 2009). The E4 and E5 proteins play a less well-understood role in the viral life cycle. The E4 protein may play a role in virus egress from the cell by inducing the collapse of the cytokeratin network (Doorbar et al. 1991). The E5 protein is necessary for optimal growth of the virus, possibly involving interaction with the epidermal growth factor receptor (EGFR) (Pim et al. 1992; Straight et al. 1993).

Normally, as progeny cells of the basal epithelium layer migrate toward the upper stratum they undergo terminal differentiation and exit from the cell division cycle. Exit from the cell division cycle is adverse to the productive phase of the HPV life cycle. This is because HPV does not encode any of its own replicative enzymes and is entirely dependent on host cell DNA replication machinery to amplify the viral genome. In order to keep infected host cells in a replication competent state as they migrate to the epithelial surface, the high-risk HPV genome encodes two oncoproteins, E6 and E7, which disrupt host cell cycle control, preventing exit from the cell division cycle and promoting activation or re-expression of cellular replication factors necessary for viral replication (Hebner and Laimins 2006). Viral genome amplification and productive HPV infection is tightly regulated and only occurs in differentiated cells located within the suprabasal epithelium layer (Longworth and Laimins 2004). Differentiation-dependent expression of viral capsid proteins in these cells promotes the assembly of infectious virions. Eventually, infected desquamated cells are shed off the upper epithelial layer, promoting spread of viral progeny.

12.4 Function of High-Risk HPV Oncoproteins

High-risk HPVs express two oncoproteins, E6 and E7, which function to deregulate the host cell cycle in order to promote amplification of the viral genome. Long-term expression of HPV E6 and E7 oncoproteins are known to both extend the life span of primary human cells and facilitate their immortalization whereas transformation is rare (Munger et al. 1989).

Despite the high prevalence of HPV infection in sexually active women, most HPV infections are self-limiting and transient. However, rarely and several years or even decades after infection with high-risk HPV types, SCCs may form. Carcinomas develop following integration of the high-risk HPV genome into host chromosomes (Baker et al. 1987). Integration of the viral genome terminates the productive life cycle of the virus. Viral genome integration can occur throughout the host genome, but most frequently is found at common fragile sites (Thorland et al. 2000).

HPV genome integration often results in deletion of a large segment of the viral genome, usually containing the viral transcriptional regulators E1 and E2 (Baker et al. 1987). The E6 and E7 ORFs, however, may remain intact, along with the LCR, which lies upstream of the integration site within the viral genome. Disruption of the E1 and E2 viral genes deregulates viral gene transcription and promotes higher than normal levels of both E6 and E7 gene transcription. Unrestrained expression of the E6 and E7 oncoproteins induces both genetic and epigenetic changes in the cellular genome that promotes malignant progression of host cells. However, overexpression of the HPV-16 E6 and E7 genes is not necessary for induction of genomic instability, suggesting that the high-risk HPV E6 and E7 oncoproteins may promote an increased risk of malignant progression even when their expression is tightly controlled (Duensing et al. 2001).

12.5 HPV-16 E6 Oncoprotein

The major function of the HPV-16 E7 oncoprotein is to promote a permissive cellular state for viral replication by disrupting normal cell cycle control mechanisms. The HPV-16 E6 oncoprotein has evolved to complement the function of the high-risk E7 oncoprotein by preventing the induction of cellular apoptosis due to aberrant cell cycle regulation. The high-risk E6 oncoprotein prevents apoptosis, in part, by mediating the degradation of p53 through redirecting a host cell HECT domain containing E3 ubiquitin ligase, E6-associated protein (E6AP) (Scheffner et al. 1993). Low-risk HPV types, such as HPV-6 and HPV-11, also encode an E6 protein. Low-risk E6 proteins can bind to E6AP but cannot mediate the degradation of p53 (Scheffner et al. 1993). The inability of low-risk E6 to degrade p53 may partially explain the reduced ability of low-risk HPV types to promote carcinogenesis (Halbert et al. 1992).

The high-risk E6 oncoprotein, but not low-risk E6 proteins, contain a PDZ-domain binding motif, X-(S/T)-X-(V/I/L)-COOH, which may also contribute to malignancy through interaction with PDZ-domain containing proteins, such as hDLG, MUPP-1, MAG1-3, and hScrib, resulting in degradation of the PDZ-protein (Kiyono et al. 1997; Gardiol et al. 1999; Lee et al. 2000; Massimi et al. 2008). Targeted inactivation of these proteins by oncogenic HPV E6 may disrupt cell junctions, induce loss of cell polarity, and promote cellular transformation. In support of this notion, it has been shown that the C-terminal PDZ-binding domain of high-risk E6 is necessary for the mediation of suprabasal cell proliferation (Nguyen et al. 2003).

High-risk E6 oncoprotein expression also contributes to cellular immortalization through the transcriptional upregulation of hTERT, the catalytic subunit of human telomerase, and can contribute to telomere maintenance (Klingelhutz et al. 1996). Enhancement of hTERT expression can occur through several mechanisms including physical interaction of the high-risk E6 oncoprotein with the transcription factor, c-MYC, its cofactor Max, and the transcription factor Sp-1, leading to the transcriptional activation of the hTERT promoter (Liu et al. 2009; Katzenellenbogen et al. 2009). Further upregulating hTERT activity, the high-risk E6 oncoprotein can promote E6AP-dependent degradation of NFX1-91, a putative transcriptional repressor of the hTERT promoter (Gewin et al. 2004).

12.6 HPV-16 E7 Oncoprotein

The major function of the HPV-16 E7 oncoprotein is to promote the viral life cycle by disrupting cell cycle control mainly through inactivation of the G_1/S-phase cell cycle checkpoint on multiple levels. The HPV-16 E7 oncoprotein accomplishes this, in part, through binding to and promoting the ubiquitin-mediated proteasomal degradation of hypophosphorylated RB (pRB), and the related pocket protein family members p107 and p130, thus releasing the inhibitory complex formed between pRB and E2Fs and allowing for transcription of S-phase relevant genes (Dyson et al. 1989; Matsushime et al. 1994; Resnitzky et al. 1994). HPV-16 E7 promotes the degradation of pRB through association with CUL2-based E3 ubiquitin ligase complexes, acting as a substrate-recognition subunit of the E3 ubiquitin ligase complex (Dyson et al. 1989; Gonzalez et al. 2001; Huh et al. 2007). Degradation of pRB occurs in a non-stoichiometric manner allowing even low levels of E7 protein to promote degradation of pRB. High-risk HPV E7 associates with pRB and its family members through a Leu-X-Cys-X-Glu (LXCXE) motif located within the CR2 homology domain (Dyson et al. 1989). Additional sequences located in the amino-terminal CR1 homology domain are necessary for pRB degradation (Huh et al. 2007). High-risk HPV-16 E7 has also been shown to inactivate p600, a pRB-associated protein, contributing to anchorage-independent growth and cellular transformation (Frisch and Screaton 2001; Huh et al. 2005).

An important functional difference between low-risk HPV E7 proteins and high-risk HPV E7 oncoproteins lies in their ability to bind and degrade pRB, p107, and p130. High-risk HPV-16 E7 binds with a higher affinity to pRB-family members than do low-risk HPV-6 E7 proteins. In the case of pRB, this difference maps to a single amino acid change within the pRB-binding domain that confers high-affinity binding (Heck et al. 1992). Low-risk HPV-6 also binds p107 and p130 with a lower affinity; however, this difference does not map to the same residue as pRB-binding efficiency (Zhang et al. 2006). Moreover, low-risk HPV-6 E7 has recently been shown to destabilize p130, but not pRB or p107, suggesting that disruption of signaling pathways controlled by p130 are necessary for the productive stage of the viral life cycle and that pRB and/or p107 degradation are important for carcinogenesis (Zhang et al. 2006).

Besides degradation of pRB, and its related family members, the HPV-16 E7 oncoprotein profoundly disrupts the pRB-signaling axis through several additional mechanisms to favor replication of the viral genome. High-risk HPV-16 E7 can directly alter E2F-dependent cellular transcription through physical interaction with E2F1. This interaction results in the pRB-independent enhancement of E2F-mediated gene transcription (Hwang et al. 2002). However, the promoter of E2F6, a transcriptional repressor responsible for directing cell cycle exit, is also E2F-responsive (Lyons et al. 2006). The HPV-16 E7 oncoprotein has therefore evolved to directly associate with E2F6 resulting in inactivation of its transcriptional repression and maintenance of a replication competent cellular state (McLaughlin-Drubin et al. 2008).

Aberrant activation of CDK2 by high-risk HPV-16 E7 further contributes to alteration of the host cellular environment. The high-risk HPV E7 oncoprotein functions to disrupt the tight control of CDK2 activation on multiple levels including upregulation of the E2F-responsive regulatory subunits cyclin A and cyclin E, direct association with CDK2/cyclin E complexes, enhancing the enzymatic activities of these complexes and inhibition of the cyclin-dependent kinase (CDK) inhibitors p21^{Cip1} and p27^{Kip1} (Funk et al. 1997; Jones et al. 1997; Nguyen and Munger 2008).

Recently, it has become increasingly evident that high-risk E7 oncoprotein expression alters components of the host cell epigenetic control machinery. The first evidence for this came from experiments demonstrating that the HPV-16 and HPV-31 E7 oncoproteins are capable of associating with histone deacetylases type-1 and -2 (HDAC-1, and -2) (Brehm et al. 1999; Longworth and Laimins 2004). HDACs function as transcriptional repressors by reversing acetyl modifications of lysine residues on histones. The indirect interaction between oncogenic HPV-16 E7 and HDACs is mediated by Mi2β, a component of the NURD histone deacetylase complex (Brehm et al. 1999). This interaction is dependent on the integrity of the two Cys-X-X-Cys motifs in the HPV E7 oncoprotein carboxy-terminus and results in increased E2F-mediated gene transcription from HDAC responsive promoters (Brehm et al. 1999). Additionally, high-risk HPV-16 E7 was also found to associate with histone acetyl transferases (HATs), such as p300 and pCAF, which function to activate transcription and stimulate cellular proliferation (Avvakumov et al. 2003; Bernat et al. 2003).

High-risk HPV-16 E7 oncoprotein expression can further disrupt host cell epigenetic control through modulation of histone methylation. Transcriptional induction of the KDM6A and KDM6B histone 3 lysine 27-specific demethylases was shown to occur following, and to be dependent on, HPV-16 E7 expression (McLaughlin-Drubin et al. 2011). The induction of these two demethylases reduces the H3K27me3 mark that is necessary for the binding of polycomb repressive complexes, which in turn enhances the transcription of KDM6A- and KDM6B-regulated *HOX* gene expression, including the cervical carcinoma biomarker p16^{INK4A}, promoting epigenetic reprogramming of host cells at the histone methylation level. Importantly, the enhancement of KDM6A and KDM6B expression was shown to be independent of pRB-inactivation (McLaughlin-Drubin et al. 2011).

Ultimately, the HPV-16 E7 oncoprotein profoundly disrupts host cell transcriptional control in order to stimulate cellular proliferation and suppress programmed cell death. This occurs in a multifaceted manner: broadly by promoting E2F-activation in both a direct and indirect manner and more precisely through regulation of the host transcriptome promoted by modifications induced at the chromatin level.

12.7 HPV Oncoproteins and Genomic Instability

High-risk HPV oncoprotein induced disruption of host cell cycle control and induction of aberrant S-phase-specific gene expression not only promotes the viral life cycle but also induces genomic instability, an event which may promote carcinogenic progression (Duensing and Munger 2004). This is accomplished through disruption of the two major tumor suppressors, p53 and pRB, by the two high-risk HPV oncoproteins E6 and E7, respectively (Scheffner et al. 1993; Dyson et al. 1989). Utilizing high-risk HPV oncoproteins as tools, novel cellular signaling pathways affected by disruption of p53 and pRB may be determined. As disruption of these two tumor suppressors is also a hallmark of many non-HPV-associated carcinomas, these studies may ultimately lead to the discovery of new targets to prevent malignant progression in both non-HPV and HPV-associated neoplasms.

Mounting evidence suggests that genomic instability may be an important co-factor in promoting malignant progression. As previously mentioned, multipolar, specifically tripolar, mitoses are a hallmark of high-risk HPV-associated carcinomas (Duensing et al. 2000). Such a disruption of spindle polarity may consequently promote chromosome missegregation and ultimately aneuploidy (Boveri 2008). How multipolar spindle poles may arise in HPV-associated lesions will be discussed below. However, besides aneuploidy, numerical and structural chromosomal instability are also a critical factor for malignant progression. Cytogenic analyses of HPV-associated lesions have revealed recurring patterns of chromosome gains and losses. In particular, gain of chromosome 3q has been linked to the transition to invasiveness in high-risk HPV-associated lesions (Heselmeyer et al. 1996). Several other common structural alterations observed in HPV-associated neoplasms include gains of genetic material on chromosomes 1q, 5p, 6p, and 20q and losses mapped to chromosomes 2q, 3p, 4, 8p, and 13q (Wilting et al. 2006; Bibbo et al. 1989; Hashida and Yasumoto 1991).

The role of structural chromosomal abnormalities in high-risk HPV-associated malignant progression is highlighted by an increased incidence of anaphase bridges in high-risk HPV-associated lesions (Duensing and Munger 2002). Anaphase bridges may form through chromosome fusions at telomeres or double-stranded DNA breaks (Acilan et al. 2007) and unrepaired, broken DNA can promote gene translocations or gene amplifications/deletions, which may provide a growth advantage to cells through gain of oncogenes or loss of tumor suppressors.

Additionally, genomic instability is an early event in HPV-associated malignant progression being observed in pre-invasive high-risk HPV-associated lesions (Duensing et al. 2001). Several lines of evidence show that expression of HPV-16 E6 and E7 by themselves can independently induce structural chromosomal instability. Further, both the HPV-16 E6 and E7 oncoproteins have independently been shown to relax mitotic checkpoints (Thomas and Laimins 1998), which may be activated in response to altered DNA structures (Mikhailov et al. 2002), promoting polyploidization and possibly predisposing cells to aneuploidy (Heilman et al. 2009; Southern et al. 2004).

Centrosome abnormalities have been detected in a wide range of malignant tumors including breast, prostate, colon, and cervical cancer, and compelling evidence suggests that centrosome abnormalities can disrupt mitotic spindle polarity, drive progressive loss of genomic stability, and promote malignant progression. Centrosome overduplication has been observed in cells expressing episomal HPV-16 genomes (Duensing et al. 2001). This observation strongly underscores that viral integration and overexpression of the high-risk HPV oncoproteins E6 and E7 is not required for the induction of centrosome abnormalities and that centrosome aberrations are early alterations in high-risk HPV-associated malignant progression.

Determining the pathways that are activated by HPV oncoprotein expression leading to centrosome overduplication, cell division errors, and ultimately aneuploidy will be important in understanding and preventing the earliest steps in malignant progression.

12.8 The Centrosome

The centrosome is a crucial cellular organelle, responsible for organizing the microtubule network in most mammalian cells and coordinating bipolar spindle pole formation during mitosis (Schatten 2008). The centrosome consists of two centrioles, barrel-shaped microtubule-based cylinders, embedded in a pericentriolar material (Loncarek et al. 2008). The two centrioles are distinct in age and composition, consisting of an older maternal centriole characterized by distal and subdistal appendage proteins which function to anchor and nucleate microtubules, and a younger daughter centriole which has not yet associated with appendage proteins (Azimzadeh and Bornens 2007). In order to generate two spindle poles, the single centrosome of a non-dividing cell must duplicate precisely once, and only once, prior to mitosis in order to ensure faithful cell division.

Centrosome duplication begins during late mitosis/early G_1-phase of the cell division cycle, when the two pre-existing centrioles of the single centrosome disengage through the action of polo kinase 1 (PLK1) and separase, and move into a near parallel position (Tsou et al. 2009). This step is followed by recruitment of polo-like kinase 4 (PLK4) to the wall of the maternal centriole at the site of daughter centriole synthesis (Habedanck et al. 2005; Kleylein-Sohn et al. 2007).

(a) Normal centriole duplication

(b) HPV-16 E7 — Centriole multiplication

(c) HPV-16 E6 — Centriole accumulation

(d) Normal | 'centriole flower'

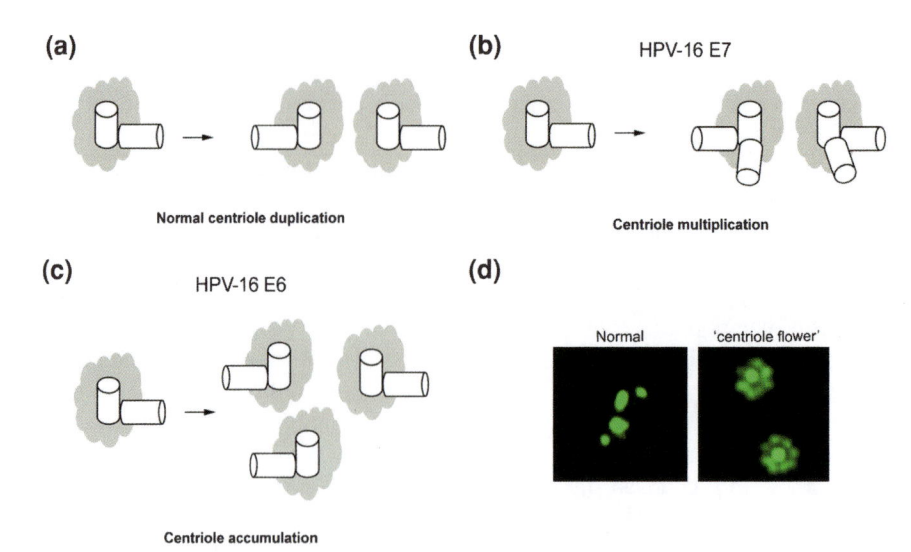

Fig. 12.1 High-risk HPV oncoproteins and centriole alterations. a Normal centriole duplication is characterized by the assembly of only one new centriole (daughter) with the pre-existing centriole (mother). b High-risk HPV-16 E7 expression induces centriole multiplication due to disruption of the centriole duplication cycle and is characterized by the presence of multiple daughter centrioles at a single maternal centriole. c Expression of high-risk HPV-16 E6 promotes centriole accumulation due to induction of cytokinesis defects or other cell division errors and results in cells containing two or more maternal centrioles. d Fluorescence microscopic analysis of normal duplicated centrioles and 'centriole flower' (centriole multiplication) phenotype induced by PLK4 overexpression in U-2 OS/centrin-GFP cells

Subsequently, structural proteins are recruited to the nascent pro-centriole to stabilize and elongate the newly formed daughter centriole. Centrosome duplication completes during the late G_2-phase of the cell cycle, when the two fully formed centriole pairs separate to form the mitotic spindle poles (Azimzadeh and Bornens 2007). A more detailed description of centrosome duplication can be found elsewhere in this book.

In principal, there are two mechanisms by which centriole amplification may occur in tumor cells: centriole overduplication and centriole accumulation. These two phenotypes can be distinguished by the number of older, mature centrioles present in a cell (Guarguaglini et al. 2005). Genuine centriole overduplication is characterized by one or two mature maternal centrioles in the presence of multiple immature daughter centrioles. In contrast, centriole accumulation is defined by multiple maternal centrioles with a normal complement of daughter centrioles (Guarguaglini et al. 2005). It is important to recognize the distinction between centriole overduplication and accumulation because cells exhibiting centriole accumulation may arise due to abortive mitoses or cytokinesis errors and may not be able to produce viable progeny (Fig. 12.1). Conversely, cells which exhibit a genuine centriole overduplication defect are, in general, less genomically altered and hence are more likely to give rise to genomically unstable daughter cells.

12.9 Models of HPV Oncoprotein-Induced Centrosome Abnormalities and Malignant Progression

In vitro studies have demonstrated that the HPV-16 E7 oncoprotein disrupts genomic integrity by directly interfering with centrosome duplication control (Duensing et al. 2000). High-risk HPV-16 E7 expression rapidly produces abnormal centriole numbers in otherwise normal cells prior to the onset of genomic instability. In contrast, high-risk HPV-16 E6 expressing cells exhibit centrosome accumulation in cells which are already genomically unstable, often expressing markers of cellular senescence, and are unlikely to remain in the proliferative pool or contribute to tumor development (Duensing et al. 2000).

The in vivo role of HPV-16 E7-induced centrosome abnormalities in malignant progression is highlighted in a transgenic mouse model of cervical carcinogenesis. Transgenic mice expressing HPV-16 E7 driven by a cytokeratin 14 promoter and treated with low doses of estrogen develop numerical centrosome abnormalities in the cervical mucosa which progress to invasive carcinomas (Riley et al. 2003). In contrast, HPV-16 E6 expressing transgenic mice display a comparable level of numerical centrosome aberrations but develop only low grade cervical lesions that do not progress to malignant tumors (Riley et al. 2003). These results suggest that centrosome aberrations in the context of HPV-16 E7 expression are associated with a greater risk of malignant progression than in HPV-16 E6 expressing cells.

12.10 The Centriole Multiplication Pathway

Following high-risk HPV-16 E7 expression, supernumerary centrioles appear rapidly and within a single cell division cycle, suggesting they arise due to direct disruption of centriole duplication control (Duensing et al. 2007). This was initially difficult to reconcile with the prevailing model of centriole duplication described above, where a single maternal centriole initiates the synthesis of only a single daughter centriole. Further analysis of HPV-16 E7-induced centriole abnormalities led to the discovery that the HPV-16 E7 oncoprotein rapidly induces centriole overduplication through stimulation of a novel centriole duplication pathway, referred to as centriole multiplication (Duensing et al. 2007). This pathway is characterized by a single maternal centriole initiating the simultaneous synthesis of two or more daughter centrioles. Although, multiciliated epithelial cells such as those in the trachea and oviduct can rapidly produce hundreds of centrioles during ciliogenesis through the centriole multiplication pathway, this had never before been observed in the context of an oncogenic stimulus relevant for a major human cancer. The mechanism behind the very rapid HPV-16 E7-mediated induction of centriole multiplication was unknown until recently.

The molecular players involved in the centriole multiplication pathway were initially determined following the observation that inhibition of protein degradation, through the use of a proteasome inhibitor Z-L$_3$VS, induced a large proportion of cells to exhibit centriole multiplication (Duensing et al. 2007). This observation led to functional studies to discover what cellular factors were necessary for this phenotype to occur and it was discovered that CDK2, cyclin E, and PLK4 were necessary factors for Z-L$_3$VS-induced centriole multiplication (Duensing et al. 2007). Further experiments revealed that ectopic expression of cyclin E/CDK2 alone was not sufficient to induce centriole multiplication (Korzeniewski et al. 2009). Deregulation of cyclin E/CDK2 complexes by themselves was found to promote the aberrant recruitment of PLK4 to maternal centrioles but endogenous levels of PLK4 were insufficient to induce centriole multiplication. Centriole multiplication occurred when CDK2/cyclin E complexes and PLK4 were upregulated, suggesting that endogenous levels of PLK4 are not sufficient to induce centriole multiplication and that PLK4 protein levels are rate-limiting for centriole multiplication (Korzeniewski et al. 2009).

PLK4 is an essential regulator of both centriole duplication and cell viability. In vitro studies have demonstrated that PLK4 overexpression in tissue culture results in an increase in supernumerary centrosomes and when PLK4 is depleted via RNA interference, centriole numbers are reduced with progressive loss of centrioles and the subsequent development of monopolar spindles (Habedanck et al. 2005). In vivo, PLK4 knockout is embryonic lethal; however, heterozygous mice develop normally (Ko et al. 2005; Hudson et al. 2001). Aged PLK4 heterozygous mice have an increased risk for development of spontaneous malignancies than do their littermates and exhibit abnormal chromosomal segregation and alignment defects, cytokinesis failure, and multinucleation (Ko et al. 2005). These observations suggest that a strict control of PLK4 transcript and protein levels is necessary to maintain cell viability and prevent malignant progression. Although the threshold level of PLK4 protein which induces centriole multiplication is not known, our own experiments have shown that very small changes in PLK4 protein level induces a small but significant percentage of cells to exhibit centriole multiplication (Korzeniewski et al. 2009) which may ultimately promote a tolerable level of chromosomal instability leading to malignant progression.

12.11 Cellular Proteolysis and Centriole Multiplication

The discovery that inhibition of cellular protein degradation mechanisms strongly induced centriole multiplication suggested that proteolysis was important in maintaining normal centriole duplication control (Duensing et al. 2007). PLK4 is a very unstable protein whose stability is known to be controlled by both ubiquitin and non-ubiquitin-mediated proteolysis (Korzeniewski et al. 2009; Cunha-Ferreira et al. 2009; Rogers et al. 2009). The PLK4 protein coding sequence has been reported to contain a PEST domain implicated in the mediation of rapid protein degradation by intracellular proteases and is also known to be targeted for E3-ubiquitin ligase-mediated degradation (Rechsteiner and Rogers 1996).

A substrate is targeted for degradation by the proteasome through addition of a polyubiquitin chain to the substrate. This reaction is catalyzed by a three-step enzymatic cascade involving an ubiquitin-activating enzyme (E1), a ubiquitin-targeting enzyme (E2), and a ubiquitin ligase protein (E3). The ubiquitin E3 ligase confers substrate specificity to the reaction (Nakayama and Nakayama 2006). Cullin-RING ubiquitin ligases (CRLs) are members of the largest family of eukaryotic E3-ubiquitin ligases (Petroski and Deshaies 2005). All CRLs consist of a cullin-backbone, a zinc-binding RING-domain containing protein which recruits the ubiquitin-conjugating E2 enzyme, and an adaptor protein which recruits interchangeable substrate recognition subunits. There are seven human cullin subunits (CUL1, -2, -3, -4A, -4B, -5, and -7) responsible for nucleating the assembly of unique E3-ubiquitin ligase complexes (Petroski and Deshaies 2005).

In the case of the prototypical CRL, the SKP1-CUL1-F-box (SCF) ubiquitin ligase complex, SKP1 acts as the adaptor protein interacting with substrate recognition subunits containing an F-box domain and a second protein–protein interaction domain which recognizes the specific target protein. Other CRLs contain unique adaptor proteins which recognize substrate recognition subunits with different functional motifs (Petroski and Deshaies 2005).

Core components of the SCF E3-ubiquitin ligase complex, including SKP1 and CUL1, have been found to localize to maternal centrioles which serve as assembly platforms for oncogene-induced centriole overduplication (Korzeniewski et al. 2009; Freed et al. 1999). SCF-ubiquitin ligase activity was found to be critically involved in suppressing centriole multiplication in human tumor cells by regulating PLK4 protein levels (Korzeniewski et al. 2009; Cunha-Ferreira et al. 2009; Rogers et al. 2009). This activity of the SCF ubiquitin ligase provides an important mechanism for restraining excessive daughter centriole formation at single maternal centrioles and hence centrosome-mediated cell division errors and chromosomal instability (Korzeniewski et al. 2009). Interestingly, HPV-16 E7 expression induces centriole multiplication in a phenotype reminiscent of SCF-ubiquitin ligase inactivation, suggesting that HPV-16 E7 expression may also deregulate PLK4 protein level, half-life, or localization to induce centriole multiplication.

12.12 Mechanism of HPV-16 E7-Induced Centriole Multiplication

HPV-16 E7 is known to deregulate cyclin E/CDK2 complexes; however, whether the HPV-16 E7 oncoprotein deregulates PLK4 protein expression to ultimately stimulate centriole multiplication had not been determined. When PLK4 protein expression was analyzed in HPV-16 E7 expressing cells, it was found that PLK4 aberrantly localized to maternal centrioles in the form of multiple PLK4 dots similar to the aberrant localization of PLK4 seen following deregulation of CDK2/cyclin E complexes alone (Korzeniewski et al. 2011). However, upregulation of

PLK4 protein level is also necessary for centriole overduplication to occur since it is known to be rate-limiting for this process.

The PLK4 promoter has been shown to contain E2F-responsive elements and be repressed by HDACs, two important cellular transcriptional control pathways that are deregulated following HPV-16 E7 oncoprotein expression (Li et al. 2005; McLaughlin-Drubin and Munger 2009). PLK4 promoter activation assays and real-time quantitative reverse transcriptase (qRT-PCR) analysis of PLK4 mRNA level were performed on cells expressing either wild-type or mutant HPV-16 E7 and it was found that wild-type HPV-16 E7 could both activate the PLK4 promoter and upregulate PLK4 mRNA levels (Korzeniewski et al. 2011).

Conversely, an HPV-16 E7 mutant with deletion of the amino acid region 21–24 which contains the LCXCE motif (HPV-16 E7 Δ21-24), and is incapable of pRB-binding was unable to activate the PLK4 promoter or upregulate PLK4 mRNA and, in accordance with previous studies, unable to promote centriole overduplication (Korzeniewski et al. 2011; Duensing and Munger 2003). Although the HPV-16 E7 Δ21-24 mutant is unable to bind and degrade pRB, it is still capable of interacting with HDACs (Phelps et al. 1992).

Like the HPV-16 E7 Δ21-24 mutant construct, an HDAC-interaction-deficient mutant L67R was also unable to activate the PLK4 promoter or upregulate PLK4 mRNA (Korzeniewski et al. 2011). The HPV-16 L67R mutant is still capable of interacting with pRB although it does so less efficiently and has a reduced capacity to activate E2F-dependent transcription (Avvakumov et al. 2003). This defect complicates analysis of the role of HDACs in the HPV-16 E7-induced modulation of PLK4 transcription. Analyzing PLK4 mRNA abundance in HPV-16 E7 expressing cell which is deficient in HDACs or pRB-family members would clarify the role of HDACs in the HPV-16 E7–mediated modulation of PLK4 mRNA.

A mechanism for the rapid induction of centriole multiplication by the HPV-16 E7 oncoprotein can now be postulated (Fig. 12.2). It has been previously shown that cyclin E/CDK2 complexes mediate the aberrant recruitment of PLK4 to maternal centrioles (Korzeniewski et al. 2009). However, a concurrent increase in PLK4 protein is necessary for centriole multiplication to occur (Korzeniewski et al. 2009). The aberrant recruitment of PLK4 to maternal centrioles is recapitulated in cells expressing HPV-16 E7 and these cells also contain increased PLK4 mRNA transcript levels (Korzeniewski et al. 2011). This increase in PLK4 mRNA, albeit only modest, may be necessary to promote the aberrant recruitment of excess PLK4 to maternal centrioles in the form of multiple PLK4 dots and ultimately centriole multiplication. Support for this notion comes from a previous study which determined that ongoing RNA polymerase II transcription is necessary for HPV-16 E7-induced centriole overduplication but dispensable for normal centriole duplication (Duensing et al. 2007). This is in line with our finding that increased PLK4 mRNA transcripts play a role in HPV-16 E7-induced centriole multiplication (Korzeniewski et al. 2011).

Therefore, HPV-16 E7 can interfere with two steps of centriole biogenesis in order to stimulate centriole multiplication: increase PLK4 at the gene expression level and enhance the recruitment of PLK4 to maternal centrioles. However, PLK4

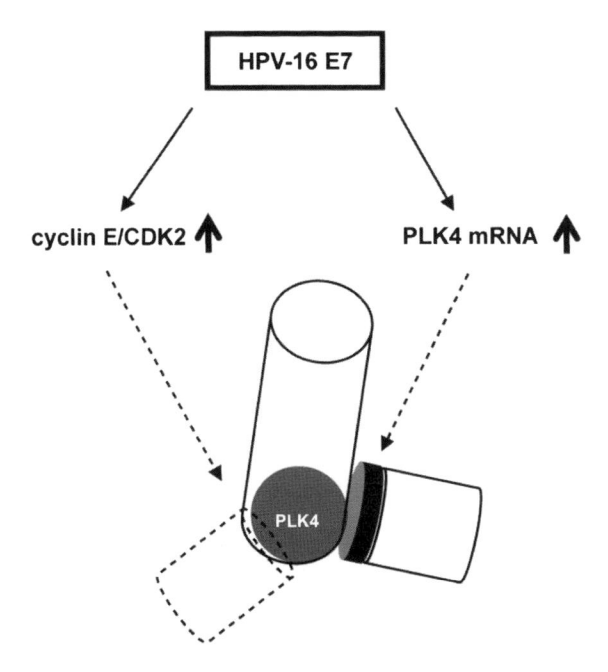

Fig. 12.2 HPV-16 E7 and the complexity of centriole duplication control. HPV-16 E7 induces the formation of an S-phase-like milieu through binding and degradation of pRB-family members, interaction with histone deacetylases (HDACs), and inactivation of the CDK inhibitors p21^{Cip1} and p27^{Kip1}, ultimately promoting the deregulation of E2F-mediated gene transcription and the aberrant activation of cyclin E/CDK2 complexes. Cyclin E/CDK2 can promote centriole multiplication through the aberrant recruitment of PLK4 protein to maternal centrioles; however, it is currently not clear whether this is through a direct effect on PLK4 or through another protein. The increase of PLK4 mRNA expression in HPV-16 E7 expressing cells is likley to be crucial for the induction of centriole multiplication. PLK4 protein level is normally restrained at maternal centrioles by cellular proteolytic mechanisms. Therefore, HPV-16 E7 expression may also interfere with the proteolytic control of PLK4 protein level to induce aberrant daughter centriole formation, a hypothesis that warrants further testing

is highly unstable and it is also possible that HPV-16 E7 may interfere with post-translational regulatory mechanisms to further increase PLK4 abundance and/or stability at maternal centrioles.

12.13 Additional Mechanisms of HPV-16 E7-Induced Centrosome Abnormalities

As mentioned earlier, HPV-16 E7 can also induce centrosome overduplication in a pRB-independent manner. The first evidence of this came from the observation that, in contrast to full-length wild-type HPV-16 E7, the HPV-16 E7 mutant construct, Δ21-24, was unable to induce centriole overduplication in both normal and pRB-family-deficient mouse embryo fibroblasts (Duensing and Munger 2003). Further experiments suggested that the ability of HPV-16 E7 to interact with γ-tubulin, a component of the pericentriolar material important for microtubule nucleation, may play a role in the pRB-independent induction of supernumerary centrosomes (Nguyen et al. 2007). Disruption of γ-tubulin may play a role in the regulation of centrosome duplication, and this interaction with HPV-16 E7 which relies on an intact LXCXE motif may hence contribute to pRB-independent overduplication induced by HPV-16 E7.

12.14 Conclusion

High-risk HPV oncoproteins have evolved to efficiently disrupt host cell cycle control on multiple levels in order to promote optimal cellular conditions for viral genome replication to occur. The collateral damage of this profound dysfunction of cellular regulation is uncontrolled cellular proliferation, genomic instability, and carcinogenic progression. Loss of centrosome duplication control is one of the earliest steps in the development of high-risk HPV-associated malignancies and a hallmark of several other non-HPV-associated cancers. By utilizing the HPV-16 E7 oncoprotein as a tool, much progress has been made in determining the biological pathways involved in disruption of centrosome duplication control. These new insights may be utilized in the future to help develop better preventive and therapeutic approaches.

References

Acilan C, Potter DM, Saunders WS (2007) DNA repair pathways involved in anaphase bridge formation. Genes Chromosomes Cancer 46:522–531

Avvakumov N, Torchia J, Mymryk JS (2003) Interaction of the HPV E7 proteins with the pCAF acetyltransferase. Oncogene 22:3833–3841

Azimzadeh J, Bornens M (2007) Structure and duplication of the centrosome. J Cell Sci 120:2139–2142

Baker CC, Phelps WC, Lindgren V, Braun MJ, Gonda MA, Howley PM (1987) Structural and transcriptional analysis of human papillomavirus type 16 sequences in cervical carcinoma cell lines. J Virol 61:962–971

Berkhout RJ, Bouwes Bavinck JN, ter Schegget J (2000) Persistence of human papillomavirus DNA in benign and (pre)malignant skin lesions from renal transplant recipients. J Clin Microbiol 38:2087–2096

Bernat A, Avvakumov N, Mymryk JS, Banks L (2003) Interaction between the HPV E7 oncoprotein and the transcriptional coactivator p300. Oncogene 22:7871–7881

Bibbo M, Dytch HE, Alenghat E, Bartels PH, Wied GL (1989) DNA ploidy profiles as prognostic indicators in CIN lesions. Am J Clin Pathol 92:261–265

Boveri T (2008) Concerning the origin of malignant tumours by Theodor Boveri. Translated and annotated by Henry Harris. J Cell Sci 121(Suppl 1):1–84

Brehm A, Nielsen SJ, Miska EA, McCance DJ, Reid JL, Bannister AJ, Kouzarides T (1999) The E7 oncoprotein associates with Mi2 and histone deacetylase activity to promote cell growth. EMBO J 18:2449–2458

Cunha-Ferreira I, Rodrigues-Martins A, Bento I, Riparbelli M, Zhang W, Laue E, Callaini G, Glover DM, Bettencourt-Dias M (2009) The SCF/Slimb ubiquitin ligase limits centrosome amplification through degradation of SAK/PLK4. Curr Biol 19:43–49

de Villiers EM, Fauquet C, Broker TR, Bernard HU, zur Hausen H (2004) Classification of papillomaviruses. Virology 324:17–27

Doorbar J, Ely S, Sterling J, McLean C, Crawford L (1991) Specific interaction between HPV-16 E1-E4 and cytokeratins results in collapse of the epithelial cell intermediate filament network. Nature 352:824–827

Duensing S, Munger K (2002) The human papillomavirus type 16 E6 and E7 oncoproteins independently induce numerical and structural chromosome instability. Cancer Res 62:7075–7082

Duensing S, Munger K (2003) Human papillomavirus type 16 E7 oncoprotein can induce abnormal centrosome duplication through a mechanism independent of inactivation of retinoblastoma protein family members. J Virol 77:12331–12335

Duensing S, Munger K (2004) Mechanisms of genomic instability in human cancer: insights from studies with human papillomavirus oncoproteins. Int J Cancer 109:157–162

Duensing S, Lee LY, Duensing A, Basile J, Piboonniyom S, Gonzalez S, Crum CP, Munger K (2000) The human papillomavirus type 16 E6 and E7 oncoproteins cooperate to induce mitotic defects and genomic instability by uncoupling centrosome duplication from the cell division cycle. Proc Natl Acad Sci U S A 97:10002–10007

Duensing S, Duensing A, Crum CP, Munger K (2001a) Human papillomavirus type 16 E7 oncoprotein-induced abnormal centrosome synthesis is an early event in the evolving malignant phenotype. Cancer Res 61:2356–2360

Duensing S, Duensing A, Flores ER, Do A, Lambert PF, Munger K (2001b) Centrosome abnormalities and genomic instability by episomal expression of human papillomavirus type 16 in raft cultures of human keratinocytes. J Virol 75:7712–7716

Duensing A, Liu Y, Perdreau SA, Kleylein-Sohn J, Nigg EA, Duensing S (2007a) Centriole overduplication through the concurrent formation of multiple daughter centrioles at single maternal templates. Oncogene 26:6280–6288

Duensing A, Liu Y, Spardy N, Bartoli K, Tseng M, Kwon JA, Teng X, Duensing S (2007b) RNA polymerase II transcription is required for human papillomavirus type 16 E7- and hydroxyurea-induced centriole overduplication. Oncogene 26:215–223

Duensing A, Chin A, Wang L, Kuan SF, Duensing S (2008) Analysis of centrosome overduplication in correlation to cell division errors in high-risk human papillomavirus (HPV)-associated anal neoplasms. Virology 372:157–164

Dyson N, Howley PM, Munger K, Harlow E (1989) The human papilloma virus-16 E7 oncoprotein is able to bind to the retinoblastoma gene product. Science 243:934–937

Freed E, Lacey KR, Huie P, Lyapina SA, Deshaies RJ, Stearns T, Jackson PK (1999) Components of an SCF ubiquitin ligase localize to the centrosome and regulate the centrosome duplication cycle. Genes Dev 13:2242–2257

Frisch SM, Screaton RA (2001) Anoikis mechanisms. Curr Opin Cell Biol 13:555–562

Funk JO, Waga S, Harry JB, Espling E, Stillman B, Galloway DA (1997) Inhibition of CDK activity and PCNA-dependent DNA replication by p21 is blocked by interaction with the HPV-16 E7 oncoprotein. Genes Dev 11:2090–2100

Gardiol D, Kuhne C, Glaunsinger B, Lee SS, Javier R, Banks L (1999) Oncogenic human papillomavirus E6 proteins target the discs large tumour suppressor for proteasome-mediated degradation. Oncogene 18:5487–5496

Gewin L, Myers H, Kiyono T, Galloway DA (2004) Identification of a novel telomerase repressor that interacts with the human papillomavirus type-16 E6/E6-AP complex. Genes Dev 18:2269–2282

Gonzalez SL, Stremlau M, He X, Basile JR, Munger K (2001) Degradation of the retinoblastoma tumor suppressor by the human papillomavirus type 16 E7 oncoprotein is important for functional inactivation and is separable from proteasomal degradation of E7. J Virol 75:7583–7591

Grassmann K, Rapp B, Maschek H, Petry KU, Iftner T (1996) Identification of a differentiation-inducible promoter in the E7 open reading frame of human papillomavirus type 16 (HPV-16) in raft cultures of a new cell line containing high copy numbers of episomal HPV-16 DNA. J Virol 70:2339–2349

Guarguaglini G, Duncan PI, Stierhof YD, Holmstrom T, Duensing S, Nigg EA (2005) The forkhead-associated domain protein Cep170 interacts with Polo-like kinase 1 and serves as a marker for mature centrioles. Mol Biol Cell 16:1095–1107

zur Hausen H (1996) Papillomavirus infections–a major cause of human cancers. Biochim Biophys Acta 1288:F55–78

Habedanck R, Stierhof YD, Wilkinson CJ, Nigg EA (2005) The Polo kinase Plk4 functions in centriole duplication. Nat Cell Biol 7:1140–1146

Halbert CL, Demers GW, Galloway DA (1992) The E6 and E7 genes of human papillomavirus type 6 have weak immortalizing activity in human epithelial cells. J Virol 66:2125–2134

Hashida T, Yasumoto S (1991) Induction of chromosome abnormalities in mouse and human epidermal keratinocytes by the human papillomavirus type 16 E7 oncogene. J Gen Virol 72(Pt 7):1569–1577

Hebner CM, Laimins LA (2006) Human papillomaviruses: basic mechanisms of pathogenesis and oncogenicity. Rev Med Virol 16:83–97

Heck DV, Yee CL, Howley PM, Munger K (1992) Efficiency of binding the retinoblastoma protein correlates with the transforming capacity of the E7 oncoproteins of the human papillomaviruses. Proc Natl Acad Sci U S A 89:4442–4446

Heilman SA, Nordberg JJ, Liu Y, Sluder G, Chen JJ (2009) Abrogation of the postmitotic checkpoint contributes to polyploidization in human papillomavirus E7-expressing cells. J Virol 83:2756–2764

Heselmeyer K, Schrock E, du Manoir S, Blegen H, Shah K, Steinbeck R, Auer G, Ried T (1996) Gain of chromosome 3q defines the transition from severe dysplasia to invasive carcinoma of the uterine cervix. Proc Natl Acad Sci U S A 93:479–484

Hudson JW, Kozarova A, Cheung P, Macmillan JC, Swallow CJ, Cross JC, Dennis JW (2001) Late mitotic failure in mice lacking Sak, a polo-like kinase. Curr Biol 11:441–446

Huh KW, DeMasi J, Ogawa H, Nakatani Y, Howley PM, Munger K (2005) Association of the human papillomavirus type 16 E7 oncoprotein with the 600-kDa retinoblastoma protein-associated factor, p600. Proc Natl Acad Sci U S A 102:11492–11497

Huh K, Zhou X, Hayakawa H, Cho JY, Libermann TA, Jin J, Harper JW, Munger K (2007) Human papillomavirus type 16 E7 oncoprotein associates with the cullin 2 ubiquitin ligase complex, which contributes to degradation of the retinoblastoma tumor suppressor. J Virol 81:9737–9747

Hwang SG, Lee D, Kim J, Seo T, Choe J (2002) Human papillomavirus type 16 E7 binds to E2F1 and activates E2F1-driven transcription in a retinoblastoma protein-independent manner. J Biol Chem 277:2923–2930

Jones DL, Alani RM, Munger K (1997) The human papillomavirus E7 oncoprotein can uncouple cellular differentiation and proliferation in human keratinocytes by abrogating p21Cip1-mediated inhibition of cdk2. Genes Dev 11:2101–2111

Katzenellenbogen RA, Vliet-Gregg P, Xu M, Galloway DA (2009) NFX1-123 increases hTERT expression and telomerase activity posttranscriptionally in human papillomavirus type 16 E6 keratinocytes. J Virol 83:6446–6456

Kiyono T, Hiraiwa A, Fujita M, Hayashi Y, Akiyama T, Ishibashi M (1997) Binding of high-risk human papillomavirus E6 oncoproteins to the human homologue of the Drosophila discs large tumor suppressor protein. Proc Natl Acad Sci U S A 94:11612–11616

Kleylein-Sohn J, Westendorf J, Le Clech M, Habedanck R, Stierhof YD, Nigg EA (2007) Plk4-induced centriole biogenesis in human cells. Dev Cell 13:190–202

Klingelhutz AJ, Foster SA, McDougall JK (1996) Telomerase activation by the E6 gene product of human papillomavirus type 16. Nature 380:79–82

Ko MA, Rosario CO, Hudson JW, Kulkarni S, Pollett A, Dennis JW, Swallow CJ (2005) Plk4 haploinsufficiency causes mitotic infidelity and carcinogenesis. Nat Genet 37:883–888

Korzeniewski N, Zheng L, Cuevas R, Parry J, Chatterjee P, Anderton B, Duensing A, Munger K, Duensing S (2009) Cullin 1 functions as a centrosomal suppressor of centriole multiplication by regulating polo-like kinase 4 protein levels. Cancer Res 69:6668–6675

Korzeniewski N, Treat B, Duensing S (2011) The HPV-16 E7 oncoprotein induces centriole multiplication through deregulation of Polo-like kinase 4 expression. Mol Cancer 10:61

Lee SS, Glaunsinger B, Mantovani F, Banks L, Javier RT (2000) Multi-PDZ domain protein MUPP1 is a cellular target for both adenovirus E4-ORF1 and high-risk papillomavirus type 18 E6 oncoproteins. J Virol 74:9680–9693

Li J, Tan M, Li L, Pamarthy D, Lawrence TS, Sun Y (2005) SAK, a new polo-like kinase, is transcriptionally repressed by p53 and induces apoptosis upon RNAi silencing. Neoplasia 7:312–323

Liu X, Dakic A, Zhang Y, Dai Y, Chen R, Schlegel R (2009) HPV E6 protein interacts physically and functionally with the cellular telomerase complex. Proc Natl Acad Sci U S A 106:18780–18785

Loncarek J, Hergert P, Magidson V, Khodjakov A (2008) Control of daughter centriole formation by the pericentriolar material. Nat Cell Biol 10:322–328

Longworth MS, Laimins LA (2004a) Pathogenesis of human papillomaviruses in differentiating epithelia. Microbiol Mol Biol Rev 68:362–372

Longworth MS, Laimins LA (2004b) The binding of histone deacetylases and the integrity of zinc finger-like motifs of the E7 protein are essential for the life cycle of human papillomavirus type 31. J Virol 78:3533–3541

Lyons TE, Salih M, Tuana BS (2006) Activating E2Fs mediate transcriptional regulation of human E2F6 repressor. Am J Physiol Cell Physiol 290:C189–C199

Massimi P, Shai A, Lambert P, Banks L (2008) HPV E6 degradation of p53 and PDZ containing substrates in an E6AP null background. Oncogene 27:1800–1804

Matsushime H, Quelle DE, Shurtleff SA, Shibuya M, Sherr CJ, Kato JY (1994) D-type cyclin-dependent kinase activity in mammalian cells. Mol Cell Biol 14:2066–2076

McLaughlin-Drubin ME, Munger K (2009a) Oncogenic activities of human papillomaviruses. Virus Res 143:195–208

McLaughlin-Drubin ME, Munger K (2009b) The human papillomavirus E7 oncoprotein. Virology 384:335–344

McLaughlin-Drubin ME, Huh KW, Munger K (2008) Human papillomavirus type 16 E7 oncoprotein associates with E2F6. J Virol 82:8695–8705

McLaughlin-Drubin ME, Crum CP, Munger K (2011) Human papillomavirus E7 oncoprotein induces KDM6A and KDM6B histone demethylase expression and causes epigenetic reprogramming. Proc Natl Acad Sci U S A 108:2130–2135

Mikhailov A, Cole RW, Rieder CL (2002) DNA damage during mitosis in human cells delays the metaphase/anaphase transition via the spindle-assembly checkpoint. Curr Biol 12:1797–1806

Munger K, Phelps WC, Bubb V, Howley PM, Schlegel R (1989) The E6 and E7 genes of the human papillomavirus type 16 together are necessary and sufficient for transformation of primary human keratinocytes. J Virol 63:4417–4421

Munoz N, Bosch FX, de Sanjose S, Herrero R, Castellsague X, Shah KV, Snijders PJ, Meijer CJ (2003) Epidemiologic classification of human papillomavirus types associated with cervical cancer. N Engl J Med 348:518–527

Nakayama KI, Nakayama K (2006) Ubiquitin ligases: cell-cycle control and cancer. Nat Rev Cancer 6:369–381

Nguyen CL, Munger K (2008) Direct association of the HPV16 E7 oncoprotein with cyclin A/CDK2 and cyclin E/CDK2 complexes. Virology 380:21–25

Nguyen ML, Nguyen MM, Lee D, Griep AE, Lambert PF (2003) The PDZ ligand domain of the human papillomavirus type 16 E6 protein is required for E6's induction of epithelial hyperplasia in vivo. J Virol 77:6957–6964

Nguyen CL, Eichwald C, Nibert ML, Munger K (2007) Human papillomavirus type 16 E7 oncoprotein associates with the centrosomal component gamma-tubulin. J Virol 81:13533–13543

Orth G (1986) Epidermodysplasia verruciformis: a model for understanding the oncogenicity of human papillomaviruses. Ciba Found Symp 120:157–174

Paz IB, Cook N, Odom-Maryon T, Xie Y, Wilczynski SP (1997) Human papillomavirus (HPV) in head and neck cancer. An association of HPV 16 with squamous cell carcinoma of Waldeyer's tonsillar ring. Cancer 79:595–604

Petroski MD, Deshaies RJ (2005) Function and regulation of cullin-RING ubiquitin ligases. Nat Rev Mol Cell Biol 6:9–20

Phelps WC, Munger K, Yee CL, Barnes JA, Howley PM (1992) Structure-function analysis of the human papillomavirus type 16 E7 oncoprotein. J Virol 66:2418–2427

Pim D, Collins M, Banks L (1992) Human papillomavirus type 16 E5 gene stimulates the transforming activity of the epidermal growth factor receptor. Oncogene 7:27–32

Rechsteiner M, Rogers SW (1996) PEST sequences and regulation by proteolysis. Trends Biochem Sci 21:267–271

Resnitzky D, Gossen M, Bujard H, Reed SI (1994) Acceleration of the G1/S phase transition by expression of cyclins D1 and E with an inducible system. Mol Cell Biol 14:1669–1679

Riley RR, Duensing S, Brake T, Munger K, Lambert PF, Arbeit JM (2003) Dissection of human papillomavirus E6 and E7 function in transgenic mouse models of cervical carcinogenesis. Cancer Res 63:4862–4871

Rogers GC, Rusan NM, Roberts DM, Peifer M, Rogers SL (2009) The SCF Slimb ubiquitin ligase regulates Plk4/Sak levels to block centriole reduplication. J Cell Biol 184:225–239

Schatten H (2008) The mammalian centrosome and its functional significance. Histochem Cell Biol 129:667–686

Scheffner M, Huibregtse JM, Vierstra RD, Howley PM (1993) The HPV-16 E6 and E6-AP complex functions as a ubiquitin-protein ligase in the ubiquitination of p53. Cell 75:495–505

Schiller JT, Lowy DR (2006) Prospects for cervical cancer prevention by human papillomavirus vaccination. Cancer Res 66:10229–10232

Schiller JT, Lowy DR (2009) Immunogenicity testing in human papillomavirus virus-like-particle vaccine trials. J Infect Dis 200:166–171

Sedman J, Stenlund A (1995) Co-operative interaction between the initiator E1 and the transcriptional activator E2 is required for replicator specific DNA replication of bovine papillomavirus in vivo and in vitro. EMBO J 14:6218–6228

Sedman J, Stenlund A (1998) The papillomavirus E1 protein forms a DNA-dependent hexameric complex with ATPase and DNA helicase activities. J Virol 72:6893–6897

Southern SA, Lewis MH, Herrington CS (2004) Induction of tetrasomy by human papillomavirus type 16 E7 protein is independent of pRb binding and disruption of differentiation. Br J Cancer 90:1949–1954

Straight SW, Hinkle PM, Jewers RJ, McCance DJ (1993) The E5 oncoprotein of human papillomavirus type 16 transforms fibroblasts and effects the downregulation of the epidermal growth factor receptor in keratinocytes. J Virol 67:4521–4532

Stubenrauch F, Laimins LA (1999) Human papillomavirus life cycle: active and latent phases. Semin Cancer Biol 9:379–386

Thomas JT, Laimins LA (1998) Human papillomavirus oncoproteins E6 and E7 independently abrogate the mitotic spindle checkpoint. J Virol 72:1131–1137

Thorland EC, Myers SL, Persing DH, Sarkar G, McGovern RM, Gostout BS, Smith DI (2000) Human papillomavirus type 16 integrations in cervical tumors frequently occur in common fragile sites. Cancer Res 60:5916–5921

Tsou MF, Wang WJ, George KA, Uryu K, Stearns T, Jallepalli PV (2009) Polo kinase and separase regulate the mitotic licensing of centriole duplication in human cells. Dev Cell 17:344–354

White AE, Livanos EM, Tlsty TD (1994) Differential disruption of genomic integrity and cell cycle regulation in normal human fibroblasts by the HPV oncoproteins. Genes Dev 8:666–677

Wilting SM, Snijders PJ, Meijer GA, Ylstra B, van den Ijssel PR, Snijders AM, Albertson DG, Coffa J, Schouten JP, van de Wiel MA, Meijer CJ, Steenbergen RD (2006) Increased gene copy numbers at chromosome 20q are frequent in both squamous cell carcinomas and adenocarcinomas of the cervix. J Pathol 209:220–230

Yang L, Mohr I, Fouts E, Lim DA, Nohaile M, Botchan M (1993) The E1 protein of bovine papilloma virus 1 is an ATP-dependent DNA helicase. Proc Natl Acad Sci U S A 90:5086–5090

Zhang B, Chen W, Roman A (2006) The E7 proteins of low- and high-risk human papillomaviruses share the ability to target the pRB family member p130 for degradation. Proc Natl Acad Sci U S A 103:437–442

Chapter 13
Centrosomes, DNA Damage and Aneuploidy

Chiara Saladino, Emer Bourke and Ciaran G. Morrison

Abstract Understanding how the genomic instability that accompanies tumour development arises has been an important question for more than a century. One potential cause of such instability is defective chromosome segregation during mitosis. A cause of mitotic defects may lie in the acquisition of multiple mitotic spindle poles, through an increase in the number of centrosomes. Cancer cells frequently possess multiple centrosomes. DNA damaging treatments, or mutations in key DNA repair genes, also lead to centrosome amplification. Here, we review current models for how cells may lose the normal controls on centrosome duplication and acquire more than the normal number of these organelles. We also discuss how genotoxic stresses may contribute to the dysregulation of centrosome duplication and how this process may be a contributory factor in cellular transformation.

13.1 Mechanisms of Aneuploidy

Aneuploidy has been described as the most common characteristic of cancer cells (Weaver and Cleveland 2006). Numerous genetic alterations have been observed in neoplastic cells, including chromosome and gene deletions, amplification and translocation. However, the presence of these alterations does not necessarily

---

C. Saladino · C. G. Morrison (✉)
School of Natural Sciences, Centre for Chromosome Biology,
National University of Ireland Galway, Galway, Ireland
e-mail: Ciaran.Morrison@nuigalway.ie

E. Bourke
Institute of Technology Sligo, Ash Lane, Sligo, Ireland

indicate that the tumour is genetically unstable. It was observed that in some haematological cancers, malignant cells were stably aneuploid, following chromosomal redistribution earlier during tumorigenesis. More often, though, aneuploid cancer cells derive from an increase in the rate of gain or loss of whole chromosomes, a condition known as chromosome instability (CIN) (Kops et al. 2005; Lengauer et al. 1998). Although aneuploidy has been often suggested as the driving force behind tumorigenesis, the rate at which chromosomes are gained or lost can cause different outcomes. While moderate levels of CIN facilitate tumour formation and development, massive changes in chromosome content can be intolerable to cancer cells (reviewed by Godinho et al. 2009). A number of studies have shown that high levels of chromosome missegregation and aneuploidy reduce cell viability in cancer cells by affecting a broad number of cellular processes (Kops et al. 2004; Thompson and Compton 2008; Williams et al. 2008). Thus, under normal circumstances, high levels of genetic instability impair cell growth, unless the mutations introduced provide a selective pressure for the accumulation of further changes, allowing cells to survive the adverse effects of aneuploidy (Holland and Cleveland 2009).

Aneuploidy or CIN can arise from defects in chromosome segregation during mitosis. Cells may gain or lose chromosomes as a result of defects in the mitotic checkpoint or in sister chromatid cohesion, of microtubule misattachments and of aberrant mitotic division (reviewed by Kops et al. 2005). The major cell cycle checkpoint ensuring the correct segregation of chromosomes between daughter cells is the spindle assembly checkpoint (SAC), which prevents metaphase–anaphase transition until all kinetochores have established a correct bi-orientation on the spindle (Musacchio and Salmon 2007). In mammalian cells, the complete inactivation of the mitotic spindle checkpoint results in cell death and early embryonic lethality due to massive chromosome missegregation (Kalitsis et al. 2000; Kops et al. 2004). However, altered expression or mutations in genes coding for components of the SAC have been observed in aneuploid human cancers (Cahill et al. 1998; Dai et al. 2004; Li et al. 2003). In these cells, the mitotic checkpoint is impaired and anaphase can begin even in the presence of unattached or misattached kinetochores, leading to chromosome missegregation and aneuploidy (Hanks et al. 2004; Sotillo et al. 2007).

Chromosome missegregation events may also occur following the generation of incorrect kinetochore–microtubule attachments. When one kinetochore interacts with microtubules coming from both spindle poles (merotelic attachment), the chromosome is attached and under tension, so that the SAC is not activated and cells can exit mitosis without any significant delay (Cimini et al. 2004; Khodjakov et al. 1997). Merotelic attachments are usually corrected before anaphase onset, although occasionally sister chromatids with merotelic attachment can missegregate, failing to move in either direction and yielding a lagging chromosome (reviewed by Salmon et al. 2005). At the end of mitosis, the lagging chromatid will be pushed into either one of the daughter cells, and upon nuclear envelope reassembly, will form a separate micronucleus (Cimini et al. 2002). A further source of CIN arises when cells enter mitosis with more than two centrosomes (Holland and Cleveland 2009).

Centrosomes play a fundamental role in the organisation of the mitotic spindle. In the presence of supernumerary centrosomes, multipolar spindles may form and contribute to aneuploidy, although how such aneuploidy arises is not yet fully understood.

13.2 Centrosome Abnormalities and Tumorigenesis

In 1902, Theodor Boveri first described the detrimental effects on organism and cell physiology of an abnormal chromosome number. Several years earlier, the pathologist David Hansemann had observed the presence of aberrant chromosome segregation during mitosis in cancer cells. These findings led Boveri to propose that aneuploidy might promote tumorigenesis. In 1914, following his studies on sea urchin embryos, Boveri observed that cells forced to undergo multipolar mitosis produced progeny with an aberrant chromosome number. The prevalence of chromosome aberrations in cancer cells led Boveri to suggest that they were the result of multipolar mitoses in cells with supernumerary centrosomes (reviewed by Boveri 2008; Godinho et al. 2009; Holland and Cleveland 2009).

Since 1914, several studies have shown that supernumerary centrosomes are common to almost all types of solid and haematological malignancies, including breast, brain, lung, colon, ovary, liver, prostate, bone, gall bladder, head and neck cancers as well as lymphoma and leukaemia (Gustafson et al. 2000; Kramer et al. 2003; Kuo et al. 2000; Lingle et al. 1998; Nitta et al. 2006; Pihan et al. 1998; Pihan et al. 2001; Sato et al. 1999; Weber et al. 1998). Furthermore, in cancer cells, aberrations in centrosome number are often associated with structural irregularities such as increased centrosome size and alterations in the expression and phosphorylation status of PCM components (Lingle et al. 2002). Aberrant centrosomes often exhibit aberrant recruitment of gamma-TuRCs and defects in microtubule nucleation, which, in turn, affect the cellular architecture (Lingle et al. 2002; Lingle and Salisbury 2001). Furthermore, it has been shown that centrosome abnormalities correlate with increased levels of multipolar mitosis and aneuploidy in cancer cells (Ghadimi et al. 2000; Gisselsson et al. 2004). Several studies showed that in highly invasive cancers and in situ carcinoma, centrosomal defects are often associated with chromosomal aberrations, which occur at a later stage in tumor progression (Gisselsson et al. 2004; Lingle et al. 2002; Pihan et al. 2001). However, centrosome defects were also identified in cancers at an early stage in animal models and were shown to become more severe with tumour progression (D'Assoro et al. 2002a; Duensing et al. 2001; Goepfert et al. 2002; Shono et al. 2001). Although extra centrosomes failed to generate large-scale genome instability in *Drosophila*, likely due to the low proliferative index of mature *Drosophila* cells, serial transplantation in the abdomen of adult flies of larval brain cells carrying mutations in genes that encode centrosomal regulators and showing centrosome amplification, generated both benign and malignant hyperplasia, demonstrating that centrosome amplification can initiate tumorigenesis in flies (Castellanos et al. 2008). While these observations supported the theory that amplified centrosomes represent a cause of aneuploidy, they did not establish how the two phenomena are related or whether centrosome amplification was a cause or a consequence of cancer progression (reviewed by D'Assoro et al. 2002b; Nigg 2002).

13.3 Mechanisms of Aneuploidy that Involve Centrosome Amplification

Conceptually, the simplest mechanism of aneuploidy to arise from centrosome aberrations is that multipolar mitoses occur through the formation of multiple spindle poles and cause aneuploidy through unequal distribution of chromosomes between daughter cells (Fukasawa 2005). However, recent time-lapse video microscopy studies demonstrated that cultured human cells containing amplified centrosomes efficiently cluster their extra centrosomes and divide in a bipolar fashion. Only a small fraction of cells with extra centrosomes underwent multipolar division and the progeny originating from such divisions was mostly non-viable (Ganem et al. 2009), consistent with the view that massive aneuploidy induced by multipolar cell division is lethal. Similarly, analysis of *Drosophila* lines in which around 60 % of the cells possessed supernumerary centrosomes revealed a delay in mitosis due to the formation of a transient multipolar intermediate, but the cells ultimately divided in a bipolar fashion (Basto et al. 2008).

Centrosome clustering appears to be the major strategy that human cells employ to minimise the impact of multiple centrosomes (Quintyne et al. 2005), although there exist several other approaches, such as inactivation or sequestration of extra centrosomes (Gergely and Basto 2008; Godinho et al. 2009). However, even though centrosome clustering prevents lethality caused by multipolar division, centrosome amplification, nevertheless, leads to chromosome missegregation and instability (Ganem et al. 2009; Silkworth et al. 2009). A recent study suggested a novel potential mechanism for how supernumerary centrosomes cause chromosome aberrations and aneuploidy. Pellman and colleagues showed that cells with amplified centrosomes go through a transient multipolar state during spindle formation, before clustering their centrosomes (Ganem et al. 2009). This intermediate state predisposes cells to develop aberrant merotelic attachments with high frequency. Unresolved merotelic attachments impair chromosome segregation by causing lagging chromosomes during anaphase (Cimini et al. 2001; Gregan et al. 2011), so that this model provides an explanation for how multiple centrosomes can lead to chromosome abnormalities, without causing multipolar divisions. Therefore, how cells acquire multiple centrosomes is an important question in understanding how genome stability is normally maintained.

13.4 Centrosome Pathways

There are two key pathways by which centrosomes can arise: the normal, templated pathway in which the pre-existing centrioles serve as the scaffolding for new centriole formation during S-phase, and a de novo pathway (Loncarek and Khodjakov 2009). However, these are not distinct in terms of the controlling activities, but differ in the sense that the existing mother centrioles serve as regulators of the 'templated' process (Rodrigues-Martins et al. 2007b).

In experiments where centrosomes were removed from monkey kidney cells by micromanipulation (Hinchcliffe et al. 2001; Maniotis and Schliwa 1991) or laser microsurgery (Khodjakov et al. 2000; Khodjakov and Rieder 2001), the centrosomes did not regenerate. However, subsequent work that examined what happened when the centrosomes were removed from S-phase arrested CHO cells by laser ablation (Khodjakov et al. 2002), or from *Chlamydomonas* cells by a mutation that causes a fraction of the daughter cells to have no centrioles (Marshall et al. 2001), demonstrated that cells can form centrosomes de novo. These observations were further supported by the finding of de novo centriole assembly in transformed (La Terra et al. 2005) and normal (Uetake et al. 2007) human cells. A p53-dependent cell cycle arrest in late G1 phase is caused by the loss or damage of centrosomes (Mikule et al. 2007; Srsen et al. 2006), which suggests a reason why the potentiation of de novo centrosome formation was only observed when cells were treated after this point. The formation of the de novo structures and the maturation of these centrioles require passage through an entire cycle (Khodjakov et al. 2002; La Terra et al. 2005). Once activated, this de novo pathway allows cells to produce multiple centrosomes, suggesting that numerical control of the centrosome resides in the existing centrosomes (Khodjakov et al. 2002; La Terra et al. 2005). Together, these findings indicate a general pathway of de novo centrosome formation that is normally inhibited by the presence of existing centrioles (La Terra et al. 2005) but which, upon activation or loss of inhibition, can generate large numbers of centrioles.

An evolutionarily conserved series of proteins govern the process by which centrioles normally duplicate (Carvalho-Santos et al. 2010). A key polo box-containing kinase, PLK4 in human (Habedanck et al. 2005; Kleylein-Sohn et al. 2007), SAK in *Drosophila melanogaster* (Bettencourt-Dias et al. 2005) is recruited to the centrosome by the coiled-coil protein SPD2/CEP192, which also directs the recruitment of the pericentriolar material (PCM) to the nascent centriole (Kemp et al. 2004; Pelletier et al. 2004; Zhu et al. 2008). PLK4/SAK is required for the recruitment of the coiled-coil proteins, SAS-4 (CPAP/CENP-J in human cells) and SAS-6, which specify the base of the forming centriole, direct the elongation of its microtubules and are required for centriole duplication (Kirkham et al. 2003; Leidel et al. 2005; Leidel and Gonczy 2003; Pelletier et al. 2006; Rodrigues-Martins et al. 2007a; Strnad et al. 2007). A further coiled-coil component of the *Caenorhabditis elegans* centriole regulatory apparatus, SAS-5 (Ana2 in *Drosophila*), is also required for centriole duplication (Pelletier et al. 2006; Stevens et al. 2010). ZYG-1 plays a role similar to SAK/PLK4 in *C. elegans* (O'Connell et al. 2001) and is required for the localisation of SAS-4, SAS-5 and SAS-6 (Pelletier et al. 2006). Asterless/CEP152 has recently been described as a Plk4-interactor that is required for centriole duplication, being required for SAS-6 localisation to centrioles (Dzhindzhev et al. 2010; Guernsey et al. 2010; Hatch et al. 2010; Varmark et al. 2007).

Overexpression of SAK/PLK4, SAS-4 or SAS-6 causes centriole overduplication (Bettencourt-Dias et al. 2005; Habedanck et al. 2005; Kleylein-Sohn et al. 2007; Peel et al. 2007; Strnad et al. 2007), although the structure of centrioles formed through SAS-4 and SAS-6 overexpression may be abnormal (Kohlmaier et al. 2009; Rodrigues-Martins et al. 2007a). Notably, the centriole overduplication induced by overexpression of these key regulators involves the formation of multiple daughters around a single mother in a distinctive 'rosette' arrangement, rather than the general initiation of de novo centrosome assembly (Kleylein-Sohn et al. 2007; Strnad et al. 2007). However, in cells where there are no centrosomes, such as unfertilised *Drosophila* eggs, such overexpression does lead to de novo centriole assembly (Peel et al. 2007). These data indicate a limitation of centriole number that is imposed by a pre-existing mother.

Another element involved in the control of centriole number is the PCM. Establishment of a PCM cloud is a relatively early event in the de novo centriole duplication process, after which centrioles arise within the cloud (Khodjakov et al. 2002). Induction of an expanded PCM in cells with centrioles by overexpression of pericentrin led to the appearance of multiple daughter centrioles independently of any spatial or numerical control from the mother centrioles (Loncarek et al. 2008). This observation prompted the hypothesis that the mother centriole's principal role in centriole assembly is the regulation and specification of a PCM scaffold, rather than the provision of a template (Loncarek et al. 2008). In either case, the control of centriole duplication resides in the extant structure.

13.5 Centrosome Amplification

Changes in the coordination of the chromosome and centrosome cycles lead to centrosome amplification, which has been noted when key cell cycle regulators or regulatory components of the centrosome are aberrantly expressed or suppressed (Hergovich et al. 2007; Hochegger et al. 2007; Leidel et al. 2005; McDermott et al. 2006; Mussman et al. 2000; Swanton et al. 2007; Tachibana et al. 2005). The altered expression of cell cycle regulators is a frequently observed phenomenon in human cancers, so this may represent one source of centrosome abnormalities. Alternatively, the dysregulation of such regulatory genes may occur as a consequence of ongoing genome instability during tumour development.

A further activity that disconnects the chromosome and centrosome cycles appears to be a controlled response to genotoxic stress. Abnormal amplification of centrosomes has also been observed following DNA damage induced by irradiation (Dodson et al. 2007; Sato et al. 2000a, b) or DNA replication stress (Balczon et al. 1995; Meraldi et al. 2002). Amplification of the centrosome occurs in cells that carry mutations in DNA repair or checkpoint genes (Bertrand et al. 2003; Dodson et al. 2004; Fukasawa et al. 1996; Griffin et al. 2000; Kraakman-van der Zwet et al. 2002; Mantel et al. 1999; Tutt et al. 2002; Yamaguchi-Iwai et al. 1999), express mutant forms of telomerase (Guiducci et al. 2001) or express viral oncogenes (Duensing et al. 2006; Duensing et al. 2000; Duensing and Munger 2003; Watanabe et al. 2000). Although these examples cover a broad range of genotoxic insults, it is clear that centrosome amplification is a potential consequence of DNA damage.

13.6 Mechanisms that Permit Centrosome Amplification

Multiple centrosomes can be observed in cells that suffer failure in cytokinesis due to altered expression of cell cycle and checkpoint regulators such as p53, BRCA2 and Aurora A (Daniels et al. 2004; Meraldi et al. 2002). In general, DNA-damaging treatments do not lead to tetraploidisation, so cytokinesis failure is not sufficient to explain how centrosome amplification occurs after genotoxic stress. Additional models are required, which we consider below.

Given the importance of ensuring the right number of centrosomes, normal centrosome duplication occurs in a manner that is strictly co-ordinated with the cell cycle (Delattre and Gonczy 2004; Hinchcliffe and Sluder 2001; Nigg 2007). This coordination is ensured by at least two controls:

i. A requirement for cyclin-dependent kinase activity in centrosome duplication. A specific link between the chromosome and centrosome cycles is CDK2, which requires heterodimerisation with cyclin A or cyclin E for activity and which is necessary for the centrosome duplication that occurs during extended S-phase arrest in mammalian cells (Hinchcliffe et al. 1999; Lacey et al. 1999; Matsumoto et al. 1999; Meraldi et al. 1999), but not in chicken DT40 cells (Bourke et al. 2010). Cdk2 is also necessary for the centriole overduplication that is induced by expression of the human papillomavirus (HPV) type 16 E7 oncoprotein or by proteasome inhibition (Duensing et al. 2007; Duensing et al. 2006). However, Cdk2 is not required in mouse or chicken cells for normal centrosome duplication and it is likely that other kinases can compensate for its absence in the cell cycle (Adon et al. 2010; Duensing et al. 2006; Hochegger et al. 2007).

ii. A 'licensing' of centrosome duplication through centriole disengagement, which is mediated by Polo-like kinase 1 and separase, a protease that is activated through anaphase promoting complex/cyclosome activity at the metaphase–anaphase transition (Tsou and Stearns 2006; Tsou et al. 2009). This licencing requirement normally limits when cells can duplicate their centrosomes, even within a cytoplasm that contains the requisite Cdk activity (Wong and Stearns 2003).

Therefore, for a cell to acquire multiple centrosomes, the following conditions must be fulfilled. Centrosomes must acquire a license for reduplication and the cell cycle regulators that drive centrosome duplication must be activated. As the generation of a centriole takes time, an additional condition may be added: the cell must not divide for a sufficient period to allow centriole duplication.

Taking one particular example of where centrosome amplification is induced experimentally, all these conditions are fulfilled. Extended S-phase arrest of many mammalian cells by hydroxyurea (HU) treatment allows the appearance of multiple centrosomes (Balczon et al. 1995; Prosser et al. 2009). The acquisition of multiple centrosomes during this arrest is dependent on Cdk activity (Prosser et al. 2009), and numerous reports have implicated Cdk2 as the particular kinase involved, acting predominantly with cyclin E (Hinchcliffe et al. 1999; Lacey et al. 1999; Matsumoto et al. 1999; Meraldi et al. 1999). Centrosomes are licensed in this cell cycle stage, Cdk2 is activated and cells do not progress through mitosis.

High levels of centrosome amplification are observed in p53-deficient mice and cells (Fukasawa et al. 1996). This is believed to arise from the dysregulation of Cdk2 activity when p53 is absent and p53-independent cell cycle arrest to provide sufficient time for amplification (Fukasawa 2008). Upregulation of Cdk2 activity through overexpression of cyclin E led to centrosome amplification in p53-deficient mouse cells, but little impact was seen in wild-type rat or mouse fibroblasts (Mussman et al. 2000; Spruck et al. 1999). Similarly, in human tumour cells, cyclin E overexpression induced centrosome amplification, but only in the absence of p53 function (Kawamura et al. 2004). Control of cyclin E levels has been cited as a mechanism by which Krüppel-like factor 4 influences centrosome amplification after irradiation (Yoon et al. 2005). HPV oncoprotein-induced centrosome amplification requires both the disabling of p53 and the loss of normal CDK2 regulation (Duensing and Munger 2002). The key downstream targets of CDK2 in centrosome duplication described to date include nucleophosmin (B23) (Okuda et al. 2000), Mps1 kinase (Fisk and Winey 2001) and CP110 (Chen et al. 2002). Interestingly, overactivation of a centrosomal nucleophosmin interactor, ROCK II kinase, has recently been shown to drive centrosome amplification in CDK2-deficient cells (Hanashiro et al. 2011), indicating a possible effector of CDK2 signalling in centrosome control.

13.7 DNA Damage and Centrosome Amplification

Early studies conducted on mouse cells showed that ionising radiation (IR) causes the amplification of microtubule-organising centres (MTOCs). Electron microscopy analysis of these MTOCs revealed structures which did not contain the paired centrioles and PCM typical of normal centrosomes (Sato et al. 1983). In Chinese hamster ovary (CHO) cells, incomplete DNA replication due to HU treatment caused mitotic centrosome fragmentation (Hut et al. 2003). A similar finding was made in *Drosophila* embryos, where it was shown that damaged DNA caused centrosome fragmentation, along with errors in chromosome segregation and cell death (Sibon et al. 2000). Premature centriole splitting has also been observed after IR in various human cell types (Saladino et al. 2009). Such splitting may indicate the disengagement of centrioles, providing licensed templates for centrosome reduplication (Tsou and Stearns 2006; Tsou et al. 2009). It should be noted that IR actually blocks the separation of duplicated centrosomes that accompanies normal entry into M phase, through an ATM-dependent, Plk1-mediated inhibition of the Nek2 kinase (Fletcher et al. 2004; Zhang et al. 2005), so that the precise impact of IR on the centriole cohesion machinery is not yet clear. Furthermore, although cell fusion experiments have indicated that irradiation is required for G2 phase centrosomes to acquire a licence for duplication (Inanc et al. 2010), it is not known what effect IR has on the principal licencing activity, separase, or its centrosomal target(s) (Tsou et al. 2009). DNA damage signalling actually inhibits the other known licencing signal, that of Plk1 activation (Smits et al. 2000; van Vugt et al. 2001; Zhang et al. 2005). In any case, as individual centrioles can organise spindle poles (Keryer et al. 1984; Ring et al. 1982; Sluder and Rieder 1985), aberrantly disengaged centrioles still retain their ability to nucleate microtubules and may contribute to multipolar spindle formation (Fig. 13.1).

Although the process of DNA damage-induced licencing of centrosome duplication is not yet understood, additional time sufficient for duplication is provided by the cell cycle delays that arise as part of the DNA damage response. The MTOC amplification seen in human cells after irradiation (Sato et al. 2000a, b) was confirmed by light and electron microscopy as being due to centrosome amplification in a wide range of transformed and non-transformed cell lines from mammals and chickens (Bourke et al. 2007; Dodson et al. 2004; Saladino et al. 2009). Other forms of DNA-damaging treatment also induced centrosome amplification (Robinson et al. 2007; Saladino et al. 2009). Importantly, the principal signalling components of the DNA damage response that blocks cell cycle progress after genotoxic stress are required to permit centrosome overduplication. Loss of the apical DNA damage-responsive kinase, ATM, greatly impedes centrosome amplification after IR or DNA damage resulting from the absence of the Rad51 recombinase (Dodson et al. 2004), and IR-induced centrosome amplification is entirely abrogated by loss of the downstream Chk1 kinase (Bourke et al. 2007). IR-induced centrosome amplification occurs independently of p53 status (Dodson et al. 2007), even though the extent of G2-to-M arrest occasioned by IR is an important factor that has implicated p53 in some studies (Kawamura et al. 2006), suggesting that the process is not governed by the same mechanisms that alter centrosome numbers during extended S-phase arrest. Furthermore, the model proposed for how S-phase arrested cells overduplicate their centrosomes, in which multiple, immature daughter centrioles assemble around a single mother (Duensing et al. 2007; Guarguaglini et al. 2005), is not sufficient to explain what happens after IR, when amplification leads to centrosome splitting and/or the duplication of single daughters per mother and the majority of centrosomes carry the maturation marker CEP170 (Bourke et al. 2007; Saladino et al. 2009). The recent demonstration that IR-induced centrosome amplification can occur outside S phase (Inanc et al. 2010) provides further evidence

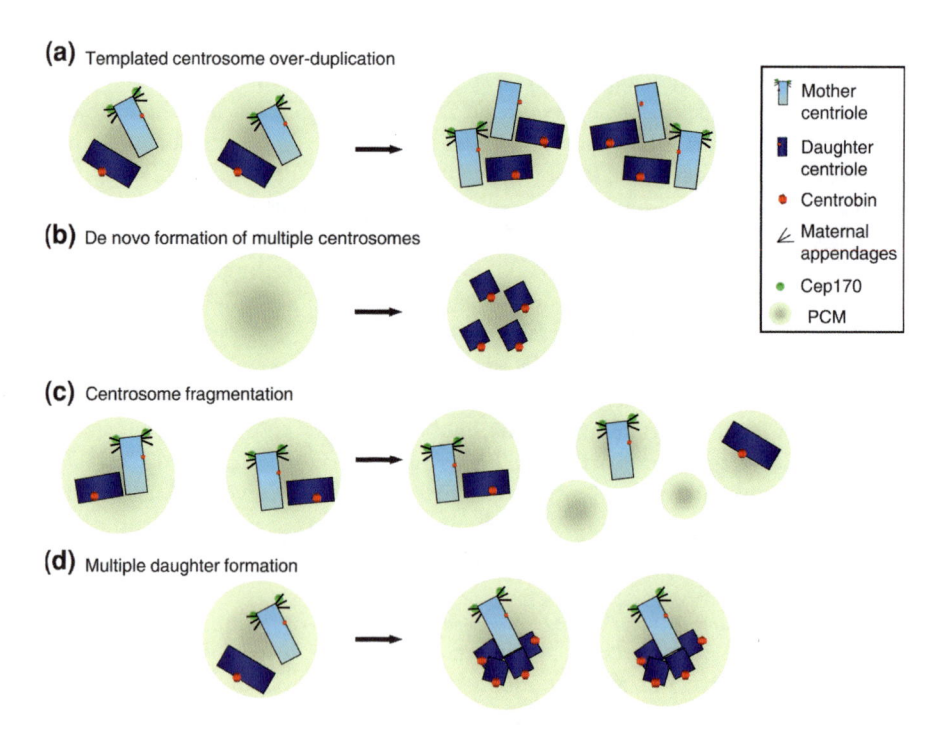

Fig. 13.1 Current models for centrosome amplification pathways: multiple centrosomes may be generated **a** during a prolonged cell cycle arrest, by templated centrosome duplication; **b** upon loss of existing centrioles, by a de novo pathway; **c** following DNA-damaging treatment by centrosome fragmentation; and **d** through the formation of multiple daughters from a single mother. Parental centrioles are schematically represented by rectangles coloured *light blue* for the mother, and *dark blue* for daughter centrioles. *Red spots* indicate centrobin association with centrioles during the cell cycle (Zou et al. 2005) and *green spots* indicate Cep170 localisation at the mature centriole (Guarguaglini et al. 2005; Saladino et al. 2009)

that an arrest in G2 phase after IR is permissive for the overduplication of the centrosome, as we have proposed (Dodson et al. 2004).

Although an extended G2 phase delay is necessary for DNA damage-induced centrosome overduplication, a question that remains is whether such an arrest is sufficient. Chk1 localises to the centrosome, along with many other elements of the DNA damage response (Loffler et al. 2006; Oricchio et al. 2006), so that it has been technically challenging to address this issue. Inhibition of CDK1 by pharmacological means or by the use of an analogue-sensitive mutant causes a robust cell cycle arrest at the transition to mitosis, without any DNA damage signal, which is accompanied by high levels of centrosome amplification (Hochegger et al. 2007). However, CDK1 inhibition also elevates the activity of CDK2 (Bourke et al. 2010). Notably, IR also causes the activation of CDK2 activity in a subset of cell types (Bourke et al. 2010), so that of the conditions for centrosome amplification that we have outlined, DNA damage leads to the fulfilment of several at once.

In several instances given above (Bettencourt-Dias et al. 2005; Habedanck et al. 2005; Kleylein-Sohn et al. 2007; Peel et al. 2007; Strnad et al. 2007), overexpression of the key upstream regulators of centriole duplication causes the formation of multiple daughter centrioles. However, in a preliminary study on a subset of centrosomal candidates, we found no evidence for significant upregulation of centrosome protein-coding genes after irradiation of non-transformed human cells (Saladino 2010), suggesting that increasing the levels of the structural components of new centrioles is not how IR drives centrosome amplification. Another question that arises is how additional centrioles assemble after irradiation, once the conditions that allow their overduplication have been met. Time-lapse microscopy of CHO cells during extended S-phase arrest has indicated that multiple centrosomes can assemble around the pre-existing mother (Guarguaglini et al. 2005; Kuriyama et al. 2007), or from nuclear aggregates of centrin (Prosser et al. 2009). Multiple daughters are also induced by peptide vinyl sulfone proteasome inhibitor Z-L(3)VS treatment (Duensing et al. 2007) or by the HPV16 E7 oncoprotein (Duensing et al. 2006). However, apart from centriole splitting and fragmentation, which may reflect initial steps in centrosome reduplication, it appears that IR induces the duplication of the entire centrosome in the form of paired mother–daughter centrioles (Bourke et al. 2007; Dodson et al. 2007).

13.8 IR Impact on the Cell Cycle and on Cells

IR and other forms of DNA damage kill cells through caspase-dependent apoptosis or mitotic catastrophe (Blagosklonny 2007; Jonathan et al. 1999; Okada and Mak 2004; Roninson et al. 2001). Mitotic catastrophe is a consequence of a mitotic delay in which cells with incompletely replicated genomes or unrepaired DNA damage enter mitosis and undergo apoptosis during M phase (reviewed by Vakifahmetoglu et al. 2008). While the G2 checkpoint normally averts mitotic entry under such circumstances, problems with this checkpoint can allow cells to initiate premature mitosis, suffer mitotic delay through activation of the SAC and ultimately, die (Johnson et al. 1999; Mikhailov et al. 2002; Nitta et al. 2004; Shin et al. 2003; Vogel et al. 2005). Centrosome amplification, a response to DNA damage that occurs during a checkpoint-mediated delay, will also cause a mitotic delay and compromise cell viability during such a delay (Ganem et al. 2009; Inanc et al. 2010; Loffler et al. 2006). In support of the notion that centrosome amplification contributes to the death of cells with DNA damage, live-cell imaging analysis of human tumour cells demonstrated that the vast majority of irradiated cells with multiple centrosomes fail in mitosis, but also that >60 % of cells undergoing mitotic catastrophe have multiple centrosomes (Dodson et al. 2007).

As noted in a recent review of how aneuploidy arises, it is not yet clear whether centrosome amplification is a cause or a consequence of genome instability, or both (Chandhok and Pellman 2009). It is clear that a deficiency in the DNA damage response is likely to lead toward cancer. Recent data have demonstrated the activation of the DNA damage response in pre-cancerous lesions in a range of human tissues

(Bartkova et al. 2005; Gorgoulis et al. 2005). This activation of the DNA damage response constrains tumourigenesis by inducing cell cycle delay, cell death or senescence, so that cells that no longer respond normally to DNA damage signals have a selective advantage in tumour development (Bartkova et al. 2005, 2006; Braig et al. 2005; Gorgoulis et al. 2005). However, the potential contribution of centrosome amplification to aneuploidy might make it a rather hazardous component of the normal DNA damage response or mechanism of cell death. Nevertheless, it is interesting to speculate that inducing centrosome amplification might be a means by which the killing effects of DNA damaging treatments could be potentiated.

References

Adon AM, Zeng X, Harrison MK, Sannem S, Kiyokawa H, Kaldis P, Saavedra HI (2010) Cdk2 and Cdk4 regulate the centrosome cycle and are critical mediators of centrosome amplification in p53-null cells. Mol Cell Biol 30(3):694–710

Balczon R, Bao L, Zimmer WE, Brown K, Zinkowski RP, Brinkley BR (1995) Dissociation of centrosome replication events from cycles of DNA synthesis and mitotic division in hydroxyurea-arrested Chinese hamster ovary cells. J Cell Biol 130(1):105–115

Bartkova J, Horejsi Z, Koed K, Kramer A, Tort F, Zieger K, Guldberg P, Sehested M, Nesland JM, Lukas C, Orntoft T, Lukas J, Bartek J (2005) DNA damage response as a candidate anti-cancer barrier in early human tumorigenesis. Nature 434(7035):864–870

Bartkova J, Rezaei N, Liontos M, Karakaidos P, Kletsas D, Issaeva N, Vassiliou LV, Kolettas E, Niforou K, Zoumpourlis VC, Takaoka M, Nakagawa H, Tort F, Fugger K, Johansson F, Sehested M, Andersen CL, Dyrskjot L, Orntoft T, Lukas J, Kittas C, Helleday T, Halazonetis TD, Bartek J, Gorgoulis VG (2006) Oncogene-induced senescence is part of the tumorigenesis barrier imposed by DNA damage checkpoints. Nature 444(7119):633–637

Basto R, Brunk K, Vinadogrova T, Peel N, Franz A, Khodjakov A, Raff JW (2008) Centrosome amplification can initiate tumorigenesis in flies. Cell 133(6):1032–1042

Bertrand P, Lambert S, Joubert C, Lopez BS (2003) Overexpression of mammalian Rad51 does not stimulate tumorigenesis while a dominant-negative Rad51 affects centrosome fragmentation, ploidy and stimulates tumorigenesis, in p53-defective CHO cells. Oncogene 22(48):7587–7592

Bettencourt-Dias M, Rodrigues-Martins A, Carpenter L, Riparbelli M, Lehmann L, Gatt MK, Carmo N, Balloux F, Callaini G, Glover DM (2005) SAK/PLK4 is required for centriole duplication and flagella development. Curr Biol 15(24):2199–2207

Blagosklonny MV (2007) Mitotic arrest and cell fate. Cell Cycle 6(1):e1–e5

Bourke E, Brown JA, Takeda S, Hochegger H, Morrison CG (2010) DNA damage induces Chk1-dependent threonine-160 phosphorylation and activation of Cdk2. Oncogene 29(4):616–624

Bourke E, Dodson H, Merdes A, Cuffe L, Zachos G, Walker M, Gillespie D, Morrison CG (2007) DNA damage induces Chk1-dependent centrosome amplification. EMBO Report 8(6):603–609

Boveri T (2008) Concerning the origin of malignant tumours by Theodor Boveri. Translated and annotated by Henry Harris. J Cell Sci 121(Suppl 1):1–84

Braig M, Lee S, Loddenkemper C, Rudolph C, Peters AH, Schlegelberger B, Stein H, Dorken B, Jenuwein T, Schmitt CA (2005) Oncogene-induced senescence as an initial barrier in lymphoma development. Nature 436(7051):660–665

Cahill DP, Lengauer C, Yu J, Riggins GJ, Willson JK, Markowitz SD, Kinzler KW, Vogelstein B (1998) Mutations of mitotic checkpoint genes in human cancers. Nature 392(6673):300–303

Carvalho-Santos Z, Machado P, Branco P, Tavares-Cadete F, Rodrigues-Martins A, Pereira-Leal JB, Bettencourt-Dias M (2010) Stepwise evolution of the centriole-assembly pathway. J Cell Sci 123(Pt 9):1414–1426

Castellanos E, Dominguez P, Gonzalez C (2008) Centrosome dysfunction in Drosophila neural stem cells causes tumors that are not due to genome instability. Curr Biol 18(16):1209–1214

Chandhok NS, Pellman D (2009) A little CIN may cost a lot: revisiting aneuploidy and cancer. Curr Opin Genet Dev 19(1):74–81

Chen Z, Indjeian VB, McManus M, Wang L, Dynlacht BD (2002) CP110, a cell cycle-dependent CDK substrate, regulates centrosome duplication in human cells. Dev Cell 3(3):339–350

Cimini D, Cameron LA, Salmon ED (2004) Anaphase spindle mechanics prevent mis-segregation of merotelically oriented chromosomes. Curr Biol 14(23):2149–2155

Cimini D, Fioravanti D, Salmon ED, Degrassi F (2002) Merotelic kinetochore orientation versus chromosome mono-orientation in the origin of lagging chromosomes in human primary cells. J Cell Sci 115(Pt 3):507–515

Cimini D, Howell B, Maddox P, Khodjakov A, Degrassi F, Salmon ED (2001) Merotelic kinetochore orientation is a major mechanism of aneuploidy in mitotic mammalian tissue cells. J Cell Biol 153(3):517–527

D'Assoro AB, Barrett SL, Folk C, Negron VC, Boeneman K, Busby R, Whitehead C, Stivala F, Lingle WL, Salisbury JL (2002a) Amplified centrosomes in breast cancer: a potential indicator of tumor aggressiveness. Breast Cancer Res Treat 75(1):25–34

D'Assoro AB, Lingle WL, Salisbury JL (2002b) Centrosome amplification and the development of cancer. Oncogene 21(40):6146–6153

Dai W, Wang Q, Liu T, Swamy M, Fang Y, Xie S, Mahmood R, Yang YM, Xu M, Rao CV (2004) Slippage of mitotic arrest and enhanced tumor development in mice with BubR1 haploinsufficiency. Cancer Res 64(2):440–445

Daniels MJ, Wang Y, Lee M, Venkitaraman AR (2004) Abnormal cytokinesis in cells deficient in the breast cancer susceptibility protein BRCA2. Science 306(5697):876–879

Delattre M, Gonczy P (2004) The arithmetic of centrosome biogenesis. J Cell Sci 117(Pt 9):1619–1630

Dodson H, Bourke E, Jeffers LJ, Vagnarelli P, Sonoda E, Takeda S, Earnshaw WC, Merdes A, Morrison C (2004) Centrosome amplification induced by DNA damage occurs during a prolonged G2 phase and involves ATM. EMBO J 23(19):3864–3873

Dodson H, Wheatley SP, Morrison CG (2007) Involvement of centrosome amplification in radiation-induced mitotic catastrophe. Cell Cycle 6(3):364–370

Duensing A, Liu Y, Perdreau SA, Kleylein-Sohn J, Nigg EA, Duensing S (2007) Centriole overduplication through the concurrent formation of multiple daughter centrioles at single maternal templates. Oncogene 26(43):6280–6288

Duensing A, Liu Y, Tseng M, Malumbres M, Barbacid M, Duensing S (2006) Cyclin-dependent kinase 2 is dispensable for normal centrosome duplication but required for oncogene-induced centrosome overduplication. Oncogene 25(20):2943–2949

Duensing S, Duensing A, Crum CP, Munger K (2001) Human papillomavirus type 16 E7 oncoprotein-induced abnormal centrosome synthesis is an early event in the evolving malignant phenotype. Cancer Res 61(6):2356–2360

Duensing S, Lee LY, Duensing A, Basile J, Piboonniyom S, Gonzalez S, Crum CP, Munger K (2000) The human papillomavirus type 16 E6 and E7 oncoproteins cooperate to induce mitotic defects and genomic instability by uncoupling centrosome duplication from the cell division cycle. Proc Natl Acad Sci U S A 97(18):10002–10007

Duensing S, Munger K (2002) The human papillomavirus type 16 E6 and E7 oncoproteins independently induce numerical and structural chromosome instability. Cancer Res 62(23):7075–7082

Duensing S, Munger K (2003) Human papillomavirus type 16 E7 oncoprotein can induce abnormal centrosome duplication through a mechanism independent of inactivation of retinoblastoma protein family members. J Virol 77(22):12331–12335

Dzhindzhev NS, Yu QD, Weiskopf K, Tzolovsky G, Cunha-Ferreira I, Riparbelli M, Rodrigues-Martins A, Bettencourt-Dias M, Callaini G, Glover DM (2010) Asterless is a scaffold for the onset of centriole assembly. Nature 467(7316):714–718

Fisk HA, Winey M (2001) The mouse Mps1p-like kinase regulates centrosome duplication. Cell 106(1):95–104

Fletcher L, Cerniglia GJ, Nigg EA, Yend TJ, Muschel RJ (2004) Inhibition of centrosome separation after DNA damage: a role for Nek2. Radiat Res 162(2):128–135

Fukasawa K (2005) Centrosome amplification, chromosome instability and cancer development. Cancer Lett 230(1):6–19

Fukasawa K (2008) P53, cyclin-dependent kinase and abnormal amplification of centrosomes. Biochim Biophys Acta 1786(1):15–23

Fukasawa K, Choi T, Kuriyama R, Rulong S, Vande Woude GF (1996) Abnormal centrosome amplification in the absence of p53. Science 271(5256):1744–1747

Ganem NJ, Godinho SA, Pellman D (2009) A mechanism linking extra centrosomes to chromosomal instability. Nature 460(7252):278–282

Gergely F, Basto R (2008) Multiple centrosomes: together they stand, divided they fall. Genes Dev 22(17):2291–2296

Ghadimi BM, Sackett DL, Difilippantonio MJ, Schrock E, Neumann T, Jauho A, Auer G, Ried T (2000) Centrosome amplification and instability occurs exclusively in aneuploid, but not in diploid colorectal cancer cell lines, and correlates with numerical chromosomal aberrations. Genes Chromosom Cancer 27(2):183–190

Gisselsson D, Palsson E, Yu C, Mertens F, Mandahl N (2004) Mitotic instability associated with late genomic changes in bone and soft tissue tumours. Cancer Lett 206(1):69–76

Godinho SA, Kwon M, Pellman D (2009) Centrosomes and cancer: how cancer cells divide with too many centrosomes. Cancer Metastasis Rev 28(1–2):85–98

Goepfert TM, Adigun YE, Zhong L, Gay J, Medina D, Brinkley WR (2002) Centrosome amplification and overexpression of aurora A are early events in rat mammary carcinogenesis. Cancer Res 62(14):4115–4122

Gorgoulis VG, Vassiliou LV, Karakaidos P, Zacharatos P, Kotsinas A, Liloglou T, Venere M, Ditullio RA Jr, Kastrinakis NG, Levy B, Kletsas D, Yoneta A, Herlyn M, Kittas C, Halazonetis TD (2005) Activation of the DNA damage checkpoint and genomic instability in human precancerous lesions. Nature 434(7035):907–913

Gregan J, Polakova S, Zhang L, Tolić-Nørrelykke IM, Cimini D (2011) Merotelic kinetochore attachment: causes and effects. Trends Cell Biol 21(6):374–381

Griffin CS, Simpson PJ, Wilson CR, Thacker J (2000) Mammalian recombination-repair genes XRCC2 and XRCC3 promote correct chromosome segregation. Nat Cell Biol 2(10):757–761

Guarguaglini G, Duncan PI, Stierhof YD, Holmstrom T, Duensing S, Nigg EA (2005) The forkhead-associated domain protein Cep170 interacts with Polo-like kinase 1 and serves as a marker for mature centrioles. Mol Biol Cell 16(3):1095–1107

Guernsey DL, Jiang H, Hussin J, Arnold M, Bouyakdan K, Perry S, Babineau-Sturk T, Beis J, Dumas N, Evans SC, Ferguson M, Matsuoka M, Macgillivray C, Nightingale M, Patry L, Rideout AL, Thomas A, Orr A, Hoffmann I, Michaud JL, Awadalla P, Meek DC, Ludman M, Samuels ME (2010) Mutations in centrosomal protein CEP152 in primary microcephaly families linked to MCPH4. Am J Hum Genet 87(1):40–51

Guiducci C, Cerone MA, Bacchetti S (2001) Expression of mutant telomerase in immortal telomerase-negative human cells results in cell cycle deregulation, nuclear and chromosomal abnormalities and rapid loss of viability. Oncogene 20(6):714–725

Gustafson LM, Gleich LL, Fukasawa K, Chadwell J, Miller MA, Stambrook PJ, Gluckman JL (2000) Centrosome hyperamplification in head and neck squamous cell carcinoma: a potential phenotypic marker of tumor aggressiveness. Laryngoscope 110(11):1798–1801

Habedanck R, Stierhof YD, Wilkinson CJ, Nigg EA (2005) The Polo kinase Plk4 functions in centriole duplication. Nat Cell Biol 7(11):1140–1146

Hanashiro K, Brancaccio M, Fukasawa K (2011) Activated ROCK II by-passes the requirement of the CDK2 activity for centrosome duplication and amplification. Oncogene 30(19):2188–2197

Hanks S, Coleman K, Reid S, Plaja A, Firth H, Fitzpatrick D, Kidd A, Mehes K, Nash R, Robin N, Shannon N, Tolmie J, Swansbury J, Irrthum A, Douglas J, Rahman N (2004) Constitutional aneuploidy and cancer predisposition caused by biallelic mutations in BUB1B. Nat Genet 36(11):1159–1161

Hatch EM, Kulukian A, Holland AJ, Cleveland DW, Stearns T (2010) Cep152 interacts with Plk4 and is required for centriole duplication. J Cell Biol 191(4):721–729

Hergovich A, Lamla S, Nigg EA, Hemmings BA (2007) Centrosome-associated NDR kinase regulates centrosome duplication. Mol Cell 25(4):625–634

Hinchcliffe EH, Li C, Thompson EA, Maller JL, Sluder G (1999) Requirement of Cdk2-cyclin E activity for repeated centrosome reproduction in Xenopus egg extracts. Science 283(5403):851–854

Hinchcliffe EH, Miller FJ, Cham M, Khodjakov A, Sluder G (2001) Requirement of a centrosomal activity for cell cycle progression through G1 into S phase. Science 291(5508):1547–1550

Hinchcliffe EH, Sluder G (2001) "It takes two to tango": understanding how centrosome duplication is regulated throughout the cell cycle. Genes Dev 15(10):1167–1181

Hochegger H, Dejsuphong D, Sonoda E, Saberi A, Rajendra E, Kirk J, Hunt T, Takeda S (2007) An essential role for Cdk1 in S phase control is revealed via chemical genetics in vertebrate cells. J Cell Biol 178(2):257–268

Holland AJ, Cleveland DW (2009) Boveri revisited: chromosomal instability, aneuploidy and tumorigenesis. Nat Rev Mol Cell Biol 10(7):478–487

Hut HM, Lemstra W, Blaauw EH, Van Cappellen GW, Kampinga HH, Sibon OC (2003) Centrosomes split in the presence of impaired DNA integrity during mitosis. Mol Biol Cell 14(5):1993–2004

Inanc B, Dodson H, Morrison CG (2010) A centrosome-autonomous signal that involves centriole disengagement permits centrosome duplication in G2 phase after DNA damage. Mol Biol Cell 21(22):3866–3877

Johnson PA, Clements P, Hudson K, Caldecott KW (1999) A mitotic spindle requirement for DNA damage-induced apoptosis in Chinese hamster ovary cells. Cancer Res 59(11):2696–2700

Jonathan EC, Bernhard EJ, McKenna WG (1999) How does radiation kill cells? Curr Opin Chem Biol 3(1):77–83

Kalitsis P, Earle E, Fowler KJ, Choo KH (2000) Bub3 gene disruption in mice reveals essential mitotic spindle checkpoint function during early embryogenesis. Genes Dev 14(18):2277–2282

Kawamura K, Izumi H, Ma Z, Ikeda R, Moriyama M, Tanaka T, Nojima T, Levin LS, Fujikawa-Yamamoto K, Suzuki K, Fukasawa K (2004) Induction of centrosome amplification and chromosome instability in human bladder cancer cells by p53 mutation and cyclin E overexpression. Cancer Res 64(14):4800–4809

Kawamura K, Morita N, Domiki C, Fujikawa-Yamamoto K, Hashimoto M, Iwabuchi K, Suzuki K (2006) Induction of centrosome amplification in p53 siRNA-treated human fibroblast cells by radiation exposure. Cancer Sci 97(4):252–258

Kemp CA, Kopish KR, Zipperlen P, Ahringer J, O'Connell KF (2004) Centrosome maturation and duplication in *C. elegans* require the coiled-coil protein SPD-2. Dev Cell 6(4):511–523

Keryer G, Ris H, Borisy GG (1984) Centriole distribution during tripolar mitosis in Chinese hamster ovary cells. J Cell Biol 98(6):2222–2229

Khodjakov A, Cole RW, McEwen BF, Buttle KF, Rieder CL (1997) Chromosome fragments possessing only one kinetochore can congress to the spindle equator. J Cell Biol 136(2):229–240

Khodjakov A, Cole RW, Oakley BR, Rieder CL (2000) Centrosome-independent mitotic spindle formation in vertebrates. Curr Biol 10(2):59–67

Khodjakov A, Rieder CL (2001) Centrosomes enhance the fidelity of cytokinesis in vertebrates and are required for cell cycle progression. J Cell Biol 153(1):237–242

Khodjakov A, Rieder CL, Sluder G, Cassels G, Sibon O, Wang CL (2002) De novo formation of centrosomes in vertebrate cells arrested during S phase. J Cell Biol 158(7):1171–1181

Kirkham M, Muller-Reichert T, Oegema K, Grill S, Hyman AA (2003) SAS-4 is a *C. elegans* centriolar protein that controls centrosome size. Cell 112(4):575–587

Kleylein-Sohn J, Westendorf J, Le Clech M, Habedanck R, Stierhof YD, Nigg EA (2007) Plk4-induced centriole biogenesis in human cells. Dev Cell 13(2):190–202

Kohlmaier G, Loncarek J, Meng X, McEwen BF, Mogensen MM, Spektor A, Dynlacht BD, Khodjakov A, Gonczy P (2009) Overly long centrioles and defective cell division upon excess of the SAS-4-related protein CPAP. Curr Biol 19(12):1012–1018

Kops GJ, Foltz DR, Cleveland DW (2004) Lethality to human cancer cells through massive chromosome loss by inhibition of the mitotic checkpoint. Proc Natl Acad Sci U S A 101(23):8699–8704

Kops GJ, Weaver BA, Cleveland DW (2005) On the road to cancer: aneuploidy and the mitotic checkpoint. Nat Rev Cancer 5(10):773–785

Kraakman-van der Zwet M, Overkamp WJ, van Lange RE, Essers J, van Duijn-Goedhart A, Wiggers I, Swaminathan S, van Buul PP, Errami A, Tan RT, Jaspers NG, Sharan SK, Kanaar R, Zdzienicka MZ (2002) Brca2 (XRCC11) deficiency results in radioresistant DNA synthesis and a higher frequency of spontaneous deletions. Mol Cell Biol 22(2):669–679

Kramer A, Schweizer S, Neben K, Giesecke C, Kalla J, Katzenberger T, Benner A, Muller-Hermelink HK, Ho AD, Ott G (2003) Centrosome aberrations as a possible mechanism for chromosomal instability in non-Hodgkin's lymphoma. Leukemia 17(11):2207–2213

Kuo KK, Sato N, Mizumoto K, Maehara N, Yonemasu H, Ker CG, Sheen PC, Tanaka M (2000) Centrosome abnormalities in human carcinomas of the gallbladder and intrahepatic and extrahepatic bile ducts. Hepatology 31(1):59–64

Kuriyama R, Terada Y, Lee KS, Wang CL (2007) Centrosome replication in hydroxyurea-arrested CHO cells expressing GFP-tagged centrin2. J Cell Sci 120(Pt 14):2444–2453

La Terra S, English CN, Hergert P, McEwen BF, Sluder G, Khodjakov A (2005) The de novo centriole assembly pathway in HeLa cells: cell cycle progression and centriole assembly/maturation. J Cell Biol 168(5):713–722

Lacey KR, Jackson PK, Stearns T (1999) Cyclin-dependent kinase control of centrosome duplication. Proc Natl Acad Sci U S A 96(6):2817–2822

Leidel S, Delattre M, Cerutti L, Baumer K, Gonczy P (2005) SAS-6 defines a protein family required for centrosome duplication in *C. elegans* and in human cells. Nat Cell Biol 7(2):115–125

Leidel S, Gonczy P (2003) SAS-4 is essential for centrosome duplication in C elegans and is recruited to daughter centrioles once per cell cycle. Dev Cell 4(3):431–439

Lengauer C, Kinzler KW, Vogelstein B (1998) Genetic instabilities in human cancers. Nature 396(6712):643–649

Li GQ, Li H, Zhang HF (2003) Mad2 and p53 expression profiles in colorectal cancer and its clinical significance. World J Gastroenterol 9(9):1972–1975

Lingle WL, Barrett SL, Negron VC, D'Assoro AB, Boeneman K, Liu W, Whitehead CM, Reynolds C, Salisbury JL (2002) Centrosome amplification drives chromosomal instability in breast tumor development. Proc Natl Acad Sci U S A 99(4):1978–1983

Lingle WL, Lutz WH, Ingle JN, Maihle NJ, Salisbury JL (1998) Centrosome hypertrophy in human breast tumors: implications for genomic stability and cell polarity. Proc Natl Acad Sci U S A 95(6):2950–2955

Lingle WL, Salisbury JL (2001) Methods for the analysis of centrosome reproduction in cancer cells. Methods Cell Biol 67:325–336

Loffler H, Lukas J, Bartek J, Kramer A (2006) Structure meets function–centrosomes, genome maintenance and the DNA damage response. Exp Cell Res 312(14):2633–2640

Loncarek J, Hergert P, Magidson V, Khodjakov A (2008) Control of daughter centriole formation by the pericentriolar material. Nat Cell Biol 10(3):322–328

Loncarek J, Khodjakov A (2009) Ab ovo or de novo? Mechanisms of centriole duplication. Mol Cells 27(2):135–142

Maniotis A, Schliwa M (1991) Microsurgical removal of centrosomes blocks cell reproduction and centriole generation in BSC-1 cells. Cell 67(3):495–504

Mantel C, Braun SE, Reid S, Henegariu O, Liu L, Hangoc G, Broxmeyer HE (1999) p21(cip-1/waf-1) deficiency causes deformed nuclear architecture, centriole overduplication, polyploidy, and relaxed microtubule damage checkpoints in human hematopoietic cells. Blood 93(4):1390–1398

Marshall WF, Vucica Y, Rosenbaum JL (2001) Kinetics and regulation of de novo centriole assembly. Implications for the mechanism of centriole duplication. Curr Biol 11(5):308–317

Matsumoto Y, Hayashi K, Nishida E (1999) Cyclin-dependent kinase 2 (Cdk2) is required for centrosome duplication in mammalian cells. Curr Biol 9(8):429–432

McDermott KM, Zhang J, Holst CR, Kozakiewicz BK, Singla V, Tlsty TD (2006) p16(INK4a) prevents centrosome dysfunction and genomic instability in primary cells. PLoS Biol 4(3):e51

Meraldi P, Honda R, Nigg EA (2002) Aurora-A overexpression reveals tetraploidization as a major route to centrosome amplification in p53-/- cells. EMBO J 21(4):483–492

Meraldi P, Lukas J, Fry AM, Bartek J, Nigg EA (1999) Centrosome duplication in mammalian somatic cells requires E2F and Cdk2-cyclin A. Nat Cell Biol 1(2):88–93

Mikhailov A, Cole RW, Rieder CL (2002) DNA damage during mitosis in human cells delays the metaphase/anaphase transition via the spindle-assembly checkpoint. Curr Biol 12(21):1797–1806

Mikule K, Delaval B, Kaldis P, Jurcyzk A, Hergert P, Doxsey S (2007) Loss of centrosome integrity induces p38-p53-p21-dependent G1-S arrest. Nat Cell Biol 9(2):160–170

Musacchio A, Salmon ED (2007) The spindle-assembly checkpoint in space and time. Nat Rev Mol Cell Biol 8(5):379–393

Mussman JG, Horn HF, Carroll PE, Okuda M, Tarapore P, Donehower LA, Fukasawa K (2000) Synergistic induction of centrosome hyperamplification by loss of p53 and cyclin E overexpression. Oncogene 19(13):1635–1646

Nigg EA (2002) Centrosome aberrations: cause or consequence of cancer progression? Nat Rev Cancer 2(11):815–825

Nigg EA (2007) Centrosome duplication: of rules and licenses. Trends Cell Biol 17(5):215–221

Nitta M, Kobayashi O, Honda S, Hirota T, Kuninaka S, Marumoto T, Ushio Y, Saya H (2004) Spindle checkpoint function is required for mitotic catastrophe induced by DNA-damaging agents. Oncogene 23(39):6548–6558

Nitta T, Kanai M, Sugihara E, Tanaka M, Sun B, Nagasawa T, Sonoda S, Saya H, Miwa M (2006) Centrosome amplification in adult T-cell leukemia and human T-cell leukemia virus type 1 Tax-induced human T cells. Cancer Sci 97(9):836–841

O'Connell KF, Caron C, Kopish KR, Hurd DD, Kemphues KJ, Li Y, White JG (2001) The *C. elegans* zyg-1 gene encodes a regulator of centrosome duplication with distinct maternal and paternal roles in the embryo. Cell 105(4):547–558

Okada H, Mak TW (2004) Pathways of apoptotic and non-apoptotic death in tumour cells. Nat Rev Cancer 4(8):592–603

Okuda M, Horn HF, Tarapore P, Tokuyama Y, Smulian AG, Chan PK, Knudsen ES, Hofmann IA, Snyder JD, Bove KE, Fukasawa K (2000) Nucleophosmin/B23 is a target of CDK2/cyclin E in centrosome duplication. Cell 103(1):127–140

Oricchio E, Saladino C, Iacovelli S, Soddu S, Cundari E (2006) ATM is activated by default in mitosis, localizes at centrosomes and monitors mitotic spindle integrity. Cell Cycle 5(1):88–92

Peel N, Stevens NR, Basto R, Raff JW (2007) Overexpressing centriole-replication proteins in vivo induces centriole overduplication and de novo formation. Curr Biol 17(10):834–843

Pelletier L, O'Toole E, Schwager A, Hyman AA, Muller-Reichert T (2006) Centriole assembly in *Caenorhabditis elegans*. Nature 444(7119):619–623

Pelletier L, Ozlu N, Hannak E, Cowan C, Habermann B, Ruer M, Muller-Reichert T, Hyman AA (2004) The Caenorhabditis elegans centrosomal protein SPD-2 is required for both pericentriolar material recruitment and centriole duplication. Curr Biol 14(10):863–873

Pihan GA, Purohit A, Wallace J, Knecht H, Woda B, Quesenberry P, Doxsey SJ (1998) Centrosome defects and genetic instability in malignant tumors. Cancer Res 58(17): 3974–3985

Pihan GA, Purohit A, Wallace J, Malhotra R, Liotta L, Doxsey SJ (2001) Centrosome defects can account for cellular and genetic changes that characterize prostate cancer progression. Cancer Res 61(5):2212–2219

Prosser SL, Straatman KR, Fry AM (2009) Molecular dissection of the centrosome overduplication pathway in S-phase arrested cells. Mol Cell Biol&&&AQ

Quintyne NJ, Reing JE, Hoffelder DR, Gollin SM, Saunders WS (2005) Spindle multipolarity is prevented by centrosomal clustering. Science 307(5706):127–129

Ring D, Hubble R, Kirschner M (1982) Mitosis in a cell with multiple centrioles. J Cell Biol 94(3):549–556

Robinson HM, Bratlie-Thoresen S, Brown R, Gillespie DA (2007) Chk1 is required for G2/M checkpoint response induced by the catalytic topoisomerase II inhibitor ICRF-193. Cell Cycle 6(10):1265–1267

Rodrigues-Martins A, Bettencourt-Dias M, Riparbelli M, Ferreira C, Ferreira I, Callaini G, Glover DM (2007a) DSAS-6 organizes a tube-like centriole precursor, and its absence suggests modularity in centriole assembly. Curr Biol 17(17):1465–1472

Rodrigues-Martins A, Riparbelli M, Callaini G, Glover DM, Bettencourt-Dias M (2007b) Revisiting the role of the mother centriole in centriole biogenesis. Science 316(5827):1046–1050

Roninson IB, Broude EV, Chang BD (2001) If not apoptosis, then what? Treatment-induced senescence and mitotic catastrophe in tumor cells. Drug Resist Updat 4(5):303–313

Saladino C (2010) The impact of DNA damage on centrosomes. PhD Thesis, Biochemistry, National University of Ireland Galway, Galway

Saladino C, Bourke E, Conroy PC, Morrison CG (2009) Centriole separation in DNA damage-induced centrosome amplification. Environ Mol Mutagen 50(8):725–732

Salmon ED, Cimini D, Cameron LA, DeLuca JG (2005) Merotelic kinetochores in mammalian tissue cells. Philos Trans R Soc Lond B Biol Sci 360(1455):553–568

Sato C, Kuriyama R, Nishizawa K (1983) Microtubule-organizing centers abnormal in number, structure, and nucleating activity in X-irradiated mammalian cells. J Cell Biol 96(3):776–782

Sato N, Mizumoto K, Nakamura M, Nakamura K, Kusumoto M, Niiyama H, Ogawa T, Tanaka M (1999) Centrosome abnormalities in pancreatic ductal carcinoma. Clin Cancer Res 5(5):963–970

Sato N, Mizumoto K, Nakamura M, Tanaka M (2000a) Radiation-induced centrosome overduplication and multiple mitotic spindles in human tumor cells. Exp Cell Res 255(2):321–326

Sato N, Mizumoto K, Nakamura M, Ueno H, Minamishima YA, Farber JL, Tanaka M (2000b) A possible role for centrosome overduplication in radiation-induced cell death. Oncogene 19(46):5281–5290

Shin HJ, Baek KH, Jeon AH, Park MT, Lee SJ, Kang CM, Lee HS, Yoo SH, Chung DH, Sung YC, McKeon F, Lee CW (2003) Dual roles of human BubR1, a mitotic checkpoint kinase, in the monitoring of chromosomal instability. Cancer Cell 4(6):483–497

Shono M, Sato N, Mizumoto K, Maehara N, Nakamura M, Nagai E, Tanaka M (2001) Stepwise progression of centrosome defects associated with local tumor growth and metastatic process of human pancreatic carcinoma cells transplanted orthotopically into nude mice. Lab Invest 81(7):945–952

Sibon OC, Kelkar A, Lemstra W, Theurkauf WE (2000) DNA-replication/DNA-damage-dependent centrosome inactivation in Drosophila embryos. Nat Cell Biol 2(2):90–95

Silkworth WT, Nardi IK, Scholl LM, Cimini D (2009) Multipolar spindle pole coalescence is a major source of kinetochore mis-attachment and chromosome mis-segregation in cancer cells. PLoS ONE 4(8):e6564

Sluder G, Rieder CL (1985) Centriole number and the reproductive capacity of spindle poles. J Cell Biol 100(3):887–896

Smits VA, Klompmaker R, Arnaud L, Rijksen G, Nigg EA, Medema RH (2000) Polo-like kinase-1 is a target of the DNA damage checkpoint. Nat Cell Biol 2(9):672–676

Sotillo R, Hernando E, Diaz-Rodriguez E, Teruya-Feldstein J, Cordon-Cardo C, Lowe SW, Benezra R (2007) Mad2 overexpression promotes aneuploidy and tumorigenesis in mice. Cancer Cell 11(1):9–23

Spruck CH, Won KA, Reed SI (1999) Deregulated cyclin E induces chromosome instability. Nature 401(6750):297–300

Srsen V, Gnadt N, Dammermann A, Merdes A (2006) Inhibition of centrosome protein assembly leads to p53-dependent exit from the cell cycle. J Cell Biol 174(5):625–630

Stevens NR, Dobbelaere J, Brunk K, Franz A, Raff JW (2010) Drosophila Ana2 is a conserved centriole duplication factor. J Cell Biol 188(3):313–323

Strnad P, Leidel S, Vinogradova T, Euteneuer U, Khodjakov A, Gonczy P (2007) Regulated HsSAS-6 levels ensure formation of a single procentriole per centriole during the centrosome duplication cycle. Dev Cell 13(2):203–213

Swanton C, Marani M, Pardo O, Warne PH, Kelly G, Sahai E, Elustondo F, Chang J, Temple J, Ahmed AA, Brenton JD, Downward J, Nicke B (2007) Regulators of mitotic arrest and ceramide metabolism are determinants of sensitivity to paclitaxel and other chemotherapeutic drugs. Cancer Cell 11(6):498–512

Tachibana KE, Gonzalez MA, Guarguaglini G, Nigg EA, Laskey RA (2005) Depletion of licensing inhibitor geminin causes centrosome overduplication and mitotic defects. EMBO Report 6(11):1052–1057

Thompson SL, Compton DA (2008) Examining the link between chromosomal instability and aneuploidy in human cells. J Cell Biol 180(4):665–672

Tsou MF, Stearns T (2006) Mechanism limiting centrosome duplication to once per cell cycle. Nature 442(7105):947–951

Tsou MF, Wang WJ, George KA, Uryu K, Stearns T, Jallepalli PV (2009) Polo kinase and separase regulate the mitotic licensing of centriole duplication in human cells. Dev Cell 17(3):344–354

Tutt AN, van Oostrom CT, Ross GM, van Steeg H, Ashworth A (2002) Disruption of Brca2 increases the spontaneous mutation rate in vivo: synergism with ionizing radiation. EMBO Report 3(3):255–260

Uetake Y, Loncarek J, Nordberg JJ, English CN, La Terra S, Khodjakov A, Sluder G (2007) Cell cycle progression and de novo centriole assembly after centrosomal removal in untransformed human cells. J Cell Biol 176(2):173–182

Vakifahmetoglu H, Olsson M, Zhivotovsky B (2008) Death through a tragedy: mitotic catastrophe. Cell Death Differ 15(7):1153–1162

van Vugt MA, Smits VA, Klompmaker R, Medema RH (2001) Inhibition of Polo-like kinase-1 by DNA damage occurs in an ATM- or ATR-dependent fashion. J Biol Chem 276(45):41656–41660

Varmark H, Llamazares S, Rebollo E, Lange B, Reina J, Schwarz H, Gonzalez C (2007) Asterless is a centriolar protein required for centrosome function and embryo development in Drosophila. Curr Biol 17(20):1735–1745

Vogel C, Kienitz A, Muller R, Bastians H (2005) The mitotic spindle checkpoint is a critical determinant for topoisomerase-based chemotherapy. J Biol Chem 280(6):4025–4028

Watanabe N, Yamaguchi T, Akimoto Y, Rattner JB, Hirano H, Nakauchi H (2000) Induction of M-phase arrest and apoptosis after HIV-1 Vpr expression through uncoupling of nuclear and centrosomal cycle in HeLa cells. Exp Cell Res 258(2):261–269

Weaver BA, Cleveland DW (2006) Does aneuploidy cause cancer? Curr Opin Cell Biol 18(6):658–667

Weber RG, Bridger JM, Benner A, Weisenberger D, Ehemann V, Reifenberger G, Lichter P (1998) Centrosome amplification as a possible mechanism for numerical chromosome aberrations in cerebral primitive neuroectodermal tumors with TP53 mutations. Cytogenet Cell Genet 83(3–4):266–269

Williams BR, Prabhu VR, Hunter KE, Glazier CM, Whittaker CA, Housman DE, Amon A (2008) Aneuploidy affects proliferation and spontaneous immortalization in mammalian cells. Science 322(5902):703–709

Wong C, Stearns T (2003) Centrosome number is controlled by a centrosome-intrinsic block to reduplication. Nat Cell Biol 5(6):539–544

Yamaguchi-Iwai Y, Sonoda E, Sasaki MS, Morrison C, Haraguchi T, Hiraoka Y, Yamashita YM, Yagi T, Takata M, Price C, Kakazu N, Takeda S (1999) Mre11 is essential for the maintenance of chromosomal DNA in vertebrate cells. EMBO J 18(23):6619–6629

Yoon HS, Ghaleb AM, Nandan MO, Hisamuddin IM, Dalton WB, Yang VW (2005) Kruppel-like factor 4 prevents centrosome amplification following gamma-irradiation-induced DNA damage. Oncogene 24(25):4017–4025

Zhang W, Fletcher L, Muschel RJ (2005) The role of Polo-like kinase 1 in the inhibition of centrosome separation after ionizing radiation. J Biol Chem 280(52):42994–42999

Zhu F, Lawo S, Bird A, Pinchev D, Ralph A, Richter C, Muller-Reichert T, Kittler R, Hyman AA, Pelletier L (2008) The mammalian SPD-2 ortholog Cep192 regulates centrosome biogenesis. Curr Biol 18(2):136–141

Zou C, Li J, Bai Y, Gunning WT, Wazer DE, Band V, Gao Q (2005) Centrobin: a novel daughter centriole-associated protein that is required for centriole duplication. J Cell Biol 171(3):437–445

Chapter 14
Centrosome Regulation and Breast Cancer

Zeina Kais and Jeffrey D. Parvin

Abstract Chromosomal instability and aneuploidy are commonly observed in breast tumor cells. Loss of function of the breast- and ovarian-specific tumor suppressor gene, BRCA1, results in supernumerary centrosomes that are likely to contribute to the genetic instability and tumorigenesis in breast cancer cells. Other DNA repair proteins also contribute to the regulation of the centrosome along with several oncogenic and tumor suppressor proteins. A number of centrosome regulators that are known to be involved in breast cancer will be discussed with a focused discussion of BRCA1 and its ubiquitin ligase activity in the regulation of centrosome number.

14.1 Centrosome Abnormalities and Breast Cancer

Centrosome abnormalities are a major cause of genomic instability resulting in aneuploidy that is commonly seen in tumors (Brinkley and Goepfert 1998). Many tumors, including breast tumors, have centrosome abnormalities characterized by centrosome amplifications, more than two centrosomes per cell, and centrosomal hypertrophy (Carroll et al. 1999; Lingle et al. 1998). Supernumerary centrosomes can promote the formation of multipolar spindles that may ultimately cluster into a pseudo-bipolar mitotic spindle during anaphase. This will lead to the formation of a defective kinetochore-microtubule attachment that results in the missegregation of chromosomes that will ultimately facilitate the formation of malignant tumors (Ganem et al. 2009). Centrosome abnormalities in breast cancer are mostly due to

Z. Kais · J. D. Parvin (✉)
Department of Biomedical Informatics,
Ohio State University Comprehensive Cancer Center,
Columbus OH 43210, USA
e-mail: Jeffrey.Parvin@osumc.edu

genetic alterations including mutations, activations, or deletions of oncogenes, tumor suppressors, or cell cycle regulators that affect centrosome duplication and function (Salisbury 2001).

14.2 Regulation of Centrosomes

14.2.1 Disruption of Nucleophosmin-Mediated Centrosome Control in Breast Cancer

Centrosome duplication is a process that takes place in S phase in coordination with DNA duplication. Both events are triggered by the activation of CDK2-cyclin E (Meraldi et al. 1999; Mussman et al. 2000). Several CDK2-cyclin E target proteins have been identified to be involved in centrosome regulation including nucleophosmin (NPM) a protein that functions as a chaperone in several cellular events and is frequently mutated in cancer cells (Okuda et al. 2000; Tokuyama et al. 2001). NPM localizes between the paired centrioles and is thought to function in centriole pairing (Shinmura et al. 2005). NPM dissociates from the centrosome upon phosphorylation by CDK2-cyclin E leading to the separation of the two centrioles, an initial step in the centrosome duplication cycle. The centrioles are tightly paired throughout the cell cycle except during the initiation of duplication. Abrogation of mechanisms underlying centriole pairing, such as depletion of NPM, results in the generation of extra centrosomes (Grisendi et al. 2005). In addition, it has been shown that after being phosphorylated, NPM binds to ROCK2, a member of the Rho-associated coiled containing protein kinase family, which is overexpressed in many cancers including breast cancer. Binding of phosphorylated NPM to ROCK2 activates it at the centrosomes leading to the initiation of centrosome duplication (Ma et al. 2006). Recently, NPM and ROCK2 have been shown to form a complex with BRCA2. This complex functions in maintaining the integrity of centrosome duplication. Inhibition of the NPM–BRCA2 interaction results in supernumerary centrosomes, a phenotype that would ultimately cause genetic instability and tumorigenesis. In addition, the loss of association between NPM and BRCA2 might play an important role in familial breast carcinogenesis since many missense mutations have been reported in the NPM binding region of BRCA2 in the hereditary breast and/or ovarian cancer families (Wang et al. 2011).

14.2.2 Control of Centrosomes by Cyclin A in Cancer

In addition to cyclin E, CDK2 forms a complex with cyclin A, which has also been implicated in the regulation of centrosome duplication. It has been shown that cyclin A promotes centriole overduplication in S phase-arrested cells through a continued stimulation of CDK2 activity (Duensing et al. 2006). CDK2-cyclin A complex cannot, however, initiate duplication of centrosomes during the normal cell cycle but in prolonged S phase-arrested cells, cyclin A promotes the reduplication of centrosomes (Duensing et al. 2007). Abnormal levels of cyclin A are frequently detected in human cancers and these abnormal levels may have a crucial role in centriole overduplication and amplification, which eventually might be a contributing factor to tumorigenesis (Duensing et al. 2007; Faivre et al. 2002).

Since CDK-cyclin controls the duplication of centrosomes, many proteins that control the activity of the CDK2/Cyclin A complex may also play a role in the regulation of centrosome duplication (Fukasawa 2007). One example is p21, which is a negative regulator of the CDK2 activity. p21 is regulated by p53 which is one of the proteins that have been implicated in the development of centrosome defects seen in human breast tumors (Salisbury 2001). It has been shown that cells lacking p53 are capable of undergoing centrosome reduplication even when DNA synthesis is inhibited resulting in centrosome amplification (Balczon et al. 1995; Fukasawa et al. 1996). Inhibition of DNA synthesis occurs under physiological stress irrespective of the p53 status (Fukasawa 2007). Normally, p53 is stabilized under such conditions through the inhibition of MDM2, an E3 ubiquitin ligase that promotes the degradation of the p53 protein. Stabilization of p53 will result in the upregulation of p21 that will inhibit the CDK2-cyclin complexes thus inhibiting the initiation of centrosome reduplication. Cells lacking p53, however, will have unrestrained activation of CDK2 which in turn will trigger centrosome reduplication leading to the formation of extra centrosomes.

In addition to regulating centrosome duplication through the p53–p21 pathway, p53 has been shown to localize to the centrosome where it participates in the regulation of duplication independent of its transactivation function through a mechanism that is yet to be identified (Fukasawa 2007; Morris et al. 2000; Shinmura et al. 2007). Mutant p53 protein incapable of transactivating target genes can still control centrosome duplication if the centrosome localization is retained. Thus, regulation of p53 stability in general is important in the regulation of centrosome duplication. Overexpression of MDM2, as seen in cancers, for example, will promote the degradation of p53 causing centrosome amplification (Carroll et al. 1999; Vargas et al. 2003).

14.2.3 Polo-like Kinases and Aurora Kinases in Centrosome Regulation

Several kinases involved in the regulation of the cell cycle are also known to be involved in the regulation of centrosome duplication and these include: the polo-like kinases (Plks), Aurora kinases, and NIMA-related kinases (NEK2). Plks are centrosomal kinases that control the entry into mitosis (Barr et al. 2004). Plks are known to be overexpressed in cancers, and their overexpression correlates with increased aggressiveness of the disease (Sankaran and Parvin 2006; Takai et al.

2005). Plk1 plays an important role in the maturation of centrosomes. Inhibition of this kinase results in the formation of significantly smaller centrosomes with a decreased gamma-tubulin localization (Lane and Nigg 1996). Plk2 and 4 localize to the centrosome and their depletion suppresses the initiation of centrosome duplication (Habedanck et al. 2005; Warnke et al. 2004). Similarly, Plk3 is found at the centrosome during interphase, it then co-migrates with duplicated centrosomes at all stages of the cell cycle (Wang et al. 2002). Aurora kinases (A, B, and C) are serine/threonine kinases whose expression is known to be elevated in many human cancers (Li and Li 2006). Their activity levels are highest during the G2-M phase of the cell cycle.

Aurora kinase A (AURKA) localizes to the centrosomes and is essential for mitotic progression. AURKA interacts with several centrosomal and centrosome regulating proteins and its overexpression results in supernumerary centrosomes (Meraldi et al. 2002) whereas its inhibition causes the formation of monopolar spindles (Glover et al. 1995). Thus, AURKA plays a crucial role in the regulation of centrosome separation and duplication. AURKC also localizes to the centrosome and its overexpression results in polyploidy cells with more than two centrosomes (Dutertre et al. 2005). Nek2, a member of the Nek (NIMA)-related kinases, has also been implicated in the regulation of centrosomes. Nek2 phosphorylates centrosomal Nek2-associated protein 1 (C-Nap1) triggering the dissociation of C-Nap1 from the centrosomes to mediate the centrosomal separation (Fry et al. 1998a; Mayor et al. 2002). Overexpression of Nek2 results in premature splitting of the centrosomes (Fry et al. 1998b).

14.2.4 DNA Damage Repair Proteins and the Regulation of the Centrosome

Recent studies have indicated the involvement of several DNA damage and repair proteins in the maintenance of the centrosome (Fukasawa et al. 1996; Griffin et al. 2000; Kraakman-van der Zwet et al. 2002; Shimada et al. 2009; Yamaguchi-Iwai et al. 1999). In addition, several studies have stressed on the importance of homologous recombination (HR) repair proteins in the regulation of the centrosome (Shimada and Komatsu 2009; Shimada et al. 2009). HR is a major pathway through which DNA double-stranded breaks are repaired. Several repair proteins take part in this pathway and these include: ATM/ATR kinases, NBS1-MRE11-RAD50 complex, BRCA1, and BRCA2. ATM is a serine/threonine kinase that is activated by autophosphorylation as an initial step of the DNA damage response. ATM localizes to the centrosome and regulates centrosome number through the cell cycle checkpoint (Oricchio et al. 2006; Shimada and Komatsu 2009). ATR also localizes to the centrosome and depletion of this protein leads to centrosome amplification (Collis et al. 2008).

BRCA1, breast cancer-associated gene 1, a major DNA repair factor, is also known to be involved in the regulation of centrosome number. BRCA1 localizes to the centrosome and its depletion results in the formation of extra centrosomes, a phenotype that is specific to mammary epithelial cells suggesting the dependency of mammary cells on the activity of BRCA1 in the prevention of centrosome overduplication (Ko et al. 2006; Starita et al. 2004). The regulation of the centrosome by BRCA1 will be discussed further in the following section. BRCA2, breast cancer-associated gene 2, is another key player in HR that is known to control centrosome number. BRCA2 localizes to the centrosome and its depletion results in centrosome overduplication strengthening the relationship between a defect in DNA repair and abnormal centrosome numbers (Nakanishi et al. 2007). Another DNA repair factor that localizes to the centrosome is NBS1. At the centrosome, NBS1 regulates the BRCA1-mediated ubiquitination of gamma-tubulin through its interaction with ATR. Similar to the phenotype caused by BRCA1 and ATR depletion, NBS1 depletion leads to centrosome amplification (Shimada et al. 2009). RAD51 is another HR repair factor whose depletion results in centrosomal abnormalities (Bertrand et al. 2003). Centrosome reduplication upon RAD51 depletion occurs in a prolonged G2 arrest due to the activation of the G2-M checkpoint in response to DNA damage that accumulates as a consequence of a defective DNA repair system. Thus, a defect in the DNA repair machinery will arrest cells in G2 due to the activation of a G2-M checkpoint and this would eventually cause centrosome amplification.

14.3 BRCA1 and the Centrosome

BRCA1 is a multifunctional protein involved in several cellular processes including DNA repair, cell cycle checkpoints, transcription control, and maintenance of the centrosome (Sankaran et al. 2006; Venkitaraman 2002). BRCA1 heterodimerizes with BARD1 to form a complex with an E3 ubiquitin ligase activity that catalyzes the transfer of ubiquitin moiety to a target protein. Similar to other E3 ubiquitin ligases, BRCA1/BARD1 can either monoubiquitinate or polyubiquitinate targeted proteins. Polyubiquitination on lysine 48 will target the protein for proteasomal degradation. Monoubiquitination of a protein also has several effects including intracellular trafficking or simply blocking the modification of an important lysine. Inhibition of BRCA1 by expressing a protein fragment that binds to its carboxy terminus or by RNA interference (RNAi) results in supernumerary centrosomes in mammary epithelial cells (Starita et al. 2004). BRCA1 and BARD1 both localize to the centrosome at all stages of the cell cycle (Hsu and White 1998; Sankaran et al. 2006) and the ubiquitin ligase activity of the protein regulates both centrosome number and function (Kais and Parvin 2008). Starita et al. 2004 found that the BRCA1/BARD1 complex ubiquitinates gamma-tubulin, a major centrosomal protein, at gamma-tubulin residue lysine 48. Expression of a mutant form of gamma-tubulin with a lysine to arginine substitution at residue 48, and thus unable to accept the ubiquitin moiety, results in centrosome amplification indicating the importance of BRCA1 ubiquitin ligase activity in the regulation of centrosome duplication (Sankaran et al. 2006; Starita et al. 2004). Interestingly, while inhibition of BRCA1 only affects centrosome number in cell lines derived from breast tissue, expression of the mutated gamma-tubulin gene results in centrosome amplification in non-breast cells (Starita et al. 2004). This result suggests that all cell types need to regulate gamma-tubulin by modification of residue lysine-48, and in breast cells this process is dependent on the BRCA1 ubiquitin ligase.

Centrosomes function by nucleating microtubule formation (Moritz et al. 1995). An enzymatically functional BRCA1 inhibits the microtubule nucleation function of the centrosome in vitro (Sankaran et al. 2005). Thus, BRCA1 E3 ubiquitin ligase activity inhibits centrosome function but is required for the regulation of centrosome duplication. The BRCA1 enzymatic activity is also important for the localization of gamma-tubulin to the centrosome for it has been shown that in addition to ubiquitinating gamma-tubulin itself, BRCA1/BARD1 ubiquitinates a protein in the PCM that docks gamma-tubulin to the centrosome (Sankaran et al. 2007b).

Given that BRCA1 is located at the centrosome at all stages of the cell cycle, and since the BRCA1 enzymatic activity inhibits microtubule nucleation, it needs to be explained how the centrosomes are capable of establishing the mitotic spindle at all. Ko et al. 2006 found that BRCA1 function at the centrosome is mostly critical during S-G2 phases of the cell cycle to block the reduplication of the already duplicated centrosomes. At G2, AURKA localizes to the centrosome and phosphorylates BRCA1 thus inhibiting its ubiquitin ligase activity and enabling the microtubule nucleation activity (Sankaran et al. 2007a) (Fig. 14.1). AURKA is known to be overexpressed in more than 60 % of breast cancers (Miyoshi et al. 2001) and this suggests that the overexpression of the protein will cause its localization at the centrosomes during S phase which results in the inhibition of BRCA1 ubiquitin ligase activity when the BRCA1 is needed most to inhibit the reduplication of centrosomes. Thus, either overexpression of AURKA (Fig. 14.1c) or inhibition of BRCA1 (Fig. 14.1b) results in the same phenotype of supernumerary centrosomes a phenotype that might contribute to chromosomal instability, aneuploidy, and tumorigenesis.

Since BRCA1 is associated with familial cases of breast cancer, the likelihood of a woman to inherit a mutated *BRCA1* allele in such a case is high. In such a scenario, the possibility of having a mutation during the lifetime in the second *BRCA1* allele rendering the protein defective in at least one mammary epithelial cell is common. Individuals with a family history of breast cancer may get their *BRCA1* gene sequenced. Although many of the characterized mutations result in a frameshift or a stop codon that results in a truncated protein, some women present with missense mutations whose disease association is still unclear due to their low prevalence. Such mutations therefore present a diagnostic dilemma because it is still unknown how these mutations affect the function of the protein and whether they render the protein defective or not. A set of BRCA1 point mutants in the amino terminus of the protein were studied for their effect on the HR process as an initial step in determining whether a variant of unknown function may predispose

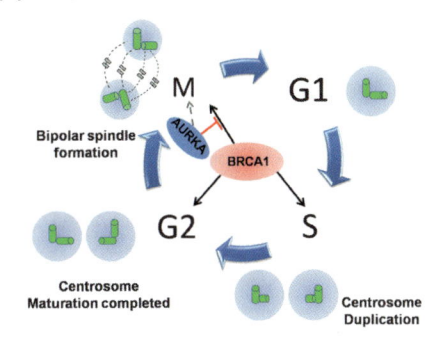

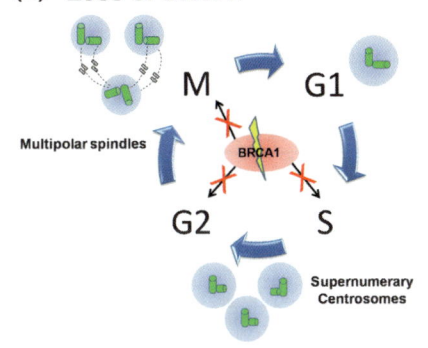

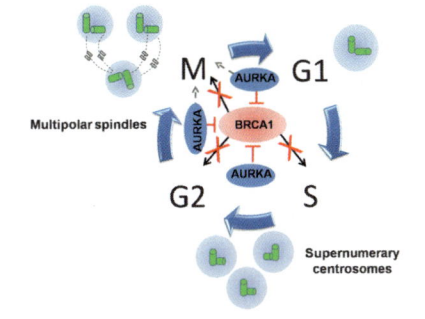

Fig. 14.1 Model for the regulation of the centrosome by BRCA1 and AURKA. **a** Normal centrosome duplication cycle (centrioles are surrounded by the pericentriolar material PCM). In the presence of normal levels of BRCA1 and AURKA and during S-G2 phases of the cell cycle BRCA1 ubiquitinates the already duplicated centrosomes to inhibit reduplication. At M phase AURKA phosphorylates BRCA1 inhibiting its ubiquitin ligase activity and stimulating microtubule nucleation. **b** Loss of BRCA1: effects on the centrosome. Loss of BRCA1 results in supernumerary centrosomes during S phase. These supernumerary centrosomes may result in the formation of mutipolar spindles that would ultimately cause abnormal mitosis. **c** Model for the effect of AURKA overexpression on the centrosome. Overexpression of AURKA mimics the effects of loss of BRCA1. Overexpressed AURKA associates with the centrosomes during S-G2 phase and blocks BRCA1 function when its activity is most critical. In addition, overexpressed AURKA overrides the spindle checkpoint contributing to an abnormal mitosis

to breast cancer development (Ransburgh et al. 2010). A similar assay that checks the effects of the different mutants on the function of BRCA1 in the control of centrosome number is currently being developed and results from both assays, the HR and the centrosome assay, with genetic and clinical analysis of the point mutants will be of great importance in counseling women carrying such mutations and ultimately it will improve the prognosis of the disease.

14.4 Identifying Proteins that Collaborate with BRCA1 in the Regulation of the Centrosome

Many genomes have been sequenced and now that large-scale microarray databases and protein–protein interaction databases are publicly available online, informatics methods can be used to identify genes/proteins that interact with a protein of interest. Using such a bioinformatics approach and making use of publicly available microarray data, proteins were identified that might collaborate with BRCA1, BRCA2, ATM, and CHK2 (Pujana et al. 2007). Among the 164 genes identified, HMMR stood out as a gene that is highly correlated with BRCA1. Depletion of HMMR resulted in centrosome amplification, a phenotype seen upon BRCA1 depletion. Interestingly, depleting both genes suppressed the phenotype. In addition, SNPs linked to HMMR were associated with an increased prevalence of breast cancer in certain populations (Pujana et al. 2007). Finding a gene whose expression is correlated with BRCA1 and that functions in a BRCA1 controlled pathway such as centrosome regulation adds confidence to such bioinformatic approaches. Therefore, using bioinformatics to find other genes that collaborate with BRCA1 in the regulation of centrosomes will be of great importance in the future knowing the importance of centrosome regulation in breast tumorigenesis.

14.5 Concluding Remarks

Regulation of the duplication and the function of the centrosome are critical for the adequate transmission of genetic material to daughter cells. Defects in the mechanisms that control centrosome number and/or function result in centrosome abnormalities that will promote genetic instability and ultimately tumorigenesis. The centrosome is regulated by many oncogenes and tumor suppressors. Disrupting the function of these genes will cause centrosome abnormalities that will contribute to the development and progression of breast cancer. BRCA1 E3 ubiquitin ligase activity is essential for the regulation of centrosome number in mammary cells. Loss of BRCA1, as seen in many breast tumors, will result in supernumerary centrosomes. AURKA is overexpressed in 62 % of human breast cancer (Miyoshi et al. 2001). Since AURKA inhibits BRCA1 enzymatic activity, such an overexpression will also result in the same phenotype seen upon loss of BRCA1: centrosome amplification. Since BRCA1 is a major player in breast cancer, screening BRCA1 mutants using functional assays such as the centrosome assay will definitely be of great importance in the genetic counseling process for individuals with family history of breast cancer. In addition, finding genes that collaborate with BRCA1 in the control of centrosome number will also be essential in finding new players that might contribute to the pathogenesis of the disease.

Acknowledgments This work was supported by funding from the NIH/NCI (R01 CA141090 and R01 CA111480) to J.D.P.

References

Balczon R, Bao L, Zimmer WE, Brown K, Zinkowski RP, Brinkley BR (1995) Dissociation of centrosome replication events from cycles of DNA synthesis and mitotic division in hydroxyurea-arrested Chinese hamster ovary cells. J Cell Biol 130:105–115

Barr FA, Sillje HH, Nigg EA (2004) Polo-like kinases and the orchestration of cell division. Nat Rev Mol Cell Biol 5:429–440

Bertrand P, Lambert S, Joubert C, Lopez BS (2003) Overexpression of mammalian Rad51 does not stimulate tumorigenesis while a dominant-negative Rad51 affects centrosome fragmentation, ploidy and stimulates tumorigenesis, in p53-defective CHO cells. Oncogene 22:7587–7592

Brinkley BR, Goepfert TM (1998) Supernumerary centrosomes and cancer: Boveri's hypothesis resurrected. Cell Motil Cytoskelet 41:281–288

Carroll PE, Okuda M, Horn HF, Biddinger P, Stambrook PJ, Gleich LL, Li YQ, Tarapore P, Fukasawa K (1999) Centrosome hyperamplification in human cancer: chromosome instability induced by p53 mutation and/or Mdm2 overexpression. Oncogene 18:1935–1944

Collis SJ, Ciccia A, Deans AJ, Horejsi Z, Martin JS, Maslen SL, Skehel JM, Elledge SJ, West SC, Boulton SJ (2008) FANCM and FAAP24 function in ATR-mediated checkpoint signaling independently of the Fanconi anemia core complex. Mol Cell 32:313–324

Duensing A, Liu Y, Spardy N, Bartoli K, Tseng M, Kwon JA, Teng X, Duensing S (2007) RNA polymerase II transcription is required for human papillomavirus type 16 E7- and hydroxyurea-induced centriole overduplication. Oncogene 26:215–223

Duensing A, Liu Y, Tseng M, Malumbres M, Barbacid M, Duensing S (2006) Cyclin-dependent kinase 2 is dispensable for normal centrosome duplication but required for oncogene-induced centrosome overduplication. Oncogene 25:2943–2949

Dutertre S, Hamard-Peron E, Cremet JY, Thomas Y, Prigent C (2005) The absence of p53 aggravates polyploidy and centrosome number abnormality induced by Aurora-C overexpression. Cell Cycle 4:1783–1787

Faivre J, Frank-Vaillant M, Poulhe R, Mouly H, Jessus C, Brechot C, Sobczak-Thepot J (2002) Centrosome overduplication, increased ploidy and transformation in cells expressing endoplasmic reticulum-associated cyclin A2. Oncogene 21:1493–1500

Fry AM, Mayor T, Meraldi P, Stierhof YD, Tanaka K, Nigg EA (1998a) C-Nap1, a novel centrosomal coiled-coil protein and candidate substrate of the cell cycle-regulated protein kinase Nek2. J Cell Biol 141:1563–1574

Fry AM, Meraldi P, Nigg EA (1998b) A centrosomal function for the human Nek2 protein kinase, a member of the NIMA family of cell cycle regulators. EMBO J 17:470–481

Fukasawa K (2007) Oncogenes and tumour suppressors take on centrosomes. Nat Rev Cancer 7:911–924

Fukasawa K, Choi T, Kuriyama R, Rulong S, Vande Woude GF (1996) Abnormal centrosome amplification in the absence of p53. Science 271:1744–1747

Ganem NJ, Godinho SA, Pellman D (2009) A mechanism linking extra centrosomes to chromosomal instability. Nature 460:278–282

Glover DM, Leibowitz MH, McLean DA, Parry H (1995) Mutations in aurora prevent centrosome separation leading to the formation of monopolar spindles. Cell 81:95–105

Griffin CS, Simpson PJ, Wilson CR, Thacker J (2000) Mammalian recombination-repair genes XRCC2 and XRCC3 promote correct chromosome segregation. Nat Cell Biol 2:757–761

Grisendi S, Bernardi R, Rossi M, Cheng K, Khandker L, Manova K, Pandolfi PP (2005) Role of nucleophosmin in embryonic development and tumorigenesis. Nature 437:147–153

Habedanck R, Stierhof YD, Wilkinson CJ, Nigg EA (2005) The Polo kinase Plk4 functions in centriole duplication. Nat Cell Biol 7:1140–1146

Hsu LC, White RL (1998) BRCA1 is associated with the centrosome during mitosis. Proc Natl Acad Sci U S A 95:12983–12988

Kais Z, Parvin JD (2008) Regulation of centrosomes by the BRCA1-dependent ubiquitin ligase. Cancer Biol Ther 7:1540–1543

Ko MJ, Murata K, Hwang DS, Parvin JD (2006) Inhibition of BRCA1 in breast cell lines causes the centrosome duplication cycle to be disconnected from the cell cycle. Oncogene 25:298–303

Kraakman-van der Zwet M, Overkamp WJ, van Lange RE, Essers J, van Duijn-Goedhart A, Wiggers I, Swaminathan S, van Buul PP, Errami A, Tan RT et al (2002) Brca2 (XRCC11) deficiency results in radioresistant DNA synthesis and a higher frequency of spontaneous deletions. Mol Cell Biol 22:669–679

Lane HA, Nigg EA (1996) Antibody microinjection reveals an essential role for human polo-like kinase 1 (Plk1) in the functional maturation of mitotic centrosomes. J Cell Biol 135:1701–1713

Li JJ, Li SA (2006) Mitotic kinases: the key to duplication, segregation, and cytokinesis errors, chromosomal instability, and oncogenesis. Pharmacol Ther 111:974–984

Lingle WL, Lutz WH, Ingle JN, Maihle NJ, Salisbury JL (1998) Centrosome hypertrophy in human breast tumors: implications for genomic stability and cell polarity. Proc Natl Acad Sci U S A 95:2950–2955

Ma Z, Kanai M, Kawamura K, Kaibuchi K, Ye K, Fukasawa K (2006) Interaction between ROCK II and nucleophosmin/B23 in the regulation of centrosome duplication. Mol Cell Biol 26:9016–9034

Mayor T, Hacker U, Stierhof YD, Nigg EA (2002) The mechanism regulating the dissociation of the centrosomal protein C-Nap1 from mitotic spindle poles. J Cell Sci 115:3275–3284

Meraldi P, Honda R, Nigg EA (2002) Aurora-A overexpression reveals tetraploidization as a major route to centrosome amplification in p53-/-cells. EMBO J 21:483–492

Meraldi P, Lukas J, Fry AM, Bartek J, Nigg EA (1999) Centrosome duplication in mammalian somatic cells requires E2F and Cdk2-cyclin A. Nat Cell Biol 1:88–93

Miyoshi Y, Iwao K, Egawa C, Noguchi S (2001) Association of centrosomal kinase STK15/BTAK mRNA expression with chromosomal instability in human breast cancers. Int J Cancer 92:370–373

Moritz M, Braunfeld MB, Sedat JW, Alberts B, Agard DA (1995) Microtubule nucleation by gamma-tubulin-containing rings in the centrosome. Nature 378:638–640

Morris VB, Brammall J, Noble J, Reddel R (2000) p53 localizes to the centrosomes and spindles of mitotic cells in the embryonic chick epiblast, human cell lines, and a human primary culture: An immunofluorescence study. Exp Cell Res 256:122–130

Mussman JG, Horn HF, Carroll PE, Okuda M, Tarapore P, Donehower LA, Fukasawa K (2000) Synergistic induction of centrosome hyperamplification by loss of p53 and cyclin E overexpression. Oncogene 19:1635–1646

Nakanishi A, Han X, Saito H, Taguchi K, Ohta Y, Imajoh-Ohmi S, Miki Y (2007) Interference with BRCA2, which localizes to the centrosome during S and early M phase, leads to abnormal nuclear division. Biochem Biophys Res Commun 355:34–40

Okuda M, Horn HF, Tarapore P, Tokuyama Y, Smulian AG, Chan PK, Knudsen ES, Hofmann IA, Snyder JD, Bove KE et al (2000) Nucleophosmin/B23 is a target of CDK2/cyclin E in centrosome duplication. Cell 103:127–140

Oricchio E, Saladino C, Iacovelli S, Soddu S, Cundari E (2006) ATM is activated by default in mitosis, localizes at centrosomes and monitors mitotic spindle integrity. Cell Cycle 5:88–92

Pujana MA, Han JD, Starita LM, Stevens KN, Tewari M, Ahn JS, Rennert G, Moreno V, Kirchhoff T, Gold B et al (2007) Network modeling links breast cancer susceptibility and centrosome dysfunction. Nat Genet 39:1338–1349

Ransburgh DJ, Chiba N, Ishioka C, Toland AE, Parvin JD (2010) Identification of breast tumor mutations in BRCA1 that abolish its function in homologous DNA recombination. Cancer Res 70:988–995

Salisbury JL (2001) The contribution of epigenetic changes to abnormal centrosomes and genomic instability in breast cancer. J Mammary Gland Biol Neoplasia 6:203–212

Sankaran S, Crone DE, Palazzo RE, Parvin JD (2007a) Aurora-A kinase regulates breast cancer associated gene 1 inhibition of centrosome-dependent microtubule nucleation. Cancer Res 67:11186–11194

Sankaran S, Crone DE, Palazzo RE, Parvin JD (2007b) BRCA1 regulates gamma-tubulin binding to centrosomes. Cancer Biol Ther 6:1853–1857

Sankaran S, Parvin JD (2006) Centrosome function in normal and tumor cells. J Cell Biochem 99:1240–1250

Sankaran S, Starita LM, Groen AC, Ko MJ, Parvin JD (2005) Centrosomal microtubule nucleation activity is inhibited by BRCA1-dependent ubiquitination. Mol Cell Biol 25:8656–8668

Sankaran S, Starita LM, Simons AM, Parvin JD (2006) Identification of domains of BRCA1 critical for the ubiquitin-dependent inhibition of centrosome function. Cancer Res 66:4100–4107

Shimada M, Komatsu K (2009) Emerging connection between centrosome and DNA repair machinery. J Radiat Res (Tokyo) 50:295–301

Shimada M, Sagae R, Kobayashi J, Habu T, Komatsu K (2009) Inactivation of the Nijmegen breakage syndrome gene leads to excess centrosome duplication via the ATR/BRCA1 pathway. Cancer Res 69:1768–1775

Shinmura K, Bennett RA, Tarapore P, Fukasawa K (2007) Direct evidence for the role of centrosomally localized p53 in the regulation of centrosome duplication. Oncogene 26:2939–2944

Shinmura K, Tarapore P, Tokuyama Y, George KR, Fukasawa K (2005) Characterization of centrosomal association of nucleophosmin/B23 linked to Crm1 activity. FEBS Lett 579:6621–6634

Starita LM, Machida Y, Sankaran S, Elias JE, Griffin K, Schlegel BP, Gygi SP, Parvin JD (2004) BRCA1-Dependent Ubiquitination of {gamma}-Tubulin Regulates Centrosome Number. Mol Cell Biol 24:8457–8466

Takai N, Hamanaka R, Yoshimatsu J, Miyakawa I (2005) Polo-like kinases (Plks) and cancer. Oncogene 24:287–291

Tokuyama Y, Horn HF, Kawamura K, Tarapore P, Fukasawa K (2001) Specific phosphorylation of nucleophosmin on Thr(199) by cyclin-dependent kinase 2-cyclin E and its role in centrosome duplication. J Biol Chem 276:21529–21537

Vargas DA, Takahashi S, Ronai Z (2003) Mdm2: a regulator of cell growth and death. Adv Cancer Res 89:1–34

Venkitaraman AR (2002) Cancer susceptibility and the functions of BRCA1 and BRCA2. Cell 108:171–182

Wang HF, Takenaka K, Nakanishi A, Miki Y (2011) BRCA2 and nucleophosmin coregulate centrosome amplification and form a complex with the Rho effector kinase ROCK2. Cancer Res 71:68–77

Wang Q, Xie S, Chen J, Fukasawa K, Naik U, Traganos F, Darzynkiewicz Z, Jhanwar-Uniyal M, Dai W (2002) Cell cycle arrest and apoptosis induced by human polo-like kinase 3 is mediated through perturbation of microtubule integrity. Mol Cell Biol 22:3450–3459

Warnke S, Kemmler S, Hames RS, Tsai HL, Hoffmann-Rohrer U, Fry AM, Hoffmann I (2004) Polo-like kinase-2 is required for centriole duplication in mammalian cells. Curr Biol 14:1200–1207

Yamaguchi-Iwai Y, Sonoda E, Sasaki MS, Morrison C, Haraguchi T, Hiraoka Y, Yamashita YM, Yagi T, Takata M, Price C et al (1999) Mre11 is essential for the maintenance of chromosomal DNA in vertebrate cells. EMBO J 18:6619–6629

Further readings

Note: Two papers relevant to this chapter have been recently published and are placed further readings

Kais Z, Barsky SH, Mathsyaraja H, Zha A, Ransburgh DJ, He G, Pilarski RT, Shapiro CL, Huang K, Parvin JD. (2011) KIAA0101 Interacts with BRCA1 and regulates centrosome number. Mol Cancer Res 9(8):1091–1099

Kais Z, Chiba N, Ishioka C, Parvin JD. (2011) Functional differences among BRCA1 missense mutations in the control of centrosome duplication. Oncogene 31(6):799–804

Chapter 15
The Role of Centrosomes in Multiple Myeloma

Benedict Yan and Wee-Joo Chng

Abstract Multiple Myeloma (MM), a neoplastic proliferation of plasma cells, displays complex genetic aberrations. This genetic complexity is due in part to centrosomal abnormalities, which are well-documented in MM. The exact mechanisms by which such abnormalities develop in MM are still not fully characterized, although various pathways and molecules, particularly the G1 cyclin-CDK/Rb pathway, receptor of hyalorunan-mediated motility, and Aurora-A molecules, have been implicated. The identification of centrosome abnormalities in MM patients is of potential clinical utility, both in the prognostic and therapeutic setting. A high centrosome index is associated with poorer prognosis in MM patients, and predicts for greater in vitro sensitivity of myeloma cell lines to certain therapeutics such as Aurora kinase inhibitors. Future studies into the mechanisms leading to centrosome abnormalities in MM may reveal novel candidates and strategies for therapeutic intervention.

B. Yan
Department of Pathology, National University Health System, Singapore

W.-J. Chng (✉)
Department of Haematology and Oncology, National University Cancer Institute of Singapore, Singapore
e-mail: mdccwj@nus.edu.sg

W.-J. Chng
National University Health System, NUHS Tower Block, Level 7, 1E Lower Kent Ridge Road, Singapore 119228, Singapore

W.-J. Chng
Department of Medicine, Yong Loo Lin School of Medicine, National University of Singapore, Singapore

W.-J. Chng
Cancer Science Institute of Singapore, National University of Singapore, Singapore

15.1 Multiple Myeloma

Multiple Myeloma (MM) is a neoplastic clonal proliferation of plasma cells occurring multifocally within the bone marrow (Swerdlow et al. 2008). At present, MM accounts for approximately 1 % of cancers, 15 % of hematopoietic neoplasms, and 20 % of deaths from hematologic malignancies (Jemal et al. 2010). Clinical manifestations include the presence of serum M-protein, hypercalcemia, renal insufficiency, anemia, and bone lesions (Kyle and Rajkumar 2004). Despite therapeutic advances, the disease remains incurable at present with a median survival of around 4 years (Kumar et al. 2008).

Monoclonal gammopathy of undetermined significance (MGUS) is a premalignant state of MM with an estimated prevalence of 3.2 % in patients above 50 years of age (Wadhera and Rajkumar 2010). MGUS is characterized by a monoclonal plasma cell proliferation within the bone marrow without end-organ damage (International Myeloma Working Group 2003). The risk of progression of MGUS to MM or related disorders is about 1 % per year (Kyle et al. 2002).

15.2 Genetic Abnormalities in Multiple Myeloma

Complex genetic aberrations in the form of either numerical and/or structural chromosomal abnormalities are ubiquitous in MM (Avet-Loiseau et al. 1999; Nishida et al. 1997; Smadja et al. 1998, 2001; Mohamed et al. 2007; Drach et al. 1995). In fact, several genetic and molecular subtypes of MM with distinct clinicopathological features have been identified (Fonseca et al. 2009). At the top hierarchical level, MM can be divided into hyperdiploid and non-hyperdiploid subtypes (Smadja et al. 1998, 2001; Carrasco et al. 2006). Hyperdiploid MM (H-MM) is characterized by trisomies of chromosomes 3, 5, 7, 9, 11, 15, 19 and 21, and has a low prevalence of primary immunoglobulin heavy chain (IgH) translocations. This is in contrast to non-hyperdiploid MM (NH-MM), which encompasses hypodiploid, pseudodiploid, and near tetraploid MM and is strongly associated with IgH translocations (Fonseca et al. 2003). The identification of unique gene expression signatures associated with these genetic subtypes provides further support that the H-MM and NH-MM categories are biologically distinct (Bergsagel et al. 2005; Zhan et al. 2006). This dichotomy into H-MM and NH-MM is also seen in MGUS (Chng et al. 2005; Brousseau et al. 2007), indicating that these distinct pathogenetic pathways arise early in the course of the disease.

15.3 Centrosomal Abnormalities as a Mechanism for Aneuploidy in Multiple Myeloma

The high prevalence of aneuploidy and evolving genetic complexity during disease progression seen in MM (Wu et al. 2007) strongly suggest that the MM genome is unstable, and specifically points to the presence of chromosomal instability (CIN) in MM pathogenesis. CIN is a phenomenon in which cells persistently demonstrate a high rate of loss and gain of whole chromosomes, and one mechanism leading to CIN is numerical centrosome anomalies (Thompson et al. 2010).

Centrosomes are the primary microtubule-organizing center (MTOC) in animal cells, and facilitate organization of the spindle poles during mitosis (Bettencourt-Dias and Glover 2007). A causal association between centrosome abnormalities and cancer was first proposed by Boveri in the early 1900s (Boveri 1914, 2008). Centrosome aberrations are frequent in various cancer types (Pihan et al. 1998) and are already present in some early premalignant lesions (Pihan et al. 2003).

There is recent evidence for a causal link between centrosome anomalies and numerical chromosomal abnormalities (Nigg and Raff 2009; Thompson et al. 2010). One proposed mechanism by which supernumerary centrosomes promote tumorigenesis is through aberrant spindle formation during cell division, resulting in CIN and aneuploidy. Cells with amplified centrosomes may form tripolar mitotic spindles and undergo cytokinesis to generate viable but highly aneuploid daughters (Fukasawa 2008).

Mitotic spindles with more than three poles may also be formed, resulting in cytokinesis failure. In the presence of p53, this failure to undergo cytokinesis triggers the checkpoint response, leading ultimately to cell death (Fukasawa 2008). Mitotic clustering of centrosomes with pseudo-bipolar spindle formation is one mechanism by which multipolar divisions in cancer cells are suppressed (Kwon et al. 2008; Quintyne et al. 2005). Centrosome amplification can still lead to chromosome missegregation due to an increased rate of merotely, where a single sister kinetochore becomes simultaneously attached to two spindle poles (Silkworth et al. 2009; Ganem et al. 2009).

MM also exhibits genomic instability in the form of structural chromosomal abnormalities such as deletions and translocations. There is at present little evidence to suggest that centrosomes play a significant causative role in the development of these abnormalities.

15.4 Centrosome Abnormalities in MM

Three studies have examined centrosome abnormalities in MM (Maxwell et al. 2005; Dementyeva et al. 2010; Chng et al. 2006). To evaluate structural and numerical centrosome abnormalities, Maxwell et al. (Maxwell et al. 2005) performed multicolor immunofluorescence on archived bone marrow core biopsies using antibodies against two recognized protein components of the centrosome, pericentrin, and gamma-tubulin. Centrosomal volumes were determined by 3D rendering of confocal z-stacks labeled with gamma-tubulin. They found that centrosome abnormalities, including the mean number of centrosomes per cell and mean total centrosome volume, were highly correlated. Centrosome abnormalities were significantly higher in MM compared to MGUS or control plasma cells from marrow of lymphoma patients.

In another study examining the clinical implications of centrosome amplification in plasma cell neoplasms, we employed immunofluorescence staining of centrin [another well-established centrosomal protein (Errabolu et al. 1994)] in combination with staining of clonal cytoplasmic light chains to identify centrosome abnormalities in clonal plasma cells (Chng et al. 2006). We found that although the prevalence of centrosome abnormalities was fairly similar from MGUS to MM (approximately two-thirds of patients), the percentage of malignant cells with centrosome abnormalities increased progressively from MGUS to MM. We also observed that centrosomal structural abnormalities such as altered shape and configuration were predominantly seen in MM. Overall, our results suggest that centrosome abnormalities, of which centrosome amplification is the most prominent, occur early in MM pathogenesis and increase with disease progression.

In the most recent study of centrosome abnormalities in MM, Dementyeva et al. performed immunofluorescence staining of centrin in bone marrow B-cells and plasma cells. They similarly identified an increased prevalence of centrosome amplification in plasma cells from MM patients. Interestingly, they also found an increased population of B-cells with centrosome amplification in MM patients compared to healthy donors (Dementyeva et al. 2010). This finding alludes to a longstanding hypothesis proposing that a population of B-cells are closely related to and might even represent precursors of MM (Zojer et al. 2002).

15.5 Clinical and Biological Implications of Centrosome Abnormalities in MM

In our study, we found that a gene expression-based index (centrosome index, CI) comprising the expression of genes encoding the main centrosomal proteins centrin, pericentrin and gamma-tubulin correlated very strongly with centrosome amplification (Chng et al. 2006). Using a high CI as a surrogate marker for centrosome amplification, it was found that centrosome amplification correlated with poor prognostic features such as a high plasma cell labeling index and high-risk genetic aberrations including chromosome 13 deletion, t(4; 14), and t(14; 16). A high CI was associated with significantly poorer survival in patients treated with chemotherapy, newly diagnosed patients treated with autologous stem cell transplantation, and relapsed patients treated with bortezomib (Chng et al. 2008). A high CI was found to be an independent prognostic factor on multivariate analysis encompassing other known prognostic factors. In our study therefore, a high CI identified a cohort of patients with poor prognosis regardless of treatment modalities, phase of presentation, and international staging system (ISS) stage. In a separate study by Hose et al., a significant correlation between CI and survival was not demonstrated, although sample size might account for the difference (Hose et al. 2009).

With regards to the molecular phenotype, no correlation between centrosome amplification and ploidy categories was identified (Chng et al. 2006). Gene expression profiling (GEP) analysis comparing high versus low CI MM revealed an overexpression of genes coding for proteins associated with the centrosome (*TUBG1*, *CETN2*, *TACC3*, *NEK2*, *PRKRA*, *STK6*, *AURKB* and *PLK4*), cell cycle (*CCNB1*, *CCNB2*, *CCND2*, *E2F2*, *CDC* gene family, *CDK5*, *CDK6*, *CDKN2C*), proliferation (RAN, CSK1B, *TOP2A*, *TTK*, *TYMS*, *MCM* gene family, *ASPM*), DNA repair/G2 cell cycle checkpoints (*BRCA1*, *CHEK1*, *CHEK2*, *MAD2L1*, *BUB1*, *BUB1B*, *FANCD2*, *REV1L*), and kinetochore and microtubule attachment (*BIRC5*, *CENPA*, *CENPE*, *CENPH*, *ZWINT*) (Chng et al. 2008). Centrosome amplification is therefore associated with abnormalities of the cell cycle, proliferation, and DNA repair in MM, findings very similar to a separate study on centrosome aberrations in acute myeloid leukemia (Neben et al. 2004).

15.6 Possible Mechanisms Leading to Centrosome Amplification in MM

Proteins that participate in the regulation of chromosomal numerical integrity belong to one of three functional groups: cell-cycle regulation, DNA-damage response/repair, and nucleocytoplasmic transport. Mutations involving these proteins lead to supernumerary centrosomes (Fukasawa 2007).

To date, direct mechanisms leading to centrosome amplification in MM have not been identified. Table 15.1 provides a list of molecules that are known to participate in centrosome function/regulation of centrosome number, and that have been separately implicated in MM pathogenesis as well. We discuss in-depth the G1 cyclin-CDK/Rb pathway, receptor of hyalorunan-mediated motility (RHAMM), and Aurora-A molecules which we feel represent important causative factors underlying centrosome amplification in MM.

15.7 G1 Cyclin-CDK/Rb Pathway

The centrosome duplication cycle is tightly linked to the cell division cycle, and the G1 cyclins (D and E) and their associated kinases (Lee and Yang 2003; Sherr and Roberts 2004; Giacinti and Giordano 2006) play integral roles in both processes (Bettencourt-Dias and Glover 2007). D-type cyclins associate with CDK4 or CDK6 and function early in G1-phase, while cyclin E associates with CDK2 and functions

Table 15.1 Genes with known roles in centrosome function/regulation of centrosome number implicated in multiple myeloma oncogenesis

Gene	Known role in centrosome function or regulation of centrosome number	Role in myeloma oncogenesis
AKT1	Active AKT1 expression induces supernumerary centrosomes (Plo and Lopez 2009)	Knockdown of Akt1 impairs survival of Akt-dependent MM lines (Zollinger et al. 2008)
Aurora-A kinase	Overexpression of Aurora-A induces centrosome amplification in vitro (Zhou et al. 1998); Aurora A gives rise to extra chromosomes through defects in cell division and consequent tetraploidization (Meraldi et al. 2002)	Aurora kinase inhibitors inhibits myeloma growth at nanomolar concentration (Shi et al. 2007); Aurora-A kinase RNAi induces apoptotic death in myeloma cells (Evans et al. 2008); VX680 induces apoptosis in HMCL (Hose et al. 2009); MLN8237, a small molecule Aurora A kinase inhibitor, inhibits cell proliferation in HMCL (Gorgun et al. 2010)
BRCA1	BRCA1 localizes to centrosomes (Hsu and White 1998); cells from mice lacking full-length BRCA1 show a high frequency of centrosome amplification (Xu et al. 1999); BRCA1-dependent ubiquitination of gamma-tubulin regulates centrosome number (Starita et al. 2004)	BRCA1 gene expression upregulated in high CI MM (Chng et al. 2008)
BRCA2	Loss of BRCA2 results in centrosome amplification (Tutt et al. 1999); exogenous expression of NPM-binding region of BRCA2 results in aberrant centrosome amplification (Wang et al. 2011b)	BRCA2 is located on 13q12; 13q deletions common in MM (Mohamed et al. 2007; Fonseca et al. 2009; Fonseca et al. 2004; Christensen et al. 2007; Yuregir et al. 2009; Wu et al. 2007); no direct evidence for BRCA2 involvement in myeloma pathogenesis at present
β-catenin	β-catenin is a component of the intercentrosomal linker and participates in centrosome separation (Bahmanyar et al. 2008); β-catenin participates in centrosome amplification, and mutations in β-catenin might contribute to the formation of abnormal centrosomes seen in cancer (Bahmanyar et al. 2010)	β-catenin small interfering RNA suppresses MM progression in a xenograft mouse model (Ashihara et al. 2009); β-catenin knockdown and overexpression in HMCL led to a significant decrease and increase in proliferation respectively (Dutta-Simmons et al. 2009)
CDK2-Cyclin E	CDK2 & CDK4 are critical mediators of centrosome amplification in p53-null cells (Adon et al. 2010)	Although there is no direct evidence that CDK2-Cyclin E is important for MM pathogenesis, p27, a CDK2-Cyclin E inhibitor, is implicated in MM; see below
CDK2-Cyclin A CDK4/CDK6-Cyclin D	Cyclin D1 overexpression induces centrosome amplification in hepatocytes and human breast epithelial cells (Nelsen et al. 2005)	Cyclin D dysregulation is an early unifying pathogenic event in MM (Bergsagel et al. 2005); CDK4-Cyclin D1 & CDK6-Cyclin D2 inactivates Rb in MM and promotes cell cycle dysregulation (Ely et al. 2005)

(continued)

Table 15.1 (continued)

Gene	Known role in centrosome function or regulation of centrosome number	Role in myeloma oncogenesis
ILK	ILK regulates centrosome clustering and prevents multipolar divisions (Fielding et al. 2011)	ILK inhibition leads to apoptosis in HMCL (Wang et al. 2011a)
KRAS	The Ras oncogene signals centrosome amplification in mammary epithelial cells through cyclin D1/Cdk4 and Nek2; KRAS^{G12D} initiates centrosome amplification in mammary precursor lesions (Zeng et al. 2010)	KRAS mutations are present in 6 % of MM cases, and are significantly associated with shorter overall survival and progression free survival (Chng et al. 2008)
LATS2	Aurora A phosphorylates and targets LATS2 to the centrosome (Toji et al. 2004); Cells from *Lats2*-deficient mice show mitotic defects associated with centrosome fragmentation (McPherson et al. 2004; Yabuta et al. 2007)	Homozygous deletion of LATS2 observed in MM (Dickens et al. 2010)
MDM2	Forced expression of MDM2 in cells harboring wild-type p53 efficiently induces centrosome amplification (Carroll et al. 1999)	MDM2 genomic amplification present in MM (Elnenaei et al. 2003)
MET	Constitutive active MET induces supernumerary centrosomes in vitro (Nam et al. 2010)	Serum HGF (hepatocyte growth factor; ligand for c-MET) levels are elevated in MM patients (Seidel et al. 1998); MET knockdown/depletion results in decreased MM cell survival (Phillip et al. 2009; Stellrecht et al. 2007)
NPM1	Probably functions in centrosome pairing (Shinmura et al. 2005); NPM controls numeral integrity of centrosomes (Grisendi et al. 2005)	NPM1 is overexpressed in hyperdiploid multiple myeloma due to a gain of chromosome 5 (Weinhold et al. 2010)
Plk	Overexpression of Plk1 in HeLa cells results in extra copies of centrosomes (Meraldi et al. 2002)	Plk inhibitor BI 2536 exhibits potent activity against malignant plasma cells (Stewart et al. 2011)
p16^{INK4a}	Loss of p16^{INK4a} generates supernumerary centrosomes through centriole pair splitting (McDermott et al. 2006)	p16 methylation observed in MGUS, SMM, and MM; conflicting results regarding association between p16 methylation status and prognosis (Kramer et al. 2002; Gonzalez-Paz et al. 2007; Dib et al. 2007; Park et al. 2011; Wong et al. 1998; Ribas et al. 2005; Mateos et al. 2002); hemizygous p16^{INK4A} deletion reported in MM (Kramer et al. 2002)

(continued)

Table 15.1 (continued)

Gene	Known role in centrosome function or regulation of centrosome number	Role in myeloma oncogenesis
p18	No direct link to centrosome regulation documented, although it inhibits CDK4/CDK6 (Hirai et al. 1995)	Prevalence of bi-allelic p18 deletion in MM is about 2 %; exogenous p18 expression inhibits growth of HMCL with low/absent endogenous p18 expression (Dib et al. 2006)
p27	p27 suppresses centrosome amplification after DNA damage (Sugihara et al. 2006)	Low p27 is an adverse prognostic factor in patients with MM (Filipits et al. 2003)
RAD51	Expression of dominant-negative RAD51 and conditional repression of RAD51 leads to centrosome amplification (Bertrand et al. 2003); reduced expression or loss of RAD51B, RAD51C, RAD51D, XRCC2, and XRCC3 all induce centrosome amplification (Renglin et al. 2007; Smiraldo et al. 2005; Griffin et al. 2000; Date et al. 2006)	One study comparing MM cells in one patient with normal plasma cells in her identical twin reported downregulation of RAD51 in MM cells by gene expression analysis (Munshi et al. 2004)
Rb	Inactivation of Rb by expression of E7 results in centrosome amplification (Duensing et al. 2000); conditional Rb loss in mice results in centrosome amplification (Iovino et al. 2006; Balsitis et al. 2003)	Heterozygous deletion of Rb present in MM (Juge-Morineau et al. 1997; Kramer et al. 2002)
RHAMM	RHAMM localizes to the centrosome (Maxwell et al. 2003)	RHAMM overexpression in vitro in HMCL results in centrosomal defects (Maxwell et al. 2005)
TP53	p53-deficient mice show a high frequency of centrosome amplification (Fukasawa and Wiener 1997; Fukasawa et al. 1996); efficient centrosome reduplication occurs only when p53 is mutated or lost (Tarapore et al. 2001); The absence of p53 favors the accumulation of cells with extra centrosomes (Meraldi et al. 2002)	TP53 mutation prevalence 0–40 % in MM (Fonseca et al. 2004; Chng et al. 2007); TP53 mutations associated with very poor survival (Chng et al. 2007)

from mid to late G1-phase (Lee and Yang 2003). Their activation leads to phosphorylation of the retinoblastoma (Rb) protein, the main molecule responsible for the G1 checkpoint, and entry into S-phase (Giacinti and Giordano 2006).

The CDK2-cyclin E kinase complex is an important initiator of centrosome duplication (Lacey et al. 1999; Hinchcliffe et al. 1999), and its centrosomal target proteins include nucleophosmin (Okuda et al. 2000), Mps1 kinase (Fisk and Winey 2001), and CP110 (Chen et al. 2002). Cyclin E overexpression alone does not lead to centrosomal amplification in vitro (Spruck et al. 1999). CDK2 has recently been demonstrated to be important for centrosome amplification in p53-null mouse embryonic fibroblasts (Adon et al. 2010). At present however, there is no demonstrated role for the CDK2-cyclin E complex in MM centrosomal amplification or pathogenesis.

CDK4 has also been implicated in centrosome amplification in p53-null cells (Adon et al. 2010), and cyclin D overexpression induces centrosome amplification in hepatocytes (Nelsen et al. 2005). Abnormalities of the cyclin D-Rb axis with increased and/or dysregulated expression of cyclins D1, D2, or D3 are seen in nearly all genetic subtypes of MM (Bergsagel et al. 2005). t(11; 14) and t(6; 14) lead directly to overexpression of cyclins D1 and D3 respectively. t(4; 14) MM shows cyclin D2 overexpression. As cyclin D2 is a transcriptional target of *maf*, *maf*-translocated MM also shows cyclin D2 overexpression (Hurt et al. 2004). H-MM shows overexpression of cyclins D1 and/or D2. These abnormalities of the cyclin D-Rb axis lead to inactivation of Rb and facilitate cell cycle dysregulation in MM (Ely et al. 2005).

Loss of Rb function also results in centrosome amplification (Iovino et al. 2006; Duensing et al. 2000; Balsitis et al. 2003). Rb is located on chromosome 13q14.2, and chromosome 13 abnormalities are detected in 50 % of MM cases. Eighty-five percentage of these abnormalities are monosomy, and the remainder are interstitial deletions (Fonseca et al. 2009). Heterozygous deletions of Rb have been reported in MM (Juge-Morineau et al. 1997; Kramer et al. 2002). However, a direct link between loss of Rb function and centrosome amplification in MM has not been demonstrated.

Other abnormalities of the Rb pathway are also seen in MM. CDK inhibitors such as p15, p16, p17, and p18 specifically suppress cyclin D kinase activity, while p21 and p27 act on other cyclin/CDK complexes (Giacinti and Giordano 2006). Methylation of the p15 and p16 genes is observed in around 20–30 % of MGUS/MM and in most human myeloma cell lines (HMCL) (Fonseca et al. 2004, 2009). Also, p16 expression is low to absent in most MM independent of methylation status (Gonzalez-Paz et al. 2007; Dib et al. 2007).

Various lines of evidence suggest that loss of function of other CDK inhibitors occurs in MM. Bi-allelic deletion of p18, a gene that is important for normal B-cell development (Morse et al. 1997; Ashihara et al. 2009; Schrantz et al. 2000), has been observed in HMCL and primary MM cases, and exogenous p18 expression inhibits growth of HMCL with low/absent endogenous p18 expression (Dib et al. 2006). Low p27 expression is reportedly an independent adverse prognostic factor in MM patients (Filipits et al. 2003); interestingly, loss of p27 function might be related to centrosomal abnormalities in MM as p27 has been shown to suppress centrosome amplification following DNA damage (Sugihara et al. 2006).

While abnormalities affecting the Rb pathway are almost universal and early events in MM, there is at present no convincing evidence that such abnormalities are causal for centrosome amplification in MM, and centrosome amplification is not seen in all MM patients. However, these abnormalities may create a permissive environment for centrosome amplification.

15.8 Receptor of Hyalorunan-Mediated Motility

RHAMM is a cell-motility molecule first described by Turley and colleagues (Turley 1992; Turley and Torrance 1985; Turley et al. 1987, 1991), and the human full-length cDNA was cloned in 1996 (Wang et al. 1996). RHAMM overexpression in fibroblasts induces transformation (Hall et al. 1995), and RHAMM-hyaluronan interactions have been shown to mediate the activation of oncogenic kinases such as Src, extracellular-regulated kinases (Erk), and protein kinase C (Turley et al. 2002). RHAMM also localizes to the centrosome and functions in the maintenance of spindle integrity (Maxwell et al. 2003).

Existing evidence suggests that RHAMM plays an important role in MM pathogenesis. MM plasma cells were first observed to express RHAMM by Turley et al. in 1993 (Turley et al. 1993). Novel RHAMM variants were later identified and shown to be overexpressed in MM plasma cells relative to normal B cells (Crainie et al. 1999). Increasing RHAMM expression in MM strongly correlated with osteolytic bone lesions (Zhan et al. 2002), poor event-free, and overall survival (Maxwell et al. 2004).

As mentioned previously, structural and numerical centrosomal abnormalities in MM cells were first identified by Maxwell et al. (2005). They also found that structural centrosomal abnormalities correlated with elevated RHAMM expression in MM. RHAMM overexpression in vitro resulted in altered centrosome size and structure. Although the mechanism by which RHAMM affected centrosome structure was not elucidated in detail, a link between RHAMM, TPX2, and Aurora A (see next section) was proposed because RHAMM was found to colocalize and coimmunoprecipitate TPX2 in a cell cycle-dependent manner. Further evidence for a RHAMM-TPX2-Aurora A pathway is seen in a later study which showed that RHAMM overexpression and silencing in vitro in a MM cell line resulted in enhanced and decreased sensitivity respectively to treatment with Aurora kinase-specific inhibitors (Shi et al. 2007).

The reason for elevated RHAMM in MM is also uncertain. The gene encoding RHAMM, HMMR, is located on chromosome 5q33. Although chromosome 5 trisomy is common and might account for the elevated expression of RHAMM in H-MM, RHAMM overexpression is more strongly associated with NH-MM. Also H-MM is associated with a better prognosis, while increasing RHAMM expression levels correlate with poorer outcomes. Hence, further work is needed to understand the mechanisms underlying RHAMM overexpression in MM.

15.9 Aurora-A Kinase

The Aurora protein kinases, comprising Aurora-A, Aurora-B, and Aurora-C, are a family of serine-threonine kinases that regulate many processes during cell division (Vader and Lens 2008). Aurora-A is recognized to play an important role in centrosome maturation, centrosome separation, and bipolar spindle assembly (Carmena et al. 2009; Carmena and Earnshaw 2003; Dutertre et al. 2002).

During centrosome maturation, several proteins accumulate at the centrosome, resulting in growth of the pericentriolar material (PCM) and enhanced centrosomal microtubule nucleation activity. Aurora-A participates in this process by recruiting pericentriolar material (PCM) proteins including centrosomin, LATS2, NDEL1, and TACC (Mori et al. 2007; Abe et al. 2006; Toji et al. 2004; Hannak et al. 2001). Following maturation, the centrosomes migrate to opposite poles of the bipolar mitotic spindle in late G2; Aurora-A is also necessary for this process of centrosome separation, as seen by abnormal monopolar spindle formation when Aurora-A is inhibited in vitro (Glover et al. 1995; Liu and Aurora 2006).

The role of Aurora-A kinase in oncogenesis is well-documented. The Aurora-A gene, AURKA, is located on chromosome 20q13, an amplification hotspot in many tumors (Staaf et al. 2010; Tsukamoto et al. 2008; Scotto et al. 2008). Aurora-A genomic amplification has been identified in several tumor types including breast, colon, ovarian, and prostate cancers (Zhou et al. 1998; Sakakura et al. 2001; Bischoff et al. 1998). Overexpression of Aurora-A leads to tumorigenesis and centrosomal amplification in vitro (Zhou et al. 1998; Meraldi et al. 2002).

Several studies have examined Aurora-A expression in MM (Chng et al. 2008; Hose et al. 2009; Dutta-Simmons et al. 2009; Gorgun et al. 2010). We and other investigators have found that Aurora-A expression correlates with the CI (Chng et al. 2008; Hose et al. 2009). Given its known role in centrosome amplification (Zhou et al. 1998), it seems likely that Aurora-A contributes to centrosome amplification in MM as well.

15.10 Postulated Mechanisms of Centrosome Amplification in Multiple Myeloma with Immunoglobulin Translocations

One postulated mechanism for centrosome amplification in MM with immunoglobulin translocations is the disruption of genomic loci encoding proteins with centrosome-related functions as a consequence of such translocations (Maxwell and Pilarski 2005). Also, other genomic aberrations such as chromosome 13q deletions may be present

Table 15.2 Genes involved in centrosome regulation that are located at deletion sites or in the vicinity of translocation breakpoints in myeloma

Gene	Locus	MM deletion/ translocation	Protein function pertaining to centrosomes
ORC1	1p32	1p32-32 deletion	Controls centrosome copy number in human cells (Hemerly et al. 2009)
TACC3	4p16	t(4; 14)(p16; q32)	Localizes to centrosomes during mitosis (Mori et al. 2007; Gergely et al. 2000); required for bipolar spindle assembly and chromosome alignment (Fu et al. 2010)
PIM1	6p21	t(6; 14)(p21; q32)	Overexpression leads to centrosome amplification (Roh et al. 2003)
STK38	6p12	t(6;14)(p21;q32)	Regulates centrosome duplication (Hergovich et al. 2007)
NuMA	11q13	t(11; 14)(q13; q32)	Required for maintenance of spindle poles (Silk et al. 2009)
PAK	11q13	t(11; 14)(q13; q32)	Localizes to the centrosome (Zhao et al. 2005; Li et al. 2002)
PPP1CA	11q13	t(11; 14)(q13; q32)	Localizes to the centrosome (Andreassen et al. 1998); regulates centrosome splitting (Mi et al. 2007)
SCYL1	11q13	t(11; 14)(q13; q32)	Localizes to the centrosome and is associated with centrosomal amplification (Gong et al. 2009)
CENPJ	13q12	13 del	Required for maintenance of centrosome integrity and normal spindle morphology during cell division (Cho et al. 2006)
LATS2	13q12	13 del	Localizes to centrosome and maintains centrosome integrity (Toji et al. 2004; McPherson et al. 2004; Yabuta et al. 2007)
VAC14/ TAX1BP2	16q22	t(14; 16)(q32; q22)	Modulates centrosome number (Ching et al. 2006)
ID1	20q11	t(14; 20)(q32; q12)	Modulates centrosome number (Manthey et al. 2010)
TPX2	20q11	t(14; 20)(q32; q12)	Required for normal spindle morphology and centrosome integrity during cell division (Garrett et al. 2002)
AURKA	20q13	t(14; 20)(q32; q12)	Important centrosome-associated kinase involved in centrosome maturation and separation (Vader and Lens 2008); regulates centrosome number (Zhou et al. 1998; Meraldi et al. 2002)

in association with IgH translocation-positive MM (Fonseca et al. 2003), representing an additional mechanism for centrosome dysregulation. Table 15.2 provides a list of possible candidates located at both translocation and deletion sites.

15.11 Therapeutic Implications of Centrosome Amplification in Multiple Myeloma

Besides being a prognostic marker, centrosome amplification (or a surrogate like the CI) may also serve as a therapeutic marker in MM. As proof of concept, we showed that HMCL with higher CI were more sensitive to a small molecular Aurora kinase inhibitor (MLN8054) (Chng et al. 2008). Although it can be argued that a more direct marker for sensitivity to Aurora A inhibition might be expression levels of the Aurora-A gene/protein itself, it is noteworthy that Aurora A is regulated both at the expression level as well as via the phosphorylation of T288 on its activation loop (Dar et al. 2010). Therefore, expression levels of Aurora-A alone as a predictive therapeutic marker might not be optimal; in theory, a downstream marker of Aurora-A activity, such as centrosome amplification, might be of greater utility.

Centrosome amplification in MM might also be a therapeutic marker for other established or potential targets, such as TOP2A or PLK4, both of which were upregulated in high CI MM in our study (Chng et al. 2008). The Polo-like kinases (Plks, of which Plk4 is a member) are particularly attractive in this regard, as they have known roles in centrosome function (Dai et al. 2002), and overexpression of Plk1 in vitro also results in numerical centrosomal abnormalities (Meraldi et al. 2002). The role of the Plks in MM pathogenesis is still not well-characterized, but at least one study has shown activity of a Plk small molecular inhibitor against MM cell lines (Stewart et al. 2011).

Another recently proposed therapeutic strategy involves disrupting centrosome clustering (Fielding et al. 2011), which as previously mentioned is a protective mechanism preventing multipolar mitoses and cell death in cells with supernumerary centrosomes (Kwon et al. 2008; Quintyne et al. 2005). In a proof of concept study, Fielding et al. showed that integrin-linked kinase (ILK) is required for centrosome clustering in breast and prostate cancer cells. Furthermore, pharmacological inhibition of ILK led to multipolar divisions and cell death, and the sensitivity of the cell lines to ILK inhibition correlated with centrosome number (Fielding et al. 2011). Interestingly, another recent study reported that pharmacological ILK inhibition led to apoptosis in HMCL (Wang et al. 2010). Further studies investigating a correlation between sensitivity to ILK inhibition and centrosome abnormalities in MM are warranted.

15.12 Future Perspectives

Although some progress has been made concerning the role of centrosomes in MM biology and pathogenesis, the molecular pathways and mechanisms by which centrosome amplification and clustering occur in MM remain poorly characterized. Table 15.1 provides a list of molecules that may represent candidates for further study in the biology of centrosome abnormalities in MM. Also, in view of our previous findings that centrosome amplification correlated with high-risk genetic aberrations, such as chromosome 13 deletion, t(4; 14), and t(14; 16), further studies are needed to understand if centrosome abnormalities contribute to the formation of structural chromosomal abnormalities.

Given the existing evidence which suggests an important role for centrosome abnormalities in oncogenesis in general, we envisage that a deeper understanding of centrosome biology in MM will, in time, reveal novel candidates and strategies for therapy, and subsequently translate into enhanced patient care.

References

Abe Y, Ohsugi M, Haraguchi K, Fujimoto J, Yamamoto T (2006) LATS2-Ajuba complex regulates gamma-tubulin recruitment to centrosomes and spindle organization during mitosis. FEBS Lett 580:782–788

Adon AM, Zeng X, Harrison MK, Sannem S, Kiyokawa H, Kaldis P, Saavedra HI (2010) Cdk2 and Cdk4 regulate the centrosome cycle and are critical mediators of centrosome amplification in p53-null cells. Mol Cell Biol 30:694–710

Andreassen PR, Lacroix FB, Villa-Moruzzi E, Margolis RL (1998) Differential subcellular localization of protein phosphatase-1 alpha, gamma1, and delta isoforms during both interphase and mitosis in mammalian cells. J Cell Biol 141:1207–1215

Ashihara E, Kawata E, Nakagawa Y, Shimazaski C, Kuroda J, Taniguchi K, Uchiyama H, Tanaka R, Yokota A, Takeuchi M, Kamitsuji Y, Inaba T, Taniwaki M, Kimura S, Maekawa T (2009) beta-catenin small interfering RNA successfully suppressed progression of multiple myeloma in a mouse model. Clin Cancer Res 15:2731–2738

Avet-Loiseau H, Brigaudeau C, Morineau N, Talmant P, Lai JL, Daviet A, Li JY, Praloran V, Rapp MJ, Harousseau JL, Facon T, Bataille R (1999) High incidence of cryptic translocations involving the Ig heavy chain gene in multiple myeloma, as shown by fluorescence in situ hybridization. Genes Chromosom Cancer 24:9–15

Bahmanyar S, Kaplan DD, Deluca JG, Giddings TH Jr, O'Toole ET, Winey M, Salmon ED, Casey PJ, Nelson WJ, Barth AI (2008) beta-Catenin is a Nek2 substrate involved in centrosome separation. Genes Dev 22:91–105

Bahmanyar S, Guiney EL, Hatch EM, Nelson WJ, Barth AI (2010) Formation of extra centrosomal structures is dependent on beta-catenin. J Cell Sci 123:3125–3135

Balsitis SJ, Sage J, Duensing S, Munger K, Jacks T, Lambert PF (2003) Recapitulation of the effects of the human papillomavirus type 16 E7 oncogene on mouse epithelium by somatic Rb deletion and detection of pRb-independent effects of E7 in vivo. Mol Cell Biol 23:9094–9103

Bergsagel PL, Kuehl WM, Zhan F, Sawyer J, Barlogie B, Shaughnessy J Jr (2005) Cyclin D dysregulation: an early and unifying pathogenic event in multiple myeloma. Blood 106:296–303

Bertrand P, Lambert S, Joubert C, Lopez BS (2003) Overexpression of mammalian Rad51 does not stimulate tumorigenesis while a dominant-negative Rad51 affects centrosome fragmentation, ploidy and stimulates tumorigenesis, in p53-defective CHO cells. Oncogene 22:7587–7592

Bettencourt-Dias M, Glover DM (2007) Centrosome biogenesis and function: centrosomics brings new understanding. Nat Rev Mol Cell Biol 8:451–463

Bischoff JR, Anderson L, Zhu Y, Mossie K, Ng L, Souza B, Schryver B, Flanagan P, Clairvoyant F, Ginther C, Chan CS, Novotny M, Slamon DJ, Plowman GD (1998) A homologue of Drosophila aurora kinase is oncogenic and amplified in human colorectal cancers. EMBO J 17:3052–3065

Boveri T (1914) Zur frage der entstehung maligner tumoren. Gustav Fishcer, Jena

Boveri T (2008) Concerning the origin of malignant tumours by Theodor Boveri. Translated and annotated by Henry Harris. J Cell Sci 121(1):1–84

Brousseau M, Leleu X, Gerard J, Gastinne T, Godon A, Genevieve F, Dib M, Lai JL, Facon T, Zandecki M (2007) Hyperdiploidy is a common finding in monoclonal gammopathy of undetermined significance and monosomy 13 is restricted to these hyperdiploid patients. Clin Cancer Res 13:6026–6031

Carmena M, Earnshaw WC (2003) The cellular geography of aurora kinases. Nat Rev Mol Cell Biol 4:842–854

Carmena M, Ruchaud S, Earnshaw WC (2009) Making the Auroras glow: regulation of Aurora A and B kinase function by interacting proteins. Curr Opin Cell Biol 21:796–805

Carrasco DR, Tonon G, Huang Y, Zhang Y, Sinha R, Feng B, Stewart JP, Zhan F, Khatry D, Protopopova M, Protopopov A, Sukhdeo K, Hanamura I, Stephens O, Barlogie B, Anderson KC, Chin L, Shaughnessy JD Jr, Brennan C, Depinho RA (2006) High-resolution genomic profiles define distinct clinico-pathogenetic subgroups of multiple myeloma patients. Cancer Cell 9:313–325

Carroll PE, Okuda M, Horn HF, Biddinger P, Stambrook PJ, Gleich LL, Li YQ, Tarapore P, Fukasawa K (1999) Centrosome hyperamplification in human cancer: chromosome instability induced by p53 mutation and/or Mdm2 overexpression. Oncogene 18:1935–1944

Chen Z, Indjeian VB, McManus M, Wang L, Dynlacht BD (2002) CP110, a cell cycle-dependent CDK substrate, regulates centrosome duplication in human cells. Dev Cell 3:339–350

Ching YP, Chan SF, Jeang KT, Jin DY (2006) The retroviral oncoprotein Tax targets the coiled-coil centrosomal protein TAX1BP2 to induce centrosome overduplication. Nat Cell Biol 8:717–724

Chng WJ, Van Wier SA, Ahmann GJ, Winkler JM, Jalal SM, Bergsagel PL, Chesi M, Trendle MC, Oken MM, Blood E, Henderson K, Santana-Davila R, Kyle RA, Gertz MA, Lacy MQ, Dispenzieri A, Greipp PR, Fonseca R (2005) A validated FISH trisomy index demonstrates the hyperdiploid and nonhyperdiploid dichotomy in MGUS. Blood 106:2156–2161

Chng WJ, Ahmann GJ, Henderson K, Santana-Davila R, Greipp PR, Gertz MA, Lacy MQ, Dispenzieri A, Kumar S, Rajkumar SV, Lust JA, Kyle RA, Zeldenrust SR, Hayman SR, Fonseca R (2006a) Clinical implication of centrosome amplification in plasma cell neoplasm. Blood 107:3669–3675

Chng WJ, Ketterling RP, Fonseca R (2006b) Analysis of genetic abnormalities provides insights into genetic evolution of hyperdiploid myeloma. Genes Chromosom Cancer 45:1111–1120

Chng WJ, Price-Troska T, Gonzalez-Paz N, Van Wier S, Jacobus S, Blood E, Henderson K, Oken M, Van Ness B, Greipp P, Rajkumar SV, Fonseca R (2007) Clinical significance of TP53 mutation in myeloma. Leukemia 21:582–584

Chng WJ, Braggio E, Mulligan G, Bryant B, Remstein E, Valdez R, Dogan A, Fonseca R (2008a) The centrosome index is a powerful prognostic marker in myeloma and identifies a cohort of patients that might benefit from aurora kinase inhibition. Blood 111:1603–1609

Chng WJ, Gonzalez-Paz N, Price-Troska T, Jacobus S, Rajkumar SV, Oken MM, Kyle RA, Henderson KJ, Van Wier S, Greipp P, Van Ness B, Fonseca R (2008b) Clinical and biological significance of RAS mutations in multiple myeloma. Leukemia 22:2280–2284

Cho JH, Chang CJ, Chen CY, Tang TK (2006) Depletion of CPAP by RNAi disrupts centrosome integrity and induces multipolar spindles. Biochem Biophys Res Commun 339:742–747

Christensen JH, Abildgaard N, Plesner T, Nibe A, Nielsen O, Sorensen AG, Kerndrup GB (2007) Interphase fluorescence in situ hybridization in multiple myeloma and monoclonal gammopathy of undetermined significance without and with positive plasma cell identification: analysis of 192 cases from the Region of Southern Denmark. Cancer Genet Cytogenet 174:89–99

Crainie M, Belch AR, Mant MJ, Pilarski LM (1999) Overexpression of the receptor for hyaluronan-mediated motility (RHAMM) characterizes the malignant clone in multiple myeloma: identification of three distinct RHAMM variants. Blood 93:1684–1696

International Myeloma Working Group (2003) Criteria for the classification of monoclonal gammopathiesmultiple myeloma and related disorders: a report of the international myeloma working group. Br J Haematol 121:749–757

Dai W, Wang Q, Traganos F (2002) Polo-like kinases and centrosome regulation. Oncogene 21:6195–6200

Dar AA, Goff LW, Majid S, Berlin J, El-Rifai W (2010) Aurora kinase inhibitors–rising stars in cancer therapeutics? Mol Cancer Ther 9:268–278

Date O, Katsura M, Ishida M, Yoshihara T, Kinomura A, Sueda T, Miyagawa K (2006) Haploinsufficiency of RAD51B causes centrosome fragmentation and aneuploidy in human cells. Cancer Res 66:6018–6024

Dementyeva E, Nemec P, Kryukov F (2010) Centrosome amplification as a possible marker of mitotic disruptions and cellular carcinogenesis in multiple myeloma. Leuk Res 34:1007–1011

Dib A, Peterson TR, Raducha-Grace L, Zingone A, Zhan F, Hanamura I, Barlogie B, Shaughnessy J Jr, Kuehl WM (2006) Paradoxical expression of INK4c in proliferative multiple myeloma tumors: bi-allelic deletion vs increased expression. Cell Div 1:23

Dib A, Barlogie B, Shaughnessy JD Jr, Kuehl WM (2007) Methylation and expression of the p16INK4A tumor suppressor gene in multiple myeloma. Blood 109:1337–1338

Dickens NJ, Walker BA, Leone PE, Johnson DC, Brito JL, Zeisig A, Jenner MW, Boyd KD, Gonzalez D, Gregory WM, Ross FM, Davies FE, Morgan GJ (2010) Homozygous deletion mapping in myeloma samples identifies genes and an expression signature relevant to pathogenesis and outcome. Clin Cancer Res 16:1856–1864

Drach J, Schuster J, Nowotny H, Angerler J, Rosenthal F, Fiegl M, Rothermundt C, Gsur A, Jager U, Heinz R et al (1995) Multiple myeloma: high incidence of chromosomal aneuploidy as detected by interphase fluorescence in situ hybridization. Cancer Res 55:3854–3859

Duensing S, Lee LY, Duensing A, Basile J, Piboonniyom S, Gonzalez S, Crum CP, Munger K (2000) The human papillomavirus type 16 E6 and E7 oncoproteins cooperate to induce mitotic defects and genomic instability by uncoupling centrosome duplication from the cell division cycle. Proc Natl Acad Sci U S A 97:10002–10007

Dutertre S, Descamps S, Prigent C (2002) On the role of aurora-A in centrosome function. Oncogene 21:6175–6183

Dutta-Simmons J, Zhang Y, Gorgun G, Gatt M, Mani M, Hideshima T, Takada K, Carlson NE, Carrasco DE, Tai YT, Raje N, Letai AG, Anderson KC, Carrasco DR (2009) Aurora kinase A is a target of Wnt/beta-catenin involved in multiple myeloma disease progression. Blood 114:2699–2708

Elnenaei MO, Gruszka-Westwood AM, A'Hernt R, Matutes E, Sirohi B, Powles R, Catovsky D (2003) Gene abnormalities in multiple myeloma; the relevance of TP53, MDM2, and CDKN2A. Haematologica 88:529–537

Ely S, Di Liberto M, Niesvizky R, Baughn LB, Cho HJ, Hatada EN, Knowles DM, Lane J, Chen-Kiang S (2005) Mutually exclusive cyclin-dependent kinase 4/cyclin D1 and cyclin-dependent kinase 6/cyclin D2 pairing inactivates retinoblastoma protein and promotes cell cycle dysregulation in multiple myeloma. Cancer Res 65:11345–11353

Errabolu R, Sanders MA, Salisbury JL (1994) Cloning of a cDNA encoding human centrin, an EF-hand protein of centrosomes and mitotic spindle poles. J Cell Sci 107(Pt 1):9–16

Evans R, Naber C, Steffler T, Checkland T, Keats J, Maxwell C, Perry T, Chau H, Belch A, Pilarski L, Reiman T (2008) Aurora A kinase RNAi and small molecule inhibition of Aurora kinases with VE-465 induce apoptotic death in multiple myeloma cells. Leuk Lymphoma 49:559–569

Fielding AB, Lim S, Montgomery K, Dobreva I, Dedhar S (2011) A critical role of integrin-linked kinase, ch-TOG and TACC3 in centrosome clustering in cancer cells. Oncogene 30:521–534

Filipits M, Pohl G, Stranzl T, Kaufmann H, Ackermann J, Gisslinger H, Greinix H, Chott A, Drach J (2003) Low p27Kip1 expression is an independent adverse prognostic factor in patients with multiple myeloma. Clin Cancer Res 9:820–826

Fisk HA, Winey M (2001) The mouse Mps1p-like kinase regulates centrosome duplication. Cell 106:95–104

Fonseca R, Debes-Marun CS, Picken EB, Dewald GW, Bryant SC, Winkler JM, Blood E, Oken MM, Santana-Davila R, Gonzalez-Paz N, Kyle RA, Gertz MA, Dispenzieri A, Lacy MQ, Greipp PR (2003a) The recurrent IgH translocations are highly associated with nonhyperdiploid variant multiple myeloma. Blood 102:2562–2567

Fonseca R, Blood E, Rue M, Harrington D, Oken MM, Kyle RA, Dewald GW, Van Ness B, Van Wier SA, Henderson KJ, Bailey RJ, Greipp PR (2003b) Clinical and biologic implications of recurrent genomic aberrations in myeloma. Blood 101:4569–4575

Fonseca R, Barlogie B, Bataille R, Bastard C, Bergsagel PL, Chesi M, Davies FE, Drach J, Greipp PR, Kirsch IR, Kuehl WM, Hernandez JM, Minvielle S, Pilarski LM, Shaughnessy JD Jr, Stewart AK, Avet-Loiseau H (2004) Genetics and cytogenetics of multiple myeloma: a workshop report. Cancer Res 64:1546–1558

Fonseca R, Bergsagel PL, Drach J, Shaughnessy J, Gutierrez N, Stewart AK, Morgan G, Van Ness B, Chesi M, Minvielle S, Neri A, Barlogie B, Kuehl WM, Liebisch P, Davies F, Chen-Kiang S, Durie BG, Carrasco R, Sezer O, Reiman T, Pilarski L, Avet-Loiseau H (2009) International Myeloma Working Group molecular classification of multiple myeloma: spotlight review. Leukemia 23:2210–2221

Fu W, Tao W, Zheng P, Fu J, Bian M, Jiang Q, Clarke PR, Zhang C (2010) Clathrin recruits phosphorylated TACC3 to spindle poles for bipolar spindle assembly and chromosome alignment. J Cell Sci 123:3645–3651

Fukasawa K (2007) Oncogenes and tumour suppressors take on centrosomes. Nat Rev Cancer 7:911–924

Fukasawa K (2008) P53, cyclin-dependent kinase and abnormal amplification of centrosomes. Biochim Biophys Acta 1786:15–23

Fukasawa K, Wiener F (1997) Vande Woude GF, Mai S. Genomic instability and apoptosis are frequent in p53 deficient young mice. Oncogene 15:1295–1302

Fukasawa K, Choi T, Kuriyama R, Rulong S, Vande Woude GF (1996) Abnormal centrosome amplification in the absence of p53. Science 271:1744–1747

Ganem NJ, Godinho SA, Pellman D (2009) A mechanism linking extra centrosomes to chromosomal instability. Nature 460:278–282

Garrett S, Auer K, Compton DA, Kapoor TM (2002) hTPX2 is required for normal spindle morphology and centrosome integrity during vertebrate cell division. Curr Biol 12:2055–2059

Gergely F, Karlsson C, Still I, Cowell J, Kilmartin J, Raff JW (2000) The TACC domain identifies a family of centrosomal proteins that can interact with microtubules. Proc Natl Acad Sci U S A 97:14352–14357

Giacinti C, Giordano A (2006) RB and cell cycle progression. Oncogene 25:5220–5227

Glover DM, Leibowitz MH, McLean DA, Parry H (1995) Mutations in aurora prevent centrosome separation leading to the formation of monopolar spindles. Cell 81:95–105

Gong Y, Sun Y, McNutt MA, Sun Q, Hou L, Liu H, Shen Q, Ling Y, Chi Y, Zhang B (2009) Localization of TEIF in the centrosome and its functional association with centrosome amplification in DNA damage, telomere dysfunction and human cancers. Oncogene 28:1549–1560

Gonzalez-Paz N, Chng WJ, McClure RF, Blood E, Oken MM, Van Ness B, James CD, Kurtin PJ, Henderson K, Ahmann GJ, Gertz M, Lacy M, Dispenzieri A, Greipp PR, Fonseca R (2007) Tumor suppressor p16 methylation in multiple myeloma: biological and clinical implications. Blood 109:1228–1232

Gorgun G, Calabrese E, Hideshima T, Ecsedy J, Perrone G, Mani M, Ikeda H, Bianchi G, Hu Y, Cirstea D, Santo L, Tai YT, Nahar S, Zheng M, Bandi M, Carrasco RD, Raje N, Munshi N, Richardson P, Anderson KC (2010) A novel Aurora-A kinase inhibitor MLN8237 induces cytotoxicity and cell-cycle arrest in multiple myeloma. Blood 115:5202–5213

Griffin CS, Simpson PJ, Wilson CR, Thacker J (2000) Mammalian recombination-repair genes XRCC2 and XRCC3 promote correct chromosome segregation. Nat Cell Biol 2:757–761

Grisendi S, Bernardi R, Rossi M, Cheng K, Khandker L, Manova K, Pandolfi PP (2005) Role of nucleophosmin in embryonic development and tumorigenesis. Nature 437:147–153

Hall CL, Yang B, Yang X, Zhang S, Turley M, Samuel S, Lange LA, Wang C, Curpen GD, Savani RC, Greenberg AH, Turley EA (1995) Overexpression of the hyaluronan receptor RHAMM is transforming and is also required for H-ras transformation. Cell 82:19–26

Hannak E, Kirkham M, Hyman AA, Oegema K (2001) Aurora-A kinase is required for centrosome maturation in Caenorhabditis elegans. J Cell Biol 155:1109–1116

Hemerly AS, Prasanth SG, Siddiqui K, Stillman B (2009) Orc1 controls centriole and centrosome copy number in human cells. Science 323:789–793

Hergovich A, Lamla S, Nigg EA, Hemmings BA (2007) Centrosome-associated NDR kinase regulates centrosome duplication. Mol Cell 25:625–634

Hinchcliffe EH, Li C, Thompson EA, Maller JL, Sluder G (1999) Requirement of Cdk2-cyclin E activity for repeated centrosome reproduction in Xenopus egg extracts. Science 283:851–854

Hirai H, Roussel MF, Kato JY, Ashmun RA, Sherr CJ (1995) Novel INK4 proteins, p19 and p18, are specific inhibitors of the cyclin D-dependent kinases CDK4 and CDK6. Mol Cell Biol 15:2672–2681

Hose D, Reme T, Meissner T, Moreaux J, Seckinger A, Lewis J, Benes V, Benner A, Hundemer M, Hielscher T, Shaughnessy JD Jr, Barlogie B, Neben K, Kramer A, Hillengass J, Bertsch U, Jauch A, De Vos J, Rossi JF, Mohler T, Blake J, Zimmermann J, Klein B, Goldschmidt H (2009) Inhibition of aurora kinases for tailored risk-adapted treatment of multiple myeloma. Blood 113:4331–4340

Hsu LC, White RL (1998) BRCA1 is associated with the centrosome during mitosis. Proc Natl Acad Sci U S A 95:12983–12988

Hurt EM, Wiestner A, Rosenwald A, Shaffer AL, Campo E, Grogan T, Bergsagel PL, Kuehl WM, Staudt LM (2004) Overexpression of c-maf is a frequent oncogenic event in multiple myeloma that promotes proliferation and pathological interactions with bone marrow stroma. Cancer Cell 5:191–199

Iovino F, Lentini L, Amato A, Di Leonardo A (2006) RB acute loss induces centrosome amplification and aneuploidy in murine primary fibroblasts. Mol Cancer 5:38

Jemal A, Siegel R, Xu J, Ward E (2010) Cancer statistics 2010. CA Cancer J Clin 60:277–300

Juge-Morineau N, Harousseau JL, Amiot M, Bataille R (1997) The retinoblastoma susceptibility gene RB-1 in multiple myeloma. Leuk Lymphoma 24:229–237

Kramer A, Schultheis B, Bergmann J, Willer A, Hegenbart U, Ho AD, Goldschmidt H, Hehlmann R (2002) Alterations of the cyclin D1/pRb/p16(INK4A) pathway in multiple myeloma. Leukemia 16:1844–1851

Kumar SK, Rajkumar SV, Dispenzieri A, Lacy MQ, Hayman SR, Buadi FK, Zeldenrust SR, Dingli D, Russell SJ, Lust JA, Greipp PR, Kyle RA, Gertz MA (2008) Improved survival in multiple myeloma and the impact of novel therapies. Blood 111:2516–2520

Kwon M, Godinho SA, Chandhok NS, Ganem NJ, Azioune A, Thery M, Pellman D (2008) Mechanisms to suppress multipolar divisions in cancer cells with extra centrosomes. Genes Dev 22:2189–2203

Kyle RA, Rajkumar SV (2004) Multiple myeloma. N Engl J Med 351:1860–1873

Kyle RA, Therneau TM, Rajkumar SV, Offord JR, Larson DR, Plevak MF, Melton LJ 3rd (2002) A long-term study of prognosis in monoclonal gammopathy of undetermined significance. N Engl J Med 346:564–569

Lacey KR, Jackson PK, Stearns T (1999) Cyclin-dependent kinase control of centrosome duplication. Proc Natl Acad Sci U S A 96:2817–2822

Lee MH, Yang HY (2003) Regulators of G1 cyclin-dependent kinases and cancers. Cancer Metastasis Rev 22:435–449

Li F, Adam L, Vadlamudi RK, Zhou H, Sen S, Chernoff J, Mandal M, Kumar R (2002) p21-activated kinase 1 interacts with and phosphorylates histone H3 in breast cancer cells. EMBO Rep 3:767–773

Liu Q, Aurora.A Ruderman JV (2006) Mitotic entry, and spindle bipolarity. Proc Natl Acad Sci U S A 103:5811–5816

Manthey C, Mern DS, Gutmann A, Zielinski AJ, Herz C, Lassmann S, Hasskarl J (2010) Elevated endogenous expression of the dominant negative basic helix-loop-helix protein ID1 correlates with significant centrosome abnormalities in human tumor cells. BMC Cell Biol 11:2

Mateos MV, Garcia-Sanz R, Lopez-Perez R, Moro MJ, Ocio E, Hernandez J, Megido M, Caballero MD, Fernandez-Calvo J, Barez A, Almeida J, Orfao A, Gonzalez M (2002) San Miguel JF. Methylation is an inactivating mechanism of the p16 gene in multiple myeloma associated with high plasma cell proliferation and short survival. Br J Haematol 118:1034–1040

Maxwell CA, Pilarski LM (2005) A potential role for centrosomal deregulation within IgH translocation-positive myeloma. Med Hypotheses 65:915–921

Maxwell CA, Keats JJ, Crainie M, Sun X, Yen T, Shibuya E, Hendzel M, Chan G, Pilarski LM (2003) RHAMM is a centrosomal protein that interacts with dynein and maintains spindle pole stability. Mol Biol Cell 14:2262–2276

Maxwell CA, Rasmussen E, Zhan F, Keats JJ, Adamia S, Strachan E, Crainie M, Walker R, Belch AR, Pilarski LM, Barlogie B, Shaughnessy J Jr, Reiman T (2004) RHAMM expression and isoform balance predict aggressive disease and poor survival in multiple myeloma. Blood 104:1151–1158

Maxwell CA, Keats JJ, Belch AR, Pilarski LM, Reiman T (2005) Receptor for hyaluronan-mediated motility correlates with centrosome abnormalities in multiple myeloma and maintains mitotic integrity. Cancer Res 65:850–860

McDermott KM, Zhang J, Holst CR, Kozakiewicz BK, Singla V, Tlsty TD (2006) p16(INK4a) prevents centrosome dysfunction and genomic instability in primary cells. PLoS Biol 4:e51

McPherson JP, Tamblyn L, Elia A, Migon E, Shehabeldin A, Matysiak-Zablocki E, Lemmers B, Salmena L, Hakem A, Fish J, Kassam F, Squire J, Bruneau BG, Hande MP, Hakem R (2004) Lats2/Kpm is required for embryonic development, proliferation control and genomic integrity. EMBO J 23:3677–3688

Meraldi P, Honda R, Nigg EA (2002) Aurora-A overexpression reveals tetraploidization as a major route to centrosome amplification in p53-/- cells. EMBO J 21:483–492

Mi J, Guo C, Brautigan DL, Larner JM (2007) Protein phosphatase-1alpha regulates centrosome splitting through Nek2. Cancer Res 67:1082–1089

Mohamed AN, Bentley G, Bonnett ML, Zonder J, Al-Katib A (2007) Chromosome aberrations in a series of 120 multiple myeloma cases with abnormal karyotypes. Am J Hematol 82:1080–1087

Mori D, Yano Y, Toyo-oka K, Yoshida N, Yamada M, Muramatsu M, Zhang D, Saya H, Toyoshima YY, Kinoshita K, Wynshaw-Boris A, Hirotsune S (2007) NDEL1 phosphorylation by Aurora-A kinase is essential for centrosomal maturation, separation, and TACC3 recruitment. Mol Cell Biol 27:352–367

Morse L, Chen D, Franklin D, Xiong Y, Chen-Kiang S (1997) Induction of cell cycle arrest and B cell terminal differentiation by CDK inhibitor p18(INK4c) and IL-6. Immunity 6:47–56

Munshi NC, Hideshima T, Carrasco D, Shammas M, Auclair D, Davies F, Mitsiades N, Mitsiades C, Kim RS, Li C, Rajkumar SV, Fonseca R, Bergsagel L, Chauhan D, Anderson KC (2004) Identification of genes modulated in multiple myeloma using genetically identical twin samples. Blood 103:1799–1806

Nam HJ, Chae S, Jang SH, Cho H, Lee JH (2010) The PI3 K-Akt mediates oncogenic Met-induced centrosome amplification and chromosome instability. Carcinogenesis 31:1531–1540

Neben K, Tews B, Wrobel G, Hahn M, Kokocinski F, Giesecke C, Krause U, Ho AD, Kramer A, Lichter P (2004) Gene expression patterns in acute myeloid leukemia correlate with centrosome aberrations and numerical chromosome changes. Oncogene 23:2379–2384

Nelsen CJ, Kuriyama R, Hirsch B, Negron VC, Lingle WL, Goggin MM, Stanley MW, Albrecht JH (2005) Short term cyclin D1 overexpression induces centrosome amplification, mitotic spindle abnormalities, and aneuploidy. J Biol Chem 280:768–776

Nigg EA, Raff JW (2009) Centrioles, centrosomes, and cilia in health and disease. Cell 139:663–678

Nishida K, Tamura A, Nakazawa N, Ueda Y, Abe T, Matsuda F, Kashima K, Taniwaki M (1997) The Ig heavy chain gene is frequently involved in chromosomal translocations in multiple myeloma and plasma cell leukemia as detected by in situ hybridization. Blood 90:526–534

Okuda M, Horn HF, Tarapore P, Tokuyama Y, Smulian AG, Chan PK, Knudsen ES, Hofmann IA, Snyder JD, Bove KE, Fukasawa K (2000) Nucleophosmin/B23 is a target of CDK2/cyclin E in centrosome duplication. Cell 103:127–140

Park G, Kang SH, Lee JH, Suh C, Kim M, Park SM, Kim TY, Oh B, Min HJ, Yoon SS, Yang IC, Cho HI, Lee DS (2011) Concurrent p16 methylation pattern as an adverse prognostic factor in multiple myeloma: a methylation-specific polymerase chain reaction study using two different primer sets. Ann Hematol 90:73–79

Phillip CJ, Stellrecht CM, Nimmanapalli R, Gandhi V (2009) Targeting MET transcription as a therapeutic strategy in multiple myeloma. Cancer Chemother Pharmacol 63:587–597

Pihan GA, Purohit A, Wallace J, Knecht H, Woda B, Quesenberry P, Doxsey SJ (1998) Centrosome defects and genetic instability in malignant tumors. Cancer Res 58:3974–3985

Pihan GA, Wallace J, Zhou Y, Doxsey SJ (2003) Centrosome abnormalities and chromosome instability occur together in pre-invasive carcinomas. Cancer Res 63:1398–1404

Plo I, Lopez B (2009) AKT1 represses gene conversion induced by different genotoxic stresses and induces supernumerary centrosomes and aneuploidy in hamster ovary cells. Oncogene 28:2231–2237

Quintyne NJ, Reing JE, Hoffelder DR, Gollin SM, Saunders WS (2005) Spindle multipolarity is prevented by centrosomal clustering. Science 307:127–129

Renglin Lindh A, Schultz N, Saleh-Gohari N, Helleday T (2007) RAD51C (RAD51L2) is involved in maintaining centrosome number in mitosis. Cytogenet Genome Res 116:38–45

Ribas C, Colleoni GW, Felix RS (2005) Regis Silva MR, Caballero OL, Brait M, Bordin JO. p16 gene methylation lacks correlation with angiogenesis and prognosis in multiple myeloma. Cancer Lett 222:247–254

Roh M, Gary B, Song C, Said-Al-Naief N, Tousson A, Kraft A, Eltoum IE, Abdulkadir SA (2003) Overexpression of the oncogenic kinase Pim-1 leads to genomic instability. Cancer Res 63:8079–8084

Sakakura C, Hagiwara A, Yasuoka R, Fujita Y, Nakanishi M, Masuda K, Shimomura K, Nakamura Y, Inazawa J, Abe T, Yamagishi H (2001) Tumour-amplified kinase BTAK is amplified and overexpressed in gastric cancers with possible involvement in aneuploid formation. Br J Cancer 84:824–831

Schrantz N, Beney GE, Auffredou MT, Bourgeade MF, Leca G, Vazquez A (2000) The expression of p18INK4 and p27kip1 cyclin-dependent kinase inhibitors is regulated differently during human B cell differentiation. J Immunol 165:4346–4352

Scotto L, Narayan G, Nandula SV, Arias-Pulido H, Subramaniyam S, Schneider A, Kaufmann AM, Wright JD, Pothuri B, Mansukhani M, Murty VV (2008) Identification of copy number gain and overexpressed genes on chromosome arm 20q by an integrative genomic approach in cervical cancer: potential role in progression. Genes Chromosom Cancer 47:755–765

Seidel C, Borset M, Turesson I, Abildgaard N, Sundan A, Waage A (1998) Elevated serum concentrations of hepatocyte growth factor in patients with multiple myeloma. The Nordic Myeloma Study Group. Blood 91:806–812

Sherr CJ, Roberts JM (2004) Living with or without cyclins and cyclin-dependent kinases. Genes Dev 18:2699–2711

Shi Y, Reiman T, Li W, Maxwell CA, Sen S, Pilarski L, Daniels TR, Penichet ML, Feldman R, Lichtenstein A (2007) Targeting aurora kinases as therapy in multiple myeloma. Blood 109:3915–3921

Shinmura K, Tarapore P, Tokuyama Y, George KR, Fukasawa K (2005) Characterization of centrosomal association of nucleophosmin/B23 linked to Crm1 activity. FEBS Lett 579:6621–6634

Silk AD, Holland AJ, Cleveland DW (2009) Requirements for NuMA in maintenance and establishment of mammalian spindle poles. J Cell Biol 184:677–690

Silkworth WT, Nardi IK, Scholl LM, Cimini D (2009) Multipolar spindle pole coalescence is a major source of kinetochore mis-attachment and chromosome mis-segregation in cancer cells. PLoS ONE 4:e6564

Smadja NV, Fruchart C, Isnard F, Louvet C, Dutel JL, Cheron N, Grange MJ, Monconduit M, Bastard C (1998) Chromosomal analysis in multiple myeloma: cytogenetic evidence of two different diseases. Leukemia 12:960–969

Smadja NV, Bastard C, Brigaudeau C, Leroux D, Fruchart C (2001) Hypodiploidy is a major prognostic factor in multiple myeloma. Blood 98:2229–2238

Smiraldo PG, Gruver AM, Osborn JC, Pittman DL (2005) Extensive chromosomal instability in Rad51d-deficient mouse cells. Cancer Res 65:2089–2096

Spruck CH, Won KA, Reed SI (1999) Deregulated cyclin E induces chromosome instability. Nature 401:297–300

Staaf J, Jonsson G, Ringner M, Vallon-Christersson J, Grabau D, Arason A, Gunnarsson H, Agnarsson BA, Malmstrom PO, Johannsson OT, Loman N, Barkardottir RB, Borg A (2010) High-resolution genomic and expression analyses of copy number alterations in HER2-amplified breast cancer. Breast Cancer Res 12:R25

Starita LM, Machida Y, Sankaran S, Elias JE, Griffin K, Schlegel BP, Gygi SP, Parvin JD (2004) BRCA1-dependent ubiquitination of gamma-tubulin regulates centrosome number. Mol Cell Biol 24:8457–8466

Stellrecht CM, Phillip CJ, Cervantes-Gomez F, Gandhi V (2007) Multiple myeloma cell killing by depletion of the MET receptor tyrosine kinase. Cancer Res 67:9913–9920

Stewart HJ, Kishikova L, Powell FL, Wheatley SP, Chevassut TJ (2011) The polo-like kinase inhibitor BI 2536 exhibits potent activity against malignant plasma cells and represents a novel therapy in multiple myeloma. Exp Hematol 39(3):330–338

Sugihara E, Kanai M, Saito S, Nitta T, Toyoshima H, Nakayama K, Nakayama KI, Fukasawa K, Schwab M, Saya H, Miwa M (2006) Suppression of centrosome amplification after DNA damage depends on p27 accumulation. Cancer Res 66:4020–4029

Swerdlow SH, Campo E, Harris NL, Jaffe ES, Pileri SA, Stein H, Thiele J, Vardiman JW (2008) WHO classification of tumours of haematopoietic and lymphoid tissues. Int Agency Res Cancer 15:145–147

Tarapore P, Horn HF, Tokuyama Y, Fukasawa K (2001) Direct regulation of the centrosome duplication cycle by the p53-p21Waf1/Cip1 pathway. Oncogene 20:3173–3184

Thompson SL, Bakhoum SF, Compton DA (2010) Mechanisms of chromosomal instability. Curr Biol 20:R285–R295

Toji S, Yabuta N, Hosomi T, Nishihara S, Kobayashi T, Suzuki S, Tamai K, Nojima H (2004) The centrosomal protein Lats2 is a phosphorylation target of Aurora-A kinase. Genes Cells 9:383–397

Tsukamoto Y, Uchida T, Karnan S, Noguchi T, Nguyen LT, Tanigawa M, Takeuchi I, Matsuura K, Hijiya N, Nakada C, Kishida T, Kawahara K, Ito H, Murakami K, Fujioka T, Seto M, Moriyama M (2008) Genome-wide analysis of DNA copy number alterations and gene expression in gastric cancer. J Pathol 216:471–482

Turley EA (1992) Hyaluronan and cell locomotion. Cancer Metastasis Rev 11:21–30

Turley EA, Torrance J (1985) Localization of hyaluronate and hyaluronate-binding protein on motile and non-motile fibroblasts. Exp Cell Res 161:17–28

Turley EA, Moore D, Hayden LJ (1987) Characterization of hyaluronate binding proteins isolated from 3T3 and murine sarcoma virus transformed 3T3 cells. Biochemistry 26:2997–3005

Turley EA, Austen L, Vandeligt K, Clary C (1991) Hyaluronan and a cell-associated hyaluronan binding protein regulate the locomotion of ras-transformed cells. J Cell Biol 112:1041–1047

Turley EA, Belch AJ, Poppema S, Pilarski LM (1993) Expression and function of a receptor for hyaluronan-mediated motility on normal and malignant B lymphocytes. Blood 81:446–453

Turley EA, Noble PW, Bourguignon LY (2002) Signaling properties of hyaluronan receptors. J Biol Chem 277:4589–4592

Tutt A, Gabriel A, Bertwistle D, Connor F, Paterson H, Peacock J, Ross G, Ashworth A (1999) Absence of Brca2 causes genome instability by chromosome breakage and loss associated with centrosome amplification. Curr Biol 9:1107–1110

Vader G, Lens SM (2008) The Aurora kinase family in cell division and cancer. Biochim Biophys Acta 1786:60–72

Wadhera RK, Rajkumar SV (2010) Prevalence of monoclonal gammopathy of undetermined significance: a systematic review. Mayo Clin Proc 85:933–942

Wang X, Zhang Z, Yao C (2011a) Targeting integrin-linked kinase increases apoptosis and decreases invasion of myeloma cell lines and inhibits IL-6 and VEGF secretion from BMSCs. Med Oncol 28:1596–1600

Wang C, Entwistle J, Hou G, Li Q, Turley EA (1996) The characterization of a human RHAMM cDNA: conservation of the hyaluronan-binding domains. Gene 174:299–306

Wang HF, Takenaka K, Nakanishi A, Miki Y (2011b) BRCA2 and nucleophosmin coregulate centrosome amplification and form a complex with the Rho effector kinase ROCK2. Cancer Res 71:68–77

Weinhold N, Moreaux J, Raab MS, Hose D, Hielscher T, Benner A, Meissner T, Ehrbrecht E, Brough M, Jauch A, Goldschmidt H, Klein B, Moos M (2010) NPM1 is overexpressed in hyperdiploid multiple myeloma due to a gain of chromosome 5 but is not delocalized to the cytoplasm. Genes Chromosom Cancer 49:333–341

Wong IH, Ng MH, Lee JC, Lo KW, Chung YF, Huang DP (1998) Transcriptional silencing of the p16 gene in human myeloma-derived cell lines by hypermethylation. Br J Haematol 103:168–175

Wu KL, Beverloo B, Velthuizen SJ, Sonneveld P (2007a) Sequential analysis of chromosome aberrations in multiple myeloma during disease progression. Clin Lymphoma Myeloma 7:280–285

Wu KL, Beverloo B, Lokhorst HM, Segeren CM, van der Holt B, Steijaert MM, Westveer PH, Poddighe PJ, Verhoef GE, Sonneveld P (2007b) Abnormalities of chromosome 1p/q are highly associated with chromosome 13/13q deletions and are an adverse prognostic factor for the outcome of high-dose chemotherapy in patients with multiple myeloma. Br J Haematol 136:615–623

Xu X, Weaver Z, Linke SP, Li C, Gotay J, Wang XW, Harris CC, Ried T, Deng CX (1999) Centrosome amplification and a defective G2-M cell cycle checkpoint induce genetic instability in BRCA1 exon 11 isoform-deficient cells. Mol Cell 3:389–395

Yabuta N, Okada N, Ito A, Hosomi T, Nishihara S, Sasayama Y, Fujimori A, Okuzaki D, Zhao H, Ikawa M, Okabe M, Nojima H (2007) Lats2 is an essential mitotic regulator required for the coordination of cell division. J Biol Chem 282:19259–19271

Yuregir OO, Sahin FI, Yilmaz Z, Kizilkilic E, Karakus S, Ozdogu H (2009) Fluorescent in situ hybridization studies in multiple myeloma. Hematology 14:90–94

Zeng X, Shaikh FY, Harrison MK, Adon AM, Trimboli AJ, Carroll KA, Sharma N, Timmers C, Chodosh LA, Leone G, Saavedra HI (2010) The Ras oncogene signals centrosome amplification in mammary epithelial cells through cyclin D1/Cdk4 and Nek2. Oncogene 29:5103–5112

Zhan F, Hardin J, Kordsmeier B, Bumm K, Zheng M, Tian E, Sanderson R, Yang Y, Wilson C, Zangari M, Anaissie E, Morris C, Muwalla F, van Rhee F, Fassas A, Crowley J, Tricot G, Barlogie B, Shaughnessy J Jr (2002) Global gene expression profiling of multiple myeloma, monoclonal gammopathy of undetermined significance, and normal bone marrow plasma cells. Blood 99:1745–1757

Zhan F, Huang Y, Colla S, Stewart JP, Hanamura I, Gupta S, Epstein J, Yaccoby S, Sawyer J, Burington B, Anaissie E, Hollmig K, Pineda-Roman M, Tricot G, van Rhee F, Walker R, Zangari M, Crowley J, Barlogie B, Shaughnessy JD Jr (2006) The molecular classification of multiple myeloma. Blood 108:2020–2028

Zhao ZS, Lim JP, Ng YW, Lim L, Manser E (2005) The GIT-associated kinase PAK targets to the centrosome and regulates Aurora-A. Mol Cell 20:237–249

Zhou H, Kuang J, Zhong L, Kuo WL, Gray JW, Sahin A, Brinkley BR, Sen S (1998) Tumour amplified kinase STK15/BTAK induces centrosome amplification, aneuploidy and transformation. Nat Genet 20:189–193

Zojer N, Schuster-Kolbe J, Assmann I, Ackermann J, Strasser K, Hubl W, Drach J, Ludwig H (2002) Chromosomal aberrations are shared by malignant plasma cells and a small fraction of circulating CD19 + cells in patients with myeloma and monoclonal gammopathy of undetermined significance. Br J Haematol 117:852–859

Zollinger A, Stuhmer T, Chatterjee M, Gattenlohner S, Haralambieva E, Muller-Hermelink HK, Andrulis M, Greiner A, Wesemeier C, Rath JC, Einsele H, Bargou RC (2008) Combined functional and molecular analysis of tumor cell signaling defines 2 distinct myeloma subgroups: Akt-dependent and Akt-independent multiple myeloma. Blood 112:3403–3411

Chapter 16
Centrosomal Amplification and Related Abnormalities Induced by Nucleoside Analogs

Ofelia A. Olivero

Abstract Although very effective, anti-retroviral nucleoside analog drugs are genotoxic in humans and carcinogenic in mice. The effect of the nucleoside reverse transcriptase inhibitor (NRTI) AZT (zidovudine) as a centrosomal disruptor will be discussed. Centrosomal amplification was measured on MCF10A cells by immunocytochemistry with antibodies to centrosomal proteins pericentrin and Cep 170. Doses of 0, 10, 100, and 200 μM AZT for 24 h showed 0.9, 1.75, 2.3, and 3.1 % of cells containing centrosomal amplification (supernumerary centrosomes), respectively. Furthermore, the origin of the extra centrosomes was analyzed by addressing the maturity of the structures. Typically mature centrosomes, identified by the presence of Cep170 proteins, are the result of overduplication of pre-existing centrosomes. Conversely, it is believed that, Cep170 negative, immature centrosomes are the result of accumulation of the organelles by impaired cytokinesis. Scoring of Cep170-stained AZT-induced centrosomes revealed that 40, 50, and 53 % of the supernumerary centrosomes in cells exposed to 10, 100, and 200 μM AZT, respectively, were mature, compared to 22 % of unexposed cells centrosomes in the control. Therefore, AZT-induced centrosomal amplification is the result of both overduplication and cytokinesis impairment.

O. A. Olivero (✉)
Laboratory of Cancer Biology and Genetics, Carcinogen-DNA Interactions Section 37 Convent Dr. MSC 4255, National Cancer Institute, NIH, Bldg 37 Rm 4032, Bethesda, MD 20892-4255, USA
e-mail: oliveroo@exchange.nih.gov

16.1 Background

The centrosome is an organelle that regulates migration of chromosomes to the daughter cells, and is also a microtubule organizer. Based on the multiple proteins residing in the centrosome-associated protein matrix, centrosomes have been implicated in other cellular functions including cell cycle transitions, such as G_1 to S-phase, G_2 to mitosis, and metaphase to anaphase (D'Assoro et al. 2002; Doxsey et al. 2005). Cells normally have one centrosome, which duplicates synchronically with the phases of the cell cycle to generate two new centrosomes, each of which comprises a pair of orthogonally placed centrioles. Most human carcinomas have an abnormal centrosomal number that contributes to the genomic instability characteristic of transformed cells (Carroll et al. 1999; Lingle et al. 1998; Pihan et al. 1998; Satoh and Lindahl 1994). Additionally, it has been reported that malfunction of the centrosome induces delay in the G_1-S boundary of the cell cycle (Hinchcliffe et al. 2001; Mikule et al. 2007). Some clastogenic agents may also behave as aneugens, generating an array of both structural and numerical chromosomal aberrations as well as chromosomal instability (Lengauer et al. 1997).

Zidovudine (3'-azido-3'-deoxythymidine, AZT), the first nucleoside reverse transcriptase inhibitor (NRTI) used for HIV-1 therapy has been shown to induce micronuclei, chromosomal aberrations, mutations, and telomeric attrition in vitro and in vivo (IARC 2000; Olivero 2007). Additionally, AZT becomes incorporated into eukaryotic DNA (Diwan et al. 1999; Olivero et al. 1997) and induces cell cycle arrest with accumulation of cells in S-phase (Chandrasekaran et al. 1995; Escobar et al. 2007; Olivero et al. 2005, 2008; Viora et al. 1997).

16.2 Genotoxicity of Nucleoside Analogs

Nucleoside reverse transcriptase inhibitors (NRTIs) constitute a high widespread type of drugs used in the therapy of AIDS. Used in monotherapy or in combination with other drugs, their mechanism of action is based upon the ability to act as chain terminators and the inhibition of the nucleotide binding site of the HIV-1 reverse transcriptase (Furman et al. 1986; Huang et al. 1990; St.Clair et al. 1987).

Usually administered as prodrugs, NRTIs become active through a cascade of cytosolic phosphorylations. Activation of Zidovudine (ZDV), the first NRTI approved by the US Food and Drug Administration, to the triphosphate form is mediated by thymidine kinase, thymidylate kinase, and pyrimidine nucleoside diphosphate kinase (St. Clair et al. 1987). Zidovudine acts as a HIV-1 reverse transcriptase inhibitor able to reduce morbidity and mortality related with HIV-1 infection (Fischl et al. 1987); however, its use has been limited due to induced toxicities (Richman et al. 1987). Although highly specific for HIV-1 reverse transcriptase, NRTIs exhibit some affinity for cellular polymerases (Lim and Copeland 2001); Parker et al. 1991). In a comprehensive review, Lee et al. report that the toxic side effects of NRTIs are correlated with the kinetics of incorporation by the mitochondrial DNA polymerase (Lee et al. 2003). Others tested all FDA approved NRTIs and provided a method to screen NRTIs for potential toxicity. Furthermore, they defined a toxicity index for chain terminators to account for relative rates of incorporation versus removal (Johnson et al. 2001).

Consequences of DNA damage by NRTI incorporation and subsequent chain termination have been compiled under the following categories: micronuclei, sister chromatid exchanges, chromosomal aberrations, and telomere shortening (IARC 2000).

16.3 Zidovudine Becomes Incorporated into DNA

The widespread chronic use of Zidovudine as first line therapy for HIV generated some concern about patients who may be susceptible to drug-induced genotoxicity. It was hypothesized that despite the high affinity of Zidovudine for HIV-1 reverse transcriptase, some incorporation into eukaryotic DNA would be predicted via host polymerases (Copeland et al. 1992; Furman et al. 1986). Numerous laboratories employing diverse techniques reported to have found Zidovudine incorporated into DNA of eukaryotic cells, in vivo as well as in vitro. Among other approaches, an original ^{32}P postlabeling method to detect Zidovudine-DNA incorporation in mice (Fang and Beland 2000), incorporation of Zidovudine in bone marrow using radiolabeled Zidovudine, and subsequent analysis by high performance liquid chromatography (Sommadossi et al. 1989) and incorporation of Zidovudine into CEM T-lymphoblastoid cells (Avramis et al. 1989) have been reported. Furthermore, incorporation of Zidovudine into DNA of the chronic myelogenous leukemic cell line K562 and evidence of its removal by an exonucleolytic repair enzyme (Vazquez-Padua et al. 1990) was demonstrated in vitro. Removal of ZDV incorporated into DNA by a 3'-5' exonuclease able to remove DNA terminated by a variety of dideoxynucleosides (Skalski et al. 1995) has also been demonstrated in leukemic H9 cells (Agarwal and Olivero 1997). In 1994, Olivero et al. (Olivero et al. 1994) used a radioimmunoassay to show Zidovudine-DNA incorporation and validated the method with radiolabeled compound. Similar levels of Zidovudine-DNA incorporation were found in mouse. Although variable and depending on metabolic activation from species to species and from cell type to cell type, incorporation of Zidovudine into DNA has been extensively documented.

As a consequence of the persistent, non-repaired incorporation, the genotoxic insult can be perpetuated. Thus, the inability of the cell to effectively remove the incorporated molecules is translated as mutations and or other forms of genotoxic events.

16.4 Cell Cycle Arrest is a Consequence of Zidovudine Exposure

Inhibition of cells to enter mitosis is a measure of the potential of damage inflicted on the cell. Impairment of compliance with a cell cycle is one of the consequences of DNA damage that reflects the inability of the cell to accomplish basic biological processes.

Cell cycle delay induced by Zidovudine has been reported by Wu et al. in a human chronic myeloid leukemia cell line (Wu et al. 2004) and in HL60 cells, (Roskrow and Wickramasinghe 1990). Others (Heagy et al. 1991; Viora et al. 1997) showed decrease in proliferative response to phytohemagglutinins and delay in S-phase in peripheral blood mononuclear cells and human leukemic CEM cells. A cytostatic effect in human colon carcinoma WiDr cells was induced at 12 h, peaking at 24 h, and reversed to baseline at 72 h (Chandrasekaran et al. 1995). The authors then proposed that AZT-induced cytostasis is a transient and reversible effect, adding that similar results were observed in 8/9 tumor cell lines examined. A correlation between changes in the gene expression of cell cycle-related genes and S-phase arrest was demonstrated in HeLa cells (Olivero et al. 2005). In experiments carried out using different microarray platforms, the authors examined the correlation of up-/downregulation of some key genes in cell cycle G_1/S progression and S-phase accumulation in cells exposed to the combination Zidovudine-Lamivudine (2',3'-dideoxy-3'-thiacytidine) for 24 h. Changes in gene expression suggested that the cyclin-D-Rb pathway could be mediating the cell cycle arrest (Olivero et al. 2005). A synchrony between cell cycle and centrosomal cycle should exist to warrant proper cell division and cell homeostasis. However, when cell cycle distortions take place the centrosomal division could progress asynchronically causing centrosomal amplification.

16.5 Zidovudine Acts as a Centrosome Disruptor

Employing immunohistochemistry, Borojerdi et al. (2009) explored the ability of Zidovudine to act as a centrosome disruptor. The authors used hamster CHO cells and human NHMEC strains, exposed to Zidovudine for 24 h and showed centrosomal disruption evidenced by pericentrin signaling. Additionally, they demonstrated aberrations in tubulin polymerization in cells bearing abnormal centrosomes. A typical consequence of these abnormalities is chromosomal instability and aneuploidy. The missegregation of chromosomes to the daughter cells was documented by scoring kinetochore positive micronuclei (Borojerdi et al. 2009).

An increase in the presence of aberrant cells, including multipolar metaphases/anaphases and cells bearing lagging chromosomes was observed in correlation with increasing doses of Zidovudine in CHO cells. When compared to untreated CHO cells, Zidovudine-exposed cells showed an increase in the number of pericentrin positive signals. In normal human mammary epithelial cells (NHMECs), pericentrin positive signals were scored for the total number of cells with visible centrosomes, demonstrating a dose increase in pericentrin positive signals in treated cells.

Multipolar spindles and multiple centrosomal bodies were observed in Zidovudine-exposed NHMEC and were identified by localization of Aurora A-positive signals in NHMECs exposed cells. Western blot analysis of NHMEC lysates confirmed an increase in protein expression of Aurora A in exposed cells.

Additionally, Zidovudine genotoxicity was documented by the presence of large bodies still attached to the nucleus containing kinetochore positive signals. These types of lesions known as nuclear buds were further studied and characterized in detail by Dutra et al. (2010).

The most current treatment for HIV-1 is combination therapy, or HAART (Highly Active Antiretroviral Therapy), which typically consists of at least two NRTIs and a protease inhibitor. Zidovudine is frequently used in combination with other NRTIs such as 2',3'-dideoxy-3'-thiacytidine (3TC) and 2', 3'-dideoxyinosine (ddI). Centrosome amplification caused by 3TC, d4T, and ddI was carried out to determine if it was a common phenomenon to NRTIs since these drugs are often used together in combination. CHO cells and two NHMECs, a high and a low incorporator of Zidovudine into DNA, were used by Yu et al. (2009). Hamster and human cells exposed to any of the three NRTIs for 24 h exhibited amplification of centrosomes and human cells also exhibited disruptions in tubulin distribution, revealing that centrosome amplification as well as tubulin disruption is a common phenomenon induced by several different NRTIs.

Increase of centrosome amplification induced by thymidine was no statistically different than the controls for these strains indicating that an effect independent of nucleotide pool imbalance was taking place.

Since not all centrosomes are equal in terms of their microtubule nucleation activity, there could be multiple scenarios for amplified centrosomes (Ghadimi et al. 2000). Active centrosomes can be defined as those able to nucleate microtubules. These centrosomes will go on to form spindle poles during mitosis. Some aberrant centrosomes may be inactive and therefore unable to nucleate microtubules and form spindle poles. To determine if extranumerary centrosomes induced by NRTIs are active, microtubule nucleation of NRTI-treated cells was interrupted with nocodazole, a disruptor of tubulin polymerization. Once nocodazole was rinsed off, active centrosomes began to form microtubule asters. These experiments revealed that multiple centrosomes induced by NRTIs kept the ability to nucleate tubulin and form asters. These cells have the potential to form multipolar spindles during mitosis that lead to missegregation of chromosomes and inevitably, aneuploidy.

The origin of the new centrosomes has been a topic of interest for many investigators. To understand if the new structures have their origin as a consequence of overduplication due to cell cycle alterations or if they are just amplified independent of the cell cycle, Davila et al. (2009) measured centrosomal amplification in human hTERT transformed MCF10A cells, by immunocytochemistry with antibodies to centrosomal proteins pericentrin and Cep 170. Furthermore, the origin of the extra centrosomes was analyzed by addressing the maturity of the newly formed structures. Typically mature centrosomes, identified by the presence of Cep170 proteins, are the result of overduplication of pre-existing centrosomes. Conversely, it is believed that, Cep170 negative, immature centrosomes are the result of accumulation of the organelles by impaired cytokinesis. Scoring of Cep170-stained Zidovudine-induced centrosomes revealed that 40, 50, and 53 % of the supernumerary centrosomes in cells exposed to Zidovudine, respectively, were mature, compared to 22 % of unexposed cells centrosomes in the control. Therefore, Zidovudine-induced centrosomal amplification is the result of both overduplication and cytokinesis impairment (Davila et al. 2009).

References

Avramis VI, Markson W, Jackson RL, Gomperts E (1989) Biochemical pharmacology of zidovudine in human T-lymphoblastoid cells (CEM). AIDS 3:417–422

Agarwal RP, Olivero OA (1997) Genotoxicity and mitochondrial damage in human lymphocytic cells chronically exposed to 3'-azido-2',3'-dideoxythymidine (AZT). Mutat Res 390:223–231

Borojerdi JP, Ming J, Cooch C, Ward Y, Semino-Mora C, Yu M, Braun HM, Taylor BJ, Poirier MC, Olivero OA (2009) Centrosomal amplification and aneuploidy induced by the antiretroviral drug AZT in hamster and human cells. Mutat Res 665:67–74

Carroll PE, Okuda M, Horn HF, Biddinger P, Stambrook PJ, Gleich LL, Li YQ, Tarapore P, Fukasawa K (1999) Centrosome hyperamplification in human cancer: chromosome instability induced by p53 mutation and/or Mdm2 overexpression. Oncogene 18:1935–1944

Chandrasekaran B, Kute TE, Duch DS (1995) Synchronization of cells in the S phase of the cell cycle by 3'-azido- 3'-deoxythymidine: implications for cell cytotoxicity. Cancer Chemother Pharmacol 35:489–495

St. Clair MH, Richards CA, Spector T, Weinhold KJ, Miller WH, Langlois AJ, Furman PA (1987) 3'-Azido-3'-deoxythymidine triphosphate as an inhibitor and substrate of purified human immunodeficiency virus reverse transcriptase. Antimicrob Agents Chemother 31:1972–1977

Copeland WC, Chen MS, Wang TS (1992) Human DNA polymerases alpha and beta are able to incorporate anti-HIV deoxynucleotides into DNA. J Biol Chem 267:21459–21464

D'Assoro AB, Lingle WL, Salisbury JL (2002) Centrosome amplification and the development of cancer. Oncogene 21:6146–6153

Davila K, Yu M, Poirier MC, Olivero OA (2009) Centrosomal amplification induced by nucleoside reverse transcriptase inhibitors (NRTIs) in cultured human breast epithelial MCF 10A cells. Environ Mol Mutagen 50:546

Diwan BA, Riggs CW, Logsdon D, Haines DC, Olivero OA, Rice JM, Yuspa SH, Poirier MC, Anderson LM (1999) Multiorgan transplacental and neonatal carcinogenicity of 3'-azido-3'-deoxythymidine in mice. Toxicol Appl Pharmacol 15:82–99

Doxsey S, Zimmerman W, Mikule K (2005) Centrosome control of the cell cycle. Trends Cell Biol 15:303–311

Dutra A, Pak E, Wincovitch S, John K, Poirier MC, Olivero OA (2010) Nuclear bud formation: a novel manifestation of Zidovudine genotoxicity. Cytogenet Genome Res 128:105–110

Escobar PA, Olivero OA, Wade NA, Abrams EJ, Nesel CJ, Ness RB, Day RD, Day BW, Meng Q, O'Neill JP, Walker DM, Poirier MC, Walker VE, Bigbee WL (2007) Genotoxicity assessed by the comet and GPA assays following in vitro exposure of human lymphoblastoid cells (H9) or perinatal exposure of mother-child pairs to AZT or AZT-3TC. Environ Mol Mutagen 48:330–343

Fang JL, Beland FA (2000) Development of a novel (32)P-postlabeling method for the analysis of 3'-azido-3'-deoxythymidine. Cancer Lett 153:25–33

Fischl MA, Richman DD, Grieco MH, Gottlieb MS, Volberding PA, Laskin OL, Leedom JM, Groopman JE, Mildvan D, Schooley RT (1987) The efficacy of azidothymidine (AZT) in the treatment of patients with AIDS and AIDS-related complex. A double-blind, placebo-controlled trial. N Engl J Med 317:185–191

Furman PA, Fyfe JA, Clair MH, Weinhold K, Rideout JL, Freeman GA, Lehrman SN, Bolognesi DP, Broder S, Mitsuya H, Barry DA (1986) Phosphorylation of 3′-azido-3′-deoxythymidine and selective interaction of the 5′-triphosphate with human immunodeficiency virus reverse transcriptase. Proc Natl Acad Sci U S A 83:8333–8337

Heagy W, Crumpacker C, Lopez PA, Finberg RW (1991) Inhibition of immune functions by antiviral drugs. J Clin Invest 87:1916–1924

Ghadimi BM, Sackett DL, Difilippantonio MJ, Schrock E, Neumann T, Jauho A, Auer G, Ried T (2000) Centrosome amplification and instability occurs exclusively in aneuploid, but not in diploid colorectal cancer cell lines, and correlates with numerical chromosomal aberrations. Genes Chromosom Cancer 27:183–190

Hinchcliffe EH, Miller FJ, Cham M, Khodjakov A, Sluder G (2001) Requirement of a centrosomal activity for cell cycle progression through G1 into S phase. Science 291:1547–1550

Huang P, Farquhar D, Plunkett W (1990) Selective action of 3'-azido-3'-deoxythymidine 5'-triphosphate on viral reverse transcriptases and human DNA polymerases. J Biol Chem 265:11914–11918

IARC (2000) Monographs on the evaluation of carcinogenic risks to humans. Some antiviral and antineoplastic drugs, and other pharmaceutical agents. In: World health organization, international agency for research on cancer, vol 76. Lyon, France, pp 73–127, 1–521

Johnson AA, Ray AS, Hanes J, Suo Z, Colacino JM, Anderson KS, Johnson KA (2001) Toxicity of antiviral nucleoside analogs and the human mitochondrial DNA polymerase. J Biol Chem 276:40847–40857

Lee H, Hanes J, Johnson KA (2003) Toxicity of nucleoside analogues used to treat AIDS and the selectivity of the mitochondrial DNA polymerase. Biochemistry 42:14711–14719

Lengauer C, Kinzler KW, Vogelstein B (1997) Genetic instability in colorectal cancers. Nature 386:623–627

Lim SE, Copeland WC (2001) Differential incorporation and removal of antiviral deoxynucleotides by human DNA polymerase gamma. J Biol Chem 276:23616–23623

Lingle WL, Lutz WH, Ingle JN, Maihle NJ, Salisbury JL (1998) Centrosome hypertrophy in human breast tumors: implications for genomic stability and cell polarity. Proc Natl Acad Sci U S A 95:2950–2955

Mikule K, Delaval B, Kaldis P, Jurcyzk A, Hergert P, Doxsey S (2007) Loss of centrosome integrity induces p38-p53-p21-dependent G1-S arrest. Nat Cell Biol 9:160–170

Olivero OA (2007) Mechanisms of genotoxicity of nucleoside reverse transcriptase inhibitors. Environ Mol Mutagen 48:215–223

Olivero OA, Beland FA, Poirier MC (1994) Immunofluorescent localization and quantitation of 3′-azido-2′, 3′-dideoxythymidine (AZT) incorporated into chromosomal DNA of human, hamster and mouse cell lines. Int J Oncol 4:49–54

Olivero OA, Anderson LM, Diwan BA, Haines DC, Harbaugh SW, Moskal TJ, Jones AB, Rice JM, Riggs CW, Logsdon D, Yuspa SH, Poirier MC (1997) Transplacental effects of 3′-azido-2′,3′-dideoxythymidine (AZT): tumorigenicity in mice and genotoxicity in mice and monkeys. J Natl Cancer Inst 89:1602–1608

Olivero OA, Tejera AM, Fernandez JJ, Taylor BJ, Das S, Divi RL, Poirier MC (2005) Zidovudine induces S-phase arrest and cell cycle gene expression changes in human cells. Mutagenesis 20:139–146

Olivero OA, Ming JM, Das S, Vazquez IL, Richardson DL, Weston A, Poirier MC (2008) Human inter-individual variability in metabolism and genotoxic response to zidovudine. Toxicol Appl Pharmacol 228:158–164

Parker WB, White EL, Shaddix SC, Ross LJ, Buckheit RW Jr, Germany JM, Secrist JA III, Vince R, Shannon WM (1991) Mechanism of inhibition of human immunodeficiency virus type 1 reverse transcriptase and human DNA polymerases alpha, beta, and gamma by the 5′-triphosphates of carbovir, 3′-azido-3′-deoxythymidine, 2′,3′-dideoxyguanosine and 3′-deoxythymidine. A novel RNA template for the evaluation of antiretroviral drugs. J Biol Chem 266:1754–1762

Pihan GA, Purohit A, Wallace J, Knecht H, Woda B, Quesenberry P, Doxsey SJ (1998) Centrosome defects and genetic instability in malignant tumors. Cancer Res 58:3974–3985

Richman DD, Fischl MA, Grieco MH, Gottlieb MS, Volberding PA, Laskin OL, Leedom JM, Groopman JE, Mildvan D, Hirsch MS (1987) The toxicity of azidothymidine (AZT) in the treatment of patients with AIDS and AIDS-related complex. A double-blind, placebo-controlled trial. N Engl J Med 317:192–197

Roskrow M, Wickramasinghe SN (1990) Acute effects of 3′-azido-3′-deoxythymidine on the cell cycle of HL60 cells. Clin Lab Haematol 12:177–184

Satoh MS, Lindahl T (1994) Enzymatic repair of oxidative DNA damage. Cancer Res 54:1899s–1901s

Skalski V, Liu SH, Cheng YC (1995) Removal of anti-human immunodeficiency virus 2′,3′-dideoxynucleoside monophosphates from DNA by a novel human cytosolic 3′ → 5′ exonuclease. Biochem Pharmacol 50:815–821

Sommadossi JP, Carlisle R, Zhou Z (1989) Cellular pharmacology of 3′-azido-3′-deoxythymidine with evidence of incorporation into DNA of human bone marrow cells. Mol Pharmacol 36:9–14

Vazquez-Padua MA, Starnes MC, Cheng YC (1990) Incorporation of 3′-azido-3′-deoxythymidine into cellular DNA and its removal in a human leukemic cell line. Cancer Commun 2:55–62

Viora M, Di Genova G, Rivabene R, Malorni W, Fattorossi A (1997) Interference with cell cycle progression and induction of apoptosis by dideoxynucleoside analogs. Int J Immunopharmacol 19:311–321

Wu YW, Xiao Q, Jiang YY, Fu H, Ju Y, Zhao YF (2004) Synthesis, in vitro anticancer evaluation, and interference with cell cycle progression of N-phosphoamino acid esters of zidovudine and stavudine. Nucleosides, Nucleotides Nucleic Acids 23:1797–1811

Yu M, Ward Y, Poirier MC, Olivero OA (2009) Centrosome amplification induced by the antiretroviral nucleoside reverse transcriptase inhibitors lamivudine, stavudine, and didanosine. Environ Mol Mutagen 50:718–724

Chapter 17
Mechanisms and Consequences of Centrosome Clustering in Cancer Cells

Alwin Krämer, Simon Anderhub and Bettina Maier

Abstract Ever since initially proposed by Theodor Boveri in 1914, centrosome abnormalities have been accused to be involved in the induction of chromosomal instability and tumorigenesis. New evidence especially on a mechanism termed centrosomal clustering now again supports Boveri's idea and adds fuel to the old debate on a mechanistic link between supernumerary centrosomes and malignant transformation. On top, inhibiting centrosome clustering might well turn out to be one of the long sought after possibilities to specifically interfere with tumor cells while leaving healthy tissues untouched.

17.1 Introduction

Centrosomes are organelles that function as microtubule-organizing centers in most animal cells. Besides controlling microtubule-associated processes like cell shape and intracellular transport, they are of crucial importance for the assembly of the mitotic spindle and subsequent cell division. Centrosomes consist of two orthogonally arranged barrel-shaped centrioles which are embedded in pericentriolar material (PCM, Fig. 17.1). Centrioles themselves are composed of a central cartwheel structure surrounded by nine microtubule (MT) triplets which are oriented anti-clockwise when observed from the proximal end (Fig. 17.1) (Uzbekov and Prigent 2007). The PCM contains proteins required for MT anchorage and nucleation including γ-tubulin and pericentrin (Gould and Borisy 1977 ; Doxsey

---

A. Krämer (✉) · S. Anderhub · B. Maier
Clinical Cooperation Unit Molecular Hematology/Oncology, German Cancer Research Center and Department of Internal Medicine University of Heidelberg, Heidelberg, Germany
e-mail: a.kraemer@dkfz.de

Fig. 17.1 Centrosome structure. Centrosomes are composed of two centrioles. Each centriole is cylindrical in structure and is made up by nine microtubule triplets. The older centriole, termed mother centriole, carries distal (*purple*) as well as subdistal (*cyan*) appendages where microtubule (MT) anchorage and nucleation takes place. Both centrioles of a centrosome are embedded in an amorphous protein mass, termed pericentriolar material (PCM; *blue*) and are tethered together in most cell cycle phases (Adopted from Anderhub et al. 2012)

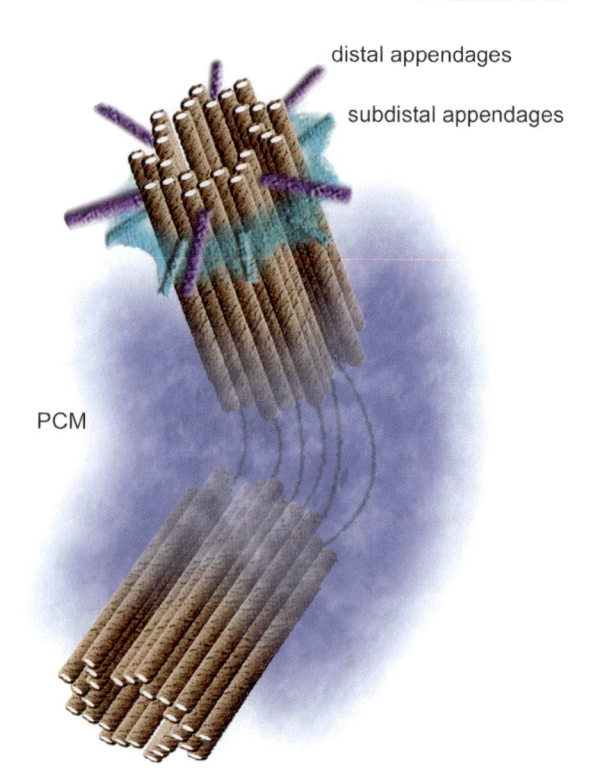

et al. 1994; Stearns et al. 1991). One centriole of each centrosome, termed the mother centriole, features distal as well as subdistal appendages which are essential for MT anchorage and nucleation (Bornens 2002). In addition, cilia formation is initiated by the mother centriole and various diseases are directly linked to defects in ciliogenesis (Nigg and Raff 2009).

Cells in the G_1-phase of the cell cycle harbor a single centrosome consisting of two loosely connected centrioles. New centrioles are formed at the proximal part of each of the two pre-existing centrioles upon transition from G_1- to S-phase. These newly formed centrioles elongate and mature during S- until late G_2-phase, giving rise to two pairs of centrioles per cell before these enter mitosis. In early mitosis, the two centrosomes are separated in order to build up a bipolar mitotic spindle array, thereby ensuring proper segregation of duplicated chromosomes. In late mitosis or early G_1-phase the tight connection between the two centrioles of each centrosome is loosened, a process termed disengagement (Kuriyama and Borisy 1981). Centrosome duplication, like DNA replication, is regulated precisely to ensure that centrosomes are duplicated only once per cell cycle. Recently, it has been shown that the formation of new procentrioles is blocked intrinsically by the engagement of centrioles. Interestingly, the activity of separase, a protein known to be important for separation of duplicated chromatids during mitosis, is also required for centriole disengagement (Tsou and Stearns 2006). More precisely, a recent study revealed that cohesin is the glue that tethers centrioles—similar to sister chromatids—together and separase, via cleavage of cohesin, is required for centriole disengagement (Schockel et al. 2011). In addition, this study also revealed a role of shugoshin, which is known to prevent premature separation of newly replicated chromatids, in the protection of centrioles against precocious disengagement (Schockel et al. 2011). Furthermore, both centrosome duplication and DNA replication are spatially and timely tightly linked to the cell cycle. Activation of cell cycle regulators like cyclin-dependent kinases (CDK) in distinct phases and distinct sites controls correct cell cycle progression as well as centrosome duplication. For instance, DNA replication and centrosome duplication are assumed to be coupled by the activity of centrosomally localized CDK2 at the G_1/S-boundary (Ferguson and Maller 2010). Similarly, entry into mitosis is dependent on active CDK1 localized at centrosomes (Jackman et al. 2003), which has recently been shown to be mediated by the centrosomal protein Cep63. Hence, depletion of Cep63 in human cells leads to polyploid cells due to mitotic skipping (Löffler et al. 2011). Activation of CDK1 at centrosomes also depends on other kinases, namely Polo-like kinase 1 (PLK1) and Aurora-A, both of which are also essential for cell cycle progression and centrosome duplication (Barr et al. 2004; Vader and Lens 2008). Hence, centrosomes and chromosomes are, at least partially, orchestrated by the same proteins to avoid untimely replication and/or separation which might lead to cells with abnormal DNA and/or centrosome content.

At the molecular level, the formation of new centrioles at the base of pre-existing centrioles is best studied in Caenorhabditis elegans (*C. elegans*). Procentriole formation is initiated by the recruitment of SPD-2 to the base of the existing centriole where it is required for the correct localization of the kinase ZYG-1. ZYG-1 in return recruits SAS-5, a protein which shuttles between the cytoplasm and the centrosome, to the newly forming centriole. Upon interaction with SAS-5, a structural protein of the central tube, SAS-6 localizes to the centriole. Following the formation of the central tube, SAS-4, which is required for centriole MT assembly at the periphery of the tube, is recruited to the centriole (Strnad and Gönczy 2008). Homologs in humans have been identified for SPD-2, ZYG-1, SAS-6, and SAS-4, termed CEP192 (Andersen et al. 2003), PLK4 (Bettencourt-Dias et al. 2005; Habedanck et al. 2005), HsSAS-6 (Leidel et al. 2005), and CPAP/CENPJ (Hung et al. 2004), respectively. Due to domain homologies between SAS-5, Ana2, the Drosophila homolog of SAS-5, and SIL/STIL, it was proposed that SIL/STIL might be the human ortholog of SAS-5 (Stevens et al. 2010). Indeed, most recently data have been published by several groups demonstrating that SIL/STIL is required for centriole replication in mammalian cells and might truly be the functional ortholog of SAS-5 in humans (Tang et al. 2011; Kitagawa et al. 2011; Vulprecht et al. 2012; Arquint et al. 2012). In mammals, centriole duplication is initiated by PLK4 activity at the pre-existing centriole and is followed by sequential recruitment of SIL/STIL, HsSAS-6, CPAP, CEP135, CEP110, and γ-tubulin to form the base of the procentriole, subsequently elongate the procentriole and to eventually nucleate MTs (Tang et al. 2011; Kitagawa et al. 2011; Vulprecht et al. 2012; Arquint et al. 2012; Kleylein-Sohn et al. 2007). Recently, the PCM protein CEP152 was attributed a function in the recruitment of

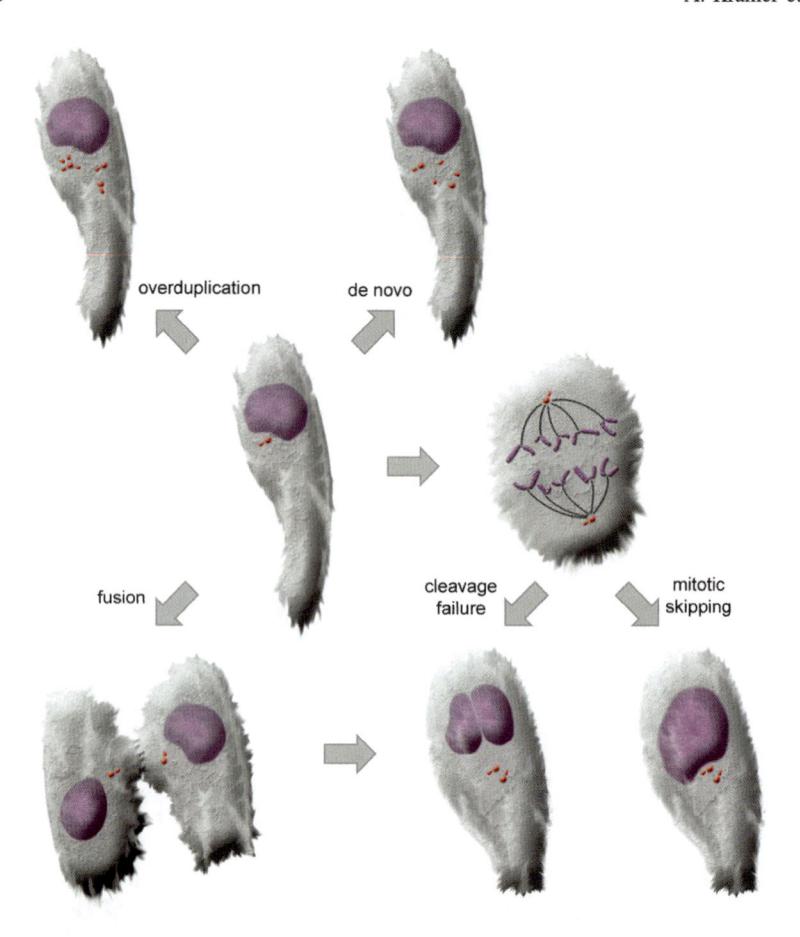

Fig. 17.2 Mechanisms leading to supernumerary centrioles/centrosomes. Centriole overduplication, and de novo formation results in cells with an undefined number of extra centrioles, whereas cell fusion, mitotic skipping as well as cleavage failure not only lead to a doubled centrosome number but also to a duplicated DNA content. In contrast to overduplication, de novo formation, mitotic skipping, and cleavage failure, cell fusion is not restricted to a defined cell cycle phase. Both overduplication and de novo centriole formation are assumed to take place during S-phase. Cleavage failure and mitotic skipping occur in G_2/M, whereas the first event follows anaphase, the latter is not limited to a specific mitotic phase (Adopted from Anderhub et al. 2012)

PLK4 and CPAP to centrioles and thus in the initiation of centriole duplication in human cells (Cizmecioglu et al. 2010; Dzhindzhev et al. 2010; Hatch et al. 2010)

Abnormalities of diverse tumor suppressors and oncogenes can cause centrosome amplification (Fukasawa 2007), which occurs through centrosome over-duplication during interphase, de novo synthesis of centrosomes or cytokinesis failure (Nigg 2002) (Fig. 17.2). Centrosome amplification is frequent in cancer, and is linked to tumorigenesis and aneuploidy (Nigg 2002; Lingle et al. 1998; Pihan et al. 1998; Neben et al. 2003; Krämer et al. 2003; Koutsami et al. 2006). The extent of centrosomal aberrations correlates with aneuploidy, defined as a state of cells with abnormal numbers of chromosomes, as well as with chromosomal instability (CIN), which describes the rate of chromosome gains and losses as well as structural chromosome aberrations in cell populations, and malignant behavior in tumor cell lines, mouse tumor models, and human tumors (Nigg 2002; Lingle et al. 1998; Neben et al. 2003; Krämer et al. 2003; Koutsami et al. 2006; Levine et al. 1991; Pihan et al. 2001)

CIN is a feature commonly observed in human cancers and was first identified almost a century ago (von Hansemann 1890). Only recently, using elaborate mouse models, it became clear that CIN does not only correlate with but probably plays a causative part in a substantial proportion of malignancies (Weaver et al. 2007; Sotillo et al. 2007; Baker et al. 2009). One major cause of CIN is mitotic checkpoint overactivation (Schvartzman et al. 2010). Here, we focus on recent discoveries related to another phenomenon intricately linked to CIN, centrosomal clustering, its emerging mechanistic basis, significance for cancer cell survival, tumor progression, role in asymmetric divisions of stem cells, and as a potential tumor-selective therapeutic target.

17.2 Centrosome Clustering and Chromosomal Instability

In mitosis, supernumerary centrosomes can form multipolar spindles, which occur in many tumor types and have long been accused of contributing to CIN and tumorigenesis (Nigg 2002; Krämer et al. 2002; Boveri 1929). However, recent findings show that multipolar divisions and the resulting CIN undermine cell viability, frequently leading to cell death (Weaver et al. 2007; Ganem et al. 2009; Brinkley 2001; Kops et al. 2004). To avoid cell death, many cancer cells cluster supernumerary centrosomes into two spindle poles thereby enabling bipolar division. The earliest observation on centrosomal clustering came from immunofluorescence studies of N115 mouse neuroblastoma cells which contain large numbers of centrioles, and yet undergo mostly bipolar divisions, with often unequal numbers of centrioles coalescing at the two spindle poles (Ring et al. 1982). Forgotten for many years, Salisbury and colleagues rediscovered the phenomenon after having noted the low frequency of abnormal mitoses despite the presence of supernumerary centrosomes in human breast cancer samples (Lingle and Salisbury 1999). Whether or not centrosomal clustering was coupled with reduced CIN was not examined at that time. Indeed, the concept that has crystallized since these pioneering studies is that centrosomal clustering enables cells to successfully divide despite the presence of supernumerary centrosomes (Nigg 2002; Brinkley 2001).

The initial description of centrosome clustering noted that cells with supernumerary centrosomes pass through a transient multipolar spindle intermediate before centrosome clustering and bipolar anaphase occurs (Ring et al. 1982). Excitingly, recent data from several laboratories demonstrate that, while passing through the transient multipolar state, merotelic kinetochore MT attachment errors, defined by the persistent

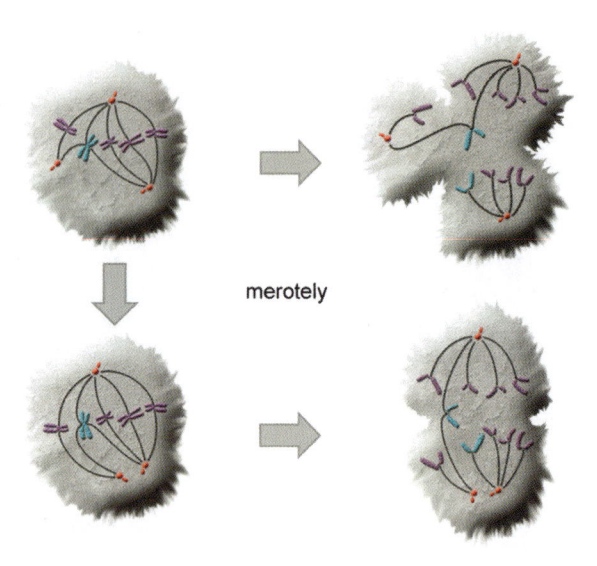

Fig. 17.3 Potential consequences of extra centrosomes. Cells with amplified centrosomes can either divide in a multipolar fashion or cluster their supernumerary centrosomes into two spindle poles. Whereas multipolar division is detrimental for cell viability likely due to gross aneuploidy, centrosome clustering may serve as a survival mechanism to compensate for centrosome amplification. Improper MT-kinetochore attachments like merotely are proposed to be enriched in cells with extra centrosomes and give rise to aneuploid progeny upon centrosomal clustering. (Adopted from Anderhub et al. 2012)

attachment of MT from both spindle poles to a single kinetochore, accumulate, and consequently increase the frequency of lagging chromosomes during bipolar anaphase after centrosomal clustering (Ganem et al. 2009; Silkworth et al. 2009) (Fig. 17.3). Importantly, this finding implies that cells with amplified centrosomes do not necessarily need to divide in a multipolar fashion to allow low-level chromosomal missegregation that can fuel tumor progression. Such interpretation also supports the emerging bimodal relationship between aneuploidy and tumorigenesis (Weaver et al. 2007): whereas moderate CIN induced tumorigenesis in mice, high-level CIN suppressed tumor formation in vivo. Suppression of tumor cell growth in this context seems to be brought about by apoptosis induction due to loss of chromosomes encoding genes required for maintenance of cell viability (Kops et al. 2004). Likewise, in patients with breast, ovarian, gastric and non-small cell lung cancer, extreme CIN is associated with improved prognosis relative to tumors with intermediate CIN levels (Birkbak et al. 2011).

As already mentioned above, several studies have recently provided substantial evidence for a causative role of CIN in malignant transformation. Similarly, the key question of whether supernumerary centrosomes are simply a passenger phenotype or can induce malignancy has now been addressed by constructing flies that overexpress SAK (also known as polo-like kinase 4 (PLK4)), a kinase important for centriole replication (Basto et al. 2008). Flies overexpressing SAK contain extra centrosomes in about 60 % of their somatic cells. Although many of the fly cells with supernumerary centrosomes initially form multipolar spindles, they ultimately cluster into bipolar arrays, resulting in only slightly increased CIN levels. Nevertheless, larval brain cells of these animals can generate metastatic tumors when transplanted into the abdomens of wild-type hosts (Basto et al. 2008).

Similar to flies, mice overexpressing the centrosomal protein ninein-like (NINL) show centrosome amplification as detected in mouse embryonic fibroblasts from the transgenic animals and develop tumors of breast, ovary, and testicles at 10–15 months of age (Shao et al. 2010). Whether or not extra centrosomes are clustered into bipolar mitoses in this system was not examined. Although NINL overexpression will certainly have additional effects other than centrosome amplification these data nevertheless indicate a role of supernumerary centrosomes in tumorigenesis in mammals as well.

The ability to cluster supernumerary centrosomes into a bipolar mitotic spindle array is not a specific trait of tumor cells. For example, during physiological hepatocyte polyploidization, primary binuclear hepatocytes—which naturally contain four centrosomes in the G_2 phase—efficiently cluster pairs of centrosomes at opposite spindle poles, leading to the generation of mononuclear 4n progeny (Guidotti et al. 2003). Recent data confirm that polyploid mouse hepatocytes in most cases reorganize their spindles into a bipolar mitotic array from an intermediate multipolar state. This process was associated with lagging chromosomes in 25–50 % of tetraploid hepatocytes undergoing bipolar anaphase and resulted in a high rate of aneuploidy (Duncan et al. 2010). Interestingly, however, a small percentage of tetraploid mouse hepatocytes underwent successful tripolar divisions, producing viable offspring. Moreover, during liver regeneration in mice, which is associated with excessive polyploidization, about 20 % of hepatocytes missegregate one or more chromosomes at each mitosis (Putkey et al. 2002), possibly also a consequence of centrosomal clustering. Furthermore, several studies showed that both non-transformed *Drosophila melanogaster (D. melanogaster)* neuroblasts and diverse types of human cells that have been manipulated to contain supernumerary centrosomes by either PLK4 overexpression or treatment with cytochalasin D to inhibit cytokinesis, can cluster multiple centrosomes into a bipolar spindle array both in vitro and in vivo (Ganem et al. 2009; Basto et al. 2008; Quintyne et al. 2005; Kwon et al. 2008; Yang et al. 2008).

Collectively, it seems that not only cancer cells but also non-transformed cell types can cluster supernumerary centrosomes into bipolar mitotic spindles. Initial evidence implicates supernumerary centrosomes and centrosomal clustering in tumorigenesis in flies. However, data generated in mouse hepatocytes show that neither centrosome amplification nor centrosome clustering or multipolar cell division with subsequent aneuploidy necessarily leads to malignant transformation in mammals. From those data it can also be concluded that multipolar divisions are not universally lethal. However, as these experiments have been performed using polyploid hepatocytes, surviving multipolar divisions might well be a peculiarity of polyploid cells which better tolerate the loss of multiple chromosomes.

It appears that more insights into the molecular and cellular basis of centrosomal clustering are needed. This topic, along with data on deregulation of the genes and proteins mechanistically involved in centrosomal clustering in diverse malignancies, are discussed in the next section.

17.3 Mechanisms of Centrosome Clustering

Mechanistically, multiple cellular systems are involved in the clustering of supernumerary centrosomes in normal and tumor cells. Three recent studies show that extra centrosomes activate a MAD2-dependent delay of anaphase onset in different cell types, which is required for centrosomal clustering and suppression of multipolar mitosis (Basto et al. 2008; Kwon et al. 2008; Yang et al. 2008). MAD2 is a central component of the spindle assembly checkpoint (SAC) that blocks disjunction of sister chromatids at metaphase until MT attachment at kinetochores is complete and spindle tension is established (Weaver et al. 2007). From those results it can be assumed that, although the SAC does not recognize abnormal spindles per se (Sluder et al. 1997), multipolarity is accompanied by improper kinetochore attachment or insufficient tension and thereby activates the SAC and leads to metaphase arrest. In line with this interpretation an RNA interference (RNAi) screen in *D. melanogaster* S2 cells suggested that knockdown of components of the actin cytoskeleton and actin-dependent cortical force generators including the formin FORM3/INF2, the myosin MYO10, the MT plus-end-tracking protein CLIP190 as well as several cell-matrix adhesion molecules (Turtle, Echinoid, CAD96CA, CG33171, FIT1) induces spindle multipolarity through interference with the interphase cell adhesion pattern (Kwon et al. 2008). When cells round up during mitosis, retraction fibers (actin-rich structures linked to the sites of former adhesion during interphase) remain attached to the extracellular substrate and promote interaction of astral spindle MTs with the cell cortex (Thery et al. 2005). Disturbance of the connection between cell-matrix adhesion proteins and the actin cytoskeleton on the one hand and spindle MT components on the other hand might therefore cause reduced spindle tension, thereby inhibiting centrosome clustering. Indeed, in elegant experiments it has been shown that O- and Y-shaped fibronectin-coated micropatterns, allowing for multidirectional distribution of retraction fiber formation, caused increased frequencies of multipolar spindles (Kwon et al. 2008). On the other hand, bipolar arrangements of adhesive contacts induced by H-shaped micropatterns promoted bipolar mitoses.

As indicated above, spindle tension is necessary for clustering of supernumerary centrosomes into a bipolar mitotic spindle array (Fig. 17.4) (Kwon et al. 2008; Leber et al. 2010). Before chromosome segregation, kinetochores of sister chromatids attach to MTs of opposite spindle poles. This configuration is achieved through a trial-and-error process in which correct attachments exert tension across the centromere, which stabilizes kinetochore-MT interactions. Incorrect attachments

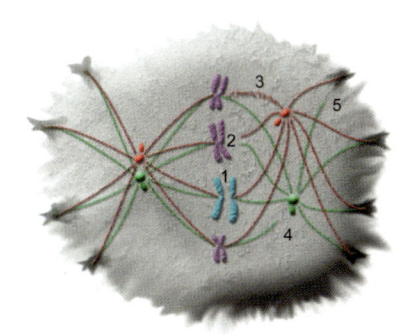

Fig. 17.4 Spindle tension is required for centrosomal clustering. Model of centrosomal clustering (*left* spindle half) and mechanisms involved in its prevention via reduction of spindle tension (*right* spindle half). Centrosome clustering is brought about by microtubule tension-dependent uniform positioning of individual centrosomes resulting in the formation of two spindle poles. Spindle tension can be disrupted by reduction of chromatid cohesion (1) (Uzbekov and Prigent 2007), disturbed microtubule–kinetochore attachment (2) (Gould and Borisy 1977), reduced microtubule generation (3) (Doxsey et al. 1994), disturbed microtubule bundling (4) (Stearns et al. 1991), or interference with the interphase cell adhesion pattern by disruption of components of the actin cytoskeleton (5) (Bornens 2002)

exert less tension and are destabilized, providing chromosomes a new opportunity to bi-orient (Liu et al. 2009). The mitotic kinase aurora B, the enzymatically active component of the chromosomal passenger complex (CPC), localizes to the inner centromere between sister kinetochores, and regulates chromosome-spindle attachments by phosphorylating kinetochore substrates, including the NDC80 MT-binding complex (Liu et al. 2009; Ruchaud et al. 2007; Wei et al. 2007). The CPC composed of aurora B and its regulatory subunits INCENP, survivin, and borealin are a key regulator of chromosome segregation and cytokinesis. Since tension across centromeres widens spatial partition of the CPC and thereby separates aurora B from its kinetochore substrates, substrate phosphorylation is reduced resulting in stabilized MT–kinetochore interactions (Liu et al. 2009).

Using genome-wide RNAi screening in human cancer cells with extra centrosomes both NDC80 complex and CPC components were found to be involved in centrosomal clustering (Leber et al. 2010). In addition, shugoshin-like 1 (SGOL1), a protein previously known to be involved in sensing spindle tension at budding yeast kinetochores (Indjeian et al. 2005), is necessary for centrosome clustering. Importantly, SGOL1 also contributes to the recruitment of the CPC to centromeres (Boyarchuk et al. 2007; Kawashima et al. 2007; Vanoosthuyse et al. 2007) while itself is loaded onto histone H2A after histone phosphorylation by BUB1 in yeast and human cells (Kawashima et al. 2010; Yamagishi et al. 2010). Fittingly, BUB1 knockdown does cause centrosome declustering as well (Sluder et al. 1997). Another recently identified centromeric recruitment factor for the CPC is haspin (Wang et al. 2010; Kelly et al. 2010). Haspin phosphorylates histone H3, thereby creating a docking site for survivin in both Xenopus and human cells. Interestingly, depletion of haspin leads to the generation of multiple spindle poles and disruption of mitotic spindle structure in U2OS and HeLa cells as a consequence of acentriolar pole formation and centriole disengagement (Dai et al. 2009).

CENPA (Hori et al. 2008), the centromere-specific histone H3 variant, CENPT (Hori et al. 2008), a component of the linker structure connecting the centromere with outer kinetochore components, sororin (also known as cell division cycle associated 5 (CDCA5)) (Schmitz et al. 2007), a protein involved in sister chromatid cohesion, and the augmin complex (Lawo et al. 2009), which promotes microtubule-dependent MT amplification within the mitotic spindle, are necessary for centrosome clustering as well (Kwon et al. 2008; Leber et al. 2010). Similar to haspin depletion, knockdown of augmin complex components leads to the formation of acentriolar spindle poles and centrosome fragmentation in addition to centrosomal declustering (Leber et al. 2010; Lawo et al. 2009; Uehara et al. 2009; Einarson et al. 2004; Wu et al. 2008). Most recently, hepatoma upregulated protein (HURP) has been shown to be required for centrosome clustering in cells with supernumerary centrosomes as well (Breuer et al. 2010). This observation is noteworthy as HURP serves as an attachment- and tension-sensitive kinetochore MT stabilizing factor during mitosis (Koffa et al. 2006; Sillje et al. 2006; Wong and Fang 2006).

Together, these findings support the notion that loss of centromere tension results in centrosome declustering. Indeed, when pulling forces are measured directly across multipolar spindles in cancer cells with supernumerary centrosomes, depletion of NDC80, CPC and augmin complexes or SGOL1 result in substantially reduced spindle tension, as indicated by shorter interkinetochore distances and BUBR1 labeling of kinetochores in multipolar metaphase cells (Leber et al. 2010; Uehara et al. 2009). Also, knockdown of haspin has been shown to reduce tension at sister kinetochores (Dai et al. 2009). However, these data also suggest that at least some of the proteins involved in the clustering of supernumerary centrosomes might contribute to centriole cohesion and bipolar spindle formation in cells with a regular centrosome content as well.

Several of the proteins of the chromosomal passenger and NDC80 complexes including aurora B, survivin, borealin, NUF2 and HEC1 as well as sororin and HURP have been found to be overexpressed in a wide variety of cancer types (Carmena and Earnshaw 2003; Bischoff et al. 1998; Adams et al. 2001; Altieri 2003; Chang et al. 2006; Hayama et al. 2006; Ferretti et al. 2010; Nguyen et al. 2010; Tsou et al. 2003). Furthermore, overexpression of highly expressed in cancer 1 (HEC1), a component of the NDC80 complex as well as aurora B have been implicated in tumor formation in mouse models (Diaz-Rodriguez et al. 2008; Nguyen et al. 2009). These findings have already led to the development of potent and selective inhibitors of aurora B kinase which are currently in early clinical trials in patients with different kinds of malignancies (Taylor and Peters 2008). Taken together, these data suggest that proteins involved in centrosome clustering in cancer cells with supernumerary centrosomes are frequently overexpressed in human cancers, suggesting that clustering of extra centrosomes into a bipolar spindle array might indeed be important for cancer cell survival and/or progression.

Several studies report that centrosome clustering also relies on MT-based motors and MT-bundling proteins that organize spindle poles in both normal and tumor cells with supernumerary centrosomes (Quintyne et al. 2005; Kwon et al. 2008; Leber et al. 2010). For example, depletion of the minus-end directed motor dynein causes declustering of centrosomes and subsequent spindle multipolarity in tumor cells as well as in non-transformed cells engineered to contain extra centrosomes (Quintyne et al. 2005; Leber et al. 2010). Mechanistically, in mitosis the dynein complex is responsible for targeting nuclear mitotic apparatus protein (NUMA) to spindle poles, where it focuses MT minus ends and tethers them to the centrosomes. However, whether delocalization of dynein from spindle MTs is responsible for the generation of multipolar spindles as initially suggested, remains controversial (Quintyne et al. 2005; Nguyen et al. 2008). Interestingly, in *D. melanogaster* S2 cells depletion of dynein does not substantially increase the frequency of multipolar mitoses, but another minus-end directed motor, non-claret disjunctional (NCD), seems to take over the role of dynein in suppressing multipolarity in fly cells (Kwon et al. 2008). In acentrosomal *D. melanogaster* oocytes NCD is necessary for efficient bundling of MTs at spindle poles. Also, mitotic centromere-associated kinesin KIF2C/MCAK, which functions as MT depolymerase and is believed to be a key component of the error correction mechanism at kinetochores, plays a role in centrosomal clustering in flies (Kwon et al. 2008). Interestingly, analogous to SGOL1 for the CPC, SGOL2 serves to recruit KIF2C/MCAK to the inner centromere (Huang et al. 2007).

Most recently integrin-linked kinase (ILK) has been shown to mediate centrosome clustering via transforming acidic coiled-coil (TACC3) and colonic and hepatic tumor overexpressed gene (ch-TOG), two centrosomal proteins involved in stabilization of MT minus ends at spindle poles (Fielding et al. 2011). In addition, depletion of protein regulator of cytokinesis 1 (PRC1), a MT-bundling protein with most prominent activity during central spindle formation that is also important for the establishment of kinetochore tension, leads to spindle multipolarity (Leber et al. 2010). These data are further evidence for the suggestion that the mechanisms responsible for holding supernumerary centrosomes together might be similar to the forces that bundle MTs into a bipolar spindle array in cells with two centrosomes or even without centrosomes.

Similar to CPC and NDC80 complex components, the majority of MT-based motors and MT-bundling proteins involved in centrosomal clustering including TACC3, ch-TOG, ILK, PRC1, KIF2C/MCAK, and the human NCD homolog KIFC1 (also known as HSET) are frequently overexpressed in different types of human cancer (Peset and Vernos 2008; Charrasse et al. 1995; Ishikawa et al. 2008; Nakamura et al. 2007; Hannigan et al. 2005; Shimo et al. 2007; Carter et al. 2006; De et al. 2009; Grinberg-Rashi et al. 2009). Expression levels of PRC1 were found to strongly correlate with aneuploidy levels, which themselves were associated with poor clinical outcome in several cancer types (Carter et al. 2006). In addition, overexpression of the kinesin KIFC1 has been shown to mediate resistance against docetaxel in breast cancer cells (De et al. 2009). Since taxanes induce spindle multipolarity at low concentrations (Chen and Horwitz 2002), high level expression of KIFC1 might counteract this effect and prevent taxane-treated cells from multipolarity-induced cell death by enabling bipolar spindle formation through centrosomal clustering.

17.4 Centrosomes and Asymmetric Stem Cell Division

Besides inducing tolerable levels of CIN, supernumerary centrosomes disrupt asymmetric stem cell division leading to expansion of the stem cell pool and tumor formation, at least in flies (Basto et al. 2008). Whereas several elegant studies have unequivocally demonstrated the link between supernumerary centrosomes and CIN in vitro, no data is available to prove that induction of CIN is the mechanism by which extra centrosomes may cause tumors in mammals. Therefore, disruption of asymmetric stem cell division by extra centrosomes should be considered as a plausible alternative mechanism of transformation in the mammalian system as well.

When stem cells divide, their daughters either self-renew stem cell identity or initiate differentiation. The balanced choice between these alternate fates is critical to maintain stem cell numbers and to rein in their potentially dangerous capacity for long-term proliferation. Symmetric division allows stem cell expansion during embryogenesis and replacement of stem cells after injury but might also harbor the risk for tumorigenesis. Recent studies have highlighted the importance of the stem cell niche as a source of local extrinsic signals that specify stem cell self-renewal. In the context of such a niche, developmentally regulated orientation of the mitotic spindle directs whether the outcome of a stem cell division is asymmetric or symmetric.

Studies of both mouse radial glia progenitors and *D. melanogaster* male germ stem cells showed that when the spindle is oriented perpendicular to the interface with the niche, upon cleavage, one daughter can maintain contact with the niche while the other is displaced away and is free to initiate differentiation. By contrast, spindle orientation parallel to the niche interface allows both daughters to inherit attachments to, and receive local self-renewal signals from the niche. Strikingly, differential labeling of mother centrosomes in both flies and mice revealed that it is always the mother centrosome that remains next to the niche in the new stem cell while the daughter centrosome enters the differentiating daughter cell (Yamashita et al. 2007; Wang et al. 2009). Why centrosome age seemingly does not impact on daughter cell fate during symmetric stem cell divisions remains to be elucidated.

As mentioned above, Basto et al. (2008) recently demonstrated that extra centrosomes can indeed initiate tumorigenesis in *D. melanogaster* overexpressing SAK. Most cells with supernumerary centrosomes initially formed multipolar spindles, but these spindles ultimately became bipolar owing to centrosomal clustering. Surprisingly, the frequency of aneuploidy was only slightly increased. Instead, spindle orientation and thereby asymmetric division of larval neural stem cells was compromised by the extra centrosomes, leading to hyperproliferation of neuroblasts and malignant transformation. A likely explanation for these findings is that amplified centrosomes interfere with asymmetric stem cell division, resulting in hyperproliferation with subsequent induction of CIN that leads to malignant transformation (Basto et al. 2008). Consistent with such a scenario is the same sequence of events in *D. melanogaster* larval neuroblasts containing mutations in genes that directly control asymmetric cell division (Caussinus and Gonzalez 2005).

The reason why asymmetric division fails in cells with supernumerary centrosomes remains unclear. One possibility is that the asymmetric features of mother versus daughter centrosomes which for example determine MT nucleation potentials are disturbed by clustering of multiple centrosomes. Alternatively, astral MT organization, which is important for the interaction between cell cortex and spindle poles and therefore inherently linked to asymmetric division might be corrupted by centrosome clustering. Third, supernumerary centrosomes seem to prevent asymmetric localization of polarity determinants like MUD (the *D. melanogaster* homolog of the human spindle pole protein NUMA) (Caussinus and Gonzalez 2005), what in turn might induce spindle positioning defects and disturbed asymmetric division.

To separately assess the contribution of centrosome defects versus CIN in tumorigenesis, Castellanos and coworkers studied the tumorigenic potential of multiple *D. melanogaster* larval brain tissue mutants defective in various aspects of centrosome biogenesis (Castellanos et al. 2008). Mutations affecting proteins required for centriole replication, pericentriolar matrix recruitment and centrosome function resulted in frequent tumor formation despite only a small fraction of cells having abnormal karyotypes. Consistently, mutations known to induce CIN, including defects in DNA replication and SAC, chromatin condensation and cytokinesis did not give rise to tumors. These results again suggest that in tissues where self-renewing asymmetric divisions are frequent, centrosome-related disturbed stem cell division rather than induction of CIN might initiate malignant transformation.

Do these thought-provoking results imply that centrosome aberrations do indeed cause cancer, however not via CIN as initially thought, but rather by perturbing stem cell division? Given the possible cell-type- and organism-specific effects, and the presence of moderate CIN along with the perturbed stem cell divisions, this conclusion seems premature. Furthermore, we urgently need insights into centrosome function in mammalian cancer stem cells. Answering the question of whether supernumerary centrosomes contribute to mammalian tumorigenesis by disruption of asymmetric division of cancer stem cells, induction of CIN, or both will be rewarding.

17.5 Inhibition of Centrosome Clustering as a Novel Anti-Cancer Treatment Strategy

Supernumerary centrosomes almost exclusively occur in a wide variety of neoplastic disorders but only rarely in non-transformed cells. Therefore, inhibition of centrosomal clustering with consequential induction of multipolar spindles and subsequent cell death would specifically target tumor cells with no effect on normal cells with a regular centrosome content (Nigg 2002; Brinkley 2001).

Recently, griseofulvin has been identified as to inhibit centrosomal clustering (Rebacz et al. 2007). Griseofulvin has been used for many years for the treatment of dermatophyte infections (Loo 2006). Mechanistically, it inhibits mitosis in sensitive fungi (Gull and Trinci 1973) and mammalian cells (Grisham et al. 1973) but whether mitotic arrest is a consequence of MT depolymerization or some other action on MTs in both fungi and human cells is still unclear (Grisham et al. 1973; Weber et al. 1976). Despite extensive studies, the mechanism by which the drug inhibits mitosis in human cells remains obscure. Although griseofulvin has been reported to bind to mammalian brain tubulin and to inhibit MT polymerization in vitro, it does so only at concentrations significantly higher than those needed for spindle multipolarity induction in cancer cells with extra centrosomes (Panda et al. 2005). Also, whether griseofulvin binds to tubulin directly or to MT associated proteins remains conflicting (Panda et al. 2005; Wehland et al. 1977; Roobol et al. 1977). Already more than 30 years ago it was reported that griseofulvin treatment induces spindle multipolarity with each mitotic center containing two centrioles in HeLa cells (Grisham et al. 1973). While at lower concentrations the drug leads to multipolar spindles with centrosomes at each pole in cells with extra centrosomes, at higher concentrations spindle multipolarity with acentrosomal spindle pole formation is induced as well, consistent with the above concept that clustering extra centrosomes in cancer cells might be similar to focusing MTs into a bipolar spindle array in normal cells. For detailed mechanistic understanding it will be important to clearly determine the sequence of events: Does the drug at low concentrations indeed cause declustering of supernumerary centrosomes with subsequent multipolar spindle formation or does spindle multipolarity occur first with successive distribution of centrosomes to each pole?

Additional evidence for an effect of griseofulvin on centrosomal clustering comes from the finding that the drug, in contrast to other MT interacting compounds, induces hepatomas in mice and rats (Epstein et al. 1967). In these animals the majority of hepatocytes are polyploid and therefore contain supernumerary centrosomes which are usually efficiently clustered into bipolar spindle arrays (Guidotti et al. 2003; Duncan et al. 2010).

Findings similar to those reported for griseofulvin have recently been described for the MT-modulating noscapinoid EM011 (Karna et al. 2011). In contrast to griseofulvin, EM011 seems to induce centrosome amplification prior to declustering, thereby potentially reducing its specificity to cancer cells with supernumerary centrosomes. Further supporting the candidacy of centrosomal clustering for a largely cancer-selective target, at low drug concentrations sufficient for spindle multipolarity induction in cancer cells, MT poisons including nocodazole and taxol induce greater cell death in tumor cells than in non-transformed cells (Brito and Rieder 2009).

Most recently, it has been described that a phenanthrene-derived poly(ADP-ribose) polymerase (PARP) inhibitor also prevents centrosome clustering and thereby leads to cell death of tumor cells with supernumerary centrosomes (Castiel et al. 2011). Poly(ADP-ribose) (PAR) is enriched in the mitotic spindle and required for bipolar spindle formation (Chang et al. 2004). In addition, several

PARP proteins localize to either centrosomes or centromeres and catalyze poly (ADP-ribosyl)ation of both centrosomal and centromeric proteins (Kanai et al. 2003; Saxena et al. 2002) arguing for a possible role of PARP proteins in the clustering process. On the other hand, centrosome clustering was apparently not affected by other potent, non-phenanthrene PARP inhibitors (Castiel et al. 2011), thereby questioning a specific role for PARP inhibition in the prevention of centrosome clustering.

In addition to drugs, siRNAs to the kinesin KIFC1/HSET, the NDC80 complex subunit HEC1, aurora B, survivin, sororin, SGOL1, the augmin complex subunits HAUS3 and FAM29A (also known as HAUS6) as well as ILK and PRC1 lead to cell death through inhibition of centrosomal clustering in tumor cells with amplified centrosomes but not in normal cells (Kwon et al. 2008; Leber et al. 2010; Fielding et al. 2011). Identified in an RNAi screen performed in *D. melanogaster* S2 cells, siRNAs to KIFC1/HSET and Myo10 increased the frequency of spindle multipolarity in human cancer cells harboring supernumerary centrosomes as well (Kwon et al. 2008). Especially, KIFC1/HSET might constitute an interesting therapeutic target as knockdown of the protein had no effect on cell division in diploid control cells but largely decreased viability of tumor cells with extra centrosomes by inducing multipolar anaphases and subsequent apoptosis (Kwon et al. 2008).

Also, small molecule inhibition of ILK, HEC1, and aurora B suppresses tumor cell growth in tissue culture as well as in animals (Huang et al. 2007; Wu et al. 2008; Wilkinson et al. 2007; Kalra et al. 2009). Therefore, induction of multipolar spindles seems to induce cell death irrespective of the underlying mechanism that induced them. By contrast, although inhibition of monopolar spindle 1 (MPS1, also known as TTK), a dual-specificity kinase required for the maintenance of SAC activation, inhibits centrosomal clustering and induces aberrant cell divisions in cells with supernumerary centrosomes, it does not cause selective cytotoxicity in cells with amplified centrosomes compared to cells with a regular centrosome content (Kwiatkowski et al. 2010). Therefore, it might be concluded that whereas SAC inhibition per se equally targets all cells, selective inhibition of centrosomal clustering through specific targeting may provide a therapeutic window to specifically target cells with supernumerary centrosomes.

Importantly, prior to cell death, cells with inhibited HEC1 or aurora B have multipolar spindles, lagging chromosomes and subsequent aneuploidy and polyploidy (Wu et al. 2008; Wilkinson et al. 2007). Therefore, a possible downside of centrosomal cluster inhibition as a potential treatment approach might be the induction of cell clones with additional chromosomal abnormalities. On the optimistic side, such a risky scenario seems relatively unlikely, as multipolar cell division mostly leads to gross CIN and cell death (Ganem et al. 2009; Kops et al. 2004).

Acknowledgment We apologize to those authors whose work is not cited because of space limitations. This work was supported by the Deutsche Forschungsgemeinschaft, the Deutsche Krebshilfe, the Deutsche José Carreras Leukämie-Stiftung and the Interdisciplinary Research Program of the National Center for Tumor Diseases Heidelberg.

References

Adams RR et al (2001) Human INCENP colocalizes with the Aurora-B/AIRK2 kinase on chromosomes and is overexpressed in tumour cells. Chromosoma 110:65–74

Altieri DC (2003) Survivin, versatile modulation of cell division and apoptosis in cancer. Oncogene 22:8581–8589

Anderhub SJ, Krämer A, Maier B (2012) Centrosome amplification in tumorigenesis. Cancer Lett. 322:8–17

Andersen JS, Wilkinson CJ, Mayor T, Mortensen P, Nigg EA, Mann M (2003) Proteomic characterization of the human centrosome by protein correlation profiling. Nature 426:570–574

Arquint C, Sonnen KF, Stierhof YD, Nigg EA (2012) Cell cycle-regulated expression of STIL controls centriole numbers in human cells. J Cell Sci 125:1342–1452

Baker DJ, Jin F, Jeganathan KB, van Deursen JM (2009) Whole chromosome instability caused by Bub1 insufficiency drives tumorigenesis through tumor suppressor gene loss of heterozygosity. Cancer Cell 16:475–486

Barr FA, Sillje HH, Nigg EA (2004) Polo-like kinases and the orchestration of cell division. Nat Rev Mol Cell Biol 5:429–440

Basto R et al (2008) Centrosome amplification can initiate tumorigenesis in flies. Cell 133:1032–1042

Bettencourt-Dias M et al (2005) SAK/PLK4 is required for centriole duplication and flagella development. Curr Biol 15:2199–2207

Birkbak NJ et al. (2011) Paradoxical relationship between chromosomal instability and survival outcome in cancer. Cancer Res. (epub ahead of print)

Bischoff JR et al (1998) A homologue of Drosophila aurora kinase is oncogenic and amplified in human colorectal cancers. EMBO J 17:3052–3065

Bornens M (2002) Centrosome composition and microtubule anchoring mechanisms. Curr Opin Cell Biol 14:25–34

Boveri T (1929) The origin of malignant tumors. Williams and Wilkins, Baltimore

Boyarchuk Y, Salic A, Dasso M, Arnaoutov A (2007) Bub1 is essential for assembly of the functional inner centromere. J Cell Biol 176:919–928

Breuer M et al (2010) HURP permits MTOC sorting for robust meiotic spindle bipolarity, similar to extra centrosome clustering in cancer cells. J Cell Biol 191:1251–1260

Brinkley BR (2001) Managing the centrosome numbers game: from chaos to stability in cancer cell division. Trends Cell Biol 11:18–21

Brito DA, Rieder CL (2009) The ability to survive mitosis in the presence of microtubule poisons differs significantly between human nontransformed (RPE-1) and cancer (U2OS, HeLa) cells. Cell Motil. Cytoskeleton 66:437–447

Carmena M, Earnshaw WC (2003) The cellular geography of aurora kinases. Nat Rev Mol Cell Biol 4:842–854

Carter SL et al (2006) A signature of chromosomal instability inferred from gene expression profiles predicts clinical outcome in multiple human cancers. Nat Genet 38:1043–1048

Castellanos E, Dominguez P, Gonzalez C (2008) Centrosome dysfunction in Drosophila neural stem cell causes tumors that are not due to genome instability. Curr Biol 18:1209–1214

Castiel A, Visochek L, Mittelman L, Dantzer F, Izraeli S, Cohen-Armon M (2011) A phenanthrene derived PARP inhibitor is an extra-centrosomes de-clustering agent exclusively eradicating human cancer cells. BMC Cancer 11:412

Caussinus E, Gonzalez C (2005) Induction of tumor growth by altered stem-cell asymmetric division in Drosophila melanogaster. Nat Genet 37:1125–1129

Chang P, Jacobson MK, Mitchison TJ (2004) Poly(ADP-ribose) is required for spindle assembly and structure. Nature 432:645–649

Chang JL et al (2006) Borealin/Dasra B is a cell cycle-regulated chromosomal passenger protein and its nuclear accumulation is linked to poor prognosis for human gastric cancer. Exp Cell Res 312:962–973

Charrasse S et al (1995) Characterization of the cDNA and pattern of expression of a new gene over-expressed in human hepatomas and colonic tumors. Eur J Biochem 234:406–413

Chen JG, Horwitz SB (2002) Differential mitotic responses to microtubule-stabilizing and -destabilizing drugs. Cancer Res 62:1935–1938

Cizmecioglu O, Arnold M, Bahtz R, Settele F, Ehret L, Haselmann-Weiss U, Antony C, Hoffmann I (2010) Cep152 acts as a scaffold for recruitment of Plk4 and CPAP to the centrosome. J Cell Biol 191:731–739

Dai J, Kataneva AV, Higgins JMG (2009) Studies of haspin-depleted cells reveal that spindle-pole integrity in mitosis requires chromosome cohesion. J Cell Sci 122:4168–4176

De S, Cipriano R, Jackson MW, Stark GR (2009) Overexpression of kinesins mediates docetaxel resistance in breast cancer cells. Cancer Res 69:8035–8042

Diaz-Rodriguez E, Sotillo R, Schvartzman J-M, Benezra R (2008) Hec1 overexpression hyperactivates the mitotic checkpoint and induces tumor formation in vivo. Proc. Natl. Acad. Sci. USA 105:16719–16724

Doxsey SJ, Stein P, Evans L, Calarco PD, Kirschner M (1994) Pericentrin, a highly conserved centrosome protein involved in microtubule organization. Cell 76:639–650

Duncan AW et al (2010) The ploidy conveyor of mature hepatocytes as a source of genetic variation. Nature 467:707–711

Dzhindzhev NS et al (2010) Asterless is a scaffold for the onset of centriole assembly. Nature 467:714–718

Einarson MB, Cukierman E, Compton DA, Golemis EA (2004) Human enhancer of invasion-cluster, a coiled-coil protein required for passage through mitosis. Mol Cell Biol 24:3957–3971

Epstein SS, Andrea J, Joshi S, Mantel N (1967) Hepatocarcinogenicity of griseofulvin following parenteral administration to infant mice. Cancer Res 27:1900–1906

Ferguson RL, Maller JL (2010) Centrosomal localization of cyclin E-Cdk2 is required for initiation of DNA synthesis. Curr Biol 20:856–860

Ferretti C et al (2010) Expression of the kinetochore protein Hec1 during the cell cycle in normal and cancer cells and its regulation by the pRb pathway. Cell Cycle 9:4147–4182

Fielding AB, Lim S, Montgomery K, Dobreva I, Dedhar S (2011) A critical role of integrin-linked kinase, ch-TOG and TACC3 in centrosome clustering in cancer cells. Oncogene 30:521–534

Fukasawa K (2007) Oncogenes and tumour suppressors take on centrosomes. Nat Rev Cancer 7:911–924

Ganem NJ, Godinho SA, Pellman D (2009) A mechanism linking extra centrosomes to chromosomal instability. Nature 460:278–282

Gould RR, Borisy GG (1977) The pericentriolar material in Chinese hamster ovary cells nucleates microtubule formation. J Cell Biol 73:601–615

Grinberg-Rashi H et al (2009) The expression of three genes in primary non-small cell lung cancer is associated with metastatic spread to the brain. Clin Cancer Res 15:1755–1761

Grisham LM, Wilson L, Bensch KG (1973) Antimitotic action of griseofulvin does not involve disruption of microtubules. Nature 244:294–296

Guidotti J-E et al (2003) Liver cell polyploidization: a pivotal role for binuclear hepatocytes. J Biol Chem 278:19095–19101

Gull K, Trinci APJ (1973) Griseofulvin inhibits fungal mitosis. Nature 244:292–294

Habedanck R, Stierhof YD, Wilkinson CJ, Nigg EA (2005) The Polo kinase Plk4 functions in centriole duplication. Nat Cell Biol 7:1140–1146

Hannigan G, Troussard AA, Dedhar S (2005) Integrin-linked kinase: a cancer therapeutic target unique among its ILK. Nat Rev Cancer 5:51–63

Hatch EM, Kulukian A, Holland AJ, Cleveland DW, Stearns T (2010) Cep152 interacts with Plk4 and is required for centriole duplication. J Cell Biol 191:721–729

Hayama S et al (2006) Activation of CDCA1-KNTC2, members of centromere protein complex, involved in pulmonary carcinogenesis. Cancer Res 66:10339–10348

Hori T et al (2008) CCAN makes multiple contacts with centromeric DNA to provide distinct pathways to the outer kinetochore. Cell 135:1039–1052

Huang H et al (2007) Tripin/hSgo2 recruits MCAK to the inner centromere to correct defective kinetochore attachments. J Cell Biol 177:413–424

Hung LY, Chen HL, Chang CW, Li BR, Tang TK (2004) Identification of a novel microtubule-destabilizing motif in CPAP that binds to tubulin heterodimers and inhibits microtubule assembly. Mol Biol Cell 15:2697–2706

Indjeian VB, Stern BM, Murray AW (2005) The centromeric protein Sgo1 is required to sense lack of tension on mitotic chromosomes. Science 307:130–133

Ishikawa K et al (2008) Mitotic centromere-associated kinesin is a novel marker for prognosis and lymph node metastasis in colorectal cancer. Br J Cancer 98:1824–1829

Jackman M, Lindon C, Nigg EA, Pines J (2003) Active cyclin B1-Cdk1 first appears on centrosomes in prophase. Nat Cell Biol 5:143–148

Kalra J et al (2009) QLT0267, a small molecule inhibitor targeting integrin-linked kinase (ILK), and docetaxel can combine to produce synergistic interactions linked to enhanced cytotoxicity, reductions in P-AKT levels, altered F-actin architecture and improved treatment outcomes in an orthotopic breast cancer model. Breast Cancer Res 11:R25

Kanai M, Tong W-M, Sugihara E, Wang Z-Q, Fukasawa K, Miwa M (2003) Involvement of poly(ADP-ribose) polymerase 1 and poly(ADP-ribosyl)ation in regulation of centrosome function. Mol Cell Biol 23:2451–2462

Karna P et al (2011) A novel microtubule-modulating noscapinoid triggers apoptosis by inducing spindle multipolarity via centrosome amplification and declustering. Cell Death Differ 18:632–644

Kawashima SA, Tsukahara T, Langegger M, Hauf S, Kitajima TS, Watanabe Y (2007) Shugoshin enables tension-generating attachment of kinetochores by loading Aurora to centromeres. Genes Dev 21:420–435

Kawashima SA, Yamagishi Y, Honda T, Ishiguro K, Watanabe Y (2010) Phosphorylation of H2A by Bub1 prevents chromosomal instability through localizing shugoshin. Science 327:172–177

Kelly AE, Ghenoiu C, Xue JZ, Zierhut C, Kimura H, Funabiki H (2010) Survivin reads phosphorylated histone H3 threonine 3 to activate the mitotic kinase aurora B. Science 330:235–239

Kitagawa D, Kohlmaier G, Keller D, Strnad P, Balestra FR, Fluckiger I, Gönczy P (2011) Spindle positioning in human cells relies on proper centriole formation and on the microcephaly proteins CPAP and STIL. J Cell Sci 124:3884–3893

Kleylein-Sohn J, Westendorf J, Le Clech M, Habedanck R, Stierhof YD, Nigg EA (2007) Plk4-induced centriole biogenesis in human cells. Dev Cell 13:190–202

Koffa MD, Casanova CM, Santarella R, Kocher T, Wilm M, Mattaj IW (2006) HURP is part of a Ran-dependent complex involved in spindle formation. Curr Biol 16:743–754

Kops GJ, Foltz DR, Cleveland DW (2004) Lethality to human cancer cells through massive chromosome loss by inhibition of the mitotic checkpoint. Proc. Natl. Acad. Sci. USA 101:8699–8704

Koutsami MK et al (2006) Centrosome abnormalities are frequently observed in non-small-cell lung cancer and are associated with aneuploidy and cyclin E overexpression. J. Pathol. 209:512–521

Krämer A, Neben K, Ho AD (2002) Centrosome replication, genomic instability and cancer. Leukemia 16:767–775

Krämer A et al (2003) Centrosome aberrations as a possible mechanism for chromosomal instability in non-Hodgkin's lymphoma. Leukemia 17:2207–2213

Kuriyama R, Borisy GG (1981) Centriole cycle in Chinese hamster ovary cells as determined by whole-mount electron microscopy. J Cell Biol 91:814–821

Kwiatkowski N et al (2010) Small-molecule kinase inhibitors provide insight into Mps1 cell cycle function. Nat Chem Biol 6:359–368

Kwon M et al (2008) Mechanisms to suppress multipolar divisions in cancer cells with extra centrosomes. Genes Dev 22:2189–2203

Lawo S et al (2009) HAUS, the 8-subunit complex, regulates centrosome and spindle integrity. Curr Biol 19:1–11

Leber B et al. Proteins required for centrosome clustering in cancer cells. Sci. Transl. Med. 2, 32ra38 (2010)

Leidel S, Delattre M, Cerutti L, Baumer K, Gönczy P (2005) SAS-6 defines a protein family required for centrosome duplication in C. elegans and in human cells. Nat Cell Biol 7:115–125

Levine DS, Sanchez CA, Rabinovitch PS, Reid BJ (1991) Formation of the tetraploid intermediate is associated with the development of cells with more than four centrioles in the elastase-simian virus 40 tumor antigen transgenic mouse model of pancreatic cancer. Proc. Natl. Acad. Sci. USA 88:6427–6431

Lingle WL, Salisbury JL (1999) Altered centrosome structure is associated with abnormal mitoses in human breast tumors. Am J Pathol 155:1941–1951

Lingle WL, Lutz WH, Ingle JN, Maihle NJ, Salisbury JL (1998) Centrosome hypertrophy in human breast tumors: implications for genomic stability and cell polarity. Proc. Natl. Acad. Sci. USA 95:2950–2955

Liu D, Vader G, Vromans MJM, Lampson MA, Lens SMA (2009) Sensing chromosome biorientation by spatial separation of aurora B kinase from kinetochore substrates. Science 323:1350–1353

Löffler H et al (2011) Cep63 recruits Cdk1 to the centrosome: implications for regulation of mitotic entry, centrosome amplification, and genome maintenance. Cancer Res 71:2129–2139

Loo DS (2006) Systemic antifungal agents: an update of established and new therapies. Adv Dermatol 22:101–124

Nakamura Y et al (2007) Clinicopathological and biological significance of mitotic centromere-associated kinesin overexpression in human gastric cancer. Br J Cancer 97:543–549

Neben K, Giesecke C, Schweizer S, Ho AD, Krämer A (2003) Centrosome aberrations in acute myeloid leukemia are correlated with cytogenetic risk profile. Blood 101:289–291

Nguyen CL, McLaughlin-Drubin ME, Münger K (2008) Delocalization of the microtubule motor dynein from mitotic spindles by the human papillomavirus E7 oncoprotein is not sufficient for induction of multipolar mitoses. Cancer Res 68:8715–8722

Nguyen HG et al (2009) Deregulated Aurora-B induced tetraploidy promotes tumorigenesis. FASEB J. 23:2741–2748

Nguyen M-H et al (2010) Phosphorylation and activation of cell division cycle associated 5 by mitogen-activated protein kinase play a crucial role in human lung carcinogenesis. Cancer Res 70:5337–5347

Nigg EA (2002) Centrosome aberrations: cause or consequence of cancer progression? Nat Rev Cancer 2:815–825

Nigg EA, Raff JW (2009) Centrioles, centrosomes, and cilia in health and disease. Cell 139:663–678

Panda D, Rathinasamy K, Santra MK, Wilson L (2005) Kinetic suppression of microtubule dynamic instability by griseofulvin: implications for its possible use in the treatment of cancer. Proc. Natl. Acad. Sci. USA 102:9878–9883

Peset I, Vernos I (2008) The TACC proteins: TACC-ling microtubule dynamics and centrosome function. Trends Cell Biol 18:379–388

Pihan GA, Purohit A, Wallace J (1998) Centrosome defects and genetic instability in malignant tumors. Cancer Res 58:3974–3985

Pihan GA, Purohit A, Wallace J, Malhotra R, Liotta L, Doxsey SJ (2001) Centrosome defects can account for cellular and genetic changes that characterize prostate cancer progression. Cancer Res 61:2212–2219

Putkey FR et al (2002) Unstable kinetochore-microtubule capture and chromosomal instability following deletion of CENP-E. Dev Cell 3:351–365

Quintyne NJ, Reing JE, Hoffelder DR, Gollin SM, Saunders WS (2005) Spindle multipolarity is prevented by centrosomal clustering. Science 307:127–129

Rebacz B et al (2007) Identification of griseofulvin as an inhibitor of centrosomal clustering in a phenotype-based screen. Cancer Res 67:6342–6350

Ring D, Hubble R, Kirschner M (1982) Mitosis in a cell with multiple centrioles. J Cell Biol 94:549–556

Roobol A, Gull K, Pogson CI (1977) Evidence that griseofulvin binds to a microtubule associated protein. FEBS Lett 75:149–153

Ruchaud S, Carmena M, Earnshaw WC (2007) Chromosomal passengers: conducting cell division. Nat Rev Mol Cell Biol 8:798–812

Saxena A, Saffery R, Wong LH, Kalitsis P, Choo A (2002) Centromere proteins CENPA, CENPB, and BUB3 interact with poly(ADP-ribose) polymerase-1 protein and are poly(ADP-ribosyl)ated. J Biol Chem 277:26921–26926

Schmitz J, Watrin E, Lénárt P, Mechtler K, Peters J-M (2007) Sororin is required for stable binding of cohesin to chromatin and for sister chromatid cohesion in interphase. Curr Biol 17:630–636

Schockel L, Mockel M, Mayer B, Boos D, Stemmann O (2011) Cleavage of cohesin rings coordinates the separation of centrioles and chromatids. Nat Cell Biol 13:966–972

Schvartzman J-M, Sotillo R, Benezra R (2010) Mitotic chromosomal instability and cancer: mouse modelling of the human disease. Nat Rev Cancer 10:102–115

Shao S et al (2010) Centrosomal Nlp is an oncogenic protein that is gene-amplified in human tumors and causes spontaneous tumorigenesis in transgenic mice. J. Clin. Invest. 120:498–507

Shimo A et al (2007) Elevated expression of protein regulator of cytokinesis 1, involved in the growth of breast cancer cells. Cancer Sci 98:174–181

Silkworth WT, Nardi IK, Scholl LM, Cimini D (2009) Multipolar spindle pole coalescence is a major source of kinetochore mis-attachment and chromosome mis-segregation. in cancer cells. PLoS ONE 4:e6564

Sillje HHW, Nagel S, Körner R, Nigg EA (2006) HURP is a Ran-importin β-related protein that stabilizes kinetochore microtubules in the vincinity of chromosomes. Curr Biol 16:731–742

Sluder G, Thompson EA, Miller FJ, Hayes J, Rieder CL (1997) The checkpoint control for anaphase onset does not monitor excess numbers of spindle poles or bipolar spindle symmetry. J Cell Sci 110:421–429

Sotillo R et al (2007) Mad2 overexpression promotes aneuploidy and tumorigenesis in mice. Cancer Cell 11:9–23

Stearns T, Evans L, Kirschner M (1991) Gamma-tubulin is a highly conserved component of the centrosome. Cell 65:825–836

Stevens NR, Dobbelaere J, Brunk K, Franz A, Raff JW (2010) Drosophila Ana2 is a conserved centriole duplication factor. J Cell Biol 188:313–323

Strnad P, Gönczy P (2008) Mechanisms of procentriole formation. Trends Cell Biol 18:389–396

Tang CJ et al (2011) The human microcephaly protein STIL interacts with CPAP and is required for procentriole formation. EMBO J 30:4790–4804

Taylor S, Peters JM (2008) Polo and Aurora kinases: lessons derived from chemical biology. Curr Opin Cell Biol 20:77–84

Thery M et al (2005) The extracellular matrix guides the orientation of the cell division axis. Nat Cell Biol 7:947–953

Tsou MF, Stearns T (2006) Mechanism limiting centrosome duplication to once per cell cycle. Nature 442:947–951

Tsou AP et al (2003) Identification of a novel cell cycle regulated gene, HURP, overexpressed in human hepatocellular carcinoma. Oncogene 22:298–307

Uehara R et al (2009) The augmin complex plays a critical role in spindle microtubule generation for mitotic progression and cytokinesis in human cells. Proc. Natl. Acad. Sci. USA 106:6998–7003

Uzbekov R, Prigent C (2007) Clockwise or anticlockwise? Turning the centriole triplets in the right direction! FEBS Lett 581:1251–1254

Vader G, Lens SM (2008) The Aurora kinase family in cell division and cancer. Biochim Biophys Acta 1786:60–72

Vanoosthuyse V, Prykhozhij S, Hardwick KG (2007) Shugoshin 2 regulates localization of the chromosomal passenger proteins in fission yeast mitosis. Mol Biol Cell 18:1657–1669

von Hansemann D (1890) Über asymmetrische Zellteilung in Epithelkrebsen und deren biologische Bedeutung. Virchows Arch. Patholog. Anat. 119, 299-326 (in German)

Vulprecht J et al (2012) SIL is required for centriole duplication in human cells. J Cell Sci 125:1353–1362

Wang X et al (2009) Asymmetric centrosome inheritance maintains neural progenitors in the neocortex. Nature 461:947–955

Wang F et al (2010) Histone H3 Thr-3 phosphorylation by haspin positions aurora B at centromeres in mitosis. Science 330:231–235

Weaver BA, Silk AD, Montagna C, Verdier-Pinard P, Cleveland DW (2007) Aneuploidy acts both oncogenically and as a tumor suppressor. Cancer Cell 11:25–36

Weber K, Wehland J, Herzog W (1976) Griseofulvin interacts with microtubules both in vitro and in vivo. J Mol Biol 102:817–829

Wehland J, Herzog W, Weber K (1977) Interaction of griseofulvin with microtubules, microtubule protein and tubulin. J Mol Biol 111:329–342

Wei RR, Al-Bassam J, Harrison SC (2007) The Ndc80/HEC1 complex is a contact point for kinetochore-microtubule attachment. Nat Struct Mol Biol 14:54–59

Wilkinson RW et al (2007) AZD1152, a selective inhibitor of Aurora B kinase, inhibits human tumor xenograft growth by inducing apoptosis. Clin Cancer Res 13:3682–3688

Wong J, Fang G (2006) HURP controls spindle dynamics to promote proper interkinetochore tension and efficient kinetochore capture. J Cell Biol 173:879–891

Wu G, Lin YT, Wei R, Chen Y, Shan Z, Lee WH (2008a) Hice1, a novel microtubule-associated protein required for maintenance of spindle integrity and chromosomal stability in human cells. Mol Cell Biol 28:3652–3662

Wu G et al (2008b) Small molecule targeting the Hec1/Nek2 mitotic pathway suppresses tumor cell growth in culture and in animal. Cancer Res 68:8393–8399

Yamagishi Y, Honda T, Tanno Y, Watanabe Y (2010) Two histone marks establish the inner centromere and chromosome bi-orientation. Science 330:239–243

Yamashita YM, Mahowald AP, Perlin JR, Fuller MT (2007) Asymmetric inheritance of mother versus daughter centrosome in stem cell division. Science 315:518–521

Yang Z, Loncarek J, Khodjakov A, Rieder CL (2008) Extra centrosomes and/or chromosomes prolong mitosis in human cells. Nat Cell Biol 10:748–751

Part IV
Centrosomes in Other Systems

Chapter 18
Re-evaluation of the Neuronal Centrosome as a Generator of Microtubules for Axons and Dendrites

Peter W. Baas and Aditi Falnikar

Abstract A typical vertebrate neuron extends a single axon and multiple dendrites, both of which are rich in highly organized arrays of microtubules that serve essential functions. In simpler cell types, microtubules are organized by their attachment to a centralized nucleating structure such as the centrosome. In axons and dendrites, however, microtubules are not attached to the centrosome or any recognizable organizing structure. Over a decade ago, we proposed that the neuronal centrosome acts as a "generator" of microtubules for the axon and dendrites. Our studies suggested that the neuronal centrosome is highly active, especially during development, nucleating and releasing microtubules into the cell body. The released microtubules are then actively transported into the axon and dendrites by molecular motor proteins. In migrating neurons, most of the microtubules are attached to the centrosome, suggesting that significant changes in the nucleation or release of microtubules from the centrosome occur as neurons cease migration and begin to form their axonal and dendritic arbors. Recent studies suggest that the centrosome eventually becomes inactive as neurons mature, and that microtubule numbers are increased by other mechanisms, such as the severing of existing microtubules. Exactly how important the centrosome is for early stages of differentiation remains unclear, and the possibility exists that the centrosome may be re-activated in more mature neurons to meet particular challenges that may arise. Here we review historical as well as contemporary data on the neuronal centrosome, with emphasis on its potential role as a generator of microtubules.

P. W. Baas (✉) · A. Falnikar
Department of Neurobiology and Anatomy,
Drexel University College of Medicine,
2900 Queen Lane, Philadelphia, PA 19129, USA
e-mail: pbaas@drexelmed.edu

18.1 Introduction

Neurons are arguably the cell type in nature with the greatest dependence upon sophisticated arrays of highly organized microtubules for their form and function. A typical vertebrate neuron extends a single axon and multiple dendrites, both of which are rich in microtubules. The microtubule arrays within these processes are essential for providing architectural support, for enabling axons and dendrites to take on different shapes and branching patterns, and for supporting bidirectional organelle transport (Baas and Buster 2004). Many of the most fundamental differences between axons and dendrites directly or indirectly result from distinct patterns of microtubule orientation in each type of process. In the axon, nearly all of the microtubules are oriented with their plus ends distal to the cell body, whereas in the dendrite, the microtubules have a mixed pattern of orientation (Baas and Lin 2011). In most textbooks, microtubules are said to be organized mainly by their attachment to microtubule-organizing centers such as the centrosome (Alberts et al. 2007), but amazingly, the highly organized microtubules in axons and dendrites are not attached to the centrosome or any recognizable organizing structure (Baas and Yu 1996). Instead, the microtubules are free at both ends, and take on various lengths within the axon and dendrites. The shortest microtubules are highly mobile, moving rapidly within the axon (Wang and Brown 2002) and perhaps the dendrite as well (Sharp et al. 1995). One of the questions that has driven our laboratory for many years is how microtubules become organized in the axon and dendrites if not via attachment to an organizing center. Another question is whether the centrosome (located in the cell body of the neuron) has any importance for generating or organizing the neuronal microtubule arrays, or alternatively, whether it is a vestigial structure with no function.

Over a decade ago, we embarked on a series of studies the results of which led us to propose that the neuronal centrosome acts as a "generator" of microtubules for the axon and dendrites (Ahmad and Baas 1995; Ahmad et al. 1994, 1998, 1999; Baas 1996; Yu et al. 1993). The premise was that the neuronal centrosome is highly active, especially during development, nucleating and releasing microtubules into the cell body. The released microtubules are then actively transported into the axon and dendrites by molecular motor proteins. The relevant motors transport the microtubules specifically with their plus or minus end leading, and thereby establish the distinct patterns of microtubule polarity orientation in each type of process (Baas and Ahmad 1993; Sharp et al. 1995, 1997; Yu et al. 1997). In this view, the centrosome does not contribute to the polarity orientation of microtubules in either type of process, except perhaps to create an initial bias of plus ends outward in the cell body as the microtubules transit away from the centrosome (Ahmad and Baas 1995). One of the main roles that we envisioned for the centrosome was to nucleate microtubules in a regulated fashion with the appropriate lattice structure, as de novo nucleation of microtubules would presumably result in a variety of different protofilament numbers comprising the lattice (Baas and Joshi 1992; Yu et al. 1993). Another role for the centrosome, as a kind of centralized "generator" of product, was to impose a level of control on the amount of microtubule polymer and the numbers of microtubules available at critical stages of neuronal development (Baas 1996). We envisioned the activity of the centrosome as being pulsatile, delivering bursts of new microtubules for example just prior to dendritic differentiation or when needed to supply a rapid increase in axonal growth.

We also noted, however, that it would be difficult to envision how the centrosome could be called upon to generate and deploy bursts of new microtubules to be used far down the length of the axon, for example, in the formation of a collateral branch. On this basis, we posited that existing microtubules in the axon or the dendrites may undergo localized severing events that could transform a single long microtubule into a population of many short ones (Joshi and Baas 1993). Each short microtubule would inherit the lattice structure of the parent microtubule, and each short microtubule would theoretically have the capability of assembling into a new long microtubule. This would render the axon or dendrites, once formed, less dependent upon or perhaps entirely independent of microtubule nucleation events at the centrosome. Since positing these ideas, we have confirmed that sites of impending branch formation do, indeed, display local severing of microtubules (Yu et al. 1994), and we have identified two different microtubule-severing proteins that participate in axonal branch formation (Qiang et al. 2010; Yu et al. 2008). Of course, this begs the question of whether microtubule severing could completely obviate the need for an active centrosome, even within the cell body.

Over the past decade, most studies on the neuronal centrosome have focused on neuronal migration, a phase of development some neurons undergo prior to axonal and dendritic development. In migrating neurons, most of the microtubules are attached to the centrosome, and this is important for pulling along the centrosome (and accompanying nucleus and cell body) as the neuron journeys to its final destination (Higginbotham and Gleeson 2007). This raises the question of what happens when the neuron ceases migration and sets forth to differentiate an axon and dendritic arbor. Is there an upregulation of microtubule severing such that all microtubules nucleated at the centrosome are now released? Does that centrosome gradually lose its nucleating potency during development, or is the ability to nucleate microtubules retained and used at key moments in development? In adult neurons, is there a slow but steady flow of new microtubules from the centrosome, or does the centrosome become quiescent in terms of manufacturing new microtubules? These questions remain unanswered but there has recently been new interest in whether or not the neuronal centrosome serves as a hub for microtubule-based activity relevant to neuronal differentiation (de Anda et al. 2005; Stiess et al. 2010; Stiess and Bradke 2011). Here, we review the older literature, summarize exciting new findings, and ponder the unanswered questions.

18.2 Location, Location, or Not Location?

The idea that axonal microtubules have a centrosomal origin is actually a rather old one. As early as 1965 before "spindle tubules" and "neurotubules" were both identified as "microtubules," Gonatas and Robbins (1965) examined the lattice structure of neurotubules in the chick embryo retina, found it to be indistinguishable from that of spindle tubules, and concluded that "neurotubules probably arise from the centrioles." Similarly, in ultrastructural studies on rabbit embryo dorsal root ganglion neuroblasts, Tennyson (1965) concluded that neurotubules "probably originate from the centriole ..." and "migrate into the neurite". Even so, a common theme of these earlier studies was that the position of the centrosome in the cell body of the neuron had no consistent correlation with the point of origin of the axon, and there was certainly no direct continuity between the microtubule array of the axon and the centrosome (Lyser 1964, 1968; Sharp et al. 1982). These observations also held true in the case of cultured rat hippocampal neurons (Baas et al. 1988; Dotti and Banker 1991) as well as various other types of neurons in culture we have studied over the years, which include rat sympathetic neurons (Yu et al. 2001) and chicken dorsal root ganglion neurons (Baas and Heidemann 1986). A lack of correlation between the position of the centrosome and the location of the axon (or dendrites) is consistent with the centrosome ejecting microtubules into the cell body that may ultimately come to reside in axons and dendrites, without the microtubules dragging the centrosome with them in the direction of the relevant axon or dendrite (see Fig. 18.1). This would distinguish a neuron that has stopped migrating and started elaborating its axonal and dendritic arbors from a migrating neuron, in which the centrosome is dragged toward the leading process by its attached microtubules.

An interesting exception to the location rule was reported in the case of cultured cerebellar granule neurons, which have a somewhat unique developmental pattern in the culture dish (Zmuda and Rivas 1998). After these neurons cease migrating, they extend an initial axon, then a secondary axon, and finally multiple dendrites. The centrosome is first positioned near where the initial axon develops and then moves to where the secondary axon develops, suggesting that the position of the centrosome is related to the development of each of the two axons. Perhaps having two axons and a single centrosome demands that the centrosome is nearest the one that is undergoing the most active phase of growth, whereas in neurons with only one axon and one centrosome, the position of the centrosome is not so important. Perhaps the centrosome is dragged toward the axon into which microtubules are being most actively transported. The same molecular motors responsible for transporting microtubules into the axon would pull on the microtubules while they are still attached to the centrosome, and thereby move the centrosome toward the relevant axon.

In a more recent paper on cultured hippocampal neurons, the laboratory of Carlos Dotti revisited his earlier result on centrosome location relative to axonal differentiation (de Anda et al. 2005). They reported that the axon consistently arose

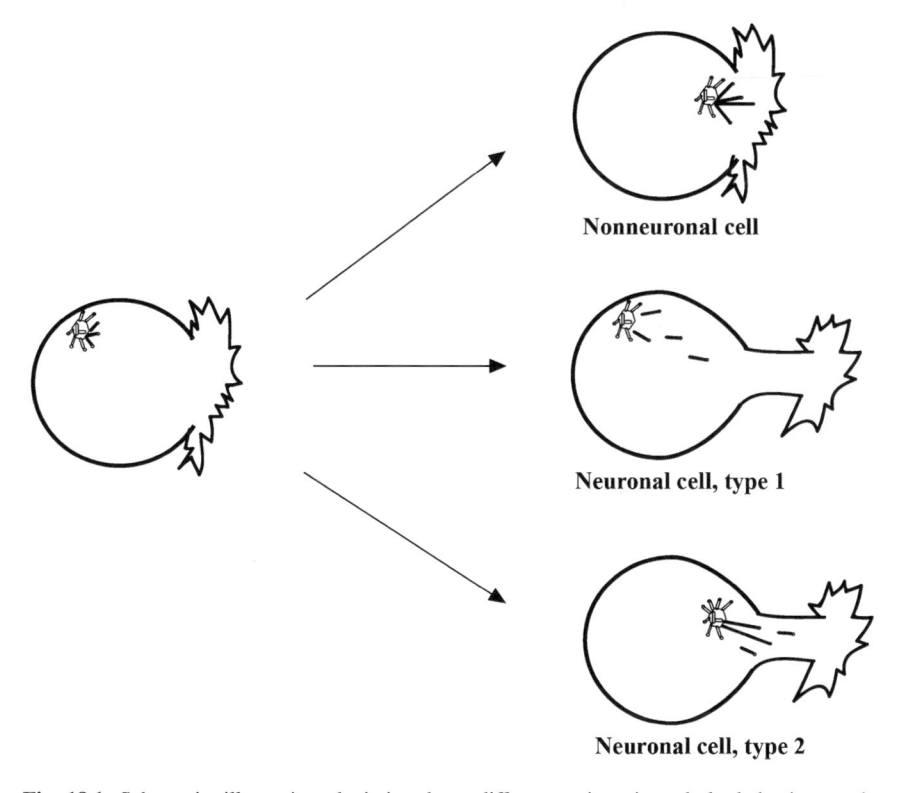

Fig. 18.1 Schematic illustration depicting how differences in microtubule behavior at the centrosome may contribute to phenotypic differences in different cell types. A pluripotent precursor cell shown on the left could give rise to either a motile non-neuronal cell or a neuron. In the case of the non-neuronal cell shown on the right at top, forces pulling on the microtubules draw the centrosome toward the leading edge of the cell as it moves. In the most typical situation of the neuron (denoted as type 1), the microtubules are released and the centrosome is not relocated. Nevertheless, the microtubules are translocated toward the leading edge, which coalesces into a growth cone. The cell body remains stationary and the microtubules translocate into the space between the cell body and the growth cone, which develops into the axon. In the case of some neurons (denoted as type 2), a subset of microtubules nucleated by the centrosome remains attached to the centrosome while others are released. The same forces that transport the released microtubules into the early axon pull on the attached microtubules, drawing the centrosome toward the axon

from the first immature neurite to form after the final mitotic division of the neuroblast, and that the Golgi and endosomes (which generally accompany the centrosome) clustered in the location where the first neurite formed. These observations are surprising in light of the earlier findings on hippocampal and other types of neurons, but are consistent with the observations on cerebellar granule neurons described by Zmuda and Rivas (1998). Interestingly, they also found that ablating the centrosome precluded normal polarization of the neuron.

One of the most enduring mysteries of neuronal polarity is why most neurons have a single axon and how it is that the formation of additional axons is suppressed. It is interesting in this regard that neurons also most typically have a single centrosome. It is tempting to propose that the singularity of the axon and the singularity of the centrosome are somehow related. In the unusual case of the cerebellar granule neurons with two axons, a single centrosome changes location to serve both. However, in the recent paper by de Anda et al. (2005), they observed a small number of hippocampal neurons with two centrosomes and such neurons consistently formed two axons. On the other hand, it should be noted that cultured sympathetic neurons initially differentiate several axons after which they re-craft their morphology into a single axon and multiple dendrites (Bruckenstein and Higgins 1988); and yet, despite initially forming several axons, they only have one centrosome (Yu et al. 1993). Unlike the case with the axon, dendrites are almost always multiple in numbers, and it would be hard to fathom that the centrosome could be so mobile in the cell body as to move from dendrite to dendrite and then back to the axon to serve each neurite one at a time. Interestingly, we reported several years ago what appears to be streams of microtubules flowing from the centrosome into developing dendrites of cultured hippocampal neurons, with a location roughly centralized among the dendrites (Sharp et al. 1995; also see Fig. 18.2). No such flow of microtubules was observed between the centrosome and the axon at this stage of development. Taken together, these several findings indicate that there is no "one size fits all" scenario for the location of the neuronal centrosome. Even so, it would certainly appear that the centrosome is an important structure in the neuron, at least for the early stages of development.

18.3 Why is the Neuronal Centrosome Important?

The centrosome is best known in eukaryotic cells as a microtubule-organizing center that organizes microtubules by virtue of its microtubule-nucleating properties. The centrosome consists of two barrel-shaped centrioles surrounded by amorphous pericentriolar material (Alberts et al. 2007). Among the components of the pericentriolar material are structures known as γ-TuRCs (gamma-tubulin ring complexes). Each γ-TuRC, which consists of gamma-tubulin together with several other proteins, is a template for nucleating a microtubule. Microtubules are nucleated from the γ-TuRCs in such a way that the plus ends of the microtubules grow away from the centrosome. Thus, if the microtubules remain attached to the centrosome, they form a radial array of uniform polarity orientation (Euteneuer and McIntosh 1981; Schiebel 2000; Teixido-Travesa et al. 2010). Such a radial array, typical of simple interphase cells, is able to direct organelle traffic by virtue of the tendency of different types of organelles to interact with specific motors that move toward either plus or minus ends of microtubules. This is why, for example, the Golgi apparatus tends to cluster at the centrosome; because membranous elements that comprise the Golgi are transported by cytoplasmic dynein toward

Fig. 18.2 Distribution of microtubules in cultured embryonic rat hippocampal neurons in the context of dendritic development. **a** A dendrite-bearing neuron immunostained for microtubules. The image is presented in a quantitative scale in which *white* indicates the highest intensity, *black* indicates the least, and shades of *gray* indicate intermediate levels. The cell body contains high levels of microtubules within a discrete region. This region is continuous with high levels of polymer within the developing dendrites. Adapted from Sharp et al. (1995). Bar, 20 µm. **b** Schematic illustration of a dendrite-bearing neuron, depicting a stream of microtubules emerging from the centrosome and flowing into the developing dendrites. The centrosome itself occupies a location that is roughly central in reference to developing dendrites

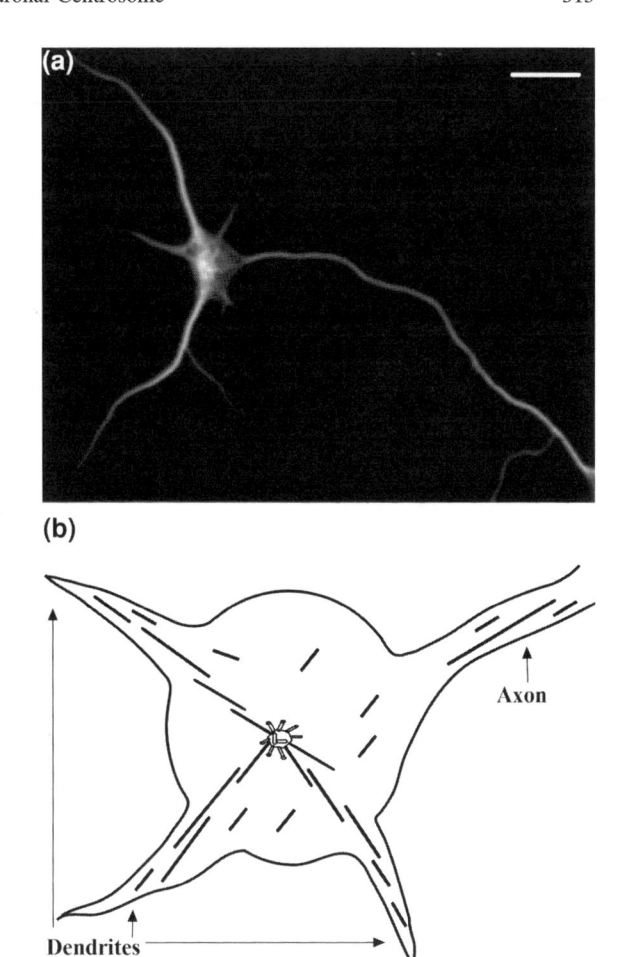

minus ends of microtubules (Corthesy-Theulaz et al. 1992). As a general but not universal principle, very little gamma-tubulin is located in cells anywhere except the centrosome (or other microtubule-organizing centers) and de novo nucleation of microtubules is suppressed in the cytoplasm relative to nucleation from such structures (Alberts et al. 2007). Nucleation from structural templates also serves the purpose of constraining the lattice of the microtubule to a consistent number of protofilaments (typically 13 in most vertebrate cells) (Evans et al. 1985) although there are other factors that influence protofilament number as well (Fourniol et al. 2010; Moores et al. 2004).

The centrosome is generally positioned in the center of the cell (hence the name centrosome) and this location is determined by a balance of forces that act upon the microtubules that emanate from the centrosome while remaining attached to it (Euteneuer and Schliwa 1992; Vallee and Stehman 2005). Without the attached microtubules, there is nothing for molecular motors to pull on in order to center the centrosome and its position becomes less centralized (Burakov et al. 2003). Based on observations from other cell types, we have posited that the reason why the centrosome relocates toward a particular neurite, if it does, is that the machinery that transports microtubules strongly favors that particular neurite at that particular moment in development (Baas 1996). Thus, the microtubules that are released from the centrosome would flow into the relevant neurite but those that are not yet released would react to the motor-driven forces by pulling the centrosome toward that particular neurite. Thus, in neurons, we would speculate that the degree to which the location of the centrosome is predictive of where an axon or dendrite emerges from the cell body probably relates to how active the centrosome is in nucleating microtubules as well as the degree to which or rate at which the microtubules are released once nucleated. In other words, if the centrosome is not very active at nucleating microtubules, it would not be relocated toward any particular neurite. If the centrosome is highly active at nucleating microtubules but most or all of the microtubules are almost immediately released upon nucleation, the centrosome would not be relocated toward any particular neurite. Also, if the relevant motors do not favor any particular neurite, the centrosome would not be relocated toward any particular neurite. These points are schematically illustrated in Figs. 18.1 and 18.2b. It is difficult to imagine a scenario by which the centrosome would relocate without being active at nucleating microtubules.

Whether or not it is functionally important that the centrosome is located where it is, in various types of neurons at particular stages of development, remains to be seen. Certainly, if there are multiple options for where the microtubules released from the centrosome could be transported, a location near the hillock of the relevant neurite would be an advantage for directing microtubules into that neurite. Another possibility is that the location of the centrosome could be functionally important but for other reasons, such as providing a flow of Golgi-derived vesicles. Support for this idea comes from work showing in cultured hippocampal neurons a particularly robust flow of membranous elements into the immature process that develops into the axon (Bradke and Dotti 1997). Yet another possibility is that the centrosome is important for reasons related to the various proteins that gather together to form the pericentriolar material. For example, the pericentriolar material is rich in kinases (Hames et al. 2005), and hence the centrosome could act as a processing center to phosphorylate functionally important proteins. Alternatively, the pericentriolar material might act as a sink for various proteins that would otherwise, and under certain circumstances, be widely distributed in the neuron. This could apply not only to proteins such as kinases, but also to proteins directly related to microtubule nucleation. In such a scenario, it may not be essential that the centrosome nucleates microtubules, but by sequestering the proteins needed for microtubule nucleation, the centrosome ensures that microtubule nucleation does not occur in other locales, where it would be problematic. Whatever the case, it may become important, as we ponder the entirety of the data on the neuronal centrosome, to think more expansively on the potential roles that it may play in organizing the cytoplasm and directing various events relevant to the axon and dendrites.

As noted earlier, our original interest in the neuronal centrosome was as a "generator" of new microtubules for axons and dendrites. However, we should note that there are exceptions to the rule of microtubule nucleation being constrained to centrosomes, as gamma-tubulin can redistribute to new locations in certain cell types (Bugnard et al. 2005). In fact, we previously proposed that the non-uniform orientation of dendritic microtubules might result from centrosomal proteins being relocated from the centrosome into dendrites at early stages of their differentiation (Baas et al. 1989). There are also examples in the literature where de novo nucleation of microtubules has been observed (Yvon and Wadsworth 1997) but not commonly, as haphazard nucleation of new microtubules would probably make more problems for cells than solutions. In studies directed at testing for de novo nucleation of microtubules in axons, we found no evidence for it, as all new assembly was observed to occur via elongation of existing polymers (see below).

18.4 Older Data on Microtubules and the Neuronal Centrosome

Our studies positing the neuronal centrosome as a generator of microtubules for the axon were conducted in the 1990s, and utilized tools that had been previously used in other cell types. The classic method for identifying sites of microtubule nucleation in cells is to depolymerize existing microtubules with nocodazole, and then remove the drug so that microtubules can reassemble from their sites of origin. This method, first used to identify the centrosome as a site of microtubule nucleation in other cell types, (De Brabander et al. 1977, 1980), was used by our laboratory to identify potential sites of microtubule nucleation in the axons of cultured rat sympathetic neurons (Baas and Ahmad 1992). After drug removal, all new microtubule assembly arose from the plus ends of the stable microtubules that resisted depolymerization. No microtubules arose independently of existing microtubules, suggesting that the plus ends of pre-existing microtubules are the exclusive sites of microtubule assembly in the axon. These findings were consistent with previous work on cultured sensory neurons demonstrating that when all microtubule polymer is pharmacologically depolymerized from isolated axons, no reassembly occurred after removal of the drug (Baas and Heidemann 1986). Based on these findings, we concluded that entirely new microtubules destined for the axon must be nucleated within the cell body. To explore the issue further, we investigated the distribution of gamma-tubulin in these neurons (Baas and Joshi 1992). Using both biochemical and immunoelectron microscopic assays, we found no evidence for gamma-tubulin in the axon. In addition, we found no appreciable levels of gamma-tubulin anywhere in the cell body except at the centrosome, suggesting that the centrosome is the sole site for the nucleation of new microtubules for the entire neuron. On the basis of these findings, we proposed that microtubules destined for the axon are nucleated at the centrosome, released, and then transported into the axon.

Electron microscopic analyses of different kinds of neurons at different developmental stages vary with regard to the appearance of the centrosome, but most studies reveal relatively few microtubules directly attached to the centrosome. In our studies on cultured sympathetic neurons, generally fewer than ten and often no microtubules were observed to be attached to the centrosome (Baas and Joshi 1992; Yu et al. 1993). These observations raised the possibility that axonal microtubules may not originate at the centrosome and that the neuronal centrosome may actually be relatively inactive. Alternatively, however, the nucleation and release of microtubules from the neuronal centrosome may be so rapid that there is insufficient time for substantial numbers of attached microtubules to accumulate at the centrosome before they are released. To address this issue, we tested the capacity of the neuronal centrosome to nucleate large numbers of microtubules, using the same drug-recovery regime that we used on the axon (Yu et al. 1993). Within a few minutes of drug removal, hundreds of microtubules reassembled in the region of the centrosome, and most of these microtubules were clearly attached to it (Fig. 18.3). Some of the microtubules were not attached to the centrosome, but were aligned side-by-side with the attached microtubules, suggesting that the unattached microtubules had been released from the centrosome after their nucleation. In addition, unattached microtubules were present in the cell body at decreasing levels with increasing distance from the centrosome. By 30 min after removing the drug, the microtubule array was indistinguishable from that of control neurons, suggesting that the hundreds of microtubules nucleated from the centrosome were subsequently released and translocated away from the centrosome.

We next tested whether microtubules derived from the centrosome are essential for the initiation and growth of the axon. Our strategy was to microinject into cultured sympathetic neurons a function-blocking antibody to gamma-tubulin previously shown to arrest microtubule nucleation at the centrosome when microinjected into other cell types (Ahmad et al. 1994). We reasoned that if centrosomally derived microtubules are required for the growth of the axon, we would expect inhibition of centrosome function to compromise or inhibit axonal growth. These experiments were tricky, however, because the cell body of the neuron is packed with microtubules that had presumably (according to our hypothesis) already been nucleated and released from the centrosome. Therefore, it was also necessary to deplete the neuron experimentally of pre-existing microtubules. After depolymerizing existing microtubules with nocodazole, the antibody was microinjected into neurons, and then the drug was rinsed from the cultures. Reassembly of microtubules over the next two hours was severely diminished under these conditions, and axonal growth was either compromised or completely abolished. These results, using an admittedly complicated experimental regime, suggested that microtubules generated from the centrosome are important for axonal growth.

Finally, we set forth to test if the microtubules nucleated at the centrosome are the same microtubules that ultimately arrive in the axon. To test this, we modified our pharmacological experiments into a kind of "pulse-chase" regime that permitted us

Fig. 18.3 Electron micrographs of cultured rat sympathetic neurons in the region of the centrosome. **a** A neuron showing centrosome consisting of two centrioles and multiple unattached microtubules. **b** A neuron treated for 6 h with 10 µg/ml nocodazole, rinsed free of the drug, and permitted to recover for 5 min. Microtubule reassembly from the centrosome is dramatic, with high levels of attached microtubules. Also apparent are other microtubules not directly attached to the centrosome. These microtubules are aligned with the attached microtubules as if they were once attached and then released from the centrosome. Analyses of every section through each centrosome were required to define and score attached and unattached microtubules. Adapted from Yu et al. (1993). Bar, 0.4 µm

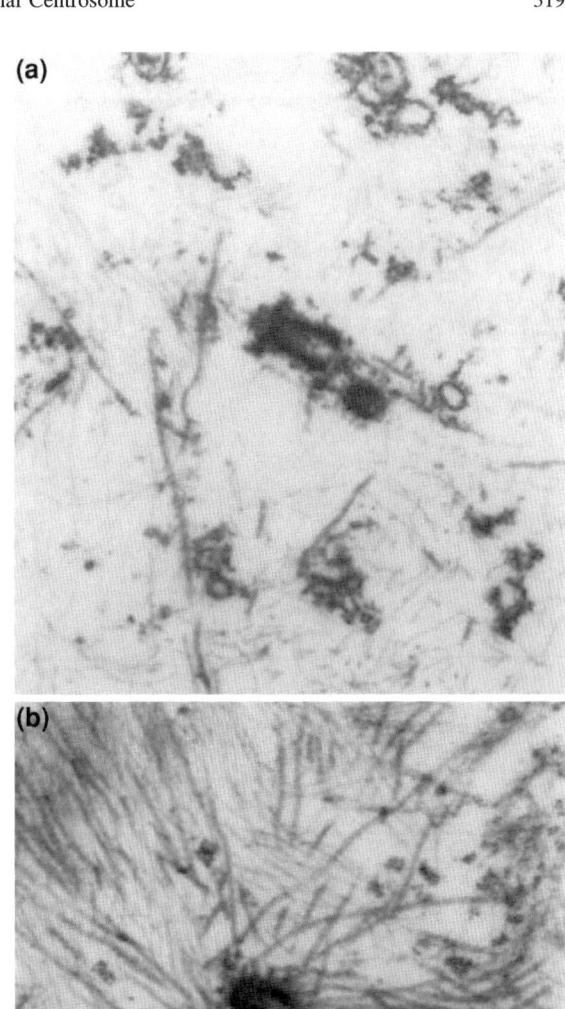

to follow the progress of a small population of microtubules nucleated at the centrosome (Ahmad and Baas 1995; also see Fig. 18.4). After drug treatment to depolymerize microtubules, and a few minutes of microtubule reassembly at the

Fig. 18.4 Schematic illustration of pharmacological strategy for revealing the progression of microtubules outward from the neuronal centrosome. Nocodazole was the first drug used to depolymerize pre-existing microtubules. Next, after a brief recovery period, vinblastine was used as an anti-microtubule drug to suppress further assembly of microtubules. Microtubules redistributed over time. Adapted from Ahmad and Baas (1995)

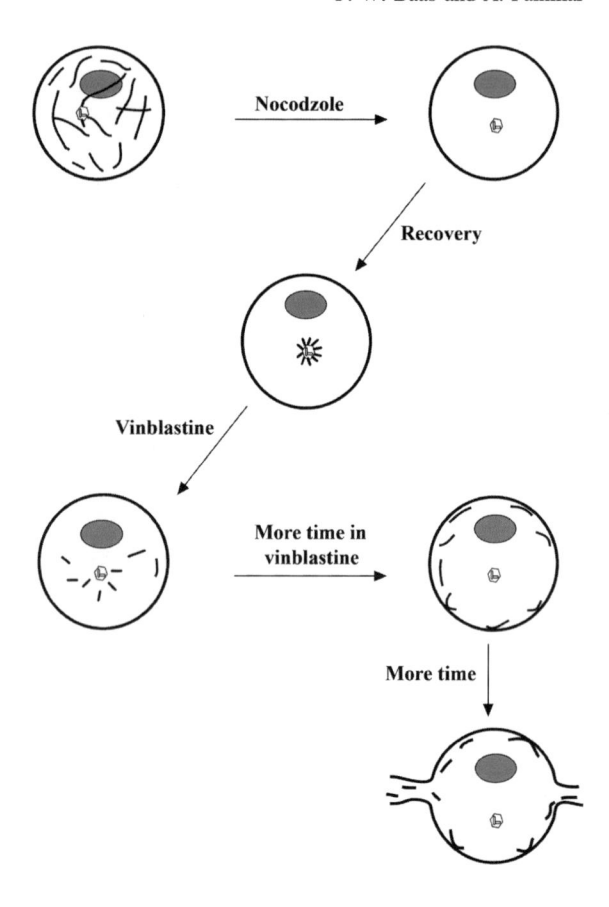

centrosome, low levels of a second anti-microtubule drug (vinblastine) were added to the cultures to suppress further microtubule assembly while not substantially depolymerizing existing microtubules. Thus, we reasoned that any alterations in the microtubule array that occur after the addition of the second drug must be the result of microtubule movements from one location in the cell to another. Consistent with this expectation, microtubule levels remained roughly the same after the addition of vinblastine, as did the lengths of individual microtubules over time. Within minutes, unattached microtubules began to appear in the cytoplasm, and by 10 min many of these had reached the periphery of the cell body. By 1 h, few or no microtubules were attached to the centrosome and most of the microtubules were concentrated at the cell periphery. In the case of the neurons that were able to grow axons under these conditions, microtubules appeared progressively further down the axons with increasing time (see Fig. 18.4). These results suggested that microtubules derived from the centrosome are transported outward from the centrosome toward cell periphery and then into and down the length of the axon.

Due to the geometry of the neuron, the density of the microtubule array, and the pool of free tubulin in neurons, we have not been able to directly visualize microtubules in living neurons moving from the centrosome into the axon. The issue arises as to whether the centrosome is actually needed under normal circumstances, or whether our pharmacologic regimes stress the system to a point where an otherwise unnecessary centrosome becomes necessary. In support of our interpretation, the active release of microtubules from the centrosome has been directly visualized in cellular extracts (Belmont et al. 1990) as well as living epithelial cells in a regime that involved no drug treatments (Keating et al. 1997). Moreover, in the case of the neuron, we have also shown that inhibition of katanin, a microtubule-severing protein, prohibits microtubule release from the centrosome, which in turn precludes the appearance of free microtubules in our pharmacologic regime (Ahmad et al. 1999).

18.5 Newer Data on Microtubules and the Neuronal Centrosome

Over a decade after our spate of papers on the neuronal centrosome, the laboratory of Frank Bradke has recently challenged the idea of the neuronal centrosome acting as a generator of microtubules for axons and dendrites (Stiess and Bradke 2010; Stiess et al. 2010). They favor the alternative view that the centrosome is dismantled during neuronal development such that its microtubule-nucleating duties are spread to new locations in the neuron, such as within the axon and dendrite themselves. This scenario would be similar to what has been shown for muscle cells, in which gamma-tubulin and its associated microtubule-nucleating properties are redistributed to the nuclear membrane and other sites within the cytoplasm (Bugnard et al. 2005). In fact, consistent with our original speculation for how a non-uniform microtubule polarity pattern might arise in dendrites (Baas et al. Baas et al. 1988, 1989), at least one pericentriolar protein has been shown to be present in dendrites but not axons (Ferreira et al. 1993). It is also provocative with regard to the centrosome dismantling hypothesis for neurons that Leask and colleagues reported a steady diminution in gamma-tubulin from the centrosome as dorsal root ganglion neurons mature, which would be consistent with a gradual redistribution of their pericentriolar proteins (Leask et al. 1997). Bradke's group found a similar diminution of gamma-tubulin levels as well as another key protein component of the γ-TuRCs, consistent with the idea that the capacity of the neuronal centrosome to act as a generator of microtubules wanes as the neuron matures. In addition, they found with hippocampal neurons that the nocodazole recovery regime resulted in a burst of microtubules from the centrosome early in development (Dotti and Banker 1991), but this was not the case later in development. In cultures that were several days old, neurons bearing dendrites showed no specific recovery of microtubules from the centrosome after nocodazole treatment and removal. Instead, the microtubules reassembled from sites throughout the cell body (Stiess et al. 2010).

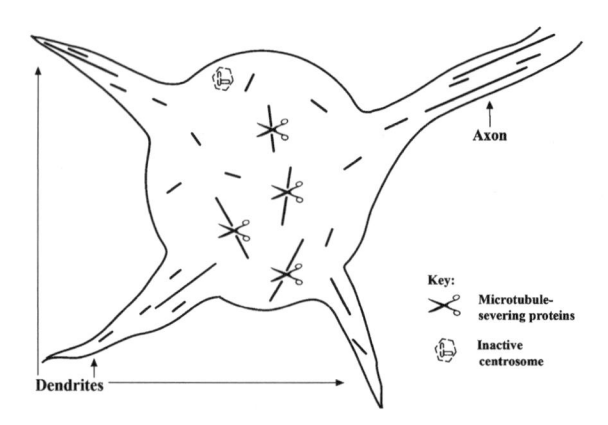

Fig. 18.5 Schematic illustration of a mature neuron showing an inactive (dismanted) centrosome and the severing of microtubules. As the neuron matures, the ability of the centrosome to nucleate microtubules diminishes. Microtubule number is increased by severing of pre-existing long microtubules into short mobile pieces, followed by the transport of short microtubules into the axon or the dendrites. These short microtubules can serve as seeds for assembly of longer microtubules

The authors posited, based on these results, that microtubule-nucleating capacity becomes de-centralized as the neuron develops such that nucleation of microtubules can occur throughout the cell body and potentially even within axons and dendrites themselves. We would agree, assuming that nucleation in this context means assembly from pre-existing microtubules, even very short fragments that are able to act as seeds for new assembly (Baas and Ahmad 1992; Baas and Black 1990; Baas and Heidemann 1986; Black et al. 1984; Brady et al. 1984; Morris and Lasek 1982). As neurons mature, it is virtually impossible to completely depolymerize the more stable microtubules, even with prolonged drug treatments, so we suspect that the reassembly of microtubules observed by Steiss and colleagues represents "nucleation" from stable microtubule fragments, but not bona fide nucleation in the de novo sense. Even so, the results reported by these authors accentuate the fact that the neuron can very ably go on "auto pilot" once a robust microtubule array has been constructed, such that a centralized factory for microtubule production can be shut down. As noted earlier (and as discussed also by Steiss and colleagues), the severing of existing microtubules in the cell body as well as in the axon and the dendrites is presumably sufficient for increasing the number of microtubules whenever and wherever needed (see Fig. 18.5).

18.6 Concluding Remarks

It may be relevant to consider that different kinds of neurons go about their business in somewhat different ways and on different timetables. For example, central and peripheral neurons may differ with regard to the importance of the centrosome, and also neurons that bear dendrites or multiple axons may differ from those that do not bear dendrites and have the more classic single axon. Migratory neurons appear to utilize their centrosome in the most traditional fashion, as the vast majority of microtubules in the migratory neuron remain attached to the centrosome (Tsai and Gleeson 2005), while a small fraction of the microtubules are apparently released from the centrosome so that they transit down the leading process or slide their minus ends behind the centrosome (Falnikar et al. 2011). Overall, it appears that the centrosome is most important early in neuronal development, especially during neuronal migration and early axonal differentiation. After that, the preponderance of the data suggests that neurons gradually lose their dependence on the centrosome in favor of self-sustaining mechanisms for maintaining the microtubule arrays of the axon and the dendrites.

Whether or not the neuron needs a centrosome for the development of proper axons or dendrites remains a debatable point, as merely being able to form an axon or dendrite in culture may be a very different thing than being able to form the appropriate axon or dendrite within the context of a functional nervous system. In addition, it is pertinent to keep in mind that the biology community continues to be surprised by the plethora of transgenic animals that are viable in the absence of proteins believed to play important roles in cellular functions. Oftentimes, the importance of a particular protein (or in this case, an organelle) is gleaned only after cells or entire organisms are challenged in particular ways. This may be the case developmentally, and also in more mature neurons in which the centrosome appears to have become vestigial. For example, perhaps under certain circumstances, the centrosome is re-activated to enable the neuron to meet a particular challenge, such as restructuring of the dendritic arbor in response to learning or disease, or regeneration of an injured axon.

Acknowledgments The work in our laboratory is supported by grants from the National Institutes of Health and the National Science Foundation.

References

Ahmad FJ, Baas PW (1995) Microtubules released from the neuronal centrosome are transported into the axon. J Cell Sci 108(Pt 8):2761–2769

Ahmad FJ, Joshi HC, Centonze VE, Baas PW (1994) Inhibition of microtubule nucleation at the neuronal centrosome compromises axon growth. Neuron 12:271–280

Ahmad FJ, Echeverri CJ, Vallee RB, Baas PW (1998) Cytoplasmic dynein and dynactin are required for the transport of microtubules into the axon. J Cell Biol 140:391–401

Ahmad FJ, Yu W, McNally FJ, Baas PW (1999) An essential role for katanin in severing microtubules in the neuron. J Cell Biol 145:305–315

Alberts B, Johnson A, Lewis J, Raff M. Roberts K, Walter P (2007) Molecular biology of the cell, 5th edn. Garland Science, New York

Baas PW (1996) The neuronal centrosome as a generator of microtubules for the axon. Curr Top Dev Biol 33:281–298

Baas PW, Ahmad FJ (1992) The plus ends of stable microtubules are the exclusive nucleating structures for microtubules in the axon. J Cell Biol 116:1231–1241

Baas PW, Ahmad FJ (1993) The transport properties of axonal microtubules establish their polarity orientation. J Cell Biol 120:1427–1437

Baas PW, Black MM (1990) Individual microtubules in the axon consist of domains that differ in both composition and stability. J Cell Biol 111:495–509

Baas PW, Buster DW (2004) Slow axonal transport and the genesis of neuronal morphology. J Neurobiol 58:3–17

Baas PW, Heidemann SR (1986) Microtubule reassembly from nucleating fragments during the regrowth of amputated neurites. J Cell Biol 103:917–927

Baas PW, Joshi HC (1992) Gamma-tubulin distribution in the neuron: implications for the origins of neuritic microtubules. J Cell Biol 119:171–178

Baas PW, Lin S (2011) Hooks and comets: the story of microtubule polarity orientation in the neuron. Dev Neurobiol 71:403–418

Baas PW, Yu W (1996) A composite model for establishing the microtubule arrays of the neuron. Mol Neurobiol 12:145–161

Baas PW, Deitch JS, Black MM, Banker GA (1988) Polarity orientation of microtubules in hippocampal neurons: uniformity in the axon and nonuniformity in the dendrite. Proc Natl Acad Sci U S A 85:8335–8339

Baas PW, Black MM, Banker GA (1989) Changes in microtubule polarity orientation during the development of hippocampal neurons in culture. J Cell Biol 109:3085–3094

Belmont LD, Hyman AA, Sawin KE, Mitchison TJ (1990) Real-time visualization of cell cycle-dependent changes in microtubule dynamics in cytoplasmic extracts. Cell 62:579–589

Black MM, Cochran JM, Kurdyla JT (1984) Solubility properties of neuronal tubulin: evidence for labile and stable microtubules. Brain Res 295:255–263

Bradke F, Dotti CG (1997) Neuronal polarity: vectorial cytoplasmic flow precedes axon formation. Neuron 19:1175–1186

Brady ST, Tytell M, Lasek RJ (1984) Axonal tubulin and axonal microtubules: biochemical evidence for cold stability. J Cell Biol 99:1716–1724

Bruckenstein DA, Higgins D (1988) Morphological differentiation of embryonic rat sympathetic neurons in tissue culture. I. Conditions under which neurons form axons but not dendrites. Dev Biol 128:324–336

Bugnard E, Zaal KJ, Ralston E (2005) Reorganization of microtubule nucleation during muscle differentiation. Cell Motil Cytoskelet 60:1–13

Burakov A, Nadezhdina E, Slepchenko B, Rodionov V (2003) Centrosome positioning in interphase cells. J Cell Biol 162:963–969

Corthesy-Theulaz I, Pauloin A, Pfeffer SR (1992) Cytoplasmic dynein participates in the centrosomal localization of the Golgi complex. J Cell Biol 118:1333–1345

de Anda FC, Pollarolo G, Da Silva JS, Camoletto PG, Feiguin F, Dotti CG (2005) Centrosome localization determines neuronal polarity. Nature 436:704–708

De Brabander M, De Mey J, Joniau M, Geuens S (1977) Immunocytochemical visualization of microtubules and tubulin at the light- and electron-microscopic level. J Cell Sci 28:283–301

De Brabander M, Geuens G, Nuydens R, Willebards R, De Mey J (1980) The microtubule nucleating and organizing activities of kinetochores and centrosomes in living PtK2 cells. In: De Brabander M and De Mey J (eds) Microtubules and microtubule inhibitors. North-Holland, Amsterdam, pp 255–268

Dotti CG, Banker G (1991) Intracellular organization of hippocampal neurons during the development of neuronal polarity. J Cell Sci Suppl 15:75–84

Euteneuer U, McIntosh JR (1981) Polarity of some motility-related microtubules. Proc Natl Acad Sci U S A 78:372–376

Euteneuer U, Schliwa M (1992) Mechanism of centrosome positioning during the wound response in BSC-1 cells. J Cell Biol 116:1157–1166

Evans L, Mitchison T, Kirschner M (1985) Influence of the centrosome on the structure of nucleated microtubules. J Cell Biol 100:1185–1191

Falnikar A, Tole S, Baas PW (2011) Kinesin-5, a mitotic microtubule-associated motor protein, modulates neuronal migration. Mol Biol Cell 22:1561–1574

Ferreira A, Palazzo RE, Rebhun LI (1993) Preferential dendritic localization of pericentriolar material in hippocampal pyramidal neurons in culture. Cell Motil Cytoskelet 25:336–344

Fourniol FJ, Sindelar CV, Amigues B, Clare DK, Thomas G, Perderiset M, Francis F, Houdusse A, Moores CA (2010) Template-free 13-protofilament microtubule-MAP assembly visualized at 8 A resolution. J Cell Biol 191:463–470

Hames RS, Crookes RE, Straatman KR, Merdes A, Hayes MJ, Faragher AJ, Fry AM (2005) Dynamic recruitment of Nek2 kinase to the centrosome involves microtubules, PCM-1, and localized proteasomal degradation. Mol Biol Cell 16:1711–1724

Higginbotham HR, Gleeson JG (2007) The centrosome in neuronal development. Trends Neurosci 30:276–283

Joshi HC, Baas PW (1993) A new perspective on microtubules and axon growth. J Cell Biol 121:1191–1196

Keating TJ, Peloquin JG, Rodionov VI, Momcilovic D, Borisy GG (1997) Microtubule release from the centrosome. Proc Natl Acad Sci U S A 94:5078–5083

Leask A, Obrietan K, Stearns T (1997) Synaptically coupled central nervous system neurons lack centrosomal gamma-tubulin. Neurosci Lett 229:17–20

Lyser KM (1964) Early differentiation of motor neuroblasts in the chick embryo as studied by electron microscopy I. Gen Aspects Dev Biol 10:433–466

Lyser KM (1968) An electron-microscopic study of centrioles in differentiating motor neuroblasts. J Embryol Exp Morphol 20:343–354

Moores CA, Perderiset M, Francis F, Chelly J, Houdusse A, Milligan RA (2004) Mechanism of microtubule stabilization by doublecortin. Mol Cell 14:833–839

Morris JR, Lasek RJ (1982) Stable polymers of the axonal cytoskeleton: the axoplasmic ghost. J Cell Biol 92:192–198

Qiang L, Yu W, Liu M, Solowska JM, Baas PW (2010) Basic fibroblast growth factor elicits formation of interstitial axonal branches via enhanced severing of microtubules. Mol Biol Cell 21:334–344

Schiebel E (2000) Gamma-tubulin complexes: binding to the centrosome, regulation and microtubule nucleation. Curr Opin Cell Biol 12:113–118

Sharp GA, Weber K, Osborn M (1982) Centriole number and process formation in established neuroblastoma cells and primary dorsal root ganglion neurones. Eur J Cell Biol 29:97–103

Sharp DJ, Yu W, Baas PW (1995) Transport of dendritic microtubules establishes their nonuniform polarity orientation. J Cell Biol 130:93–103

Sharp DJ, Yu W, Ferhat L, Kuriyama R, Rueger DC, Baas PW (1997) Identification of a microtubule-associated motor protein essential for dendritic differentiation. J Cell Biol 138:833–843

Stiess M, Bradke F (2011) Neuronal polarization: the cytoskeleton leads the way. Dev Neurobiol 71:430–444

Stiess M, Maghelli N, Kapitein LC, Gomis-Ruth S, Wilsch-Brauninger M, Hoogenraad CC, Tolic-Norrelykke IM, Bradke F (2010) Axon extension occurs independently of centrosomal microtubule nucleation. Science 327:704–707

Teixido-Travesa N, Villen J, Lacasa C, Bertran MT, Archinti M, Gygi SP, Caelles C, Roig J, Luders J (2010) The gammaTuRC revisited: a comparative analysis of interphase and mitotic human gammaTuRC redefines the set of core components and identifies the novel subunit GCP8. Mol Biol Cell 21:3963–3972

Tsai LH, Gleeson JG (2005) Nucleokinesis in neuronal migration. Neuron 46:383–388

Vallee RB, Stehman SA (2005) How dynein helps the cell find its center: a servomechanical model. Trends Cell Biol 15:288–294

Wang L, Brown A (2002) Rapid movement of microtubules in axons. Curr Biol 12:1496–1501

Yu W, Centonze VE, Ahmad FJ, Baas PW (1993) Microtubule nucleation and release from the neuronal centrosome. J Cell Biol 122:349–359

Yu W, Ahmad FJ, Baas PW (1994) Microtubule fragmentation and partitioning in the axon during collateral branch formation. J Neurosci 14:5872–5884

Yu W, Sharp DJ, Kuriyama R, Mallik P, Baas PW (1997) Inhibition of a mitotic motor compromises the formation of dendrite-like processes from neuroblastoma cells. J Cell Biol 136:659–668

Yu W, Ling C, Baas PW (2001) Microtubule reconfiguration during axogenesis. J Neurocytol 30:861–875

Yu W, Qiang L, Solowska JM, Karabay A, Korulu S, Baas PW (2008) The microtubule-severing proteins spastin and katanin participate differently in the formation of axonal branches. Mol Biol Cell 19:1485–1498

Yvon AM, Wadsworth P (1997) Non-centrosomal microtubule formation and measurement of minus end microtubule dynamics in A498 cells. J Cell Sci 110(Pt 19):2391–2401

Zmuda JF, Rivas RJ (1998) The Golgi apparatus and the centrosome are localized to the sites of newly emerging axons in cerebellar granule neurons in vitro. Cell Motil Cytoskelet 41:18–38

Chapter 19
Centrosomes and Cell Division in Apicomplexa

Leandro Lemgruber, Marek Cyrklaff and Freddy Frischknecht

Abstract Apicomplexans are curious single-celled organisms. Belonging to the group of chromalveolates, life for an apicomplexan can be parasitic and some species can cause diseases such as malaria or toxoplasmosis. No apicomplexan is alike, although they share some common features such as being highly polar cells with unique apical organelles. They often change the cells of their metazoan hosts. When they move, apicomplexans do not crawl but glide; when they divide, apicomplexans go through mechanisms matched in cell biological bizarreness only by their names. They undergo schizogony or endodyogeny, processes that are usually not part of a regular molecular cell biology textbook; but they should, as their uniqueness might lead to insights into what proteins and processes are truly essential to make progeny. Here we highlight some of our current knowledge of centrosome and microtubule biology of selected apicomplexan parasites for the yeast and metazoan cell biologist to contemplate.

19.1 Introduction

Most molecular and cell biological studies of any kind are conducted on vertebrate or metazoan model organisms, all belonging to the Opisthokonta, while some research is also performed on plants. In comparison, little attention is being paid to the vast majority of biological life forms outside these two groups. This is curious considering

---

L. Lemgruber (✉) · M. Cyrklaff · F. Frischknecht
Department of Infectious Diseases, Parasitology,
University of Heidelberg Medical School,
Im Neuenheimer Feld 324, 69120
Heidelberg, Germany
e-mail: leandro.lemgruber@med.uni-heidelberg.de

the incredible divergence of these organisms with the associated wide open space for fundamental discoveries and the fact that some of them are causing devastating disease in humans and farm animals. The neglect is neatly illustrated by a PubMed search with "microtubules and yeast" retrieving 1871 articles, while only 38 hits were found for "microtubules and malaria" at the time of writing. Indeed, just the phylum Apicomplexa within the group of chromalveolates comprises an incredibly heterogeneous group of intracellular parasites that includes species of relevance for humans due to their health and/or economical effects (Tenter et al. 2002). Some are: *Plasmodium* sp., the causative agents of malaria; *Toxoplasma gondii*, causing toxoplasmosis (a worldwide distributed disease, especially important for pregnant women and immunecompromised patients); *Cystoisospora belli*, causing isosporiasis and *Cryptosporidium* that is responsible for cryptosporidiosis (the latter ones being intestinal parasites, affecting especially immunodeficient people). Also, the livestock economy has to cope with severe losses due to disease impact and prevention costs for some infectious agents, like *Eimeria* (poultry), *Babesia*, and *Theileria* (cattle).

These protozoa present an intracellular organization that is superficially similar to metazoan cells. They have a nucleus, an endoplasmic reticulum, a Golgi complex and only one mitochondrion (Joiner and Roos 2002). The group Apicomplexa is formed by organisms that present a somewhat peculiar and exclusive structure-the apical complex, composed of cytoskeletal components and secretory organelles (micronemes and rhoptries) that are essential for invasion of the host cell (Morrissette and Sibley 2002a; Dubremetz 2007; Carruthers and Tomley 2008) and dense granules, which modify the parasitophorous vacuole surrounding the intracellular stages of these parasites (Cesbron-Delauw et al. 2008). Curious here to note are the polar rings and conoids that are present in some but not all apicomplexans (Fig. 19.1a). The conoid is formed by a unique open microtubule-like structure with the highest bend observed so far for a tubulin "tube" (Hu et al. 2002). This structure is important for host cell invasion by *Toxoplasma* but curiously absent in parasites causing malaria. The polar rings, appear to organize microtubules in a cage (Hu et al. 2002). Only recently, the first component of this structure, a protein with no homology to anything outside the apicomplexans, and not even across all apicomplexans, was identified (Tran et al. 2010). In addition to the above features, apicomplexans contain a unique organelle derived from a secondary endosymbiotic event, the apicoplast (Köhler et al. 1997; Ralph et al. 2004). Curiously, this organelle was documented for decades mainly by electron microscopy studies, but only in the late 1990s was it identified as a unique organelle (Köhler et al. 1997). The apicoplast is a non-photosynthetic plastid responsible for the type II fatty acid synthesis, isoprenoids biosynthesis, and carbohydrate metabolism (Gleeson 2000; Mazumbar et al. 2006; Fleige et al. 2007). Recently, Yeh and DeRisi (2011) demonstrated by adding isopentenyl pyrophosphate to parasites lacking the apicoplast that in blood stages of *Plasmodium* only the isoprenoid biosynthesis pathway is essential. Intriguingly, *Theileria* lacks the fatty acid biosynthesis enzymes, and *Cryptosporidium* may have lost the apicoplast along its evolution (Goodman and McFadden 2007), attesting the diversity of this group of organisms. Apicomplexan parasites present other unique cytoplasmic inclusions including enigmatic virus-like (Lemgruber and Lupetti 2011).

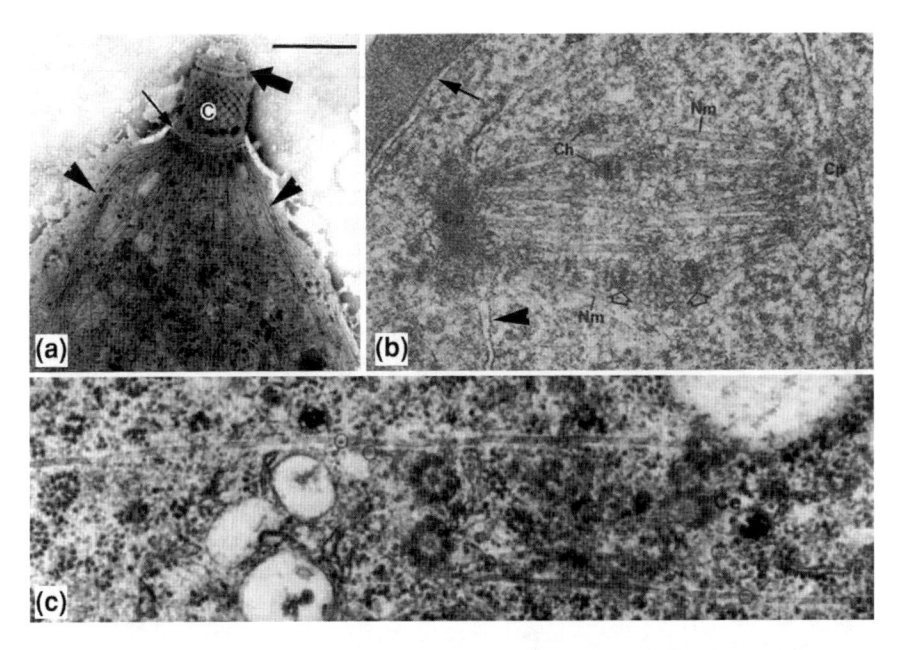

Fig. 19.1 Examples of microtubular structures in Apicomplexa as revealed by electron microscopy. **a** Negative-stained *Toxoplasma* after membrane extraction showing the conoid (C), subpellicular microtubules (*arrowheads*) and the *upper* (*thick arrow*) and *lower* (*thin arrow*) polar rings (Modified after Morrissette et al. 1997; with permissions of The Company of Biologists). **b** A thin section of *Plasmodium* undergoing nuclear division, showing centriolar plaques (Cp), intranuclear microtubules (Nm), and electron-dense structures (Ch and *open arrows*). *Arrow* points to the parasite plasma membrane and the *arrowhead* to the nuclear envelope (Modified after Aikawa and Beaudoin 1968; with permissions of The Rockefeller University Press). **c** *Centrioles of Eimeria* with the unusual "9+1" arrangement. In the vicinity, extranuclear microtubules are observed (After Dubremetz and Elsner 1979; with permissions of the John Wiley and Sons)

In order to successfully infect a host and be able to be transferred from one host to another, these parasites must proliferate. As obligatory intracellular pathogens, the apicomplexans duplicate and divide within a host cell. The nuclear divisions in apicomplexans are achieved by cryptomitosis, where the nuclear membrane continues to be present during the different phases of the cell division process, with no chromosomal condensation during karyokinesis (Striepen et al. 2007). The replication process occurs in three different manners: endodyogeny, where two daughter cells are formed within a mother cell; schizogony, in which sequential nuclear divisions are followed by budding of new parasites; or by endopolygeny, with sequential DNA replication without nuclear division (Fig. 19.2). Endodyogeny allows the parasite to remain infective even during the division process, since the mother cell still preserves its apical structures. In schizogony, however, the apical complex and the cytoskeletal system break down. They are eventually being formed de novo just before the budding of progeny cells occurs. This process

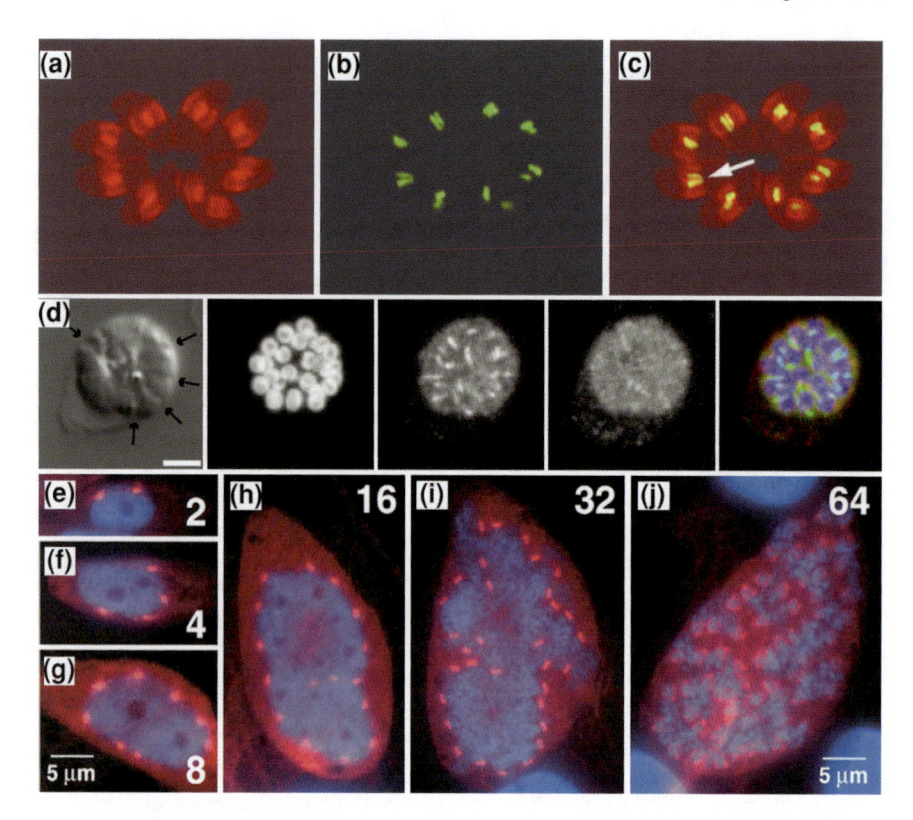

Fig. 19.2 Fluorescence light microscopy images of examples of the different division processes in apicomplexa: endodyogeny (**a–c**), schizogony (**d**), and endopolygeny (**e–j**). **a–c** *Toxoplasma* expressing an apicoplast molecule tagged with GFP (in *green*), and stained with an antibody against the inner membrane complex (in *red*). The *arrow* in **c** points to the "U" shape of the apicoplast during division (After Striepen et al. 2000; with permissions of The Rockefeller University Press). **d** *Plasmodium* parasites (*arrows*) at the final stage of mitosis. The nuclei are stained in *blue*, the MTOCs in *red* and the microtubules in *green* (After Gerald et al. 2011; with permissions of the American Society for Microbiology). **e–j**: Nuclear division and cytokinesis in *Sarcocystis*. The α-tubulin is stained in *red* and the DNA in *blue*. Numbers indicate the number of nuclei (After Vaishnava et al. 2005; with permissions of The Company of Biologists)

must be well regulated in order to form new cells containing the correct set of organelles and nuclear material. Endopolygeny is best characterized in *Sarcocystis neurona*, a parasite of horses. During this process the DNA replication, nuclear division, and cytokinesis processes are dissociated from one another, with five cycles of DNA replication occurring prior to nuclear division, generating a 32N nucleus (Vaishnava et al. 2005). A final division generates 64 haploid daughter cells. Curiously, the intranuclear spindle persists throughout the cell cycle. As will be outlined below for other parasites, the apicoplast is associated with the centrosomes and thus equally distributed to the daughter cells after replication.

19.2 Centrosomes in Apicomplexa

In most metazoans, the centrosome functions as the main microtubule-organizing center (MTOC) during interphase as well as during cell division, where it is responsible for mitotic spindle formation and chromosome segregation after disassembly of the nuclear membrane. Centrosomes are formed around a pair of centrioles that are arranged perpendicular to each other. Each centriole is a barrel-shaped structure with nine triplets of microtubules forming the barrel wall (Azimzadeh and Marshall 2010). In apicomplexans, like in yeast, the nuclear membrane continues to be present throughout the cell cycle. The spindle structure and the centrosomes are formed from an electron-dense plaque named centrocone or spindle pole plaque (Aikawa and Beaudoin 1968; Kelley and Hammond 1972; Schrevel et al. 1977; Dubremetz and Elsner 1979; Sibert and Speer 1981; Morrissette and Sibley 2002a; Gerald et al. 2011) (Fig. 19.1b). Also, the centrioles present a unique "9 + 1" microtubule structure, with a sole central microtubule within the nine microtubule triplets (Dubremetz and Elsner 1979) (Fig. 19.1c). Curiously, it has been observed in intranuclear spindles of *Eimeria* that some microtubules span between and connect both centrocones, while other microtubules terminate at kinetochores (Sibert and Speer 1981). However, in another parasitic protozoan, *Lecudina tuzetae*, a gregarine living in the intestine of a marine polychaete annelid, only a few or no astral microtubules were observed (Kuriyama et al. 2005). Because no cytokinesis occurs, the spindle may not require astral microtubules for mitosis. Curiously, *Lecudina* induces microtubule assemblies in different developmental stages that originate from different MTOCs containing γ-tubulin, pericentrin, Cep135, and mitosis-specific phosphoproteins (Kuriyama et al. 2005).

In contrast to spindle microtubules, the subpellicular microtubules that subtend the pellicle of the parasite are mostly not susceptible to microtubule-depolymerizing drugs. These microtubules are often cited as being responsible for maintaining parasite shape or might be involved in vesicular transport (Bannister et al. 2000; Morrissette and Sibley 2002a; Schrével et al. 2008). They are located ca. 20 nm beneath the inner leaflet of the inner membrane complex (IMC), a membrane structure formed from flattened vesicles and unique to this clade of organisms. The IMC encircles the entire parasite at a distance of about 30 nm from the plasma membrane. The space between plasma membrane and IMC is containing the actin-myosin machinery that drives parasites through tissues and into host cells (Heintzelman 2006; Baum et al. 2008). The microtubules may be connected to the IMC with linker proteins (Morrissette et al. 1997; Kudryashev et al. 2010) and do not seem to undergo continuous assembly and disassembly steps. Curiously, these subpellicular microtubules in some apicomplexan parasites contain what could be a protein associated with the luminal part of the microtubule wall (Cyrklaff et al. 2007). This could be important for the unusual stability of these microtubules.

After host cell invasion the subpellicular microtubules slowly disassemble. During formation of the parasite progeny when the subpellicular microtubules are being assembled, they are prone to some drugs that depolymerize microtubules. For example, treatment with oryzalin or colchicine at lower concentrations in *Toxoplasma* does not interrupt the division process but the forming cells do not present subpellicular microtubules and are not capable to infect new host cells (Morrissette and Sibley 2002b). Oryzalin belongs to a group of compounds that selectively disrupt microtubules in several protozoa and in plants, with little effect on host cell microtubules. Generation of oryzalin-resistant *Toxoplasma* lines showed that the drug specifically targets α-tubulin (Morrissette et al. 2004; Ma et al. 2007, 2010). Also in *Plasmodium* erythrocytic stages several drugs were shown to affect microtubules (Pouvelle et al. 1994; Schrével et al. 1994, Sinou et al. 1996, 1998; Fowler et al. 1998; Fennell et al. 2006).

Most of the studies on the replication process performed so far focused on the more prominent members of apicomplexa, *Toxoplasma*, and *Plasmodium*, where molecular tools are available to study the cell division process, e.g. by gene replacement. In others organisms, like *Eimeria* and *Theileria*, most of the studies were limited to observation by light or transmission electron microscopy. Nevertheless, these studies show that the apicomplexa are a diverse group of organisms, each one presenting an intriguing variation of unique cell biological processes.

Here, we summarize what is known about the division process of *Toxoplasma*, *Plasmodium*, and *Theileria* without being exhaustive in the hope to encourage more readers to work on these intriguing and understudied organisms.

19.2.1 Plasmodium

Malaria is an infectious disease caused by members of the genus *Plasmodium* and transmitted by *Anopheles* mosquitoes. In developing countries it represents an important health and economical problem. Despite declining disease incidence, there are still approximately 225 million annual infections, resulting in nearly one million deaths, mostly children (WHO report 2010). The parasite undergoes a complex life cycle between its mosquito vector and vertebrate host with hundreds of species being transmitted by different species of *Anopheles* to vertebrates as different as lizards, birds, and humans. The clinical symptoms of malaria in humans are caused by the blood stages, when parasites infect and undergo asexual replication in red blood cells.

Parasite transmission occurs when an infected female mosquito bites a host in order to probe for blood (Fig. 19.3a). Prior to sucking blood, the mosquito injects saliva into the skin and with the saliva a small number of parasites. These start to migrate immediately in the skin to ultimately enter the blood stream (Vanderberg and Frevert 2004; Amino et al. 2006). Through the blood circulatory system the parasites reach the liver, where they leave the blood stream to infect a hepatocyte (Prudêncio et al. 2006). Within the hepatocyte, the sporozoite forms a parasitophorous vacuole (PV), where it transforms into a stage that undergoes a sequential cell division process, resulting in the formation of thousands of progeny

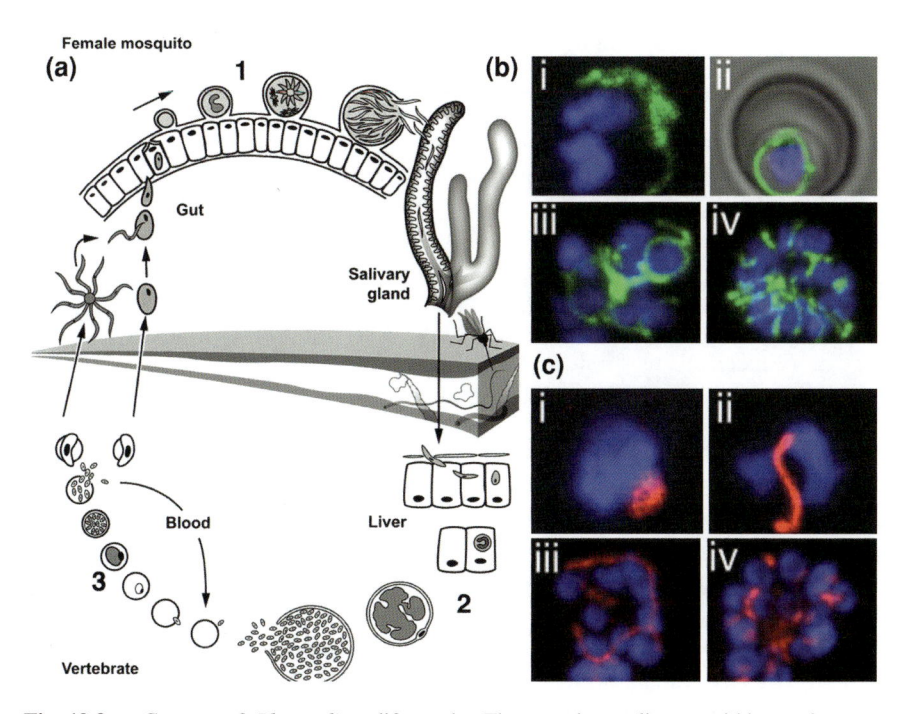

Fig. 19.3 a Cartoon of *Plasmodium* life cycle. The parasite replicates within erythrocytes through schizogony. Some differentiate into sexual stages that are transmitted to the mosquito during a blood meal, generating a motile egg cell that transverses the mosquito midgut, forming a cyst. Within the cyst, the parasite replicates through schizogony, forming the infective stages that are transmitted back to the host. These forms enter the blood circulation, ending up in hepatocytes, where they again multiply, generating thousands of parasites that will ultimately enter the erythrocytes. The numbers indicate the different stages of the cycle where the division process occurs: inside the cyst in the mosquito (1), in the hepatocyte (2), and within the red-blood cell (3). **b** and **c** Morphological modifications occur in the apicoplast and mitochondrion during schizogony. Here are fluorescence light microscopy images of a *Plasmodium* strain expressing an mitochondrion molecule tagged with GFP (in *green*) and a apicoplast protein tagged with DsRed (in *red*) during the division process inside an erythrocyte (After van Dooren et al. 2005; with permissions of the John Wiley and Sons)

parasites. This rapid division process where cell division is uncoupled from genome replication is called schizogony (Striepen et al. 2007; Gerald et al. 2011). Once the parasites are released into the bloodstream, they invade erythrocytes, again forming a PV. Inside the PV, the parasites replicate generating ca. 20 progeny per cell, which can infect new erythrocytes. While this cycle can go on until most of the red blood cells in the body are consumed, some parasites can form two different sexual stages. When those are picked up by a subsequent mosquito bite, the flagellated male gamete fertilizes the female gamete, forms a zygote and subsequently a diploid motile egg cell that exits from the mosquito stomach, crossing the surrounding epithelial cell layer to form a cyst. In this cyst, again, the parasite undergoes schizogony to generate the forms that are ultimately transmitted back to the host.

Plasmodium thus has different stages that undergo intense cell division, indicating that an effective cell cycle control process is present. This intense replication within a brief period of time is critical for the survival and further transmission of malaria parasites. Interestingly, the parasite multiplication in the erythrocytes generates variable numbers of daughter cells (between 18–24), probably because the different copies of the nucleus divide at different moments during schizogony (Read et al. 1993). This gives the impression that the cell cycle control works differently than the ones characterized in higher eukaryotes, which is controlled and synchronized by cyclin and cyclin-dependent kinases (CDKs) that are differently expressed throughout the division process. In *Plasmodium*, the regulatory mechanisms of the CDKs in the cell cycle are still largely not understood (Doering et al. 2008). Inhibitors of higher eukaryotic CDKs can interfere with *Plasmodium* (PfCDK), indicating that structural features required for inhibition are conserved in these kinases (Graeser et al. 1996; Li et al. 2001; Holton et al. 2003).

Due to their roundish shape and apparent lack of polarity, the intracellular stages of *Plasmodium* are not infective during schizogony. This is in contrast to the related parasite *Toxoplasma* that maintains some of its infectiveness during the division process (discussed below). Also different from *Toxoplasma*, *Plasmodium* does not present clear centrioles but has spindle pole plaques for chromosomal division (Morrissette and Sibley 2002a; Gerald et al. 2011).

Within a liver cell the parasite undergoes sequential nuclear divisions to form a multinuclear syncytium containing thousands of nuclei. After 2–3 days, the nuclei move to the periphery of the schizont and associate with the newly assembled inner membrane complex, the subpellicular microtubules, and other apical organelles (Striepen et al. 2007). The next generation of parasites starts to be formed with a sequential invagination process of the mother-parasite membrane (Sturm et al. 2009) and is concluded when these new parasites begin to bud into the PV matrix. The parasites ultimately destroy the PV membrane and are released into the host cell cytoplasm from where they bleb off in carriers into the blood stream (Sturm et al. 2006). There are no experimental data available as to the molecular components or mechanisms governing these interesting steps of the *Plasmodium* life cycle despite the recognized medical importance of the liver stage for disease prevention (Borrmann and Matuschewski 2011).

During the early erythrocytic part of the cycle, unpolymerized tubulin is distributed diffusely within the parasite cytoplasm. Later a centrocone (alternatively named centriolar plaque or spindle pole body or spindle pole plaque) is formed at the nuclear membrane. This structure then divides into two identical parts that migrate in opposite directions and form the spindle. Long and short intranuclear microtubules connect the two halves within the nucleus; at the cytoplasmic side an electron-dense material accumulates. In later stages of schizogony, cytoplasmic microtubules associate with each nucleus forming a radial arrangement and extending to the central zone of the parasite. The nuclei divide at different moments during schizogony, which allows the observation of mitotic spindles at distinct developmental stages within a single schizont (Read et al. 1993).

However, checkpoint mechanisms may regulate the number of nuclei re-entering the cell cycle at a given time point thus determining the overall number of produced progeny (Reininger et al. 2011).

Several kinases regulate spindle dynamics and control the cell cycle in higher eukaryotes (Nigg 2001; Carmena and Earnshaw 2003), including members of the Aurora-kinase family. Pfark-1 (*Plasmodium* Aurora-related kinase 1) has been observed to be associated with the spindle during blood stage schizogony, and PfPK5 (a Cdk1-like kinase) is considered to be the main cell cycle controller in *Plasmodium* (Graeser et al. 1996; Reininger et al. 2011). Recently, it was proposed that the molecules that associate with and control the centrocone might be responsible for the asynchronous division in each nucleus. This controlled asynchronous division could thus have evolved in order to avoid destabilizing sudden even numbers of nuclei within the restricted space available for the division in the host cell cytoplasm (Arnot et al. 2011).

Centrins associate with centrosomes, form fibrous structures, and are important for centrosome duplication (Salisbury 1995). In *Plasmodium*, four centrins appear to be conserved from the ancestral alveolate (Mahajan et al. 2008). These centrins localize to spindle plaques and are differently expressed during the intraerythrocytic stage and thus likely perform diverse roles.

During schizogony, the endoplasmic reticulum transforms into a perinuclear ring before it develops into an extensive branched network in later stages of cell segmentation. Part of the endoplasmic reticulum is known to be a specialized area of the secretory pathway. It has been shown that this site is in close proximity to the spindle plaques, being postulated to produce precursors for organelles like rhoptries and Golgi during division (Bannister et al. 2000; Striepen et al. 2007). The apicoplast and the mitochondrion also develop into large branched structures during schizogony (Fig. 19.3b or Fig. 19.3b and c). These organelles are then synchronously cleaved–first the apicoplast, then the mitochondrion—creating several copies of these organelles that are transferred to the individual daughter cells. While this looks impressive during the blood stage (van Dooren et al. 2005), it is simply awe-inspiring during the liver stage, when they shatter into thousands of fragments (Stanway et al. 2011). Curiously, the apicoplast and the mitochondrion appear to be linked to each other in the blood stages possibly facilitating the transfer to the budding parasites (Aikawa and Beaudoin 1968; van Dooren et al. 2005), while they are separated from each other in the mosquito and liver stage (Kudryashev et al. 2010; Stanway et al. 2011). Why care? In *Toxoplasma*, the apicoplast is connected to the centrioles during division (see below). So perhaps the *Plasmodium* apicoplast could also be linked to the spindle plaques and might thus in turn connect the mitochondrion to the nucleus-division machinery; at least in the blood stage. This could provide the basis for targeting organelles into new progeny parasites. If so, one wonders what the differences between the blood and liver stages are.

To continue their life cycle, some parasites depart from their asexual replication to differentiate into sexual parasite forms (Fig. 19.3a). In the mosquito stomach they can sense the new environment and start to differentiate even further and form gametes, a process that requires among much else cGMP-dependent protein kinase signaling and NIMA-related kinases (Reininger et al. 2005, 2009; McRobert et al. 2008). Most impressive is the formation of the male gamete, within the host cell cytoplasm. After three rapid rounds of genome replication, followed by three endomitoses, eight flagellated gametes are formed in a process that somehow requires a specific actin isoform, actin II (Siden-Kiamos et al. 2011). The flagella in gametes consist of alpha tubulin II, which like actin II is specifically expressed at this stage. The small differences between the two isoforms of actin and tubulin thus may be essential for their specific role in exflagellation and gamete motility (Rawlings et al. 1992). Male-gametes of *Toxoplasma* also exflagellate in the sexual phase in cats. However, *Toxoplasma* only has one copy of alpha and beta tubulin (Nagel and Boothroyd 1988). However, due to the intrinsic difficulty to work with *Toxoplasma* sexual forms (in cats), little is known about the molecules and mechanisms of male-gamete formation. The axonemes of plasmodium flagella are derived from eight kinetosomes (acting like basal bodies) that in turn originate from a single, amorphous MTOC, located at a nuclear pore (Sinden et al. 1976, 2010). Important for this is an atypical mitogen-activated protein kinase, required to start cytokinesis and axoneme motility (Tewari et al. 2005). Eventually the gametes escape from the host cell and, propelled by their flagella, zoom at high speed through the mosquito stomach cavity, to find and fuse with a female gamete.

19.2.2 Toxoplasma

Toxoplasma gondii is an obligate intracellular pathogen able to infect and replicate in most nucleated cells of warm-blooded animals. It is the etiologic agent of toxoplasmosis, one of the most common parasitic infections in humans, with a worldwide distribution, chronically infecting approximately one-third of the world population. In most adults toxoplasmosis is asymptomatic. But in some individuals, it can cause blindness and in embryos infected maternally during pregnancy even mental retardation and death.

The life cycle of *Toxoplasma* has an asexual phase, which takes place in most animals, and a sexual phase in felids (cats), with three distinct infective forms.

The asexual cycle comprises two distinct parasite forms. The rapidly replicating form is responsible for the clinical manifestations of acute toxoplasmosis (Fig. 19.4a). With the activation of the inflammatory response by the host, the parasite differentiates into a tissue cyst form, where both multiplication and metabolism are slowed down.

Both the faster and the slower replicating forms divide by the same process, called endodyogeny, a specialized system of reproduction where two daughter cells are formed within a mother cell (Fig. 19.4b – d). In contrast to the schizogony of *Plasmodium*, this unique form of division allows the parasite to remain infective during the division process. At the beginning of endodyogeny, the parasite elongates and widens the Golgi apparatus and apicoplast. This is followed by the

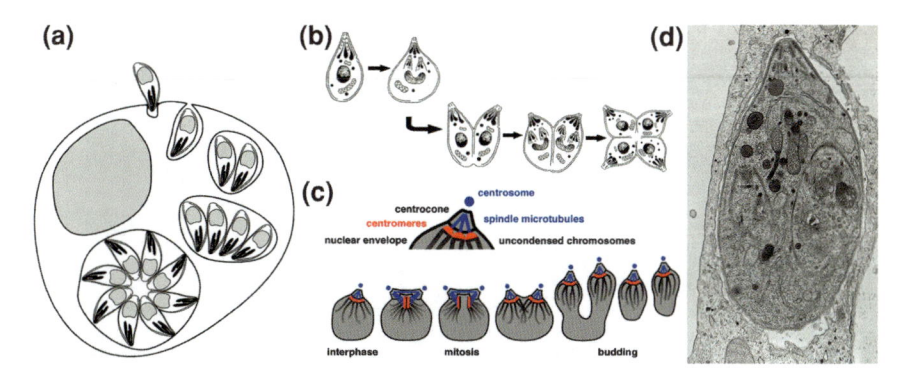

Fig. 19.4 **a** Cartoon of the *Toxoplasma* replication process within a parasitophorous vacuole inside the host cell (Curtesy of Dr. Marc-Jan Gubbels). **b** Schematic drawing of *Toxoplasma* endodyogeny, where two daughter cells are formed within a mother cell (Modified from Martins-Duarte et al. 2008; with permissions of the John Wiley and Sons). **c** Schematic representation of centrosome and spindle microtubules during the different phases of parasite division process (Modified after Brooks et al. 2011; with permissions of the Proceedings of the National Academy of Sciences). **d** Electron micrograph of a thin section of a *Toxoplasma* parasite undergoing endodyogeny

division of the centrosome at the posterior of the nucleus and the fission of the Golgi. Two new rudimentary apical complexes appear that are formed by a new conoid and apical rings from where new subpellicular microtubules and the inner membrane complex (IMC) will originate. This IMC formed below the parasite plasma membrane delineates for some time the daughter cells within the mother cell cytoplasm. As the daughter cells extend, the single nucleus finalizes DNA replication and assumes a lobular shape, resembling a horseshoe. The chromatin is then distributed to both poles by an intranuclear spindle.

MORN repeats are believed to act as protein–protein or protein–phospholipid-binding domains, being important for the organization of membranous and cyto-skeletal structures (Ju and Huang 2004; Kunita et al. 2004; Satouh et al. 2005). In *Toxoplasma*, a protein with MORN repeats named TgMORN1 is a key player in nuclear division, daughter cell formation, organelle partitioning and basal complex assembly, and cytokinesis (Fig. 19.5) (Gubbels et al. 2006; Lorestani et al. 2010), associating with the nuclear centrocone and probably also with a component of the membrane skeleton that is linking to the IMC. Actin-myosin rings are important elements for constriction in cytokinesis (Glotzer 2005). TgMORN1 associates with IMC components forming a ring during mitosis and cell division, which moves during mitosis, constricting perpendicular to the parasite longitudinal axis, resulting in nuclear division and cytokinesis (Gubbels et al. 2006). This constriction may also require myosin C and actin. Myosin C co-localizes with actin at the posterior end of the parasite, the same position where TgMORN1 was observed (Delbac et al. 2001; Gubbels et al. 2006). In this way, TgMORN1 could be acting as a connector between the posterior end of the IMC and an internal constrictive ring formed by myosin C (Gubbels et al. 2006; Lorestani et al. 2010). Curiously,

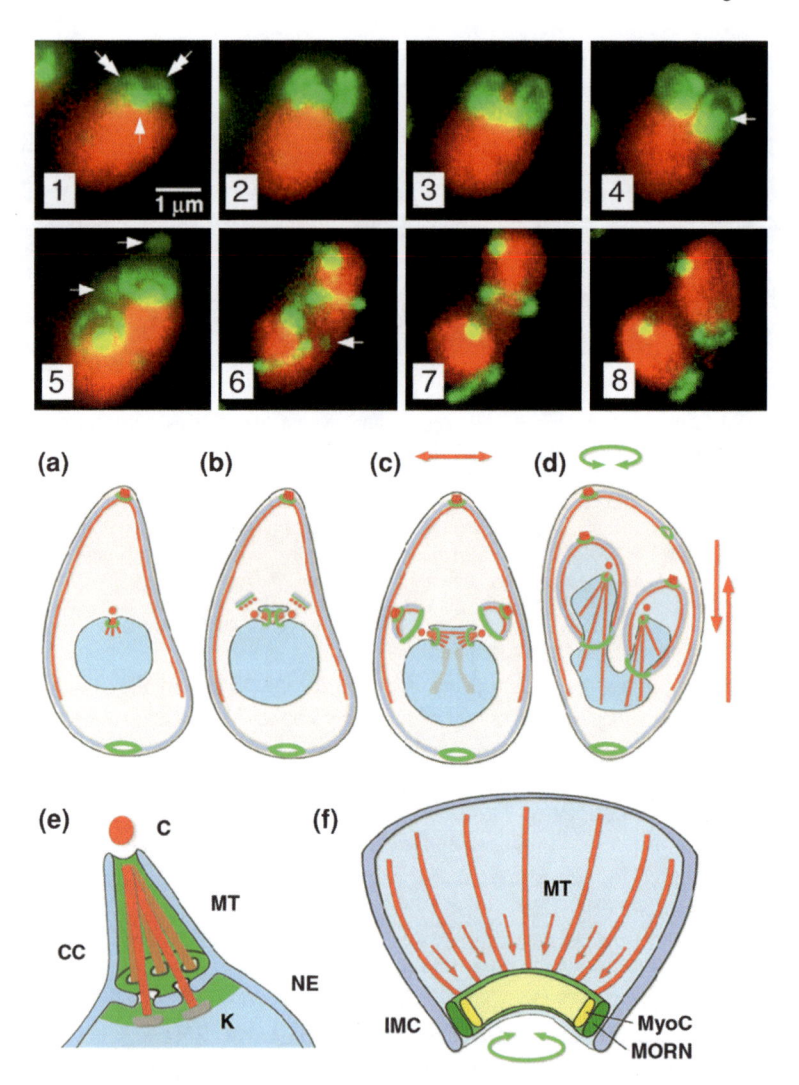

Fig. 19.5 *Upper panel* Fluorescent light microscopy images of a *Toxoplasma* strain expressing TgMORN1 coupled to YFP (in *green*) throughout the parasite division process. The parasite nucleus is stained in red. *Arrows* point to the centrocone (Modified from Gubbels et al. 2006; with permissions of The Company of Biologists). *Lower panel* Schematic model of the role played by TgMORN1 in *Toxoplasma* division. Conoid (C), subpellicular microtubules (MT), centrocone (CC) are highlighted in *red*, the nucleus in *light blue* and MORN1 in *green*. *NE* nuclear envelope, *K* kinetochore, *IMC* inner membrane complex, *MyoC* myosin C. *Red arrows* indicate microtubule driven movements and *green arrows* the constriction caused by MORN1 (Modified after Gubbels et al. 2006; with permissions of The Company of Biologists)

the TgMORN1-centrocone connection is observed throughout the *Toxoplasma* cell cycle, although spindles have been found only during mitosis. Vaishnava et al. (2005) described short spindles and centrocones through interphase in *Sarcocystis neurona*, another apicomplexa parasite of veterinary importance. In this aspect centrocone and possibly kinetochore persistence could be a general aspect of apicomplexan nuclear organization.

As the IMC associated with the subpellicular microtubules extends, the forming daughter cell includes first the centrosome and the Golgi complex followed by the remaining organelles. New sets of apical organelles important for migration and host cell invasion (rhoptries and micronemes) are formed in both apical ends of the nascent daughter cells. The mitochondrion is the last organelle to be incorporated into the forming cell (Nishi et al. 2008), although just how this works is less clear as for the other organelles. Each daughter cell continues to mature until the cytoplasm and all its contents are divided between the two daughter cells. Eventually, the inner membrane complex of the mother cell breaks down, and the plasma membrane is used to form part of the plasma membrane of the daughter cells. The cleavage to separate the daughter cells begins at the apical pole, extending through the cell bodies. During this process more plasma membrane is formed through the fusion of new membranous structures derived from Rab-11-mediated vesicular traffic (Agop-Nersesian et al. 2009, 2010). At the end of this process, there are two daughter cells and a residual body at the posterior end connecting the new cells. Within the same host cell, each daughter cell can then replicate again and again until ultimately their progeny will exit the host cell to infect neighboring cells (Fig.19.4a).

During interphase, the centrosome locates near the apical polar ring. As the division process continues the centrosome links to the spindle pole plaque or centrocone, a structure localized at the nuclear envelope and responsible for the formation of the intranuclear spindle microtubules (Fig. 19.4c). At the initial steps of endodyogeny, the centrosome is associated with the Golgi apparatus. The centrosome protein centrin is located near the Golgi during mitosis (Stedman et al. 2003). As the Golgi grows, the centrosome remains associated at one side. Afterwards, the centrosome re-localizes to the posterior pole of the mother cell nucleus. At this location, the centrosome divides, returning then to the apical part of the mother nucleus, re-connecting with the Golgi complex, with each centrosome associating with each inner end of the duplicated Golgi.

Another important role of the centrosome is the division of the apicoplast (Striepen et al. 2000). Curiously, parasites in which the apicoplast was destroyed can undergo one full cycle of replication before dying a "delayed death" (He et al. 2001). Thus, like the single mitochondrion and the Golgi, this organelle must be correctly duplicated and segregated to each forming cell. To assure this, the apicoplast division is linked to the parasite's mitotic division process by association with the dividing centrosomes. As these move away from each other, the apicoplast elongates. With the budding of the daughter cells, the apicoplast is pulled into a U shape and divided in two halves with a dynamin-related protein playing a key role during fission (van Dooren et al. 2009). Each new daughter plastid stays connected to one centrosome and migrates together with the newly-formed nucleus into the daughter cell. Thus, the association with the nucleus via the centrosome is a perfect strategy to assure that each daughter cell receives a copy of the organelle during mitosis. As discussed above, in schizogony when dozens of budding cells are formed at the same time, the apicoplast and the mitochondrion enlarge and then segment, being later directed to the budding cells. Is this apicoplast-mitochondrion-nucleus connection a common strategy to assure each daughter cell a copy of the organelle? Or perhaps each apicomplexan developed its own mechanisms? These are questions that still require further studies.

Until recently, it was poorly understood how daughter cells acquired the complete set of chromosomes after the sequential division process. Brooks et al. (2011) recently demonstrated that the parasite's chromosomes are connected to the centrosome throughout the cell cycle, allowing the correct distribution during daughter cell budding. They followed a component of the chromosomes' centromeres—CenH3—throughout the cell cycle to characterize the chromosomal architecture during mitosis. This centromere associates with the centrocone during both mitosis and interphase, allowing genome integrity during rapid DNA replication and re-location to budding cells.

Toxoplasma reorganizes host cell intracellular structures, like lysosomes, mitochondria, endoplasmic reticulum, and the Golgi in order to acquire nutrients for its intracellular development. Curiously, *Toxoplasma* also reorganizes the host cell centrosomes. The mammalian target of rapamycin signaling pathway (mTOR), a highly conserved serine/threonine kinase, is a key controller of cell growth and proliferation (Zoncu et al. 2011). It has been shown as well that mTORC2 is important for cell polarity control, actin organization, and stimulation of F-actin stress fibers (Schmidt et al. 1996; Sarbassov et al. 2005). By meddling with the protein mTORC2 the parasite interferes with host centrosome location, re-directing it to the parasitophorous vacuole and suppressing host cell migration (Wang et al. 2010). Interfering with the polarization of the host cell microtubules may lead to an arrest of the host cell in a state favorable for the intracellular development of the parasite.

19.2.3 Theileria

Theileria is another unique apicomplexan parasite, being an important pathogen for cattle in Africa and Asia. After infecting the immune system's cells (i.e. macrophages or lymphocytes), *Theileria* induces an uncontrolled proliferation of these cells by recruitment of IKK onto the parasite surface and thus activation of NFkB, effectively turning them into cancer cells (Heussler et al. 2002). The infected host cells do not undergo apoptosis and can survive for years (Küenzi et al. 2003; Heussler et al. 2006;). Infected cells migrate across tissue barriers and establish new areas of proliferation, especially in the lymphoid tissue (reviewed by Baumgartner 2011). In the laboratory, *Theileria* infected cells can be cultivated and experiments probing parasite infection are done "in reverse", not by infecting uninfected host cells, but by curing infected ones.

Unlike others Apicomplexans, *Theileria* does not reside in a parasitophorous vacuole. After it enters the host cell, it dissolves the membrane surrounding it and sits naked in the cytoplasm surrounded just by a fuzzy layer of secreted material (Shaw et al. 1991; Jura et al. 1983). It closely associates and interacts with host cell microtubules (Shaw 2003). The *Theileria*-associated host cell microtubules are unusually stable as they are not affected by microtubule-disrupting agents that cause microtubule disruption in uninfected cells (Shaw 2003).

Within the host cell cytoplasm, the parasite undergoes serial asexual nuclear multiplications and differentiates into a syncytium. The syncytium can divide and redistribute itself into the two daughter lymphocytes. During lymphocyte cytokinesis the parasite evolves to associate with the host mitotic spindle itself, thus almost passively hitching a ride. Recently, von Schubert et al. (2010) described how the parasite regulates host cell spindles to assure that each daughter cell continues to be infected. This occurs by a direct regulation of host cell Polo-like kinase (Plk) during parasite mitosis. Mitotic kinases are monitors of the cell cycle. One of these key regulators is Cdk1. After inactivation of this kinase, Plk, which is associated with kinetochores, dissociates from them and localizes to form the central spindle. Regulatory proteins, such as Plk1, Aurora B, and some Rho GTPases allow the central spindle to act as a platform that will form the plane of cleavage during cytokinesis and initiate the process of cell division (Barr and Gruneberg 2007).

At the early mitotic phases, *Theileria* schizont binds to host cell spindle pole microtubules. This allows the parasite to be directed to the equatorial region of the dividing cell, the same region where host cell chromosomes are positioned. At this stage, the parasite binds to astral as well as central microtubules by recruiting host cell Plk, allowing parasite segregation during cytokinesis into the daughter cells. Using inhibitors of the connection between the parasite and the spindle microtubules, von Schubert et al. (2010) prevented parasite segregation, showing the usurpation of the host cell mitotic machinery by *Theileria* to guarantee the continuity of itself in the daughter cells. During the abscission process, part of the parasite is trapped as a slim tube. It remains to be determined whether parasite-own structures provide its abscission or whether such signals are entirely derived from the host cell. Once incorporated into the central spindle/midbody, the parasite does not affect host cell central spindle function or abscission.

Eventually the *Theileria* syncytium undergoes a process called merogony, where several parasites bud from this central mass; they are released afterwards, infecting now erythrocytes (Shaw 2003). This budding process as well as the sequential nuclear division must be tightly regulated in order to form new mature, infective parasites, containing all the organelles and with the correct nuclear repertoire. In order to accomplish this, the organelles are tethered to the nuclear membrane that is in turn connected to the parasite plasma membrane (Shaw and Tilney 1992). Thus, although little is known about the molecules involved, all organelles are associated to each other, facilitating the distribution of the formed daughter organelles into the budding parasites.

19.3 Conclusions

We have written this chapter in the hope to raise the interest of cell biologist readers toward these curious parasites. It is our belief that the study of these organisms can result in many surprising discoveries as they go about their intracellular life in such different ways. Often considered to be close relatives of each other, as they belong to the apicomplexa phylum, already a superficial look at how they divide shows the tremendous variation of solutions they present during their distinct evolutionary paths. While they clearly share some of the key proteins with other eukaryotes they also have distinct subsets of proteins at their disposal to optimize cellular processes such as cell invasion and proliferation. Uncovering these will contribute not just to the understanding of their parasitic way of life but also to comprehend the basic principles of life itself.

Acknowledgments We thank Mirko Singer for reading the manuscript, Laboratório de Ultraestrutura Hertha Meyer (UFRJ-Brazil) for the access to their image archive and the Chica and Heinz Schaller Foundation, the Cluster of Excellence CellNetworks at the University of Heidelberg and the German Research Foundation (SPP1399) for funding. FF is a member of the European Network of Excellence EVIMalaR.

References

Agop-Nersesian C, Naissant B, Ben Rached F, Rauch M, Kretzschmar A, Thiberge S, Menard R, Ferguson DJ, Meissner M, Langsley G (2009) Rab11A-controlled assembly of the inner membrane complex is required for completion of apicomplexan cytokinesis. PLoS Pathog 5:e1000270

Agop-Nersesian C, Egarter S, Langsley G, Foth BJ, Ferguson DJ, Meissner M (2010) Biogenesis of the inner membrane complex is dependent on vesicular transport by the alveolate specific GTPase Rab11B. PLoS Pathog 6:e1001029

Aikawa M, Beaudoin RL (1968) Studies on nuclear division of a malarial parasite under pyrimethamine treatment. J Cell Biol 39:749–754

Amino R, Thiberge S, Martin B, Celli S, Shorte S, Frischknecht F, Menard R (2006) Quantitative imaging of *Plasmodium* transmission from mosquito to mammal. Nat Med 12:220–224

Arnot DE, Ronander E, Bengtsson DC (2011) The progression of the intra-erythrocytic cell cycle of *Plasmodium falciparum* and the role of the centriolar plaques in asynchronous mitotic division during schizogony. Int J Parasitol 41:71–80

Azimzadeh J, Marshall WF (2010) Building the centriole. Curr Biol 20:R816–R825

Bannister LH, Hopkins JM, Fowler RE, Krishna S, Mitchell GH (2000) A brief illustrated guide to the ultrastructure of *Plasmodium falciparum* asexual blood stages. Parasitol Today 16:427–433

Barr FA, Gruneberg U (2007) Citokinesis: placing and making the final cut. Cell 131:847–860

Baum J, Gilberger TW, Frischknecht F, Meissner M (2008) Host-cell invasion by malaria parasites: insights from *Plasmodium* and *Toxoplasma*. Trends Parasitol 24:557–563

Baumgartner M (2011) Enforcing host cell polarity: an apicomplexan parasite strategy towards dissemination. Curr Opin Microbiol 14:436–444

Borrmann S, Matuschewski K (2011) Protective immunity against malaria by 'natural immunization': a question of dose, parasite diversity, or both? Curr Opin Immunol 23:500–508

Brooks CF, Francia ME, Gissot M, Croken MM, Kim K, Striepen B (2011) *Toxoplasma gondii* sequesters centromeres to a specific nuclear region throughout the cell cycle. Proc Nat Acad Sci USA 108:3767–3772

Carmena M, Earnshaw WC (2003) The cellular geography of Aurora kinases. Nat Rev Mol Cell Biol 4:842–854
Carruthers VB, Tomley FM (2008) Microneme proteins in apicomplexans. Subcell Biochem 47:33–45
Cesbron-Delauw MF, Gendrin C, Travier L, Ruffiot P, Mercier C (2008) Apicomplexan in mammalian cells: trafficking to the parasitophorous vacuole. Traffic 9:657–664
Cyrklaff M, Kudryashev M, Leis A, Leonard K, Baumeister W, Menard R, Meissner M, Frischknecht F (2007) Cryoelectron tomography reveals periodic material at the inner side of subpellicular microtubules in apicomplexan parasites. J Exp Med 204:1281–1287
Delbac F, Sanger A, Neuhaus EM, Stratmann R, Ajioka JW, Toursel C, Herm-Gotz A, Tomavo S, Soldati T, Soldati D (2001) *Toxoplasma gondii* myosins B/C: one gene, two tails, two localizations, and a role in parasite division. J Cell Biol 155:613–623
Dubremetz JF (2007) Rhoptries are major players in *Toxoplasma gondii* invasion and host cell interaction. Cell Microbiol 9:841–848
Dubremetz JF, Elsner YY (1979) Ultrastructural study of schizogony of *Eimeria bovis* in cell cultures. J Protozool 26:367–376
Fennell BJ, Naughton JA, Dempsey E, Bell A (2006) Cellular and molecular actions of dinitroaniline and phosphorothioamidate herbicides on *Plasmodium falciparum*: tubulin as a specific antimalarial target. Mol Biochem Parasitol 145:226–238
Fleige T, Fischer K, Ferguson DJ, Gross U, Bohne W (2007) Carbohydrate metabolism in the *Toxoplasma gondii* apicoplast: localization of three glycolytic isoenzymes, the single pyruvate dehydrogenase complex, and a plastid phosphate translocator. Eukaryot Cell 6:984–996
Fowler RE, Fookes RE, Lavin F, Bannister LH, Mitchell GH (1998) Microtubules in *Plasmodium falciparum* merozoites and their importance for invasion of erythrocytes. Parasitology 117:425–433
Gerald N, Mahajan B, Kumar S (2011) Mitosis in the human malaria parasite *Plasmodium falciparum*. Eukaryot Cell 10:474–482
Gleeson MT (2000) The plastid in Apicomplexa: what use is it? Int J Parasitol 30:1053–1070
Glotzer M (2005) The molecular requirements for cytokinesis. Science 307:1735–1739
Goodman CD, McFadden GI (2007) Fatty acid biosynthesis as a drug target in apicomplexan parasites. Curr Drug Targets 8:15–30
Graeser R, Wernli B, Franklin RM, Kappes B (1996) *Plasmodium falciparum* protein kinase 5 and the malarial nuclear division cycles. Mol Biochem Parasitol 82:37–49
Gubbels MJ, Vaishnava S, Boot N, Dubremetz JF, Striepen B (2006) A morn—repeat protein is a dynamic component of the *Toxoplasma gondii* cell division apparatus. J Cell Sci 119:2236–2245
He CY, Shaw MK, Pletcher CH, Striepen B, Tilney LG, Roos DS (2001) A plastid segregation defect in the protozoan parasite *Toxoplasma gondii*. EMBO J 20:330–339
Heintzelman MB (2006) Cellular and molecular mechanics of gliding locomotion in eukaryotes. Int Rev Cytol 251:79–129
Heussler VT, Rottenberg S, Schwab R, Küenzi P, Fernandez PC, McKellar S, Shiels B, Chen ZJ, Orth K, Wallach D, Dobbelaere DA (2002) Hijacking of host cell IKK signalosomes by the transforming parasite *Theileria*. Science 298:1033–1036
Heussler V, Sturm A, Langsley G (2006) Regulation of host cell survival by intracellular *Plasmodium* and *Theileria* parasites. Parasitology 132:S49–S60
Holton S, Merckx A, Burgess D, Doerig C, Noble M, Endicott J (2003) Structures of *Plasmodium falciparum* PfPK5 test the CDK regulation paradigm and suggest mechanisms of small molecule inhibition. Structure 11:1329–1337
Hu K, Roos DS, Murray JM (2002) A novel polymer of tubulin forms the conoid of *Toxoplasma gondii*. J Cell Biol 156:1039–1050
Joiner KA, Roos DS (2002) Secretory traffic in the eukaryotic parasite *Toxoplasma gondii*: less is more. J Cell Biol 157:557–563
Ju TK, Huang FL (2004) MSAP, the meichroacidin homolog of carp (*Cyprinus carpio*), differs from the rodent counterpart in germline expression and involves flagellar differentiation. Biol Reprod 71:1419–1429

Jura WG, Brown CG, Kelly B (1983) Fine structure and invasive behaviour of the early developmental stages of *Theileria annulata* in vitro. Vet Parasitol 12:31–44

Kelley GL, Hammond DM (1972) Fine structural aspects of early development of *Eimeria ninakohlyakimovae* in cultured cells. Zeitschrift für Parasitenkunde 38:271–284

Köhler S, Delwiche C, Denny D, Tilney L, Webster P, Wilson R, Palmer J, Roos D (1997) A plastid of probable green algal origin in Apicomplexan parasites. Science 275:1485–1489

Kudryashev M, Lepper S, Stanway R, Bohn S, Baumeister W, Cyrklaff M, Frischknecht F (2010) Positioning of large organelles by a membrane- associated cytoskeleton in *Plasmodium* sporozoites. Cell Microbiol 12:362–371

Küenzi P, Schneider P, Dobbelaere DA (2003) *Theileria* parva-transformed T cells show enhanced resistance to Fas/Fas ligand-induced apoptosis. J Immunol 171:1224–1231

Kunita R, Otomo A, Mizumura H, Suzuki K, Showguchi-Miyata J, Yanagisawa Y, Hadano S, Ikeda JE (2004) Homo-oligomerization of ALS2 through its unique carboxyl-terminal regions is essential for the ALS2-associated Rab5 guanine nucleotide exchange activity and its regulatory function on endosome trafficking. J Biol Chem 279:38626–38635

Kuriyama R, Besse C, Gèze M, Omoto CK, Schrével J (2005) Dynamic organization of microtubules and microtubule-organizing centers during the sexual phase of a parasitic protozoan, *Lecudina tuzetae* (Gregarine, Apicomplexa). Cell Motil Cytoskelet 62:195–209

Lemgruber L, Lupetti P (2011) Crystalloid body, refractile body and virus-like particles in Apicomplexa: what is in there? Parasitology 138:1–9

Li Z, Le Roch K, Geyer JA, Woodard CL, Prigge ST, Koh J, Doerig C, Waters NC (2001) Influence of human p16(INK4) and p21(CIP1) on the in vitro activity of recombinant *Plasmodium falciparum* cyclin-dependent protein kinases. Biochem Biophys Res Commun 288:1207–1211

Lorestani A, Sheiner L, Yang K, Robertson SD, Sahoo N, Brooks CF, Ferguson DJ, Striepen B, Gubbels MJ (2010) A Toxoplasma MORN1 null mutant undergoes repeated divisions but is defective in basal assembly, apicoplast division and cytokinesis. PLoS ONE 5:e12302

Ma C, Li C, Ganesan L, Oak J, Tsai S, Sept D, Morrissette NS (2007) Mutations in alpha-tubulin confer dinitroaniline resistance at a cost to microtubule function. Mol Biol Cell 18:4711–47120

Ma C, Tran J, Gu F, Ochoa R, Li C, Sept D, Werbovetz K, Morrissette N (2010) Dinitroaniline activity in *Toxoplasma gondii* expressing wild-type or mutant alpha-tubulin. Antimicrob Agents Chemother 54:1453–1460

Mahajan B, Selvapandiyan A, Gerald NJ, Majam V, Zheng H, Wickramarachchi T, Tiwari J, Fujioka H, Moch JK, Kumar N, Aravind L, Nakhasi HL, Kumar S (2008) Centrins, cell cycle regulation proteins in human malaria parasite *Plasmodium falciparum*. J Biol Chem 283:31871–31883

Martins-Duarte E, de Souza W, Vommaro RC (2008) Itraconazole affects Toxoplasma gondii endodyogeny. FEMS Microbiol Lett 282:290–298

Mazumbar J, Wilson EH, Masek K, Hunter CA, Striepen B (2006) Apicoplast fatty acid synthesis is essential for organelle biogenesis and parasite survival in *Toxoplasma gondii*. Proc Nat Acad Sci USA 103:13192–13197

Morrissette N, Sibley LD (2002a) Cytoskeleton of Apicomplexan parasites. Microbiol Mol Biol Rev 66:21–38

Morrissette NS, Sibley LD (2002b) Disruption of microtubules uncouples budding and nuclear division in *Toxoplasma gondii*. J Cell Sci 115:1017–1025

Morrissette NS, Murray JM, Roos DS (1997) Subpellicular microtubules associate with an intramembranous particle lattice in the protozoan parasite *Toxoplasma gondii*. J Cell Sci 110:35–42

Morrissette NS, Mitra A, Sept D, Sibley LD (2004) Dinitroanilines bind alpha-tubulin to disrupt microtubules. Mol Biol Cell 15:1960–1968

Nagel SD, Boothroyd JC (1988) The α- and β-tubulins of *Toxoplasma gondii* are encoded by single copy genes containing multiple introns. Mol Biochem Parasitol 29:261–273

Nigg EA (2001) Mitotic kinases as regulators of cell division and its checkpoints. Nat Rev Mol Cell Biol 2:21–32

Nishi M, Hu K, Murray JM, Roos DS (2008) Organellar dynamics during the cell cycle of *Toxoplasma gondii*. J Cell Sci 121:1559–1568

Pouvelle B, Farley PJ, Long CA, Taraschi TF (1994) Taxol arrests the development of blood-stage *Plasmodium falciparum* in vitro and *Plasmodium chabaudi adami* in malaria-infected mice. J Clin Investig 94:413–417

Prudêncio M, Rodriguez A, Mota MM (2006) The silent path to thousands of merozoites: the *Plasmodium* liver stage. Nat Rev Microbiol 4:849–856

Ralph SA, van Dooren GG, Waller RF, Crawford MJ, Fraunholz MJ, Foth BJ, Tonkin CJ, Roos DS, McFadden GI (2004) Tropical infectious diseases: metabolic maps and functions of the *Plasmodium falciparum* apicoplast. Nat Rev Microbiol 2:203–216

Rawlings DJ, Fujioka H, Fried M, Keister DB, Aikawa M, Kaslow DC (1992) Alpha-tubulin II is a male-specific protein in *Plasmodium falciparum*. Mol Biochem Parasitol 56:239–250

Read M, Sherwin T, Holloway SP, Gull K, Hyde JE (1993) Microtubular organization visualized by immunofluorescence microscopy during erythrocytic schizogony in *Plasmodium falciparum* and investigation of post-translational modifications of parasite tubulin. Parasitology 106:223–232

Reininger L, Billker O, Tewari R, Mukhopadhyay A, Fennell C, Dorin-Semblat D, Doerig C, Goldring D, Harmse L, Ranford-Cartwright L, Packer J, Doerig C (2005) A NIMA-related protein kinase is essential for completion of the sexual cycle of malaria parasites. J Biol Chem 280:31957–31964

Reininger L, Tewari R, Fennell C, Holland Z, Goldring D, Ranford-Cartwright L, Billker O, Doerig C (2009) An essential role for the *Plasmodium* Nek-2 Nima-related protein kinase in the sexual development of malaria parasites. J Biol Chem 284:20858–20868

Reininger L, Wilkes JM, Bourgade H, Miranda-Saavedra D, Doerig C (2011) An essential aurora-related kinase transiently associates with spindle pole bodies during *Plasmodium falciparum* erythrocytic schizogony. Mol Microbiol 79:205–221

Salisbury JL (1995) Centrin, centrosomes, and mitotic spindle poles. Curr Opin Cell Biol 7:39–45

Sarbassov DD, Ali SM, Sabatini DM (2005) Growing roles for the mTOR pathway. Curr Opin Cell Biol 17:596–603

Satouh Y, Padma P, Toda T, Satoh N, Ide H, Inaba K (2005) Molecular characterization of radial spoke subcomplex containing radial spoke protein 3 and heat shock protein 40 in sperm flagella of the ascidian *Ciona intestinalis*. Mol Biol Cell 16:626–636

Schmidt A, Kunz J, Hall MN (1996) TOR2 is required for organization of the actin cytoskeleton in yeast. Proc Nat Acad Sci USA 93:13780–13785

Schrevel J, Asfaux-Foucher G, Bafort JM (1977) Ultrastructural study of multiple mitoses during sporogony of *Plasmodium b. berghei*. J Ultrastruct Res 59:332–350

Schrével J, Sinou V, Grellier P, Frappier F, Guénard D, Potier P (1994) Interactions between docetaxel (Taxotere) and *Plasmodium falciparum*-infected erythrocytes. Proc Nat Acad Sci U S A 91:8472–8476

Schrevel J, Asfaux-Foucher G, Hopkins JM, Robert V, Bourgouin C, Prensier G, Bannister LH (2008) Vesicle trafficking during sporozoite development in *Plasmodium berghei*: ultrastructural evidence for a novel trafficking mechanism. Parasitology 135:1–12

Shaw MK (2003) Cell invasion by *Theileria* sporozoites. Trends Parasitol 19:2–6

Shaw MK, Tilney LG (1992) How individual cells develop from a syncytium: merogony in *Theileria parva* (Apicomplexa). J Cell Sci 101:109–123

Shaw MK, Tilney LG, Musoke AJ (1991) The entry of *Theileria parva* sporozoites into bovine lymphocytes: evidence for MHC class I involvement. J Cell Biol 113:87–101

Sibert GJ, Speer CA (1981) Fine structure of nuclear division and microgametogony of *Eimeria nieschulzi* Dieben, 1924. Zeitschrift für Parsitenkunde 66:179–189

Siden-Kiamos I, Ganter M, Kunze A, Hliscs M, Steinbüchel M, Mendoza J, Sinden RE, Louis C, Matuschewski K (2011) Stage-specific depletion of myosin A supports an essential role in motility of malarial ookinetes. Cell Microbiol 13:1996–2006

Sinden RE, Talman A, Marques SR, Wass MN, Sternberg MJ (2010) The flagellum in malarial parasites. Curr Opin Microbiol 13:491–500

Sinou V, Grellier P, Schrevel J (1996) In vitro and in vivo inhibition of erythrocytic development of malarial parasites by docetaxel. Antimicrob Agents Chemother 40:358–361

Sinou V, Boulard Y, Grellier P, Schrevel J (1998) Host cell and malarial targets for docetaxel (Taxotere) during the erythrocytic development of *Plasmodium falciparum*. J Eukaryot Microbiol 45:171–183

Stedman TT, Sussmann AR, Joiner KA (2003) *Toxoplasma gondii* Rab6 mediates a retrograde pathway for sorting of constitutively secreted proteins to the Golgi complex. J Biol Chem 278:5433–5443

Striepen B, Jordan CN, Reiff S, van Dooren GG (2007) Building the perfect parasite: cell division in apicomplexa. PLoS Pathog 3:e78

Stanway RR, Mueller N, Zobiak B, Graewe S, Froehlke U, Zessin PJ, Aepfelbacher M, Heussler VT (2011) Organelle segregation into *Plasmodium* liver stage merozoites. Cell Microbiol 13:1768–1782

Striepen B, Crawford MJ, Shaw MK, Tilney LG, Seeber F, Roos DS (2000) The plastid of Toxoplasma gondii is divided by association with the centrosomes. J Cell Biol 151:1423–1434

Sturm A, Amino R, van de Sand C, Regen T, Retzlaff S, Rennenberg A, Krueger A, Pollok JM, Menard R, Heussler VT (2006) Manipulation of host hepatocytes by the malaria parasite for delivery into liver sinusoids. Science 313:1287–1290

Sturm A, Graewe S, Franke-Fayard B, Retzlaff S, Bolte S, Roppenser B, Aepfelbacher M, Janse C, Heussler V (2009) Alteration of the parasite plasma membrane and the parasitophorous vacuole membrane during exo-erythrocytic development of malaria parasites. Protist 160:51–63

Tenter AM, Barta JR, Beveridge I, Duszynski DW, Mehlhorn H, Morrison DA, Thompson RC, Conrad PA (2002) The conceptual basis for a new classification of the coccidia. Int J Parasitol 32:595–616

Tewari R, Dorin D, Moon R, Doerig C, Billker O (2005) An atypical mitogen-activated protein kinase controls cytokinesis and flagellar motility during male gamete formation in a malaria parasite. Mol Microbiol 58:1253–1263

Tran JQ, de Leon JC, Li C, Huynh MH, Beatty W, Morrissette NS (2010) RNG1 is a late marker of the apical polar ring in *Toxoplasma gondii*. Cytoskeleton (Hoboken) 67:586–598

van Dooren GG, Marti M, Tonkin CJ, Stimmler LM, Cowman AF, McFadden GI (2005) Development of the endoplasmic reticulum, mitochondrion and apicoplast during the asexual life cycle of *Plasmodium falciparum*. Mol Microbiol 57:405–419

Vaishnava S, Morrison DP, Gaji RY, Murray JM, Entzeroth R, Howe DK, Striepen B (2005) Plastid segregation and cell division in the apicomplexan parasite *Sarcocystis neurona*. J Cell Sci 118:3397–3407

Vanderberg JP, Frevert U (2004) Intravital microscopy demonstrating antibody-mediated immobilisation of *Plasmodium berghei* sporozoites injected into skin by mosquitoes. Int J Parasitol 34:991–996

von Schubert C, Xue G, Schmuckli-Maurer J, Woods KL, Nigg EA, Dobbelaere DA (2010) The transforming parasite *Theileria* co-opts host cell mitotic and central spindles to persist in continuously dividing cells. PLoS Biol 8:e1000499

Wang Y, Weiss LM, Orlofsky A (2010) Coordinate control of host centrosome position, organelle distribution, and migratory response by *Toxoplasma gondii* via host mTORC2. J Biol Chem 285:15611–15618

World Health Organization (2010) World malaria report 2010, WHO Press, pp. 1–152

Yeh E, DeRisi JL (2011) Chemical rescue of malaria parasites lacking an apicoplast defines organelle function in blood-stage *Plasmodium falciparum*. PLoS Biol 9:e1001138

Zoncu R, Efeyan A, Sabatini DM (2011) mTOR: from growth signal integration to cancer, diabetes and ageing. Nat Rev Mol Cell Biol 12:21–35

Chapter 20
The Centrosome Life Story in *Xenopus laevis*

Jacek Z. Kubiak and Claude Prigent

Abstract *Xenopus laevis* is a privileged model for the centrosome research, and cell cycle and developmental studies. Centrosomes are composed of their core components, the centrioles, surrounded by the pericentriolar material. Like in most vertebrates, with the exception of the mouse, *Xenopus* centriole is paternally inherited. During gametogenesis, spermatozoa retain centrioles, but lose most of its pericentriolar material, whereas oocytes lose their centrioles, but maintain centrosomal proteins. Upon fertilization, the sperm centriole is transmitted to the egg, where it assembles maternal proteins, such as γ-tubulin and pericentrin, to form a biparental functional centrosome. The centrosome formed in a zygote plays a crucial role in embryo development by providing a novel axis of polarity and transmission of correct number of MTOCs to all embryonic cells. In parthenogenetic embryos, which do not inherit paternal centrioles, embryonic development arrests through the formation of abnormal spindles and chaotic abortive cleavages. Centrosome assembly and maturation have been extensively studied at the molecular level in cell-free extracts obtained from *Xenopus* oocytes, eggs, and embryos. Studies on *Xenopus* centrosome have proven very useful for better understanding of many fundamental functions of centrosomes during embryo development and in cancers.

J. Z. Kubiak (✉) · C. Prigent
CNRS, UMR 6290, Institut de Génétique et Développement de Rennes,
35043 Rennes, France
e-mail: jacek.kubiak@univ-rennes1.fr

J. Z. Kubiak · C. Prigent
Faculté de Médecine, Université Rennes 1, UEB, IFR 140,
35043 Rennes, France

20.1 Centrosome Structure and Major Functions

The centrosome is a complex organelle with a crucial role in the organization of microtubule (MT) cytoskeleton, which in turn organizes many aspects of cell architecture including polarity and ciliogenesis. Typically, a centrosome is composed of a pair of centrioles surrounded by electron dense fibrillar pericentriolar material (PCM) (Fig. 20.1). In fact, the centrosome is organized by centrioles.

The centrosome controls crucial cellular processes such as cell division, polarity, motility, sensing the environment and communication with neighbor cells. These cellular centrosomal functions allow coordination of embryo development starting from oogenesis and ending with tissue differentiation. The centrosome is the major MT-organizing center (MTOC) in animal cells, even if in certain cell types, other MTOCs (secondary or diffused) may take control over MT organization. During the interphase, centrosomes nucleate MTs. The plus (+) ends of MTs are directed toward the cell periphery, while their minus (−) ends remain close to the MTOC. The centrosome controls the formation of the organized array of long cytoplasmic MTs radiating toward cell periphery from the discrete "cell center" determined by the position of the centrosome, which is usually located in the vicinity of the nucleus or in a discrete nuclear pocket. This function of the centrosome is crucial for establishment and maintenance of cell polarity (recently reviewed by Bornens 2012). Centriole duplication takes place in the S phase, thus, in the G2 phase of the cell cycle each cell possesses already four mature centrioles. At the G2/M transition, centrosomes separate to build future division spindle poles and increase their MT nucleating activity. During mitosis the centrosomes actively participate in the formation of the bipolar spindle via nucleation of MTs, very dynamic, and much shorter than in the interphase. Each of duplicated and separated centrosomes forms a spindle pole. MTs originating from centrosomes are captured by chromosomes, and more precisely by the specialized structures called kinetochores, to complete the assembly of the spindle.

The size and number of centrioles, the key constituent of a centrosome, is precisely controlled in the cell, while the volume of PCM change during the cell cycle and varies in different cell types changing the size of the centrosome. Within the centrosome, both centrioles are connected through their proximal ends during G1 and S phase. The two centrioles are functionally and structurally unequal. Due to the semi-conservative mode of centriole duplication (each centriole gives rise to a new centriole, then the pair splits and give rise to a new centrosome), one of them is always more matured than the other (Kochanski and Borisy 1990). Figure 20.2 illustrates the centrosome cycle, including centriole duplication and maturation within the centrosome in relation to the nuclear cell cycle phases G1, S, G2, and M. The centriole more advanced in maturation is called the mother centriole (centriole "père" in French, or "father", since the centriole is masculine in French) and the younger one: the daughter centriole (centriole "fils" in French, or "son") (see also Fig. 20.1). At the structural level the mother centriole possesses appendages at the distal end. The daughter centriole is slightly shorter. The

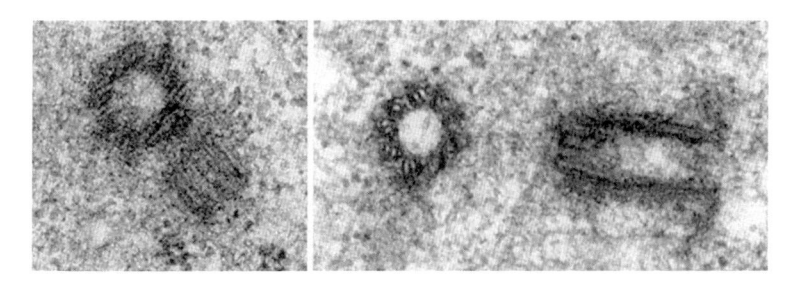

Fig. 20.1 Electron microscopy images of centrioles in *Xenopus laevis* oogonia. *Left panel*: pair of closely apposed centrioles; *Right panel*: pair of separated centrioles. In both panels a presumably mother centriole (*on the left*) is shown in cross section and a daughter centriole (*on the right*) in longitudinal section. Courtesy of Malgorzata Kloc, Houston, TX

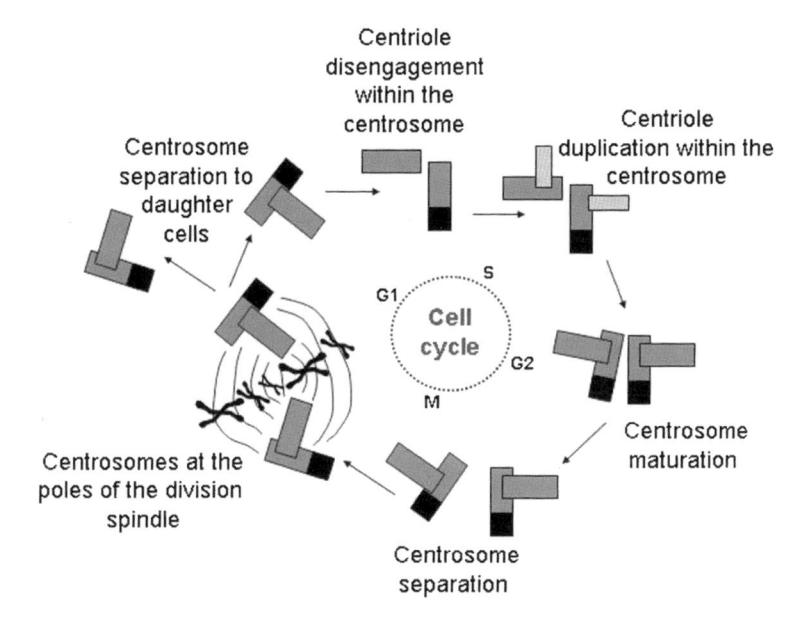

Fig. 20.2 Centrosome cycle in relation to the cell cycle progression. The mother centriole is *black* and *gray*, daughter is *gray*

two are connected by intercentriolar links embedded in the PCM. Due to slow growth and development of the daughter centriole the full reproduction cycle of the centrosome covers two cell division cycles (Fig. 20.2).

The structural cues have functional impact because the MT nucleation preferentially occurs at the mother centriole and MTs are fixed to the appendages of the mother. Also, the formation of a primary cilium belongs to exclusive capacities of the mother centriole. Thus, the polarity-delivering centrosome is intrinsically asymmetric assuring transmissibility of this cue to the cellular level.

20.1.1 Role of Pericentriolar Material (PCM) vs. Centrioles

Centrioles, through their capability to duplicate once per cell cycle, assure also the duplication of centrosomes. Similar to DNA replication the centrosome duplication must occur once and only once per cell cycle to keep their correct number. However, in contrast to DNA replication, centrosome duplication can be disconnected from cell cycle progression, despite that, physiologically, both remain tightly correlated in time (Gard et al. 1990; Hinchcliffe et al. 1999). The PCM, on the other hand, bears the functional and regulatory machinery for the nucleation of MTs. Centrosomes nucleate four kinds of MTs, the interphasic, mitotic, centriolar, and those of cilia. Not all cells nucleating MTs contain centrosomes. In vertebrates, including *Xenopus laevis*, fully grown female germ cells (oocytes) do not have centriole, but they possess functional MTOCs formed exclusively by the PCM foci. Because the centrosome is structurally defined as an organelle composed of centrioles and PCM, it precludes that oocytes have no centrosomes. However, there is no doubt that oocytes do possess MTOCs. Molecular studies of the composition of oocyte MTOC have confirmed that it corresponds to the centrosomal PCM. Thus, in oocytes the PCM alone is apparently sufficient to fulfill the major functions of centrosomes, namely MT nucleation and organization.

As mentioned above, centrosomes are self-replicating organelles and centriole duplication plays a critical role in this process. In somatic cells the restriction of centrosome duplication to only once per cell cycle is of the highest importance for their physiology (Tsou and Stearns 2006). An incorrect number of centrosomes predisposes cells to chromosome missegregation and induces aneuploidy eventually leading to carcinogenesis (Nigg 2007).

It is still unknown how in the cells lacking centrosome sensu stricto, in which MTOCs (PCM foci) nucleate MT, the information about the structure of centrosome is transmitted from a mother cell to a daughter and how the progeny of such cells are able to develop centrioles and centrosomes during development. The best example is a mouse embryo in which cell divisions during at least five initial cleavages occur without the centrioles (Gueth-Hallonet et al. 1993). How are these MTOCs re-substituted after each division remains absolutely unknown? Do they retain the structural memory of the centriole structure revealing itself when centrioles form de novo? We mention here the mouse model to illustrate the general lack of knowledge on the nature of the heredity of centrosomal structures.

Although it is now accepted that the centrosome does not contain any DNA, it was suggested that the centrosome may contain specific RNAs. These RNAs were identified in centrosomes isolated from the clam *Spisula solidissima* oocytes (Alliegro et al. 2006; Alliegro and Alliegro 2008) and they also seem to be present in *Xenopus laevis* and human centrosomes (Blower et al. 2007; Alliegro 2008). Thus, besides the proteins, which form an extremely well-organized structure, the centrosome may contain RNA. The presence of "centrosomal RNAs" is still a very controversial area. It stimulates, however, a discussion on whether such RNAs can be involved in centriole/centrosome duplication. Two major questions can be asked: (i) Do "centrosomal RNAs" serve as the templates for propagation of eventual structural information? (ii) Are these RNAs responsible for the structural memory of the centrioles? Nucleic acids are certainly better suited for this role than proteins and further understating of the role of centrosomal RNAs may have an extremely important impact on our understanding of the centrosome biology. Some RNAs are not only specifically localized on centrosomes, but also "take advantage" of highly asymmetric localization on one of the two centrosomes of the division spindle. In spirally cleaving *Ilyanassa obsoleta* mollusk some of the embryo RNAs are localized next to centrally placed spindle poles/centrosomes through a micotubule-dependent pathway. Following the cleavage these RNAs are asymmetrically inherited by the interior blastomeres (Lambert and Nagy 2002). Recent studies suggest that non-protein-coding microRNAs (miRNAs) build-up of about 22 nucleotides may regulate centrosome duplication. It is the case of miR-210, the miRNA overexpressed in renal carcinoma cells and involved in centrosome amplification (Nakada et al. 2011). Another example is delivered by miR-449 in *Xenopus laevis* and human cells. Mi9-449 accumulates in multiciliated cells, promotes centriole multiplication and multiciliogenesis via Delta/Notch pathway repression (Marcet et al. 2011). Future studies will have to verify whether miRNAs physically associate with centrosomes/centrioles and if and how it regulates centrosome development.

20.1.2 MT Nucleation and Organization in Xenopus laevis Oocytes and Cell-Free Extract

Early observations on the MTOC's role of centrosomes in *Xenopus laevis* have shown that centrosome forms MT asters when injected into the oocytes (Heidemann and Kirschner 1975; Mitchison and Kirschner 1984; Karsenti et al. 1984). Similar observations in other experimental systems suggest that MTOC activity of centrosomes is a general phenomenon (Kuriyama and Borisy 1983). The cell-free extract obtained from *Xenopus* eggs enabled to study the MT nucleation and organization in a cell-free experimental environment (Gard and Kirschner 1987) and to determine how the centrosomes are assembled via recruitment of proteins from the cytoplasm (Félix et al. 1994; Stearns and Kirschner 1994). These studies pointed to the crucial role of γ-tubulin recruitment in MT nucleation and MT asters formation. The experiments on the cell-free extract depleted of γ-tubulin showed that the recruitment of this protein from the oocyte maternal store of proteins is necessary for the paternal centrosome to gain the ability to nucleate MTs (Félix et al. 1994) (Fig. 20.3). In parallel, the pericentrin was identified as another critical component of the mature and fully active centrosome (Doxsey et al. 1994; Dictenberg et al. 1989). The incorporation of maternal proteins restores a functional centrosome nucleating and organizing MTs (Fig. 20.3). In these studies isolated sperm heads or centrioles introduced to cell-free extracts were used to follow centrosome formation and

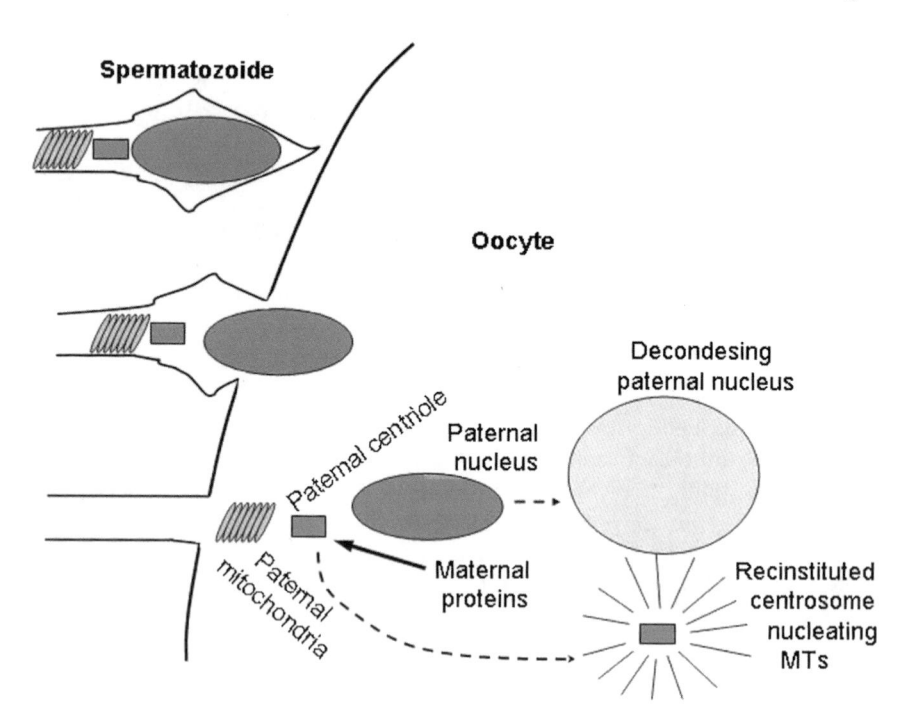

Fig. 20.3 Schematic representation of the origin of paternal centriole and maternal centrosomal proteins (such as γ-tubulin and pericentrin) and their role in reconstitution of the functional centrosome during fertilization events

functions, but similar transformation of centrosomes occurs during fertilization. The other milestone in centrosome research in *Xenopus laevis* cell-free system was the discovery that the MT self-assembly is sufficient to organize the bipolar structure of the spindle (Heald et al. 1996, 1997). These studies, in which the artificial chromosomes (beads covered by plasmid DNA) were used, demonstrated that the kinetochores, the complex chromosomal structures physiologically assuring MT attachment, are not necessary for the proper spindle assembly. Studies on mouse oocytes deprived of chromosomes showed that bipolar spindles may form also in the total absence of chromatin. This happens through the MT self assembly with the help of motor proteins organizing both MTs and PCM first into asters and then into spindle-like structures (Brunet et al. 1998). All these experiments indicate that centrioles and even chromosomes are dispensable for spindle assembly both in *Xenopus laevis* cell-free extracts and, under special conditions, in living cells of different species (Walczak et al. 1998; Varmark 2004). The optional character of the presence of centrioles was also confirmed in HeLa cells in which centriole disassembly was induced by a monoclonal antibody GT335 directed against glutamylated tubulin (Bobinnec et al. 1998). In these cells, the centriole disappearance had a severe impact on the structure of centrosome and resulted in the disaggregation of PCM.

Nevertheless, the absence of centrioles is tolerated only in appropriate physiological conditions, as confirmed by a series of studies of acentriolar MTOC and acentriolar spindles in *Drosophila melanogaster* (for further reading see e.g. Basto et al. 2006; Stevens et al. 2007; Debec et al. 2010). The discovery that flies do not need a centriole for cell division confirms that the centriole is a dispensable element of MTOC. However, the cells lacking centriole cannot generate cilia and acentrosomal *Drosophila* eventually die due to cilia deficiencies. Thus, the dominant role of centrioles and centrosomes appears to be linked to their function in ciliogenesis.

Recent study has shown that some planarians lack centrosomes (Azimzadeh et al. 2012). Centrioles and cilia are formed only in their terminally differentiating muliciliated cells. How planarian cells organize their MT network is, however, unknown. Notably, the lack of centrosomes does not alter cell division, but seems to be related to the change in the mode of cleavages during embryo development from spiral to anarchic-like (ibid.). Curiously, mammalian embryos also lacking centrosomes have much less stereotypic pattern of cleavages than invertebrates or amphibians. These two examples suggest that centrosomes may control the pattern of cleavages during early development more tightly than it was previously thought.

20.2 Centrosome Reduction During Gametogenesis

As stated above, during gametogenesis the centrosomes are reduced in both male and female gametes. The restoration of a fully functional centrosome occurs during fertilization. Theodor Boveri already suspected and wrote 100 years ago that the male gamete "sperm" was carrying a "division centre" with the "cytoplasm active substance" whereas the female gamete "oocyte" was carrying the "cytoplasm active substance" with the "division centre". After fertilization the fusion of the division center and the cytoplasm active substance would give rise to an active centrosome. It is now known that, in most cases, upon fertilization the sperm brings the centrosome and the oocyte is the source of the cytoplasm proteins.

20.2.1 Centrosome Reduction in Spermatozoa

The centrosomes reduction was reported in many species of invertebrates (insects, mollusks) suggesting that this phenomenon is the rule rather than exception during spermatogenesis (reviewed by Manandhar et al. 2005). Among vertebrates, this process was the most extensively studied in mammalian spermatozoa (Manandhar et al. 1998, 2000), and not much information is available on the centrosome reduction in spermatozoa in amphibians. Centrioles are present in *Xenopus laevis* spermatozoa (Bernardini et al. 1986). However, the fact that *Xenopus* sperm heads do not contain γ-tubulin demonstrates that centrosomes are also reduced in this species (Félix et al. 1994; Stearns and Kirschner 1994).

20.2.2 Centrosome Reduction in Oocytes

The phenomenon of centrosome reduction during oogenesis has been much better described for mouse than for *Xenopus* oocytes. The reason for that is the large size of fully grown amphibian oocytes. It makes their electron microscopy study by conventional methods very difficult and thus the direct examination of such tiny structures as potential centrosomes or PCM foci is almost impossible. The immunofluorescence localization of selected antigens has been of great help. Early studies of *Xenopus laevis* oocytes by immunolocalization of MTs suggested, as judged by the barrel shape of the meiotic spindles and the general manner of MTs' organization upon the entry into the first meiotic division, that centrioles are absent from these oocytes (Huchon et al. 1981). Later studies using confocal microscoy confirmed these observations (Gard et al. 1995a, b). The immunolocalization of centrin and electon microscopy images of very early stages of oogenesis in ovarian cysts of female froglets show the presence of centrioles (Kloc et al. 2004, and Fig. 20.1 in this paper). Thus, in *Xenopus laevis* the centrioles disappear at the beginning, or during the long-lasting phase of oocyte growth. Similarly, the presence of centrioles was also demonstrated in mouse fetal oogonia still organized in the cytocysts (Kloc et al. 2008). The centrioles were, however, absent from spindle poles of fully grown metaphase I and metaphase II mouse oocytes (Szöllosi et al. 1972). Thus, both in *Xenopus* and the mouse the disintegration of centrioles seems to occur during the oocyte growth phase, but so far the mechanism and precise timing of this phenomenon remain unknown.

20.3 Centrosome Inheritance upon Fertilization

The clear biological relevance of the reduction of centrosomes in the gametes is unknown. The morphological reduction is probably related to the functional one. The spermatozoon and the oocyte are fully differentiated cells. The reduction of centrosomes may be related to a diminution of their MTOC function in gametes. However, as these functions are of great importance for the somatic cells, fully active centrosomes must be rebuilt following fertilization.

Upon fertilization the spermatozoon and the oocyte join together and reduce paternal and maternal centrosomes (Fig. 20.3). The MTOC activity of the paternal centrosome is easily detectable due to the rapid formation of so-called sperm aster of MTs. The sperm aster's major role is to bring the male and female pronuclei together. This role of the paternal centrosome was the easiest to be visualized in transparent zygotes of marine invertebrates such as sea urchin of starfish

(Sluder et al. 1986; Zhang et al. 2004). Immunofluorescence studies confirmed that the aster is indeed composed of MTs (Schatten et al. 1986, 1987). In the sea urchin *Lytechinus pictus* zygote, the mechanical separation of male and female pronuclei (with their accompanying respective cytoplasm including the paternal centrosome next to the male pronucleus) immediately after fertilization, produces two haploid half-embryos behaving in two very distinct ways. The male pronucleus-containing half-embryo forms the sperm aster, centers its pronuclus, and then assembles a division spindle and undergoes series of regular cleavages. On the other hand, the female pronucleus-containing half-embryo undergoes nuclear envelope breakdown and reformation without either spindle formation or cleavages (Kubiak 1991). Thus, apparently, the sperm aster formed by the paternal centrosome is the prerequisite for successful embryo cleavages. The spermatozoon delivers the paternal centriole to the oocyte during fertilization and it is necessary to support the centrosome reconstitution and further duplication for the first mitotic division. We described here the example of transparent and relatively small sea urchin embryo, especially appropriate for this type of experiments, but the same role of the paternal centrosome is also attributed to the first cell cycle during *Xenopus laevis* development (Fig. 20.3). This is the best illustrated by the absence of cleavages in parthenogenetic *Xenopus* embryos (Klotz et al. 1990). The parthenogenetic activation of eggs results in cycling embryos undergoing cell cycle transitions, from interphase to mitosis and back to the interphase, without cleavages. However, when a single centriole is experimentally introduced into such a one-cell parthenogenetic embryo, the division spindle forms (Maller et al. 1976) and cleavages reappear with normal frequency and timing (Tournier et al. 1989; Klotz et al. 1990). Moreover, only the centrioles capable to duplicate have the capacity to support the parthenogenetic development of frog (Tournier et al. 1991a, b). Surprisingly, an electron microscopy study showed that the mature centrioles injected into *Xenopus laevis* parthenogenetic embryos were transformed into the juvenile centrioles showing the capacity of the cytoplasm to transform these organelles into a novel juvenile form (Nadezhdina et al. 1999). The absence of centrioles in *Xenopus* parthenogenetic embryos prevents successful development. Thus, the evolutionary conserved mechanism involved in centrosome reduction and centriole elimination in the oocytes seems to protect the species against parthenogenetic mode of reproduction. It may be, therefore, one of the key mechanisms promoting sexual reproduction in animal kingdom.

Interestingly, the mouse, which, as we already described above, also loses centrioles during oogenesis (Kloc et al. 2008; Szöllosi et al. 1972), the cleavages of early parthenogenetic embryos proceed normally (Tarkowski et al. 1970). However, in this species, during normal fertilization the sperm-delivered centriole disintegrates in the zygote and the early cleavages occur without centrioles until their de novo formation in the early blastocyste stage (Gueth-Hallonet et al. 1993). This implies that in the mouse, the mature centrosomes are exclusively of maternal origin. In this context it is not surprising that in contrast to *Xenopus*, mouse parthenogenetic and normal embryos fully support cleavages and early steps of development. However, we have to stress here that in other mammals, including humans, centrosomes are, like in *Xenopus*, of biparental origin (Sathananthan et al. 1996; Simerly et al. 1999).

The absence of centrioles in oocytes is not an absolute obstacle for successful cleavages since the centrioles may form de novo, for example after the fifth cleavage in normal mouse embryos (Gueth-Hallonet et al. 1993). De novo formation of centrioles was first observed in sea urchin embryos (Kato and Sugiyama 1971; Kallenbach 1983). Later, it was also found in rabbit parthenogenetic embryo (Szöllosi and Ozile 1991), and transformed as well as untransformed human cells (La Terra et al. 2005; Uetake et al. 2007). Recently, this pathway of centriole formation was also identified in *Xenopus laevis* embryos, and Plk4 kinase seems to be the key enzyme involved in this process (Eckerdt et al. 2011). Moreover, this paper shows that Mos/MAPK pathway suppresses Plk4-dependent centriole formation in oocytes. As Mos/MAPK pathway is activated in all oocytes studied so far, it seems to prevent maternal inheritance of centrosomes and thus protects against the parthenogenetic mode of reproduction.

20.4 Molecular Studies of Centrosome Function in *Xenopus laevis*

The function and behavior of centrosomes have been extensively studied in *Xenopus* egg extracts recapitulating cell cycle events at the molecular level. A role of CDK/cyclin complexes in MT nucleation activity of centrosome was identified in *Xenopus* egg extracts (Buendia et al. 1992). Importantly, a new role, during mitosis, of the interphase cytoplasm-nuclear transport machinery was also discovered in these extracts, as importin complexes and the small GTPase Ran involved in this process are also actively involved in spindle assembly (Nachury et al. 2001). The Aurora-A activator TPX2, which is sequestered by importin in G2 phase of the cell cycle, is released during mitosis by the small GTPase Ran to bind to Aurora-A already present on the centrosomes at the spindle poles. This association results in the activation of the Aurora-A kinase and its relocation to the MTs of the spindle poles making this kinase a peculiar player in the centrosome-driven spindle assembly (Kufer et al. 2002; Tsai et al. 2003). It is, however, not known whether TPX2 participates in centrosomal Aurora-A activation. Identifying the centrosomal activator of Aurora-A is one of the most important challenges in the field.

Aurora-A kinase gene was first identified in *Xenopus laevis* embryo as the Eg2 in a screen focused on identification of genes whose maternal mRNA was differentially polyadenylated during oocyte maturation and early embryo development (Paris and Philippe 1990). Eg2 was found to be the second gene of a series of mRNAs deadenylated following fertilization in *Xenopus laevis* embryo (Legagneux et al. 1995; Detivaud et al. 2003). In the meantime, a gene named Aurora was identified in *Drosophila*, and it was shown to code a protein kinase required for centrosome separation (Glover et al. 1995). The studies of Eg2 protein function in *Xenopus laevis* cell-free extract have shown that Eg2 is associated with centrosomes and involved in MTs' organization and spindle formation (Roghi et al. 1998). Very soon, it became clear that Eg2 was the homolog of *D. melanogaster* Aurora-A kinase (Glover et al. 1995; Giet et al. 1999a, b). Since this discovery the Aurora-A/Eg2 protein kinase was studied in parallel in *Drosophila* and *Xenopus* embryo. Its activation was soon correlated with M-phase entry upon *Xenopus laevis* oocyte entry into the meiotic maturation (Frank-Vaillant et al. 2000). The primordial role of this kinase in oocyte maturation seems to be the regulation of maternal mRNAs polyadenylation status (Mendez et al. 2000; Pascreau et al. 2005). Aurora-A has been shown to regulate MT nucleation and organization via phosphorylation of maskin/TACC (Pascreau et al. 2005; Kinoshita et al. 2005) necessary for the MT-Associated (MAP) protein XMAP215 targeting to the centrosomes (Brittle and Ohkura 2005). Aurora-A kinase was also shown to phosphorylate kinesine Eg5 distributed along MTs (Giet et al. 1999a). Because Aurora-A has been recognized as an oncogene and a tumor suppressor, depending on the cell type, the studies in *Xenopus* have an important impact on the understanding of Aurora-A and centrosome functions in carcinogenesis.

20.5 Molecular Studies of Centrosome Duplication in *Xenopus laevis*

Centrosome duplication has been extensively studied in *Xenopus* extract and embryos allowing deciphering its regulation at the molecular level. The possible disconnection between centrosome duplication and cell cycle progression was observed using *Xenopus* embryos (stage blastula) treated with cycloheximide (Gard et al. 1990).

The second important breakthrough was the discovery of the control of centrosome duplication by CDK/cyclin complexes, in particular, CDK2/cyclin E (Lacey et al. 1999) (Hinchcliffe et al. 1999). It is, however, still unknown what substrate(s) of CDK2/cyclin E is (are) necessary for this regulation. In *Xenopus* egg extract, the duplication of centrosome was found to be sensitive to calcium with a role of calmodulin and CaM kinase II in the initiation of duplication (Matsumoto and Maller 2002). As pointed out by the authors, these data provided a link between cell cycle progression, calcium waves, and centrosome duplication.

Plk4, a divergent member of the Polo-like kinase family already mentioned above, is another key molecule involved in centriole and centrosome duplication (Bettencourt-Dias et al. 2005; reviewed by Sillibourne and Bornens 2010). Little is known how Plk4 control centriole duplication but the network of genes involved in this process is steadily increasing. For instance, Hatch et al. (2010) have used recently a "kinase dead" mutant of Plk4 expressed in the cell-free extract to screen for Plk4 associated proteins. They identified seven proteins and among them Cep152, the homolog of *D. melanogaster* Asterless (*Asl* gene product), already known to be involved in centrosome regulation and centriole assembly (Dzhindzhev et al. 2010). Cep152 associates with Plk4 and is phosphorylated by this kinase in vitro suggesting that Cep52 can be a substrate of Plk4 (Hatch et al. 2010). Other kinases and phosphatases, such as CDK11 (Franck et al. 2011) and PP2A (Brownlee et al. 2011) are also involved. Precise dosage of Plk4 is required to control centriole duplication (Holland et al. 2010). Accordingly Plk4 over expression or stabilization induces centrosome amplification (Brownlee et al. 2011). This is an important link to cancer since the amplification of centrosomes is a condition found in numerous cancers (reviewed by Chan 2011). Experiments based on *Xenopus laevis* cell-free extracts depletion and reconstitution deliver unique and highly advantageous possibilities over culture cell transfection to study this aspect of cell regulation critical for carcinogenesis.

Studies of centrosome duplication revealed few features common with DNA replication: (i) centrosome duplication and DNA replication must occur once and only once per cell cycle, (ii) DNA is composed of two strands of nucleotides, the centrosome is composed of two centrioles, (iii) while DNA replication involves copying of each strand to create two new double strands, centrosome duplication implies copying of each centriole to create two new centrosomes. In addition, *Xenopus* eggs cell-free extracts studies have shown that the centrosome duplication is restricted to only one per cell cycle, and similarly to DNA replication is under a "licensing control". Notably, the centrosome duplication licensing is relieved upon the exit from mitosis by the activity of separase, also involved in degradation of cohesion required for chromatid separation. The action of separase on the centrosome results in the disengagement of centrioles allowing a new round of duplication (Tsou and Stearn 2006) increasing the similarity between the mode of centrosome and DNA duplication.

Regarding centrosome duplication per se, studies of Plk4 in *Xenopus* oocytes revealed that overexpression of the kinase induces de novo centriole formation in a CDK2-independent manner (Eckerdt et al. 2011). Thus, de novo formation of centrioles seems to be under the control of different pathways (CDK2-independent) than centriole duplication (CDK2-dependent). A solution for this intriguing paradox is actively searched nowadays.

20.6 Conclusions

The centrosome research in *Xenopus laevis* eggs has delivered a number of fundamental information on mechanisms governing not only oogenesis, fertilization, and embryo development, but also different aspects of physiology and pathology of human cells. This was possible using advantages of functional cell-free extracts and the injection of living oocytes and embryos with specific blocking antibodies or morpholino. The field of centrosome research will certainly be followed up in more detail using human cells. However, we believe the *Xenopus laevis* egg studies have not pronounced the last word yet.

Acknowledgments We are grateful to Malgorzata Kloc, Houston, TX for valuable discussions and providing unpublished electron microscopy images of centrioles in *Xenopus laevis* oogonia.

References

Alliegro MC (2008) The implications of centrosomal RNA. RNA Biol 5:198–200
Alliegro MC, Alliegro MA (2008) Centrosomal RNA correlates with intron-poor nuclear genes in Spisula oocytes. Proc Natl Acad Sci U S A 105(19):6993–6997
Alliegro MC, Alliegro MA, Palazzo RE (2006) Centrosome-associated RNA in surf clam oocytes. Proc Natl Acad Sci U S A 103(24):9034–9038
Azimzadeh J, Wong ML, Downhour DM, Sánchez Alvarado A, Marshall WF (2012) Centrosome loss in the evolution of planarians. Science 335:461–463
Basto R, Lau J, Vinogradova T, Gardiol A, Woods CG, Khodjakov A, Raff JW (2006) Flies without centrioles. Cell 125:1375–1386
Bernardini G, Stipani R, Melone G (1986) The ultrastructure of Xenopus spermatozoon. J Ultrastr Mol Struct Res 94:188–194
Bettencourt-Dias M, Rodrigues-Martins A, Carpenter L, Riparbelli M, Lehmann L, Gatt MK, Carmo N, Balloux F, Callaini G, Glover DM (2005) SAK/PLK4 is required for centriole duplication and flagella development. Curr Biol 15:2199–2207
Blower MD, Feric E, Weis K, Heald R (2007) Genome-wide analysis demonstrates conserved localization of messenger RNAs to mitotic microtubules. J Cell Biol 179:1365–1373
Bobinnec Y, Khodjakov A, Mir LM, Rieder CL, Eddé B, Bornens M (1998) Centriole disassembly in vivo and its effect on centrosome structure and function in vertebrate cells. J Cell Biol 143(6):1575–1589
Bornens M (2012) The centrosome in cells and organisms. Science 335:422–426
Brittle AL, Ohkura H (2005) Centrosome maturation: Aurora lights the way to the poles. Curr Biol 15:R880–R882
Brownlee CW, Klebba JE, Buster DW, Rogers GC (2011) The protein phosphatase 2A regulatory subunit twins stabilizes Plk4 to induce centriole amplification. J Cell Biol 195:231–243
Brunet S, Polanski Z, Verlhac MH, Kubiak JZ, Maro B (1998) Bipolar meiotic spindle formation without chromatin. Curr Biol 8:1231–1234
Buendia B, Draetta G, Karsenti E(1992) Regulation of the microtubule nucleating activity of centrosomes in Xenopus egg extracts: role of cyclin A-associated protein kinase. J Cell Biol 116(6):1431–1442
Chan JY (2011) A clinical overview of centrosome amplification in human cancers. Int J Biol Sci 7:1122–1144
Debec A, Sullivan W, Bettencourt-Dias M (2010) Centrioles: active players or passengers during mitosis? Cell Mol Life Sci 67:2173–2194
Detivaud L, Pascreau G, Karaiskou A, Osborne HB, Kubiak JZ (2003) Regulation of EDEN-dependent deadenylation of Aurora A/Eg2-derived mRNA via phosphorylation and dephosphorylation in Xenopus laevis egg extracts. J Cell Sci 116:2697–2705
Dictenberg JB, Zimmerman W, Sparks CA, Young A, Vidair C, Zheng Y, Carrington W, Fay FS, Doxsey SJ (1989) Pericentrin and γ-tubulin form a protein complex and are organized into a novel lattice at the centrosome. J Cell Biol 141(1):163–174
Doxsey SJ, Stein P, Evans L, Calarco PD, Kirschner M (1994) Pericentrin, a highly conserved centrosome protein involved in microtubule organization. Cell 76:639–650
Dzhindzhev NS, Yu QD, Weiskopf K, Tzolovsky G, Cunha-Ferreira I, Riparbelli M, Rodrigues-Martins A, Bettencourt-Dias M, Callaini G, Glover DM (2010) Asterless is a scaffold for the onset of centriole assembly. Nature 467:714–718
Eckerdt F, Yamamoto TM, Lewellyn AL, Maller JL (2011) Identification of a polo-like kinase 4-dependent pathway for de novo centriole formation. Curr Biol 21(5):428–432

Félix MA, Antony C, Wright M, Maro B (1994) Centrosome assembly in vitro: role of γ-tubulin recruitment in Xenopus sperm aster formation. J Cell Biol 124(1–2):19–31

Franck N, Montembault E, Romé P, Pascal A, Cremet JY, Giet R (2011) CDK11(p58) is required for centriole duplication and Plk4 recruitment to mitotic centrosomes. PLoS ONE 6(1):e14600

Frank-Vaillant M, Haccard O, Thibier C, Ozon R, Arlot-Bonnemains Y, Prigent C, Jessus C (2000) Progesterone regulates the accumulation and the activation of Eg2 kinase in Xenopus oocytes. J Cell Sci 113:1127–1138

Gard DL, Kirschner MW (1987) Microtubule assembly in cytoplasmic extracts of Xenopus oocytes and eggs. J Cell Biol 105(5):2191–2201

Gard DL, Hafezi S, Zhang T, Doxsey SJ (1990) Centrosome duplication continues in cycloheximide-treated Xenopus blastulae in the absence of a detectable cell cycle. J Cell Biol 110(6):2033–2042

Gard DL, Affleck D, Error BM (1995a) Microtubule organization, acetylation, and nucleation in Xenopus laevis oocytes: II. A developmental transition in microtubule organization during early diplotene. Dev Biol 168(1):189–201

Gard DL, Cha BJ, Schroeder MM (1995b) Confocal immunofluorescence microscopy of microtubules, MAPs, and MTOCs during amphibian oogenesis and early development. Curr. Top. Dev Biol 31: 383–431

Giet R, Uzbekov R, Cubizolles F, Le Guellec K, Prigent C (1999a) The Xenopus laevis aurora-related protein kinase pEg2 associates with and phosphorylates the kinesin-related protein XlEg5. J Biol Chem 274:15005–15013

Giet R, Uzbekov R, Kireev I, Prigent C (1999b) The Xenopus laevis centrosome aurora/Ipl1-related kinase. Biol Cell 91:461–470

Glover DM, Leibowitz MH, McLean DA, Parry H (1995) Mutations in aurora prevent centrosome separation leading to the formation of monopolar spindles. Cell 81:95–105

Gueth-Hallonet C, Antony C, Aghion J, Santa-Maria A, Lajoie-Mazenc I, Wright M, Maro B (1993) γ-tubulin is present in acentriolar MTOCs during early mouse development. J Cell Sci 105(Pt 1):157–166

Hatch EM, Kulukian A, Holland AJ, Cleveland DW, Stearns T (2010) Cep152 interacts with Plk4 and is required for centriole duplication. J Cell Biol 191:721–729

Heald R, Tournebize R, Blank T, Sandaltzopoulos R, Becker P, Hyman A, Karsenti E (1996) Self-organization of microtubules into bipolar spindles around artificial chromosomes in Xenopus egg extracts. Nature 382:420–425

Heald R, Tournebize R, Habermann A, Karsenti E, Hyman A (1997) Spindle assembly in Xenopus egg extracts: respective roles of centrosomes and microtubule self-organization. J Cell Biol 138(3):615–628

Heidemann SR, Kirschner MW (1975) Aster formation in eggs of Xenopus laevis. Induction by isolated basal bodies. J Cell Biol 67(1):105–117

Hinchcliffe EH, Li C, Thompson EA, Maller JL, Sluder G (1999) Requirement of Cdk2-cyclin E activity for repeated centrosome reproduction in Xenopus egg extracts. Science 283(5403):851–854

Holland AJ, Lan W, Niessen S, Hoover H, Cleveland DW (2010) Polo-like kinase 4 kinase activity limits centrosome overduplication by autoregulating its own stability. J Cell Biol 188:191–198

Huchon D, Crozet N, Cantenot N, Ozon R (1981) Germinal vesicle breakdown in the Xenopus laevis oocyte: description of a transient microtubular structure. Reprod Nutr Dev 21(1):135–148

Kallenbach RJ (1983) The induction of de novo centrioles in sea urchin eggs: a possible common mechanism for centriolar activation among parthenogenetic procedures. Eur J Cell Biol 30(2):159–166

Karsenti E, Newport J, Hubble R, Kirschner M (1984) Interconversion of metaphase and interphase microtubule arrays, as studied by the injection of centrosomes and nuclei into Xenopus eggs. J Cell Biol 98(5):1730–1745

Kato KH, Sugiyama M (1971) On the de novo formation of the centriole in the activated sea urchin egg. Dev Growth Differ 13(4):359–366

Kinoshita K, Noetzel TL, Pelletier L, Mechtler K, Drechsel DN, Schwager A, Lee M, Raff JW, Hyman AA (2005). Aurora A phosphorylation of TACC3/maskin is required for centrosome-dependent microtubule assembly in mitosis. J Cell Biol 170(7):1047–1055

Kloc M, Bilinski S, Dougherty MT, Brey EM, Etkin LD (2004) Formation, architecture and polarity of female germline cyst in Xenopus. Dev Biol 266:43–61

Kloc M, Jaglarz M, Dougherty M, Stewart MD, Nel-Themaat L, Bilinski S (2008) Mouse early oocytes are transiently polar: three-dimensional and ultrastructural analysis. Exp Cell Res 314:3245–3254

Klotz C, Dabauvalle MC, Paintrand M, Weber T, Bornens M, Karsenti E (1990) Parthenogenesis in Xenopus eggs requires centrosomal integrity. J Cell Biol 110(2):405–415

Kochanski RS, Borisy GG (1990) Mode of centriole duplication and distribution. J Cell Biol 110:1599–1605

Kubiak JZ (1991) Cleavage divisions of bisected sea urchin eggs and zygotes: implications for centrosome role and inheritance. Eur Arch Biol (Bruxelles) 102:103–109

Kufer TA, Silljé HH, Körner R, Gruss OJ, Meraldi P, Nigg EA (2002) Human TPX2 is required for targeting Aurora-A kinase to the spindle. J Cell Biol 158:617–623

Kuriyama R, Borisy GG (1983) Cytasters induced within unfertilized sea-urchin eggs. J Cell Sci 61:175–189

La Terra S, English CN, Hergert P, McEwen BF, Sluder G, Khodjakov A (2005) The de novo centriole assembly pathway in HeLa cells: cell cycle progression and centriole assembly/maturation. J Cell Biol 168(5):713–722

Lacey KR, Jackson PK, Stearns T (1999) Cyclin-dependent kinase control of centrosome duplication. Proc Natl Acad Sci U S A 96(6):2817–2822

Lambert JD, Nagy LM (2002) Asymmetric inheritance of centrosomally localized mRNAs during embryonic cleavages. Nature 420:682–686

Legagneux V, Omilli F, Osborne HB (1995) Substrate-specific regulation of RNA deadenylation in Xenopus embryo and activated egg extracts. RNA 1:1001–1008

Maller J, Poccia D, Nishioka D, Kidd P, Gerhart J, Hartman H (1976) Spindle formation and cleavage in Xenopus eggs injected with centriole-containing fractions from sperm. Exp Cell Res 99(2):285–294

Manandhar G, Sutovsky P, Joshi HC, Stearns T, Schatten G (1998) Centrosome reduction during mouse spermiogenesis. Dev Biol 203(2):424–434

Manandhar G, Simerly C, Schatten G (2000) Highly degenerated distal centrioles in rhesus and human spermatozoa. Hum Reprod 15(2):256–263

Manandhar G, Schatten H, Sutovsky P (2005) Centrosome reduction during gametogenesis and its significance. Biol Reprod 72:2–13

Marcet B, Chevalier B, Luxardi G, Coraux C, Zaragosi LE, Cibois M, Robbe-Sermesant K, Jolly T, Cardinaud B, Moreilhon C, Giovannini-Chami L, Nawrocki-Raby B, Birembaut P, Waldmann R, Kodjabachian L, Barbry P (2011) Control of vertebrate multiciliogenesis by miR-449 through direct repression of the Delta/Notch pathway. Nat Cell Biol 13:693–699

Matsumoto Y, Maller JL (2002) Calcium, calmodulin, and CaMKII requirement for initiation of centrosome duplication in Xenopus egg extracts. Science 295(5554):499–502

Mendez R, Murthy KG, Ryan K, Manley JL, Richter JD (2000) Phosphorylation of CPEB by Eg2 mediates the recruitment of CPSF into an active cytoplasmic polyadenylation complex. Mol Cell 6:1253–1259

Mitchison T, Kirschner M (1984) Microtubule assembly nucleated by isolated centrosomes. Nature 312(5991):232–237

Nachury MV, Maresca TJ, Salmon WC, Waterman-Storer CM, Heald R, Weis K (2001 Jan 12) Importin beta is a mitotic target of the small GTPase Ran in spindle assembly. Cell 104(1):95–106

Nadezhdina ES, Skoblina MN, Fais D, Chentsov YS (1999) Exclusively juvenile centrioles in Xenopus laevis oocytes injected with preparations of mature centrioles. Microsc Res Tech 44(6):430–434

Nakada C, Tsukamoto Y, Matsuura K, Nguyen TL, Hijiya N, Uchida T, Sato F, Mimata H, Seto M, Moriyama M (2011) Overexpression of miR-210, a downstream target of HIF1α, causes centrosome amplification in renal carcinoma cells. J Pathol 224:280–288

Nigg EA (2007) Centrosome duplication: of rules and licenses. Trends Cell Biol 17(5):215–221

Paris J, Philippe M (1990) Poly(A) metabolism and polysomal recruitment of maternal mRNAs during early Xenopus development. Dev Biol 140:221–224

Pascreau G, Delcros JG, Cremet JY, Prigent C, Arlot-Bonnemains Y (2005) Phosphorylation of maskin by Aurora-A participates in the control of sequential protein synthesis during Xenopus laevis oocyte maturation. J Biol Chem 280:13415–13423

Roghi C, Giet R, Uzbekov R, Morin N, Chartrain I, Le Guellec R, Couturier A, Dorée M, Philippe M, Prigent C (1998) The Xenopus protein kinase pEg2 associates with the centrosome in a cell cycle-dependent manner, binds to the spindle microtubules and is involved in bipolar mitotic spindle assembly. J Cell Sci 111:557–572

Sathananthan AH, Ratnam SS, Ng SC, Tarín JJ, Gianaroli L, Trounson A (1996) The sperm centriole: its inheritance, replication and perpetuation in early human embryos. Hum Reprod 11(2):345–356

Schatten H, Schatten G, Mazia D, Balczon R, Simerly C (1986) Behavior of centrosomes during fertilization and cell division in mouse oocytes and in sea urchin eggs. Proc Natl Acad Sci U S A 83:105–109

Schatten H, Walter M, Mazia D, Biessmann H, Paweletz N, Coffe G, Schatten G (1987) Centrosome detection in sea urchin eggs with a monoclonal antibody against Drosophila intermediate filament proteins: characterization of stages of the division cycle of centrosomes. Proc Natl Acad Sci U S A 84:8488–8492

Sillibourne JE, Bornens M (2010) Polo-like kinase 4: the odd one out of the family. Cell Div 5:25

Simerly C, Zoran SS, Payne C, Dominko T, Sutovsky P, Navara CS, Salisbury JL, Schatten G (1999) Biparental inheritance of γ-tubulin during human fertilization: molecular reconstitution of functional zygotic centrosomes in inseminated human oocytes and in cell-free extracts nucleated by human sperm. Mol Biol Cell 10(9):2955–2969

Sluder G, Miller FJ, Rieder CL (1986) The reproduction of centrosomes: nuclear versus cytoplasmic controls. J Cell Biol 103:1873–1881

Stearns T, Kirschner M (1994) In vitro reconstitution of centrosome assembly and function: the central role of γ-tubulin. Cell 76(4):623–637

Stevens NR, Raposo AA, Basto R, Johnston D, Raff JW (2007) From stem cell to embryo without centrioles. Curr Biol 17:1498–1503

Szöllosi D, Ozil JP (1991) De novo formation of centrioles in parthenogenetically activated, diploidized rabbit embryos. Biol Cell 72(1–2):61–66

Szöllosi D, Calarco P, Donahue RP (1972) Absence of centrioles in the first and second meiotic spindles of mouse oocytes. J Cell Sci 11:521–541

Tarkowski AK, Witkowska A, Nowicka J (1970) Experimental partheonogenesis in the mouse. Nature 226:162–165

Tournier F, Karsenti E, Bornens M (1989) Parthenogenesis in Xenopus eggs injected with centrosomes from synchronized human lymphoid cells. Dev Biol 136(2):321–329

Tournier F, Cyrklaff M, Karsenti E, Bornens M (1991a) Centrosomes competent for parthenogenesis in Xenopus eggs support procentriole budding in cell-free extracts. Proc Natl Acad Sci U S A 88(22):9929–9933

Tournier F, Komesli S, Paintrand M, Job D, Bornens M (1991b) The intercentriolar linkage is critical for the ability of heterologous centrosomes to induce parthenogenesis in Xenopus. J Cell Biol 113(6):1361–1369

Tsai MY, Wiese C, Cao K, Martin O, Donovan P, Ruderman J, Prigent C, Zheng Y (2003) A Ran signalling pathway mediated by the mitotic kinase Aurora A in spindle assembly. Nat Cell Biol 5:242–248

Tsou MF, Stearns T (2006) Mechanism limiting centrosome duplication to once per cell cycle. Nature 442(7105):947–951

Uetake Y, Loncarek J, Nordberg JJ, English CN, La Terra S, Khodjakov A, Sluder G (2007) Cell cycle progression and de novo centriole assembly after centrosomal removal in untransformed human cells. J Cell Biol 176(2):173–182

Varmark H (2004) Functional role of centrosomes in spindle assembly and organization. J Cell Biochem 91:904–914

Walczak CE, Vernos I, Mitchison TJ, Karsenti E, Heald R (1998) A model for the proposed roles of different microtubule-based motor proteins in establishing spindle bipolarity. Curr Biol 8:903–913

Zhang QY, Tamura M, Uetake Y, Washitani-Nemoto S, Nemoto S (2004) Regulation of the paternal inheritance of centrosomes in starfish zygotes. Dev Biol 266(1):190–200

Chapter 21
Role of the MTOC in T Cell Effector Functions

Martin Poenie, Laura Christian, Sarah Tan and Yuri Sykulev

Abstract T cells play important roles in defending the host against infections, in allergic responses, and in the destruction of tumor cells. The directed or focused delivery of effector molecules to another cell is minimally achieved by a two-step process that involves focusing of secretory vesicles around the microtubule-organizing center (MTOC) and movement of the MTOC up to the site of contact with the target cell. This chapter is focused on mechanisms involved in the movement of the MTOC to the target contact site in T cells. Modulated polarization microscopy (MPM) and several other imaging methods were employed to visualize the cytoskeleton in general and in particular, the dynamics of MTOC movement. Understanding the processes of MTOC translocation has important medical ramifications that are addressed in this chapter.

21.1 Introduction

T cells play important roles in defending the host against infections, in allergic responses, and in destruction of tumor cells. These roles can be broadly divided into two categories, those mediated by $CD8^+$ T cells, (cytotoxic T lymphocytes; CTLs) that are aimed at killing antigenic target cells, and those mediated by $CD4^+$ T cells (helper T cells) aimed at helping activate other cell types such as B cells or

M. Poenie (✉) · L. Christian · S. Tan
Molecular Cell and Developmental Biology, University of Texas at Austin,
1 University Station Mail Code C1000, Austin, TX 78712, USA
e-mail: poenie@mail.utexas.edu

Y. Sykulev
Department of Microbiology and Immunology, Kimmel Cancer Center and Jefferson Vaccine Center, Thomas Jefferson University, Philadelphia, PA, USA

macrophages. However, despite a diversity of roles, much of what T cells do can be summarized in cellular terms as the directed or focused delivery of effector molecules to another cell. This is achieved minimally by a two-step process that involves focusing of secretory vesicles around the microtubule-organizing center (MTOC) and movement of the MTOC up to the site of contact with the target cell. These two steps can apparently occur in either order but the sequence influences the speed and magnitude of effector molecule delivery (Poenie 2010; Sykulev 2010).

Movement of the MTOC to the target contact site in T cells was first described by Geiger and colleagues (Geiger et al. 1982) and extended by a series of studies from Kupfer et al. (1983, 1985, 1986), Kupfer and Dennert (1984), and Kupfer and Singer (1989). During this time, evidence was accumulating to support the idea that the directional or localized stimulation of the T cell receptor (TcR) led to directional secretion of cytokines (Kupfer et al. 1986; Takayama and Sitkovsky 1987; Poo et al. 1988) and directional killing (Kupfer et al. 1985, 1986). For example, disruption of microtubules with nocodazole reversibly blocked killing by NK cells (Kupfer et al. 1983). Other treatments that blocked MTOC translocation such as the nonenzymatic cholera toxin B subunit or heat shock also blocked killing by CTLs (Sugawara et al. 1993; Knox et al. 1991).

The importance of focused secretion might be inferred from the early study of Trenn and colleagues who artificially stimulated secretion by a combination of a calcium ionophore (ionomycin) and a PKC activator (PMA or Bryostatins) (Trenn et al. 1988). They found that the combination of these two compounds induced significant secretion of lytic vesicle contents but the effect was to increase lysis of non-antigenic targets while decreasing the lysis of antigen-specific target cells. This could be interpreted to mean that non-localized stimuli (PMA and ionomycin) triggered omnidirectional secretion. This in turn led to reduced specific killing, where secretory vesicles are normally focused and secreted directly at the target cell, and increased non-specific killing. A similar effect is seen in gunmetal mice (Stinchcombe 2001b).

More recent developments have shown that when a T cell contacts an antigen-bearing cell, a specialized junction develops which has been dubbed the immunological synapse (Fig. 21.1) (Dustin et al. 2010; Monks et al. 1998; Norcross 1984). Classically, the synapse is characterized by concentric zones containing particular groups of molecules known as supramolecular activation clusters or SMACs (Monks et al. 1998). The central or cSMAC is characterized by accumulation of signaling molecules such as the TcR, CD28, CD2, and PKC-Θ. Surrounding the central SMAC is the peripheral or pSMAC. This region is characterized by a ring of clustered LFA-1, an integrin. In addition to the pSMAC, there is a third zone known as the distal SMAC (dSMAC) (Freiberg et al. 2002) that contains CD43 and CD45, a phosphatase.

Secretion of CTL lytic vesicles appears to be directed to a zone between the pSMAC and cSMAC (Stinchcombe et al. 2001a). The mechanism of delivery clearly involves translocation of the MTOC up to the immunological synapse (Stinchcombe et al. 2006) and dynein-driven movement of secretory vesicles toward the MTOC (Mentlik et al. 2010; Poenie et al. 2004) although late stages of vesicle movement

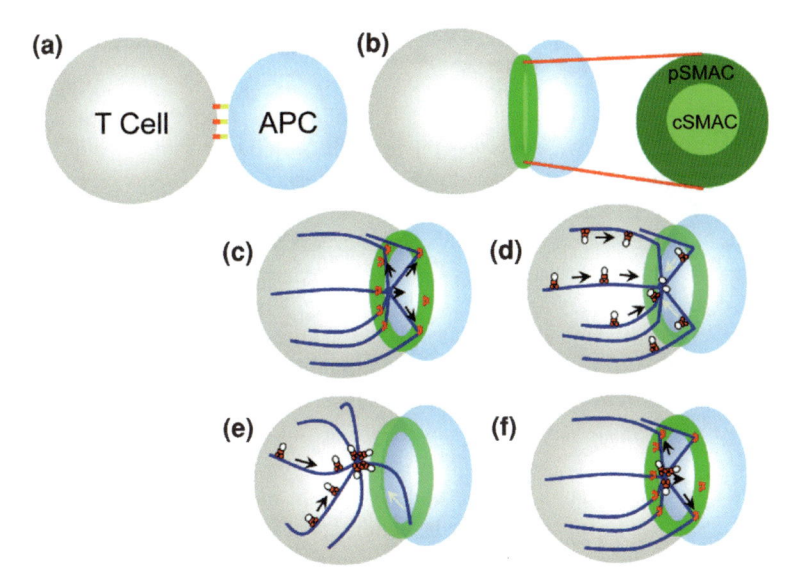

Fig. 21.1 Steps in T cell activation. **a** T cell activation begins when the T cell receptor (TcR) contacts antigen in the context of the major histocompatibility complex (MHC) of an antigen-presenting cell (APC). **b** Signaling downstream of the TcR results in the formation of the immunological synapse, which consists of several supramolecular activation clusters (SMACs). The central SMAC (cSMAC) contains signaling molecules such as the TcR. The peripheral SMAC (pSMAC) contains molecules such as the integrin LFA-1. Not shown is the distal SMAC (dSMAC), which forms a ring around the pSMAC. In helper T cells and cytotoxic T lymphocytes (CTLs) that kill their target slowly (**c, d**), the MTOC is thought to be translocated to the synapse by dynein molecular motor anchored in a ring similar to the pSMAC (**c**). This is followed by secretory vesicle transport by dynein toward the MTOC, where the vesicles are secreted (**d**). In CTLs that perform fast target cell lysis (**e, f**), secretory vesicles are thought to be concentrated by dynein around the MTOC before the MTOC is polarized (**e**). The MTOC is then translocated to the synapse where the vesicles are secreted (**f**)

could reportedly involve kinesin and/or myosin-mediated transport of secretory vesicles (Stinchcombe et al. 2001b; Sanborn et al. 2011; Sanborn et al. 2009; Andzelm et al. 2007; Kurowska et al. 2012; Burkhardt et al. 1993).

Understanding the processes of MTOC translocation has important medical ramifications. Tumor cells can escape destruction by tumor-infiltrating CTLs, despite the fact that these CTLs can be activated inside the tumor and can lyse tumor cells outside the tumor environment. One of the reasons that tumor cells can escape destruction is that something in the tumor environment blocks CTL MTOC translocation and thus delivery of cytotoxic vesicles (Frey and Monu 2008; Koneru et al. 2005; Monu and Frey 2007; Prevost-Blondel et al. 1998; Radoja et al. 2001). It is possible that by understanding how this system works, we may be able to detect exactly what fails in the tumor environment and then devise solutions to prevent this from happening.

21.2 The Mechanics of MTOC Repositioning

In an effort to visualize the cytoskeleton in general and in particular, the dynamics of MTOC movement, we developed modulated polarization microscopy (MPM) (Kuhn et al. 2001). This microscope images cytoskeletal elements based on their birefringence which allowed us to noninvasively image cells for extended lengths of time without concerns of photobleaching. MPM imaging data showed that MTOC movement was associated with development of tension in microtubules that pulled the MTOC toward the contact site (Fig. 21.2) (Kuhn and Poenie 2002). Once it reached the contact site, the MTOC oscillated laterally along the face of the contact site. It turned out that the magnitude of these oscillations was significant, about three to four microns, which is about the same as the inner diameter of the pSMAC. These oscillations were even more dramatic when two target cells were bound to a single CTL. Here, the MTOC moved repeatedly from one contact site to the other, a distance of approximately eight microns.

To explore this further, we examined numerous computerized 3D reconstructions of CTL-target pairs immunostained for tubulin and LFA-1. From this data, it became clear that the MTOC could be positioned in various regions of the cSMAC, but there were no examples where the MTOC crossed into the pSMAC (Fig. 21.3 a and b). Thus, the inner margin of the pSMAC seemed to represent a limit or barrier to oscillations of the MTOC under normal conditions. This raises the question concerning what happens when the MTOC oscillates between two target cells. Although we did not have specific examples of this in the immunostaining data, one often sees only partial rings of LFA-1. We might speculate that in cases where two target cells are engaged, there could be a partial pSMAC at each site which serves as endpoints for the oscillating MTOC.

The observation that tensioning of microtubules is associated with movement of the MTOC toward the synapse suggests that there must be a motor protein, anchored at the synapse, that reels in the microtubules. The most likely candidate is the microtubule minus end-directed motor, dynein. Given that lateral oscillations along the contact site can be in all directions (horizontal, vertical etc.), we suspected that the motor must be distributed as a ring with dimensions similar to the pSMAC. This would explain why the MTOC could only travel to the inner edge of the pSMAC.

As the MTOC moves toward the synapse, the microtubules projecting from the MTOC to the synapse take on the form roughly of a hollow cone, where the central region of the cone is devoid of microtubules (Fig. 21.3). The cone becomes progressively wider as the MTOC comes closer to the membrane. Microtubules at the edge of the cone appear to contact the cell surface or cortex in the region of the pSMAC and then bend backwards toward the rear of the cell (Fig. 21.3a–c) (Kuhn and Poenie 2002). These bend points had the appearance of "knuckles" in the microtubules where sharp bends were seen. This would make sense if motor proteins were anchored in the pSMAC region of the synapse and suggested the possibility that microtubules interact with a dynein ring that was in turn associated with sites where LFA-1 was clustered.

21 Role of the MTOC in T Cell Effector Functions

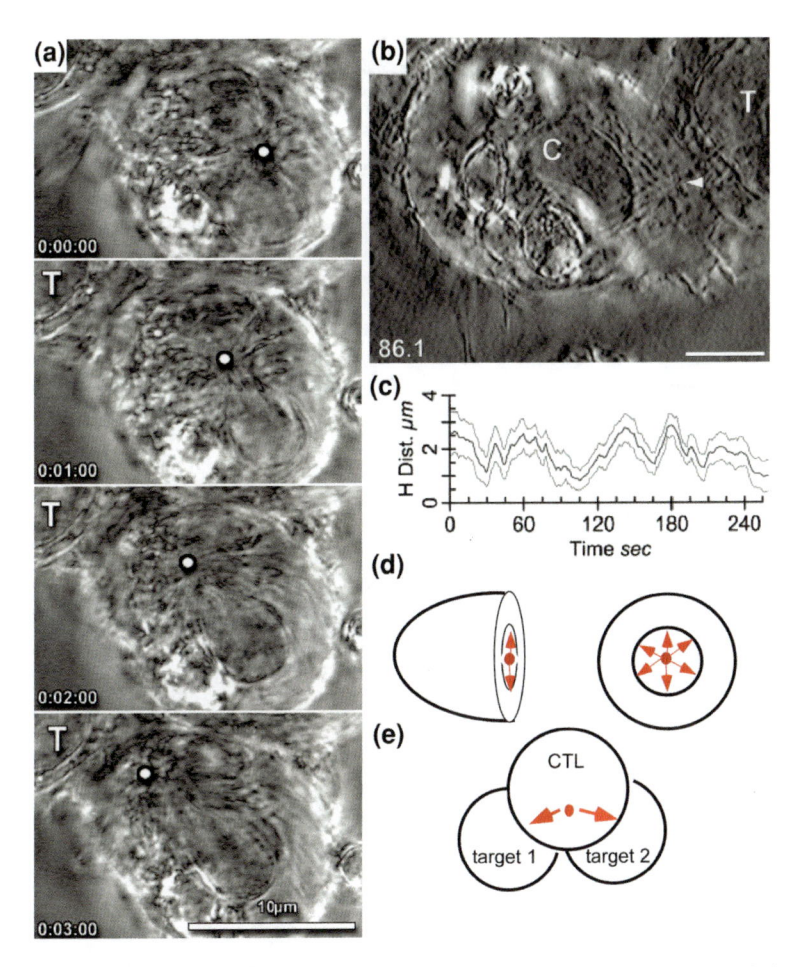

Fig. 21.2 Movement of the MTOC to the immunological synapse. **a** Modulated polarization microscopy (MPM) images of a CTL bound to an EL4.BU target cell (T) showing that the MTOC (*white circle*) translocated to the target contact site within 3 min. **b** MPM image showing a polarized MTOC (*arrowhead*) in a CTL (C) bound to a target cell (T). The scale bar is 5 microns wide. **c** The horizontal oscillations of the MTOC for the activated CTL in (**b**) are shown in relation to the target contact site *center*. The mean horizontal distances are plotted versus time. **d** Views of an activated T cell with the target cell removed. The MTOC (*red*) can oscillate in any direction within the pSMAC ring. The right image shows the synapse face-on. **e** If a T cell (*middle*) is activated by two target cells, the MTOC (*red*) can oscillate between the two target contact sites. Figures **a–c** are reprinted from Immunity 16:1 Kuhn and Poenie Dynamic polarization of the microtubule cytoskeleton during CTL-mediated killing. Copyright 2002 with permission from Elsevier

One way to test that idea was to take advantage of cases mentioned earlier where there are either partial LFA-1 rings or even just patches of LFA-1. When we compared the distribution of microtubules and regions where LFA-1 was clustered, we only saw microtubules in regions where LFA-1 was clustered. In activated T cells with partial

370   M. Poenie et al.

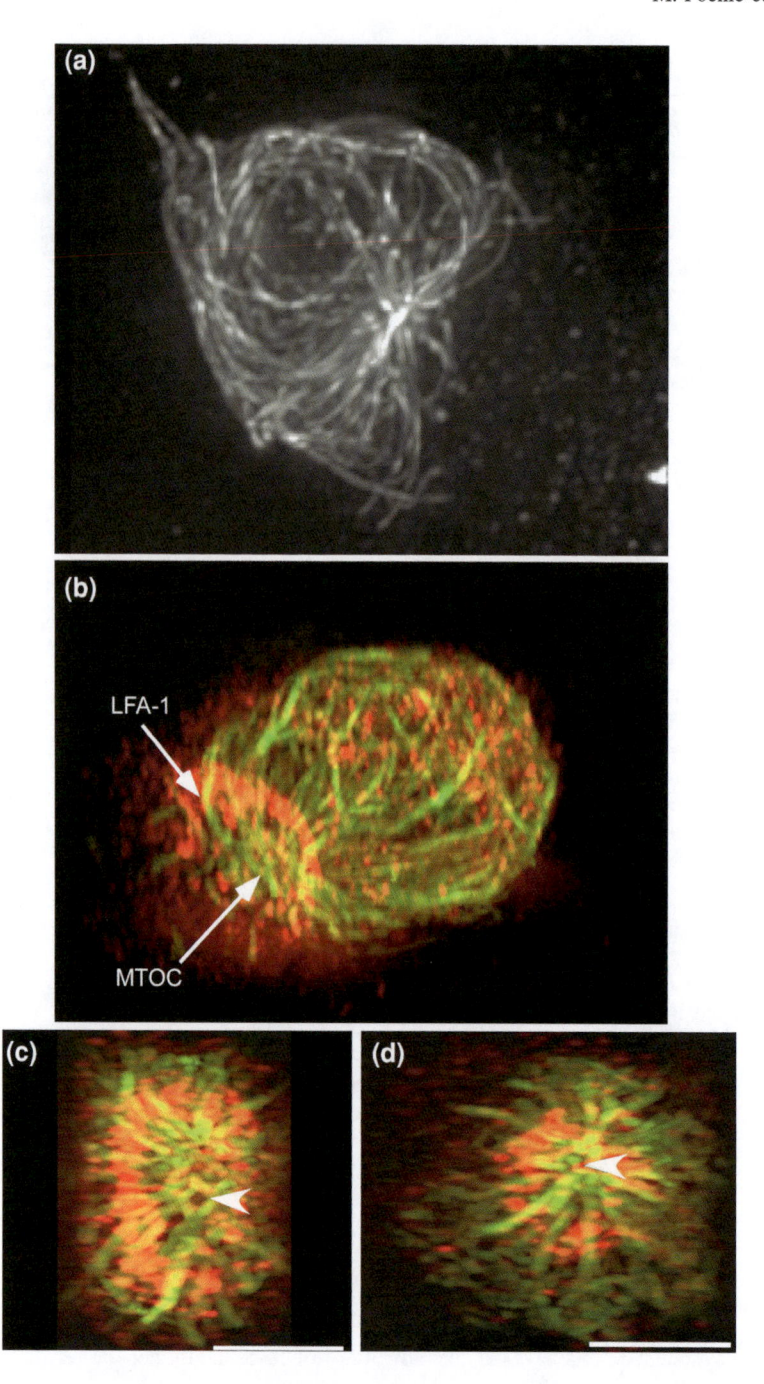

◄**Fig. 21.3** MTOC positioning at the immunological synapse. CTLs were activated by EL4.BU target cells and immunostained with tubulin (**a**) or LFA-1 (*red*) and tubulin (*green*) (**b, c, d**). The target cells were treated with colchicine to depolymerize microtubules prior to pairing for clarity. Three-dimensional reconstructions were derived from the fluorescence images. **a** In the late stage of MTOC translocation, microtubules appear to contact the pSMAC and then bend backwards away from the target cell. **b** The MTOC can be seen in the middle of the target contact site defined by the ring of LFA-1. **c, d** Cropped and rotated views of the synapse. In **c**, the MTOC is centered in the LFA-1 ring. In **d**, the MTOC is located toward the top of the LFA-1 ring image. No images obtained during these experiments showed the MTOC outside of the LFA-1 ring. Figures **a–d** are reprinted from Immunity 16:1 Kuhn and Poenie Dynamic polarization of the microtubule cytoskeleton during CTL-mediated killing. Copyright (2002) with permission from Elsevier

LFA-1 rings, the MTOC generally localized close to the partial ring, as opposed to activated T cells with full rings, where the MTOC was generally found close to the center of the ring.

The imaging data suggested that points where microtubules contact the cortex formed a ring-like pattern that correlated with the pSMAC and suggested the hypothesis that dynein might be organized similarly. To further investigate the role of dynein in MTOC translocation, we used Jurkat T cells. These cells have a number of advantages and disadvantages when compared to the mouse CTL line we used for MPM studies. The main disadvantage is that they are a T helper tumor line and less representative of normal T cells. Furthermore, the Jurkat TcR-antigen specificity is not known. As a substitute for an antigen-presenting target cell, we used Raji cells (a B cell line) coated with *Staphylococcus* enterotoxin E (SEE), a superantigen. The superantigen mimics the normal TcR-peptide-MHC interaction insofar as it binds to both the Jurkat TcR and the Raji MHC (Muller-Alouf et al. 2001).

The strong Jurkat cell TcR engagement by SEE, while an artificial stimulus, also potentially has certain advantages when it comes to trying to understand the basic mechanism of MTOC translocation. Signal transduction through the TcR is complex and a number of molecules may be important for T cell activation under conditions of weak TcR-antigen binding that might not otherwise be necessary. There are a number of molecules involved in T cell adhesion and signaling that not only affect MTOC reorientation but also impact numerous signaling events. Examples include the integrin LFA-1 (lymphocyte function-associated antigen-1) (Davignon et al. 1981; Li et al. 2009; Davignon et al. 1981; Anikeeva et al. 2005), Fyn kinase (Martin-Cofreces et al. 2006), and Vav, a guanine nucleotide exchange factor for small G proteins (Ardouin et al. 2003). When CTLs are treated with monoclonal antibodies to LFA-1, their cytolytic function is profoundly inhibited (90 %) (Davignon et al. 1981). On the other hand, when LFA-1-deficient Jurkat cells stimulated by SEE-coated Raji cells, MTOC reorientation was not greatly affected (Combs et al. 2006). Similarly, siRNA knockdown of Fyn to undetectable levels in Jurkat cells also had little effect on MTOC translocation (Tan and Poenie, unpublished observations). With respect to Vav, Ardouin and colleagues showed that for $Vav^{-/-}$ thymocytes, MTOC reorientation was reduced from 74 % to 49 %. This partial reduction of MTOC translocation correlates with impaired calcium elevation and signaling events. It was noted, however, that in $Vav-1^{-/-}$ Jurkat cells, these effects were less severe leading to the suggestion that there may be developmental defects that contribute to defects in Vav$^{-/-}$ mice (Cao et al. 2002). In any case, it does not seem that Vav is absolutely necessary for MTOC translocation.

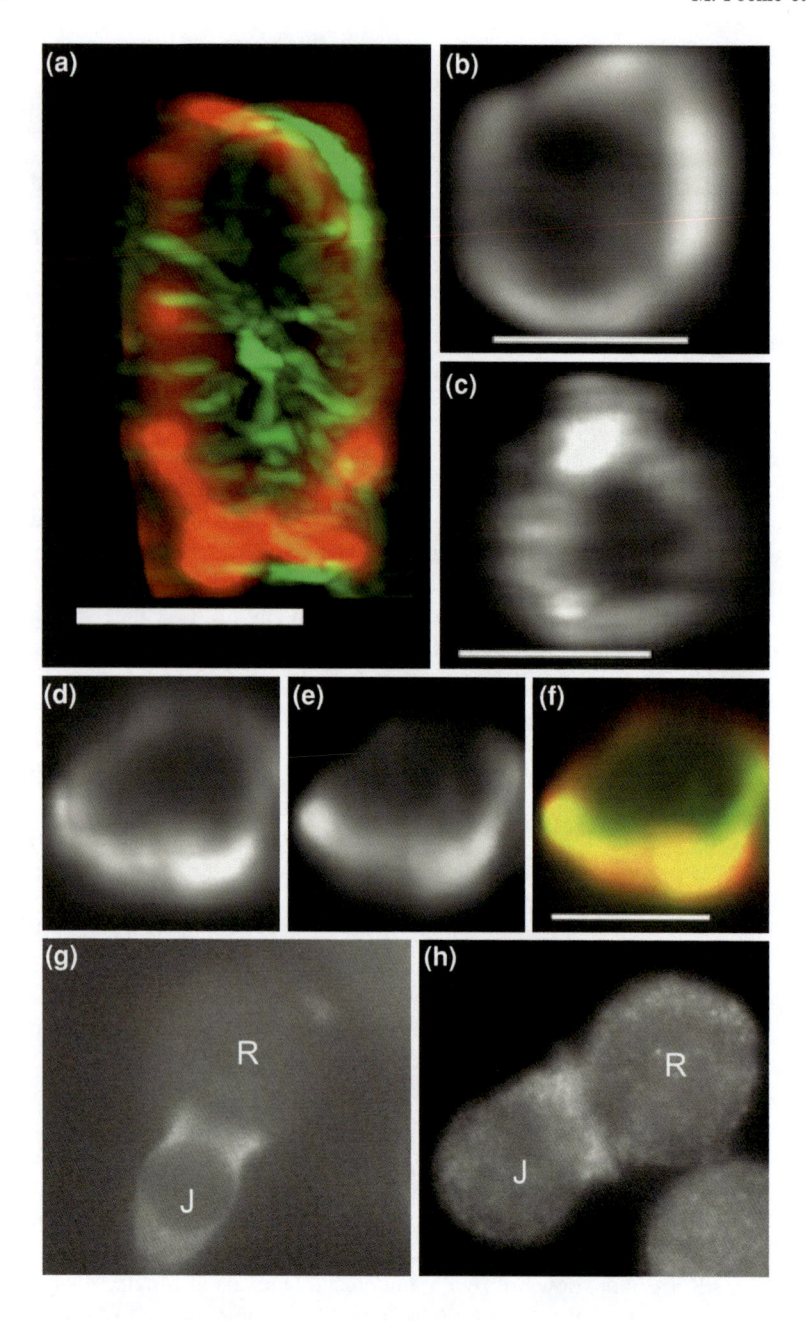

◀**Fig. 21.4** Dynein forms a ring at the immunological synapse. **a–f** Jurkat T cells were paired with Raji B cells coated with the superantigen *Staphylococcus* enterotoxin E (SEE). Cells were immunostained and the fluorescence images were processed to obtain 3D images. In **a**, mouse anti-ADAP formed a ring at the synapse with microtubules and the MTOC (*green*) in the *center*. Cell pairs were also immunostained with rabbit anti-dynein intermediate chain (DIC) (**b**) or mouse anti-DIC (**c**). Both antibodies stained a ring at the synapse. Jurkat cell pairs were stained and a merged image of **d**, anti-ADAP, and **e**, mouse anti-DIC is shown in **f**. ADAP (*red*) and DIC (*green*) colocalized in a ring at the synapse. Scale bars 5 microns. **g, h** Jurkat cells expressing GFP-DIC were activated with SEE-coated Raji cells. The GFP fluorescence alone (**g**) and immunostaining cell pairs for GFP (**h**) both show GFP-DIC accumulation at the synapse. Figures 4 (**a–f**) are reprinted from Combs et al. (2006). Copyright (2006) National Academy of Sciences, U.S.A

In studies of MTOC polarization in Jurkat cells, having found that LFA-1 was not absolutely required, the question remained as to why there was a correlation between microtubule anchor points and the pSMAC. One possibility was that microtubule contact points and possibly dynein were more closely correlated with ADAP, a molecule needed for LFA-1 clustering (Wang et al. 2009; Kliche et al. 2012). Immunostaining and computerized 3D reconstructions of ADAP and microtubules showed that ADAP forms a ring at the synapse that is closely related to where microtubules contact the pSMAC (Fig. 21.4a). Furthermore, when ADAP expression was reduced by introduction of antisense morpholino oligonucleotides, MTOC translocation was blocked (Combs et al. 2006). On the other hand, when T cells were prepared from ADAP$^{-/-}$ mice, MTOC translocation was essentially normal. The reason for this difference is not clear.

The Jurkat cell studies were extended to an examination of dynein. Immunostaining using two different antibodies against the dynein intermediate chain (70.1, 1467) showed that dynein was present as a ring at the synapse, closely related to ADAP and other markers for the pSMAC Fig. 21.4(b–f) (Combs et al. 2006). Furthermore, immunoprecipitation of dynein also pulled down ADAP suggesting that the two molecules were linked in some way. Finally, when ADAP expression was reduced using morpholino oligonucleotides, there was also a loss of dynein at the synapse. These data suggested there was a link between ADAP, dynein, and MTOC translocation in Jurkat T cells.

Although previous studies showed a link between dynein and MTOC translocation they were not directly interfering with dynein. Recent studies by Martin-Cofreces et al. showed that use of siRNA to reduce dynein expression led to a loss of MTOC translocation (Martin-Cofreces et al. 2008). They reported similar effects due to overexpression of dynactin, which disrupts dynein complexes. More recently, we have used molecular traps against the dynein intermediate chain (Varma et al. 2010) to show that when the trap is activated by dimerization, MTOC translocation is reduced by more than 50 % (Christian and Poenie, unpublished observations). Thus, several lines of evidence suggest that dynein is important for MTOC translocation.

Fig. 21.5 Model of actin expansion driving MTOC polarization. **a** Actin first accumulates at the synapse in a small patch in the area of the future cSMAC. Microtubules may be linked to actin through IQGAP or CIP4. **b** During maturation of the synapse, actin is cleared from the cSMAC area and widens out in the form of a ring. Microtubules anchored to actin would be put under tension, drawing the MTOC up to the synapse

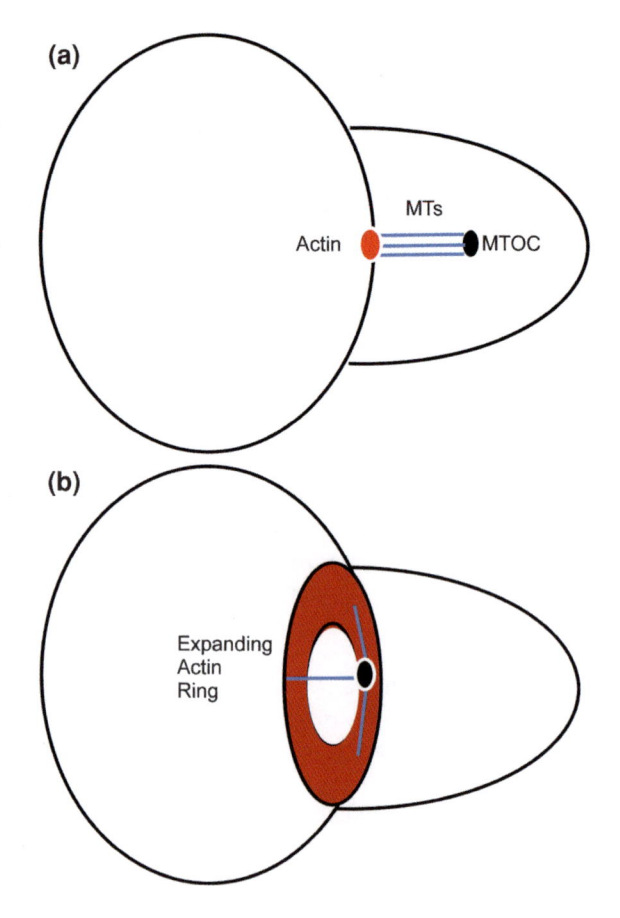

21.3 Alternative Mechanisms for MTOC Translocation

Although there are a number of lines of evidence that dynein anchored at the synapse is responsible for MTOC translocation, alternative or supplemental mechanisms have been proposed. Several studies have implicated an actin or actomyosin-based movement. One proposed mechanism is based on observations that actin is initially polymerized in the region of the cSMAC but it is then cleared out of the cSMAC as the patch of actin takes the form of an expanding ring (Stinchcombe et al. 2006). A similar observation is seen in the studies by Bunnel and colleagues who followed actin polymerization at the surface where Jurkat cells contact anti-TcR-coated coverslips. They saw that actin polymerization begins in the center of the contact zone and then widens out as an expanding ring (Bunnell et al. 2001). One could envision that if microtubule plus ends were tied to actin, then tension would develop on microtubules as the ring expanded (Fig. 21.5).

In support of this idea, previous studies had reported that Cdc42, which is necessary for triggering actin polymerization, was also needed for MTOC reorientation (Stowers et al. 1995). Based on reports that IQGAP links microtubule plus ends to the actin cortex (Fukata et al. 2002; Watanabe et al. 2004), Stinchcombe and colleagues looked at IQGAP at the synapse and found that both actin and IQGAP clear out of the synapse before the MTOC arrives. They proposed that IQGAP might link microtubules to actin. Then as the actin ring expanded, the microtubules would spread out with it generating the tension that would pull the MTOC forward (Fig. 21.5).

In a study of NK cells, Banerjee and colleagues overexpressed normal or mutant Cdc42 interacting protein (CIP4) and showed that in both cases, MTOC translocation was blocked (Banerjee et al. 2007). CIP4, like IQGAP, is also thought to link actin to microtubules. Once again this study argued for a role of Cdc42 in MTOC polarization. On the other hand, when Tskvitaria-Fuller and colleagues loaded T cells with a dominant negative Cdc42, it had only a small effect, slowing down MTOC repositioning (Tskvitaria-Fuller et al. 2006). In a study by Gomez and colleagues, Cdc42 expression was reduced using shRNA and this apparently also had no effect on MTOC translocation. They argued instead that Rac1 was necessary, perhaps though its ability to activate the formin FML1 (Gomez et al. 2007).

The notion that actin-based motility is involved with MTOC repositioning is not simple to dissect because actin dynamics are intimately linked to T cell activation. The idea of microtubules linked to an expanding actin ring is perhaps in theory, a plausible way to initially drive the MTOC toward the synapse but at the same time, it is hard to reconcile with the oscillating MTOC seen when two target cells are in contact with one CTL (Kuhn and Poenie 2002). Furthermore, the notion that this is supported by a role for Cdc42 is undermined by the papers described above. It should also be noted that Sedwick and colleagues demonstrated that the MTOC could be induced to translocate to the opposite end of the cell from where actin accumulated (Sedwick et al. 1999). Finally, we should note that we introduced an IQGAP mutant construct into Jurkat cells that has a non-functional actin binding domain (IQGAP G75Q, courtesy of David Sacks). We saw no effect on MTOC translocation (Fig. 21.6).

21.4 MTOC Polarization and T Cell Signaling

MTOC translocation is intimately tied a number of signaling events associated with T cell activation. When the TcR binds to peptide-MHC complexes, a complex series of events unfold that involve several kinases and scaffold proteins. It is thought that one of the earliest events triggered by TcR ligation is activation the Src-family kinase Lck which then phosphorylates regions of the TcR complex known as ITAMs (Smith-Garvin et al. 2009). However, this view has been challenged by other models (Davis and van der Merwe 2011). Once ITAMs are

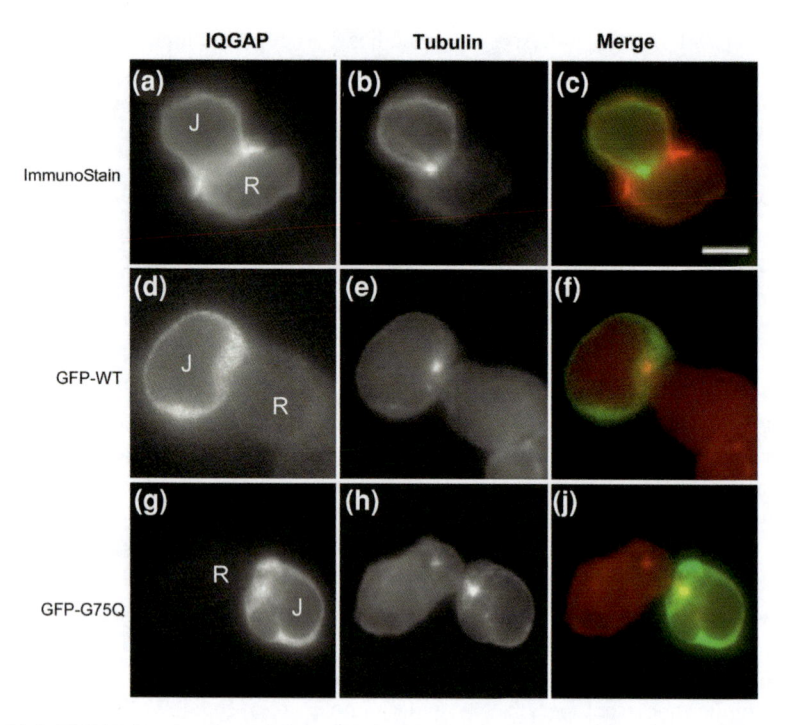

Fig. 21.6 IQGAP is not essential for MTOC polarization in activated Jurkat T cells. Either normal Jurkat T cells (**a, b, c**), Jurkat cells expressing GFP-IQGAP (**d, e, f**), or Jurkat cells expressing GFP-IQGAP G75Q (**g, h, i**) were paired with SEE-coated Raji B cells (R). Pairs were immunostained for IQGAP (**a**) and/or tubulin (**b, e, h**). IQGAP is seen at the immunological synapse (**a**). Tubulin staining marked the MTOC (*bright spot*), which was localized at the synapse in all cell lines. Since GFP-IQGAP G75Q cannot bind actin and the MTOC was still polarized in these cells, this indicates that IQGAP is not essential for MTOC polarization in this system. The merged images show IQGAP in *red* and tubulin in *green* (**c**), or GFP-IQGAP or IQGAP G75Q in green and tubulin in *red* (**f, i**). Overlap between *green* and *red* is colored *yellow*

phosphorylated, they recruit the Syk-family kinase ZAP-70 leading to its activation. ZAP-70 then phosphorylates a number of critical scaffold proteins including LAT and SLP-76 (Wang et al. 2010). Studies have shown that this cascade, especially the kinase lck, is essential for MTOC translocation (Davis and van der Merwe 2011; Lowin-Kropf et al. 1998; Kuhne et al. 2003; Tsun et al. 2011; Morgan et al. 2001; Blanchard et al. 2002).

One of the enzymes triggered through TcR activation is phospholipase Cγ. This enzyme cleaves phosphatidylinositol 4,5-biosphosphate (PIP2) into the second messengers inositol 1,4,5-trisphosphate (IP3) and diacylglycerol (DAG). These in turn trigger calcium release from the ER and activation of protein kinase C (PKC). Sustained calcium signaling is achieved by the opening of CRAC channels at the plasma membrane through the interaction of the Stim and Orai proteins (Hogan et al. 2010). Interestingly, both calcium and diacylglycerol have been implicated in MTOC translocation (Quann et al. 2009; Quintana et al. 2009; Nesic et al. 1998).

Various other signaling and adhesion molecules have been implicated in MTOC repositioning. These include the already mentioned small G proteins, Rac, Cdc42, Vav, and Fyn. In addition, the calcium-dependent tyrosine kinase Pyk-2 (RAFTK) has been implicated in MTOC movements in NK cells (Sancho et al. 2000). One of the molecules associated with Pyk-2 is paxillin which has also been implicated in MTOC repositioning (Robertson and Ostergaard 2011; Avraham et al. 2000).

At present, the most compelling evidence for a signal closely linked to MTOC repositioning is the activation of PKC. Two studies by Quann and colleagues show that localized formation of diacylglyercol by flash photolysis lead to localized accumulation of dynein at the site of photolysis followed by recruitment of the MTOC (Quann et al. 2009). In their studies, calcium was not required. This movement apparently required activation PKC-θ followed by either PKC-ε or PKC-η (Quann et al. 2011).

21.5 Movement of Secretory Vesicles Towards the MTOC

T cell, and in particular CTL, vesicles contain perforin that forms holes in the target cell membrane and granzymes that trigger apoptosis (Podack and Konigsberg 1984; Pasternack et al. 1986; Tschopp and Nabholz 1990; Podack 1992; Trapani 2001; Hoves et al. 2010). These vesicles have been characterized as "secretory lysosomes" (Blott and Griffiths 2002; Bossi and Griffiths 2005). In CTLs, trafficking of these vesicles to the synaptic interface is regulated by two principal movements (Fig. 21.1). The first is dynein-dependent movement of vesicles to the minus end of microtubules, i.e., toward the microtubule-organizing center (MTOC) (Mentlik et al. 2010), and the second is MTOC repositioning to the CTL contact surface (Stinchcombe et al. 2006; Poenie et al. 2004). Both movements occur after CTL recognition of antigen on the target cell and are initiated by proximal TCR signaling (Sykulev 2010).

As discussed in the previous section, activated phospholipase C (PLCγ) cleaves PIP$_2$ resulting in the production of DAG and IP$_3$ and activation of protein kinase C (PKC). While PKC appears to regulate MTOC movements, a rise in intracellular Ca^{2+} concentration regulates dynein-dependent granule movement toward the MTOC. Dynein motors can function in the absence of Ca^{2+} in a cell-free system (Gennerich et al. 2007) indicating that Ca^{2+} exercises its activity indirectly, and the precise nature of the downstream events is not understood. However, recent data provide evidence that the kinetics of intracellular Ca^{2+} accumulation determines how rapidly the dynein translocates the granules toward MTOC (Beal et al. 2009). This is in accord with the analysis of granule movement in another system, crustacean chromatophores, showing that a rise in intracellular Ca^{2+} concentration increases the dynein-mediated aggregation velocity of pigment granules by 4.4-fold (Ribeiro and McNamara 2007). Thus, the two movements responsible for intracellular granule trafficking are independently regulated by Ca^{2+}- and DAG-dependent signaling (Beal et al. 2009).

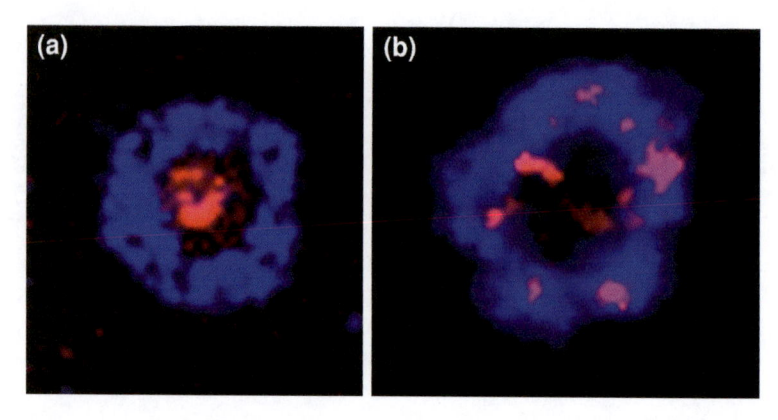

Fig. 21.7 Lytic granules are on the move. CTLs were settled onto planar lipid bilayers containing the adhesion protein ICAM-1 (*blue*) and cognate peptide-MHC ligands (*unstained*) recognizable by the CTL-TcR. Secretory vesicle contents are shown in *red*. **a** Strong TcR signaling causes vesicles (*red*) concentrate at around the MTOC prior to MTOC translocation resulting in their clustering at the center of the immunological synapse (cSMAC). **b** Weaker or indolent TcR signaling leads to MTOC translocation prior to clustering of secretory vesicles around the MTOC. Here vesicles traveling from the periphery toward the MTOC still arrive at the synapse but tend to get stuck in the pSMAC. The central location of vesicles is associated with more rapid granule release and faster destruction of target cells

These considerations led us to propose a model in which variations in the kinetics of early TCR signaling could determine the difference in temporal and spatial coordination of the two principal movements (Sykulev 2010). If the kinetics of Ca^{2+} signaling is rapid, the granules are recruited to MTOC prior to the MTOC polarization, and subsequent MTOC polarization directly delivers the granules to the secretory domain (Fig. 21.7a). The granule delivery via this short path is associated with rapid kinetics of target cell destruction by CTL. Slow kinetics of Ca^{2+} signaling allows the MTOC to polarize before the granules reach the MTOC, and the granules are redirected to the periphery of the synaptic interface (Fig. 21.7b). The granules then have to travel across the adhesion ring to be released at the center of the synaptic interface which is devoid of polymerized F-actin—the long path. The long path of granule delivery is linked to inefficient, i.e., slow, target cell destruction by CTL.

During the interaction of T cells and targets, formation of the immunological synapse is associated with polymerization of actin and a transient formation of an actin ring. Studies by Beal and colleagues indicate that the stability of this actin ring is related to sensitivity of the target cell to lysis by CTL (Beal et al. 2008). There are several possible explanations for this observation. One possibility is that actin is directly involved in driving MTOC translocation as mentioned above (Stinchcombe et al. 2006). Alternatively, it might be associated with recruitment of ADAP which interacts with actin through its Ena/VASP binding domain and has also been implicated in the recruitment of dynein (Combs et al. 2006; Obergfell et al. 2001). Finally, this actin ring might be associated with forming a seal that serves to contain the released lytic molecules at the target cell membrane (Beal et al. 2008). This may slow their inactivation and increase local concentration of these molecules at the target cell surface (Millard et al. 1984; Henkart et al. 1984; de Saint Basile et al. 2010), enhancing the effectiveness of target cell lysis.

21.6 Conclusion

In conclusion, it seems likely that dynein plays a major role in the MTOC translocation and secretory vesicle movements that result in localized secretion, which lies at the heart of T cell effector functions. The signals that link TcR ligation to activation events that trigger dynein movements appear to now have been identified. Still there is much that is unknown. At present the way that PKC and calcium might modulate dynein function are not known. Understanding how these activities are regulated may help us understand some aspects of how tumor cells avoid destruction by T cells.

Acknowledgments This work was supported by NIH grants to Y.S. (AI52812; CA131973) and NIH 1 R21 AA15437-01A1 to M.P.

References

Andzelm MM et al (2007) Myosin IIA is required for cytolytic granule exocytosis in human NK cells. J Exp Med 204(10):2285–2291

Anikeeva N et al (2005) Distinct role of lymphocyte function-associated antigen-1 in mediating effective cytolytic activity by cytotoxic T lymphocytes. Proc Nat Acad Sci USA 102(18):6437–6442

Ardouin L et al (2003) Vav1 transduces TCR signals required for LFA-1 function and cell polarization at the immunological synapse. Eur J Immunol 33(3):790–797

Avraham H et al (2000) RAFTK/Pyk2-mediated cellular signalling. Cell Signal 12(3):123–133

Banerjee PP et al (2007) Cdc42-interacting protein-4 functionally links actin and microtubule networks at the cytolytic NK cell immunological synapse. J Exp Med 204(10):2305–2320

Beal AM et al (2009) Kinetics of early T cell receptor signaling regulate the pathway of lytic granule delivery to the secretory domain. Immunity 31(4):632–642

Beal AM et al (2008) Protein kinase C theta regulates stability of the peripheral adhesion ring junction and contributes to the sensitivity of target cell lysis by CTL. J Immunol 181(7):4815–4824

Blanchard N, Di Bartolo V, Hivroz C (2002) In the immune synapse, ZAP-70 controls T cell polarization and recruitment of signaling proteins but not formation of the synaptic pattern. Immunity 17(4):389–399

Blott EJ, Griffiths GM (2002) Secretory lysosomes. Nat Rev Mol Cell Biol 3(2):122–131

Bossi G, Griffiths GM (2005) CTL secretory lysosomes: biogenesis and secretion of a harmful organelle. Semin Immunol 17(1):87–94

Bunnell SC et al (2001) Dynamic actin polymerization drives T cell receptor-induced spreading: a role for the signal transduction adaptor LAT. Immunity 14(3):315–329

Burkhardt JK et al (1993) Lytic granules from cytotoxic T cells exhibit kinesin-dependent motility on microtubules in vitro. J Cell Sci 104(Pt 1):151–162

Combs J et al (2006) Recruitment of dynein to the Jurkat immunological synapse. Proc Nat Acad Sci USA 103(40):14883–14888

Cao Y et al (2002) Pleiotropic defects in TCR signaling in a Vav-1-null Jurkat T-cell line. The EMBO J 21(18):4809–4819

Davignon D et al (1981) Lymphocyte function-associated antigen 1 (LFA-1): a surface antigen distinct from Lyt-2,3 that participates in T lymphocyte-mediated killing. Proc Nat Acad Sci USA 78(7):4535–4559

Davignon D et al (1981) Monoclonal antibody to a novel lymphocyte function-associated antigen (LFA-1): mechanism of blockade of T lymphocyte-mediated killing and effects on other T and B lymphocyte functions. J Immunol 127(2):590–595

Davis SJ, van der Merwe PA (2011) Lck and the nature of the T cell receptor trigger. Trends Immunol 32(1):1–5

Dustin ML, Chakraborty AK, Shaw AS (2010) Understanding the structure and function of the immunological synapse. Cold Spring Harbor perspectives Biol 2(10):a002311

Frey AB, Monu N (2008) Signaling defects in anti-tumor T cells. Immunol Rev 222:192–205

Freiberg BA et al (2002) Staging and resetting T cell activation in SMACs. Nat Immunol 3(10):911–917

Fukata M et al (2002) Rac1 and Cdc42 capture microtubules through IQGAP1 and CLIP-170. Cell 109(7):873–885

Geiger B, Rosen D, Berke G (1982) Spatial relationships of microtubule-organizing centers and the contact area of cytotoxic T lymphocytes and target cells. J Cell Biol 95(1):137–143

Gennerich A et al (2007) Force-induced bidirectional stepping of cytoplasmic dynein. Cell 131(5):952–965

Gomez TS et al (2007) Formins regulate the actin-related protein 2/3 complex-independent polarization of the centrosome to the immunological synapse. Immunity 26(2):177–190

Henkart PA et al (1984) Cytolytic activity of purified cytoplasmic granules from cytotoxic rat large granular lymphocyte tumors. J Exp Med 160(1):75–93

Hogan PG, Lewis RS, Rao A (2010) Molecular basis of calcium signaling in lymphocytes: STIM and ORAI. Annu Rev Immunol 28:491–533

Hoves S, Trapani JA, Voskoboinik I (2010) The battlefield of perforin/granzyme cell death pathways. J Leukoc Biol 87(2):237–243

Kliche S et al (2012) CCR7-mediated LFA-1 functions in T cells are regulated by 2 independent ADAP/SKAP55 modules. Blood 119(3):777–785

Koneru M et al (2005) Defective proximal TCR signaling inhibits CD8 + tumor-infiltrating lymphocyte lytic function. J Immunol 174(4):1830–1840

Knox JD, Mitchel RE, Brown DL (1991) Effects of hyperthermia on microtubule organization and cytolytic activity of murine cytotoxic T lymphocytes. Exp Cell Res 194(2):275–283

Kuhn JR, Wu Z, Poenie M (2001) Modulated polarization microscopy: a promising new approach to visualizing cytoskeletal dynamics in living cells. Biophys J 80(2):972–985

Kuhn JR, Poenie M (2002) Dynamic polarization of the microtubule cytoskeleton during CTL-mediated killing. Immunity 16(1):111–121

Kuhne MR et al (2003) Linker for activation of T cells, zeta-associated protein-70, and Src homology 2 domain-containing leukocyte protein-76 are required for TCR-induced microtubule-organizing center polarization. J Immunol 171(2):860–866

Kupfer A, Dennert G Singer SJ (1983) Polarization of the golgi apparatus and the microtubule-organizing center within cloned natural killer cells bound to their targets. Proc Nat Acad Sci USA 80(23):7224–7248

Kupfer A, Dennert G, Singer SJ (1985) The reorientation of the Golgi apparatus and the microtubule-organizing center in the cytotoxic effector cell is a prerequisite in the lysis of bound target cells. J Mol Cell Immunol: JMCI 2(1):37–49

Kupfer A, Dennert G (1984) Reorientation of the microtubule-organizing center and the Golgi apparatus in cloned cytotoxic lymphocytes triggered by binding to lysable target cells. J Immunol 133(5):2762–2766

Kupfer A, Singer SJ (1989) The specific interaction of helper T cells and antigen-presenting B cells. IV. Membrane and cytoskeletal reorganizations in the bound T cell as a function of antigen dose. J Exp Med 170(5):1697–1713

Kupfer A, Singer SJ, Dennert G (1986) On the mechanism of unidirectional killing in mixtures of two cytotoxic T lymphocytes. Unidirectional polarization of cytoplasmic organelles and the membrane-associated cytoskeleton in the effector cell. J Exp Med 163(3):489–498

Kupfer A et al (1986) The specific direct interaction of helper T cells and antigen-presenting B cells. Proc Nat Acad Sci USA 83(16): 6080–6083

Kurowska M et al (2012) Terminal transport of lytic granules to the immune synapse is mediated by the kinesin-1/Slp3/Rab27a complex. Blood

Li D, Molldrem JJ, Ma Q (2009) LFA-1 regulates CD8 + T cell activation via T cell receptor-mediated and LFA-1-mediated Erk1/2 signal pathways. J Biol Chem 284(31):21001–21010

Lowin-Kropf B, Shapiro VS, Weiss A (1998) Cytoskeletal polarization of T cells is regulated by an immunoreceptor tyrosine-based activation motif-dependent mechanism. J cell Biol 140(4):861–871

Martin-Cofreces NB et al (2006) Role of Fyn in the rearrangement of tubulin cytoskeleton induced through TCR. J Immunol 176(7):4201–4207

Martin-Cofreces NB et al (2008) MTOC translocation modulates IS formation and controls sustained T cell signaling. J Cell Biol 182(5):951–962

Mentlik AN et al (2010) Rapid lytic granule convergence to the MTOC in natural killer cells is dependent on dynein but not cytolytic commitment. Mol Biol Cell 21(13):2241–2256

Millard PJ et al (1984) Purification and properties of cytoplasmic granules from cytotoxic rat LGL tumors. J Immunol 132(6):3197–3204

Morgan MM et al (2001) Superantigen-induced T cell:B cell conjugation is mediated by LFA-1 and requires signaling through Lck, but not ZAP-70. J Immunol 167(10):5708–5718

Monks CR et al (1998) Three-dimensional segregation of supramolecular activation clusters in T cells. Nature 395(6697):82–86

Monu N, Frey AB (2007) Suppression of proximal T cell receptor signaling and lytic function in CD8(+) tumor-infiltrating T cells. Cancer Res 67(23):11447–11454

Muller-Alouf H et al (2001) Superantigen bacterial toxins: state of the art. Toxicon: Off J Int Soc Toxinology 39(11):1691–1701

Nesic D, Henderson S, Vukmanovic S (1998) Prevention of antigen-induced microtubule organizing center reorientation in cytotoxic T cells by modulation of protein kinase C activity. Int Immunol 10(11):1741–1746

Norcross MA (1984) A synaptic basis for T-lymphocyte activation. Annales D'Immunol 135D(2):113–134

Obergfell A et al (2001) The molecular adapter SLP-76 relays signals from platelet integrin alphaIIbbeta3 to the actin cytoskeleton. J Biol Chem 276(8):5916–5923

Pasternack MS et al (1986) Serine esterase in cytolytic T lymphocytes. Nature 322(6081):740–743

Podack ER (1992) Perforin: structure, function, and regulation. Curr Top Microbiol Immunol 178:175–184

Poo W-J, Conrad L, Janeway CA Jr (1988) Receptor-directed focusing of lymphokine release by helper T cells. Nature 332(6162):378–380

Poenie M, Kuhn J, Combs J (2004) Real-time visualization of the cytoskeleton and effector functions in T cells. Curr Opin Immunol 16(4):428–438

Poenie M (2010) Hiways and byways to the secretory synapse. Self Nonself 1(1):69–70

Prevost-Blondel A et al (1998) Tumor-infiltrating lymphocytes exhibiting high ex vivo cytolytic activity fail to prevent murine melanoma tumor growth in vivo. J Immunol 161(5):2187–2194

Podack ER, Konigsberg PJ (1984) Cytolytic T cell granules. Isolation, structural, biochemical, and functional characterization. J Exp Med 160(3):695–710

Quann EJ et al (2009) Localized diacylglycerol drives the polarization of the microtubule-organizing center in T cells. Nat Immunol 10(6):627–635

Quann EJ et al (2011) A cascade of protein kinase C isozymes promotes cytoskeletal polarization in T cells. Nat Immunol 12(7):647–654

Quintana A et al (2009) Morphological changes of T cells following formation of the immunological synapse modulate intracellular calcium signals. Cell Calcium 45(2):109–122

Radoja S et al (2001) CD8(+) tumor-infiltrating T cells are deficient in perforin-mediated cytolytic activity due to defective microtubule-organizing center mobilization and lytic granule exocytosis. J Immunol 167(9):5042–5051

Ribeiro M, McNamara JC (2007) Calcium movements during pigment aggregation in freshwater shrimp chromatophores. Pigment Cell Res/Spons Eur Soc Pigment Cell Res Int Pigment Cell Soc 20(1):70–77

Robertson LK, Ostergaard HL (2011) Paxillin associates with the microtubule cytoskeleton and the immunological synapse of CTL through its leucine-aspartic acid domains and contributes to microtubule organizing center reorientation. J Immunol 187(11):5824–5833

de Saint Basile G, Menasche G Fischer A (2010) Molecular mechanisms of biogenesis and exocytosis of cytotoxic granules. Nature Rev Immunol 10(8):568–579

Sanborn KB et al (2011) Phosphorylation of the myosin IIA tailpiece regulates single myosin IIA molecule association with lytic granules to promote NK-cell cytotoxicity. Blood 118(22):5862–5871

Sanborn KB et al (2009) Myosin IIA associates with NK cell lytic granules to enable their interaction with F-actin and function at the immunological synapse. J Immunol 182(11) 6969–6984

Sancho D et al (2000) The tyrosine kinase PYK-2/RAFTK regulates natural killer (NK) cell cytotoxic response, and is translocated and activated upon specific target cell recognition and killing. J Cell Biol 149(6):1249–1262

Sedwick CE et al (1999) TCR, LFA-1, and CD28 play unique and complementary roles in signaling T cell cytoskeletal reorganization. J Immunol 162(3):1367–1375

Smith-Garvin JE, Koretzky GA, Jordan MS (2009) T cell activation. Annu Rev Immunol 27:591–619

Stinchcombe JC et al (2001a) The immunological synapse of CTL contains a secretory domain and membrane bridges. Immunity 15(5):751–761

Stinchcombe JC et al (2006) Centrosome polarization delivers secretory granules to the immunological synapse. Nature 443(7110):462–465

Stinchcombe JC et al (2001b) Rab27a is required for regulated secretion in cytotoxic T lymphocytes. J Cell Biol 152(4):825–834

Stowers L et al (1995) Regulation of the polarization of T cells toward antigen-presenting cells by Ras-related GTPase CDC42. Proc Nat Acad Sci USA 92(11):5027–5031

Sugawara S, Kaslow HR, Dennert G (1993) CTX-B inhibits CTL cytotoxicity and cytoskeletal movements. Immunopharmacol 26(2):93–104

Sykulev Y (2010) T cell receptor signaling kinetics takes the stage. Sci Signal 3(153):pe50

Takayama H, Sitkovsky MV (1987) Antigen receptor-regulated exocytosis in cytotoxic T lymphocytes. J Exp Med 166(3):725–743

Trapani JA (2001) Granzymes: a family of lymphocyte granule serine proteases. Genome biology 2(12):REVIEWS3014

Trenn G et al (1988) Immunomodulating properties of a novel series of protein kinase C activators. Bryostatins. J Immunol 140(2):433–439

Tschopp J, Nabholz M (1990) Perforin-mediated target cell lysis by cytolytic T lymphocytes. Annu Rev Immunol 8:279–302

Tskvitaria-Fuller I et al (2006) Specific patterns of Cdc42 activity are related to distinct elements of T cell polarization. J Immunol 177(3):1708–1720

Tsun A et al (2011) Centrosome docking at the immunological synapse is controlled by Lck signaling. J Cell Biol 192(4):663–674

Varma D et al (2010) Development and application of in vivo molecular traps reveals that dynein light chain occupancy differentially affects dynein-mediated processes. Proc Nat Acad Sci USA 107(8):3493–3498

Wang H et al (2009) SLP-76-ADAP adaptor module regulates LFA-1 mediated costimulation and T cell motility. Proc Nat Acad Sci USA 106(30):12436–12441

Wang H et al (2010) ZAP-70: an essential kinase in T-cell signaling. Cold Spring Harbor Perspect Biol 2(5):a002279

Watanabe T et al (2004) Interaction with IQGAP1 links APC to Rac1, Cdc42, and actin filaments during cell polarization and migration. Dev Cell 7(6):871–83

Chapter 22
Thoughts on Progress in the Centrosome Field

Jeffrey L. Salisbury

Abstract Centrioles and centrosomes have been at the center of attention of Cell Biologists since the very beginning of the field. The approaches to the conduct of science by early investigators shaped not only the foundation of our understanding of centrosome biology, but also continue to impact its direction. What sets centrioles and centrosomes apart from membrane-bound organelles are their fascinating structure, and the intrinsic counting mechanism they employ to duplicate once in each cell cycle. The details of centriole and centrosome biogenesis, and the role that they play in ciliogenesis, cell polarity, and as a platform for cell signaling pathways ensure their central place in future investigations.

22.1 Commentary

I have been thinking a lot lately about the pace of recent progress in our understanding of the biology of centrosomes and centrioles, and the scientific excitement brought to the field with the re-emergence of the primary cilium as an extension of the centriole (Pazour et al. 2000; Pazour and Rosenbaum 2002). The cell biologist cannot help but be enamored by the centrosome and their defining functional features: the nucleation of microtubule arrays and the ability to double once in each cell cycle. The centrosome's role as a structural platform on which critical early steps in molecular signaling cascades operate is also a growing area of current interest, though the history of this notion runs deep in the field as the

---

J. L. Salisbury (✉)
Tumor Biology Program, Biochemistry and Molecular Biology, Mayo Clinic,
Rochester MN 55906, USA
e-mail: salisbury@mayo.edu

concept of the neuromotor apparatus and early thoughts on hormone signaling pass through the centrosome complex (Rees 1922; Kater 1929; Tucker et al. 1983; Salisbury 1988; Pardee 1989; Christensen et al. 2008). I feel fortunate to have experienced this recent period of astounding discovery and to have come to know personally many of the investigators who have made the latest advances possible. As is true for all progress in modern scientific achievement, the path to the breathtaking breakthroughs of today rests squarely on the shoulders of the Giants who came before us into the field (Newton 1676). Appreciation of the early pioneers, however, all too easily becomes hidden with a receding grasp on the past literature. How earlier studies developed and were formulated into the foundation of current knowledge, and how the personalities of leading workers effected this process is too often left untold in the contemporary classroom. It is therefore appropriate that we take a moment to consider our deeper history. Among many Giants in Cell Biology, a handful of investigators stand out for their exceptional contributions to the understanding of centrosome, and perhaps one, Theodor Boveri, exemplifies the species. Because there are several recent accounts of Boveri's specific contributions to the field, I will not repeat them here; rather I wish to comment more generally on his method of discovery (Metcalf 1925; Baltzer 1967; Manchester 1995; Brinkley and Goepfert 1998; Balmain 2001). Boveri was recognized as an *'extraordinary master of experimental design and objective scientific reasoning'* by his peers; see E.B. Wilson's dedication to Boveri's memory in his tome on *The Cell in Development and Heredity* (Wilson 1925). Boveri's investigations into the nature of cytoplasmic organization and the role of chromosomes as the carriers of heredity exemplify his achievement. Anyone who delves into his scientific work will discover that Boveri was not only the discoverer of fundamental facts, but also that his comprehensive insight into development and the structural basis for genetic continuity and its interruption in cancer, anticipated the very forefront of these fields today. Boveri's insight into the role of chromosomes as the carriers of the genetic traits and the concept of individuality of chromosomes was gained simply from direct observation of the behavior of chromosomes during division, and importantly through the study of abnormal divisions and their consequences in development. Boveri coined the terms *centriole* and *centrosome* during his work on the dividing *Ascaris* egg when he and his contemporary E. van Beneden observed that the astral arrays contain rounded granules that divided when new asters were formed (Boveri 1887, 1888; Van Beneden and Neyt 1887). Boveri and van Beneden both recognized that the centrosome was endowed with an *'autonomous'* behavior that included the property of 'self-replication' as seen for chromosomes. Today, we understand these features in terms of centrosome doubling and centriole duplication for which the molecular details underlying their mechanistic basis for centriole biogenesis are rapidly becoming elucidated (Kleylein-Sohn et al. 2007; Rodrigues-Martins et al. 2007; Kuriyama 2009; van Breugel et al. 2011). However, what distinguishes Boveri's conduct of science from that of the present day is a *process of discovery* that unfortunately has been diminished by the confines of the *Impact Factor*, the *Priority Score*, and the Dean's *'metrics of achievement'*.

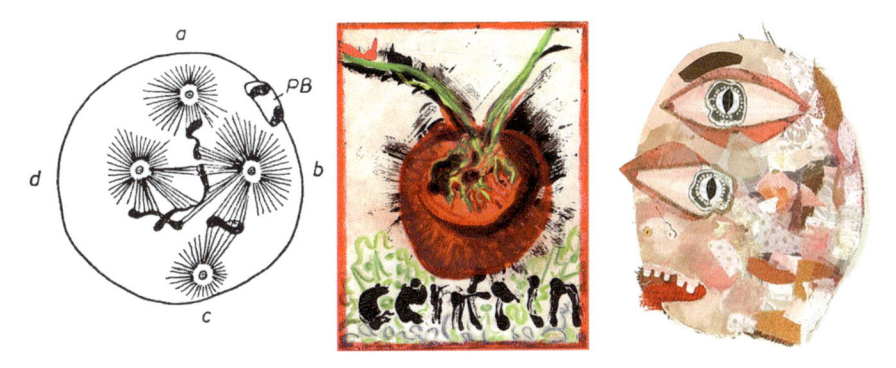

Fig. 22.1 *Left*. Drawing of a complex set of spindles in an abnormal division of an *Ascaris megalocephala* egg by Boveri (1888). *Middle*. Interpretation of a *Chlamydomonas* cell (acrylic on canvas). *Right*. Head 2010 (acrylic, paper, cloth, thread), see Portfolio at AmeliaRoseSalisbury.com

The first element of Boveri's process of discovery lies in an unencumbered freedom for the keen application of direct observation. In Boveri's time, armed only with the light microscope and the meticulous preparation of specimens, investigations into cell structure and function resulted first in a mental construct that represented the synthesis of many hours of observation. The outcome of microscopic studies was finally depicted in remarkable drawings of cytoplasmic detail that approach ultrastructural resolution (Fig. 22.1). The power of the mind to perceive beyond the practical limits of resolution of the light microscope is not unlike that of Abstract Impressionism or Cubism where multiple dimensions, times, or views are captured onto a single two-dimensional canvas. I have an artist friend who exemplifies this very style. She grew up in a scientific household before the advent of digital microscopy. During her youth on Saturday mornings instead of watching TV she spent hours at the microscope peering at histology slides that her father used to prepare for teaching. Today, her paintings reflect a vibrant palette that demonstrates the lasting impact of hematoxylin and eosin on the mind (Fig. 22.1). Contrast direct observation and unencumbered thought with the methods of today, where tagged-reporters are recorded as an expanded range of threshold intensities, projected directly on a CCD camera, and following varying degrees of computer processing onto a RGB monitor or CMYK printer. Today, the optics of the human eye and mental processing can be omitted entirely from the production of microscopic images. Obviously, power of digital microscopy cannot be denied and can be scientifically and esthetically satisfying in its own right, but we must not forget to look first to see what can be seen. The Nobel laureate and visionary Richard Feynman who thought deeply about light made this very clear in a lecture on the future direction of microscopy..."It is easy to answer many...fundamental biological questions: you *just look at the thing*!" (Feynman 1959, reprinted 1992).

A second intellectual practice that Boveri employed, and again one that is very nearly impossible or at the very least impractical today, was *not to rush to publish*. But, rather to contemplate, incubating the results, sometimes for years before a manuscript that was often a masterpiece on the subject resulted. Key to his intellectual process was to take the time to refine a more comprehensive understanding through further experimentation, and discussions with colleagues, exchange of letters and lectures on the topic, and simply to ponder. Just try and explain this method of scientific discovery to your Department Chair or Dean of Research when she asks not 'what have you done', but 'what have you done lately'? One may argue that Web-based literature forums and databases supplant the need for incubation because they make all current and past scientific thought available to the present day investigator at the click of a mouse. Undoubtedly, digital access has greatly improved the availability of information. Nonetheless, it is all too common to witness a *blind spot in citations* of important and often key studies in an area because a keyword search failed to pick up a paper that used an outmoded term for a now widely known protein or process. Lost keywords for many proteins that now carry familiar names were once referred to simply as a molecular weight or by a descriptive property, such as the 55 kD protein, the '*colchicine-binding protein*' or the '*calcium-dependent regulatory protein*'.

If Boveri was so fortunate to live to see today's stunning developments in Cell Biology he would surely be astounded by the depth of detail and progress in the field, but if he used the practice of discovery that served him so well during his career, we might see him standing on a corner carrying a cardboard sign that reads... '*Will do Microscopy for Food*'. We are the fortunate ones to have had his shoulder to rest upon and to have the opportunity to rediscover in molecular detail what he knew so well.

During the period and until the advent of modern methods in biology, progress was not simply incremental, but included many discoveries that underlie our current appreciation of centrioles and centrosomes. We must take a rather broad view of the importance of these to include fundamental studies on cilia and flagella, because, with the prescience of hindsight, we recognize that the functional relevance of centrioles lies in their supporting role for ciliogenesis and as the anchor for cilia and flagella, which was evident early on and particularly for sperm of animals and lower plants. Flagella emanating from centrioles located at the spindle poles of meiotic insect spermatocytes, the transformation of the blepharoplast in *Equisetum* to give rise to multiple flagellar bases during spermatogenesis, and the conversion of clusters of centrioles into basal bodies of ciliated epithelia led Friedrich Meves, Mihaly Lenhossek, Louis Félix Henneguy, and L.W. Sharp in particular, to solidify the concept that centrioles and basal bodies are interchangeable and indeed one-in-the-same organelle (Henneguy 1897; Lenhossék 1898; Meves 1900).

A second and still enigmatic role for centrosomes is that of organizing a cytoplasmic spatial template. The earliest hint that centrosomes embody an intrinsic structural template was seen in the giant replicating centrioles in the spermatocyte of the hagfish *Myxine*, where the orthogonal relationship between older and younger centrioles of a pair (perhaps the only macromolecular right angle known in biology) were first observed (Schreiner and Schreiner 1905). Further evidence for centrosome patterning of cytoplasmic organization is evident in the *cell axis*, a defining feature of epithelial and migrating cells discernible by a line drawn through the nucleus, intersecting the centrosome (and Golgi) and the apical membrane or leading edge of the cell (Van Beneden 1883; Wilson 1925; Bornens 2012). While the evidence suggests that establishment of centrosome positioning along the cell axis is secondary to primary cues emanating from the leading edge or apical/basal lateral membranes, its importance is nonetheless clear for maintenance or stability of cell polarity and to anchor trafficking along the microtubule array to and from the cell center (Luxton and Gundersen 2011).

With the advent of electron microscopy, studies by Giants such as Irene Manton, Keith Porter, Donald Fawcett, Daniel Mazia, and Jeremy Pickett-Heaps firmly established the centriole and centrosome as a most elegant structural feature of eukaryotic cells. Irene Manton (1904–1988) whose fine structure studies using UV and electron microscopy on reproductive cells of ferns, algae and other cryptogams, and nanoplanton gave us an early close look at the 9 + 2 structure of flagellar axonemes and the constant cytoplasmic organization characteristic of the smallest of eukaryotes (Manton and Clarke 1950; Manton 1953; Manton and Parke 1960). Electron microscopy also revealed extreme examples of nearly invariant features of cytoplasmic organization relative to the basal apparatus with a constant position of nucleus, eyespot, and other organelles seen of free-living algal cells and the regular cortical arrays of ciliates and flagellates (Dingle and Fulton 1966; Gull 1999; Sagolla et al. 2006; Paredez et al. 2011). The value of these structural features and those of the mitotic apparatus became widely recognized as an indicator of evolutionary relationships (Pickett-Heaps 1975; Stewart and Mattox 1980). We can look to these examples to find clues to the role of the centrosome in patterning organization of complex tissue types and the cytoplasm of mammalian cells, in general, albeit where the regularity of structure may be less obvious.

Finally, while much is made of the dispensability of centrioles and centrosomes for mitosis in flat worms, mutant flies, and in higher plants, it is nonetheless clear that when centrosomes become defective or 'amplified' in disease processes such as cancer, genomic continuity becomes comprised (Brinkley and Goepfert 1998). The selective pressure over nearly 2 billion years for maintaining centrioles and centrosomes in most extant eukaryotic lineages rests on two functional attributes of the organelle—the support of ciliogenesis to anchor cilia and flagella, and the definition of a 'coordinate geometry' system for mitotic and cytoplasm organization. The impact of the former became clear with the recognition of the role of defects in the primary cilium in disease processes, and that of latter is appreciated when considering the genomic instability in cancer and the definition of the plane of cleavage which, when centrosomes are absent, restricts specific developmental processes to a simple body plan as seen in the acentriolar divisions of the flat worm (Azimzadeh et al. 2012).

I have not mentioned directly 'Lesser' or burgeoning Giants, even though exceptional investigators abound. 'Lesser' only because Giants genuinely are legendary and are no longer able to dispute the title. As the field progresses we can anticipate a clarification of the role of the centrosome in the control of cytoplasmic architecture and cell polarity, and second, we can expect several distinguished players to mature into full-grown Giants for future Cell Biologists to look up to.

References

Azimzadeh J, Wong ML, Downhour DM, Alvarado AS, Marshall WF (2012) Centrosome loss in the evolution of planarians. Science 335:461–463
Balmain A (2001) Cancer genetics: from Boveri and Mendel to microarrays. Nat Rev Cancer 1(1):77–82
Baltzer F (1967) Theodor Boveri: life and work of a great biologist, University of California Press, Berkeley and Los Angeles
Bornens M (2012) The centrosome in cells and organisms. Science 335(6067):422–426
Boveri T (1887) Über Die Befruchtung Der Eier Von *Ascaris Megalocephala*. Sitz Ber Ges Morph Phys München 3:71–80
Boveri T (1888) Zellenstudien Ii. Die Befruchtung Und Teilung Des Eies Von Ascaric Megalocephala. Jene Z Naturwissen 22:685–882
Brinkley BR, Goepfert TM (1998) Supernumerary centrosomes and cancer: Boveri's hypothesis resurrected. Cell Motil Cytoskeleton 41(4):281–288
Christensen ST, Pedersen SF, Satir P, Veland IR, Schneider L (2008) The primary cilium coordinates signaling pathways in cell cycle control and migration during development and tissue repair. Curr Top Dev Biol 85:261–301
Dingle AD, Fulton C (1966) Development of the flagellar apparatus of naegleria. J Cell Biol 31(1):43–54
Feynman RP (1959) There's plenty of room at the bottom. Lecture at the 1959 Meeting of the American Physical Society at the California Institute of Technology. J Microelectromech Syst 1:60–66(reprinted 1992)
Gull K (1999) The cytoskeleton of trypanosomatid parasites. Annu Rev Microbiol 53:629–655
Henneguy L (1897) Sur Les Rapports Des Cils Vibratiles Avec Les Centrosomes. Arch. Anat. Microsc. 1:481–496
Kater J (1929) Morphology and divsion of *Chlamydomonas* with reference to the phylogeny of the flagellate neuromotor system. Univ. Calif. Publ. Zool. 33:125–168
Kleylein-Sohn J, Westendorf J, Le Clech M, Habedanck R, Stierhof YD, Nigg EA (2007) Plk4-induced centriole biogenesis in human cells. Dev Cell 13(2):190–202
Kuriyama R (2009) Centriole assembly in Cho cells expressing Plk4/Sas6/Sas4 is similar to centriogenesis in ciliated epithelial cells. Cell Motil Cytoskeleton 66(8):588–596
Lenhossék M (1898) Ueber Flimmerzellen. Verh Anat Ges 12:106
Luxton GW, Gundersen GG (2011) Orientation and function of the nuclear-centrosomal axis during cell migration. Curr Opin Cell Biol 23(5):579–588
Manchester K (1995) Theodor Boveri and the origin of malignant tumours. Trends Cell Biol 5:384–387
Manton I (1953) Number of fibrils in the cilia of green algae. Nature 171(4350):485–486
Manton I, Clarke B (1950) Electron microscope observations on the spermatozoid of fucus. Nature 166(4232):973–974
Manton I, Parke M (1960) Further observations on small green flagellates with special reference to possible relatives of *Chromulina Pusilla Butcher*. J. Mar. Biol. Assoc UK 39:275–298
Metcalf MM (1925) Boveri's Work on Cancer. JAMA 84(15):1140
Meves F (1900) Ueber Den Von La Valette St. George Entdeckten Nebenkern (Mitochondrienkörper) Der Samenzellen. Arch Mikr Anat 56:553–606

Pardee AB (1989) G1 events and regulation of cell proliferation. Science 246(4930):603–608

Paredez AR, Assaf ZJ, Sept D, Timofejeva L, Dawson SC, Wang CJ, Cande WZ (2011) An actin cytoskeleton with evolutionarily conserved functions in the absence of canonical actin-binding proteins. Proc Natl Acad Sci U S A 108(15):6151–6156

Pazour GJ, Dickert BL, Vucica Y, Seeley ES, Rosenbaum JL, Witman GB, Cole DG (2000) Chlamydomonas Ift88 and its mouse homologue, polycystic kidney disease gene Tg737, are required for assembly of cilia and flagella. J Cell Biol 151(3):709–718

Pazour GJ, Rosenbaum JL (2002) Intraflagellar transport and cilia-dependent diseases. Trends Cell Biol 12(12):551–555

Pickett-Heaps J (1975) Green algae: structure, reproduction and evolution in selected genera, Sinauer Assoc.,Stamford, CT

Rees CW (1922) The neuromotor apparatus of paramecium. Science 55(1416):184–185

Rodrigues-Martins A, Riparbelli M, Callaini G, Glover DM, Bettencourt-Dias M (2007) Revisiting the role of the mother centriole in centriole biogenesis. Science 316(5827):1046–1050

Sagolla MS, Dawson SC, Mancuso JJ, Cande WZ (2006) Three-dimensional analysis of mitosis and cytokinesis in the binucleate parasite giardia intestinalis. J Cell Sci 119(Pt 23):4889–4900

Salisbury JL (1988) The lost neuromotor apparatus of chlamydomonas: rediscovered. J Protozool 35(4):574–577

Schreiner A, Schreiner K (1905) Ueber Die Entwicklung Der Männlichen Geschlechtzellen Von *Myxine Glutinosa*. Arch Biol 21:315

Stewart KD, Mattox K (1980) Phylogeny of phytoflagellates. Phytoflagellates. ER Cox. New York, Elservier, pp 433–462

Tucker RW, Meade-Cobun KS, Jayaraman S, More NS (1983) Centrioles, primary cilia and calcium in the growth of Balb/C 3t3 cells. J Submicrosc Cytol 15(1):139–143

Van Beneden E (1883) Recherches Sur La Maturation De L'oeuf, La Fecondation El La Division Cellulaire. Arch Biol 4:265–638

Van Beneden E, Neyt A (1887) Nouvelle Rechererches Sur La Fecondation Et La Division Mitotique, Chez L'ascaride Megalocéphale. Bull Acad Roy Med. Belgique, series 3, 14: 215–295

van Breugel M, Hirono M, Andreeva A, Yanagisawa HA, Yamaguchi S, Nakazawa Y, Morgner N, Petrovich M, Ebong IO, Robinson CV, Johnson CM, Veprintsev D, Zuber B (2011) Structures of Sas-6 suggest its organization in centrioles. Science 331(6021):1196–1199

Wilson EB (1925) The cell in development and heredity. Macmillan Company, New York

About the Author

Heide Schatten is a Professor at the University of Missouri, Columbia, USA. She is a cell biologist with research focused on cytoskeletal regulation in various cell systems and on cytoskeletal dysfunctions of the centrosome–microtubule complex that play a role in disease such as cancer and in disorders such as infertility and other reproductive disorders.

She received her MS and PhD degrees from the Karl-Ruprecht University in Heidelberg, Germany, and performed her PhD research studies at the German Cancer Research Center in Heidelberg. She performed pre- and post-doctoral studies with Professor Daniel Mazia at the University of California, Berkeley, who introduced her to the fascinating field of centrosome biology and generated unlimited enthusiasm for centrosome research that she pursued in collaborations with Daniel Mazia and numerous colleagues in cell, molecular, and reproductive biology in the USA, Europe, China, and Latin America. She has given numerous presentations in these and other countries. Her studies also included collaborations with NASA scientists and experiments aboard the Space Shuttle *Endeavour* to examine the effects of spaceflight on centrosome–cytoskeletal regulation during development.

Her publications include cellular and molecular biology, cancer biology, reproductive biology, microbiology, space biology, and advanced imaging methods including novel imaging techniques that she employed in close collaborations with Professor Hans Ris at the University of Wisconsin-Madison to analyze cellular structures in their 3D network.

She is a member of the American Society for Cell Biology, American Association for the Advancement of Science, and Microscopy Society of America. She has received numerous awards including grant awards from NASA, NIH, and NSF. She has published over 190 papers, 7 book chapters, and edited several special topics journal issues and 8 books with several more in progress.

Index

A
Aneuploidy, 94, 223, 224

B
Breast cancer, 142

C
CDK2, 63, 173, 177–179, 181, 183, 188–195, 207, 212–214, 229, 230, 244, 245
Cell division, 10, 15, 19
Centrosomal amplification, 265, 280
Centrosomal dysfunctions, 56
Centrosome, 4, 59, 60, 62, 67, 134, 138, 139, 141, 142, 144, 165, 179, 190–195, 201, 209, 225–228, 233, 243, 244, 247–250, 257, 278, 299
Centrosome duplication, 4, 134, 209, 244
Centrosome protein degradation, 63
Centrosome proteolysis, 133, 138
Centrosome regulation, 67
Cyclin E, 63, 141, 161, 189
Cyclin-dependent kinases, 187
Cytoskeleton, 34

D
DNA damage, 179, 183, 189, 228, 231–233, 246, 247
Domestic cat, 50

E
Embryos, 8, 87

F
Fertility preservation, 33, 45
Fertilization, 8, 39
Flagellogenesis and cell cycle, 33

G
Gametes, 15, 54
Gametogenesis, 353
Golgi apparatus, 113–115, 118–120, 122–124, 126

H
Human centrosome, 134

I
Imaging methods, 365
Infertility, 38

M
Microtubule, 8, 118
Microtubule organizing center (MTOC), 34
Mitosis, 10
Mitotic instability, 201
Modulated polarization microscopy, 365, 368
Multiple myeloma, 256
Myeloma, 255

N
Neuronal centrosome, 310
Nuclear proteins, 63

O
Oncoproteins, 202
Oocytes, 8, 73

P
p53, 137, 143, 173, 177–179, 181–183, 191, 205, 208, 227, 229–231, 245
Papillomavirus, 202
Parasites, 328

S
Somatic cell nuclear transfer, 60
Sperm aster, 34

Spermatozoa, 34, 39, 51, 54
Stem cells, 18, 99, 102

T
T cells, 365, 369

U
Ultrastructure, 38

Printed by Printforce, the Netherlands